WITHDRAWN
WRIGHT STATE UNIVERSITY LIBRARIES

Is Arsenic an Aphrodisiac?
The Sociochemistry of an Element

Is Arsenic an Aphrodisiac?
The Sociochemistry of an Element

William R. Cullen
University of British Columbia, Vancouver BC, Canada

RSCPublishing

ISBN: 978-0-85404-363-7

A catalogue record for this book is available from the British Library

© William R. Cullen 2008

All rights reserved

Apart from fair dealing for the purposes of research for noncommercial purposes or for private study, criticism or review, as permitted under the Copyright, Designs and Patents Act 1988 and the Copyright and Related Rights Regulations 2003, this publication may not be reproduced, stored or transmitted, in any form or by any means, without the prior permission in writing of The Royal Society of Chemistry or the copyright owner, or in the case of reproduction in accordance with the terms of licences issued by the Copyright Licensing Agency in the UK, or in accordance with the terms of the licences issued by the appropriate Reproduction Rights Organization outside the UK. Enquiries concerning reproduction outside the terms stated here should be sent to The Royal Society of Chemistry at the address printed on this page.

Published by The Royal Society of Chemistry,
Thomas Graham House, Science Park, Milton Road,
Cambridge CB4 0WF, UK

Registered Charity Number 207890

For further information see our web site at www.rsc.org

Preface

In 1953 I was in my second year at the University of Otago, Dunedin, New Zealand. Dr. Ted Corbett lectured on organic chemistry but his interest in the subject was very broad. One morning he talked about the organic chemistry of arsenic, and I was hooked. From a very early age I knew that I was going to be a chemist, but I did not know what sort; so this was a very significant event. In 1956 I went to Cambridge, England, to study for a PhD with Professor H. J. Eméleus who suggested that I work on the preparation of a new class of arsenic compounds that contained fluorocarbon groups. I was happy to take up the challenge. In 1958 I accepted a position in Canada at the University of British Columbia in Vancouver, and started my independent research with a series of papers that built on my PhD work, and apart from a few side excursions, the chemistry of arsenic has never been far from my thoughts.

My objective in writing this book is not to elaborate on the vast chemistry of the element, but to try to reveal to the general reader how the element and its compounds have become embedded in our social fabric, for good and for ill. I believe no other element comes close in this regard and use the word sociochemistry to describe this interface between society and chemistry.

The average person has only one idea about arsenic – it is poison – and this reputation has a sound base. Some arsenic compounds are very toxic and have been used with criminal intent from the time of the ancient Romans to the present day (Chapter 5). This aspect of arsenic's nature has been reinforced in the fiction of authors such as Dorothy Sayers and Agatha Christie, and in Kesselring's play *Arsenic and Old Lace*. The very mention of the "A-word" promotes fear and anxiety. The accidental presence of arsenic in British beer around 1900 made thousands very ill, and prompted an inquiry by a Royal Commission leading to the first laws governing food contamination.

There is a parallel story to be told about the use of arsenic in human medicine in many cultures. This peaked in the western world in the early 20th century, with Ehrlich's discovery of salvarsan, an arsenic-based cure for syphilis, that

---

Is Arsenic an Aphrodisiac? The Sociochemistry of an Element
By William R. Cullen
© William R. Cullen 2008

gave rise to the field of chemotherapy. Salvarsan and related compounds were eventually displaced by antibiotics such as penicillin. Arsenic trioxide has staged a comeback, however, and is being used as a successful treatment for a form of leukemia (Chapter 1).

Arsenic compounds were widely used in agriculture and wood preservation during the 20th century. Although most of these applications have been abandoned or curtailed, the main market for arsenic remains as a wood preservative (Chapter 2). Much has been written about the historical use of arsenic-based pigments to colour wallpaper and the belief that it caused widespread illness and many deaths, but re-examination of these accounts suggests they are at best urban myths (Chapter 3). The myth was propagated in the 1980s in attempts to account for sudden infant death syndrome (SIDS) and the death of Napoleon Bonaparte (Chapter 4). In addition to convicted murders such as Herbert Armstrong, Madeline Smith and the Grandmothers of Nagyrev, there is also a connection between arsenic and musicians, such as Tchaikovsky, authors such as Karin Blixen, scientists such as Fritz Haber and Charles Darwin and kings such as George III.

Arsenic compounds were first successfully used as chemical-warfare agents in World War I. They were subsequently deployed against unprotected native populations in Morocco and Ethiopia. These agents were manufactured and stockpiled during WWII, but the Japanese were the only nation to use them. Some compounds saw service again in the Vietnam War. The problem of disposing of the stockpile remains with us (Chapter 6).

The topic of arsenic in the environment is discussed in Chapters 7 and 8. Arsenic is all around us: in our soil, our water, and our food, and our bodies have adapted to its presence. The arsenic in our food and water does not generally pose a problem. Although it is usually found at higher concentrations in seafood, these particular arsenic compounds are not toxic. However, the natural presence of high concentrations of arsenic in drinking water currently threatens the lives of millions of people in India, Bangladesh, Mexico and elsewhere, and the developed world has been slow to respond with aid. The situation in Bangladesh and West Bengal has been declared the "largest mass poisoning of a population in history."

Sometimes, our own activities, such as mining and pesticide manufacturing, lead to high local arsenic concentrations in soils, slag heaps and mine tailings which, when located close to human activities, can produce human health risks. In evaluating these risks we need to realise that not all arsenic compounds have the same properties, so they are not equally toxic; and also, not all arsenic compounds are equally available to human metabolic processes when ingested. Proper consideration of these aspects of arsenic's nature can greatly reduce the emotional and financial costs of dealing with the element and its compounds when the need arises.

I wish to thank the many graduate students, postdoctoral fellows, technicians and visitors who worked in my laboratories over the years and who all contributed something to this story. Special mention must be made of Vivian Lai

Preface vii

who continues to keep all our projects, including this one, on track, and Elizabeth Varty who produced many of the figures.

I am grateful to the Rockefeller Foundation for providing me with the opportunity to spend some time in Bellagio, Italy, to work in magnificent peace on this manuscript, and to the Natural Sciences and Engineering research Council of Canada for some financial support.

Thanks to Vas Aposhian of the University of Arizona, and David Thomas of the US EPA, for their past collaborations and their never-ceasing support and encouragement; thanks also to Kevin Francesconi and Walter Goessler, University of Graz; Brian Nicholson, University of Waikato; Ian Rae, University of Melbourne; Ronald Bentley, University of Pittsburgh; Helena Solo-Gabrielle, University of Miami; Joerg Feldmann, University of Aberdeen; John McArthur, University College London; Alan Storr and six volunteer graduate students, University of British Columbia for their comments on parts of the manuscript; and to the staff of the National Museum of Health and Medicine, Walter Reed Army Medical Centre, Washington DC, for supplying material for Figures 1.2 and 2.3 and other courtesies.

I am extremely grateful to Jane Bailey who provided invaluable editorial advice and assistance. I am indebted to the friendship and counsel of Ken Reimer of the Royal Military College, Kingston, my colleague and collaborator for many years.

To Sandra

Contents

Chapter 1   Medicinal Arsenic: Toxic Arsenic

    1.1   The Element   1
    1.2   Mineral Medicine   2
           1.2.1   Theophrastus Philippus Aureolus Bombastus von Hohenheim, aka Paracelsus   6
    1.3   Arsenic Eaters of Styria   7
    1.4   Fowler's Solution   13
    1.5   Acute Promyelocytic Leukemia (APL)   20
    1.6   The Organoarsenicals   22
           1.6.1   Robert Bunsen   22
           1.6.2   Paul Ehrlich   24
           1.6.3   The Golden Age of Organoarsenicals   28
           1.6.4   African Sleeping Sickness   31
    1.7   The Darker Side: Toxicity   33
    1.8   Arsenicosis and Cancer   35
    1.9   Biomarkers   38
           1.9.1   Urine   38
           1.9.2   Hair   39
           1.9.3   Finger and Toe Nails   40
           1.9.4   Saliva   40
    1.10 Animal Models   41
    1.11 Chelate Compounds and Chelating Agents   41
           1.11.1 Chelation Therapy   43
    1.12 Some Historical Connections   43
           1.12.1 Charles Darwin   43
           1.12.2 Karin Blixen aka Isak Dinesen   46

Is Arsenic an Aphrodisiac? The Sociochemistry of an Element
By William R. Cullen
© William R. Cullen 2008

	1.12.3 Alexander Borodin, Professional Organic Chemist and Amateur Composer	46
	References	47

Chapter 2 Arsenic Where You Least Expect It

2.1	Animal Feed Additives	58
2.2	Heartworm	60
2.3	Pesticides and Herbicides	61
	2.3.1 Lead and Calcium Arsenates	61
2.4	Arsenic Trioxide	63
	2.4.1 The Black Death	63
2.5	Wood Preservation	67
	2.5.1 Chromated Copper Arsenate (CCA)	67
	2.5.2 Disposal of Treated Wood	72
	2.5.3 Alternatives to CCA	74
2.6	Monomethylarsonic Acid and Dimethylarsinic Acid	75
	2.6.1 Use in the USA	75
	2.6.2 Canada	76
2.7	OBPA	78
2.8	Arsenic in Other Products and Processes	78
	2.8.1 Ironite	78
	2.8.2 Gallium Arsenide	79
	2.8.3 Glass Making	80
	2.8.4 Embalming	81
	2.8.5 Taxidermy	85
	2.8.6 Pigments	87
2.9	Some Historical Connections	88
	2.9.1 Clare Boothe Luce	88
	2.9.2 The Peale Family	89
	2.9.3 King George III	90
References		92

Chapter 3 Arsine, Scheele's Green, Gosio Gas, and Beer

3.1	Arsine	99
3.2	Scheele's Green	103
3.3	Arsenical Wallpaper	105
	3.3.1 Coal Tar Dyes and the Decline of Arsenical Colours	106
3.4	Medical Problems	107
3.5	Wallpaper Dust or Gas?	109
3.6	Gosio Gas	110
3.7	The Regulation of Arsenic, the "Verdant Assassin"	112
3.8	Other Assassins	113

Contents   xi

	3.9	Frederick Challenger	115
3.10	The Toxicity of Gosio Gas	118
3.11	Sick-Building Syndrome?	120
3.12	The Manchester Beer Incident	120
3.13	An Historical Connection. William Morris	124
References		127

Chapter 4 Arsenophobia: A Connection between the Deaths of Infants and Napoleon I

4.1	Sudden Infant Death Syndrome	130
4.2	The Toxic-Gas Hypothesis	131
	4.2.1 The Reaction	133
	4.2.2 The Turner Commission	134
	4.2.3 The Back-to-Sleep Campaign	135
4.3	The Limerick Report	137
	4.3.1 Antimony Biomethylation	138
	4.3.2 Report Summary	139
4.4	Dr. T. J. Sprott	140
	4.4.1 Sheep Skins	141
4.5	Other Proponents of the Toxic-Gas Hypothesis	143
4.6	Toxicity and Related Considerations	144
4.7	The Death of Napoleon I of France	145
	4.7.1 Was it the Arsenic in the Wallpaper?	145
	4.7.2 The Autopsy	146
	4.7.3 Arsenic Poisoning?	147
	4.7.4 The "Real" Cause of Napoleon's Death	151
	4.7.5 Who Did It?	152
4.8	Some Analytical and Chemical Problems	152
	4.8.1 The Preservation of the Corpse	152
	4.8.2 The Lethal Phase	153
	4.8.3 The Hair Analysis	153
4.9	The Overall Picture	155
	4.9.1 Sources of Arsenic	157
	4.9.2 Wine and Water	157
	4.9.3 Self-medication	158
	4.9.4 Arsenical Smoke and Preservatives	158
	4.9.5 Arsenical Straws	160
4.10	Medical Evidence	160
References		161

Chapter 5 Arsenic and Crime: The Law of Intended Consequences

| 5.1 | Introduction | 166 |
| 5.2 | Ancient Times | 167 |

5.3	European Excess: The Age of Arsenic		167
	5.3.1	Italy of the Borgias and the Medicis	168
	5.3.2	France: The Poisons Affair	170
5.4	Forensic Science		171
	5.4.1	Mary Blandy	172
	5.4.2	James Marsh	173
	5.4.3	Marie Lafarge	174
	5.4.4	The Arsenic Act of 1851	176
	5.4.5	Madeleine Smith	178
	5.4.6	Thomas Smethurst	181
	5.4.7	Florence Maybrick	183
	5.4.8	Herbert Armstrong	185
	5.4.9	Marie Besnard	187
5.5	Public Perceptions		188
	5.5.1	Arsenic and Old Lace	189
	5.5.2	Crime Fiction	190
	5.5.3	Portrait of a Poisoner	191
5.6	Some Serial Killers		192
	5.6.1	Mary Ann Cotton, Britain's First Serial Killer	192
	5.6.2	The Black Widows of Liverpool	193
	5.6.3	Vera Renczi	194
	5.6.4	Madame Popova	194
	5.6.5	Johann Hoch	195
	5.6.6	The Arsenic Gang	195
	5.6.7	The Grandmothers of Nagyrev	195
	5.6.8	Dr. Michael Swango	196
	5.6.9	Donald Harvey	197
5.7	Delivery Systems		197
	5.7.1	Food and Drink	197
	5.7.2	The Poisoned Shirt	198
	5.7.3	Application via a Prophylactic	198
	5.7.4	The Poisoned Maiden	198
	5.7.5	The Poisoned Ring	199
	5.7.6	The Poisoned Candle	199
5.8	Public Arsenic Attacks		199
	5.8.1	Japanese Curry	199
	5.8.2	Campus Coffee	200
	5.8.3	Church Picnic	201
5.9	Two Ongoing Cases		202
	5.9.1	A Political Poisoning	202
	5.9.2	Cynthia Sommer	203
5.10	Bezoars, Unicorns and Food Tasters		205
5.11	Some Historical Connections		207
	5.11.1	Zachary Taylor	207
	5.11.2	Wolfgang Amadeus Mozart	208

		5.11.3	Pyotr Ilyich Tchaikovsky	209
	References			210

Chapter 6 Arsenic at War: Mass Murder

	6.1	Introduction		215
	6.2	The First Chemical Weapons Conventions		217
	6.3	World War I: The Gas War		217
		6.3.1	Mustard Gas	221
		6.3.2	Blue Cross	223
		6.3.3	Arsenical Agents: The Second Generation	224
		6.3.4	Tactics of Chemical Warfare	225
	6.4	The US Enters the Fray		226
		6.4.1	Lewisite	227
		6.4.2	Phenarsazine Chloride	228
	6.5	Arsenical Chemical-Warfare Agents		232
	6.6	Casualties of the Chemical War		233
		6.6.1	The Combatants	233
		6.6.2	Civilian Casualties	234
	6.7	The Aftermath		234
		6.7.1	The Humane War?	234
		6.7.2	Public Reaction	235
	6.8	Living with Chemical Weapons		239
		6.8.1	The Geneva Convention	239
		6.8.2	The German Reaction	240
		6.8.3	Spain in Morocco	240
		6.8.4	Italy in Ethiopia (Abyssinia)	241
		6.8.5	Japan in China	242
	6.9	WWII – The Gas War That Never Happened		244
		6.9.1	The Buildup in Europe	244
		6.9.2	Russia	245
		6.9.3	The United States	245
		6.9.4	Canada	246
		6.9.5	The European Experience	250
		6.9.6	The War in the Pacific	251
	6.10	Human Guinea Pigs		252
		6.10.1	The Allies	252
		6.10.2	Japan and Germany	254
		6.10.3	The Nuremberg Code of 1947 and its Aftermath	254
	6.11	The Vietnam War		257
		6.11.1	Agent Blue	257
		6.11.2	Adamsite and Other Tear Gases	259
		6.11.3	Health Effects	259
		6.11.4	The Public Reaction	260
	6.12	The Chemical Weapons Convention		261

6.13	The Cleanup	263
	6.13.1 The Early Years	263
	6.13.2 Post-WWII	264
	6.13.3 Japan	266
	6.13.4 Domestic Ocean Dumping	266
6.14	Disposal of Stockpiles	267
	6.14.1 Russia	267
	6.14.2 United States	268
6.15	Some Special Problems	270
	6.15.1 Munster, Germany	270
	6.15.2 Spring Valley, US	271
	6.15.3 Bowes Moor, UK	272
	6.15.4 China	272
	6.15.5 Albania	273
	6.15.6 Other Recent Deployments of Chemical Weapons	274
6.16	Conclusions	275
References		277

Chapter 7 Arsenic and the Environment

7.1	Introduction	287
	7.1.1 Arsenic in the Atmosphere	288
	7.1.2 Arsenic in the Pedosphere	288
	7.1.3 Arsenic in the Hydrosphere	288
7.2	Arsenic in the Biosphere	289
	7.2.1 Arsenic in Seafood	289
	7.2.2 Analysis of Arsenic Species (Speciation) in Living Organisms	290
	7.2.3 Distribution of Arsenic Species in the Living Environment	294
	7.2.4 Where do Arsenosugars and Arsenobetaine Come From?	296
	7.2.5 Arsenic Accumulators and Hyperaccumulators	299
7.3	Arsenic in Our Food and Water	300
	7.3.1 Essentiality	301
	7.3.2 Arsenic Market Baskets	302
	7.3.3 The Effect of Cooking	304
	7.3.4 More on Rice	304
	7.3.5 Hijiki and Other Algal Products	305
	7.3.6 Bottled Water	308
	7.3.7 Metabolites	310
7.4	Bioavailability and Bioaccessibility	310
	7.4.1 Sequential Selective Extraction (SSE)	312
	7.4.2 Gastrointestinal Models	312

Contents xv

7.5	Arsenic in the Anthrosphere	313
	7.5.1 Gold Prospecting	314
7.6	Arsenic Trioxide and the Giant Mine, Yellowknife NT, Canada	315
	7.6.1 Giant Mine: An Underground Cleanup?	317
	7.6.2 Giant Mine: Surface Cleanup	318
7.7	American Smelting and Refining Company. Asarco	319
	7.7.1 The Everett and Tacoma Smelters	319
	7.7.2 The Globe Smelter: Some Unexpected Relief	321
7.8	A Transboundary Dispute: Teck Cominco *vs.* US EPA	321
7.9	More Woes	322
	7.9.1 Some Other Surfaces Affected by Mining	322
	7.9.2 Nickel Arsenide	323
7.10	Arsenic in Energy Sources	324
	7.10.1 Coal	324
	7.10.2 Arsenical Peppers	326
	7.10.3 The Oil Sands of Alberta	327
	7.10.4 The Sydney Tar Ponds: Arsenic as an Environmental Hammer	328
	7.10.5 Cleaning Up	329
	7.10.6 Monitored Natural Attenuation	331
7.11	Microbes and Arsenic	332
	7.11.1 More – but very Small – Arsenic Eaters	333
References		339

Chapter 8 Accidental Exposure to Arsenic: The Law of Unintended Consequences

8.1	Introduction	349
8.2	West Bengal and Bangladesh: The Devil's Water	350
	8.2.1 The Green Revolution	350
	8.2.2 Bangladesh	351
	8.2.3 The Affected	354
	8.2.4 Where does the Bangladesh Arsenic Come From?	355
8.3	Professor Dipankar Chakraborti	358
	8.3.1 Field Testing Kits	360
8.4	Arsenic Mitigation in Bangladesh	362
	8.4.1 Dhaka, Bangladesh, January 2002	362
	8.4.2 Arsenic-Mitigation Technologies	364
	8.4.3 Verification of Mitigation Technologies	365
	8.4.4 The Grainger Challenge Prize	367
	8.4.5 Nanoparticles	369
	8.4.6 Other Arsenic-Mitigation Methods	369

8.5	Where Are We Now?		370
	8.5.1 Treatment Options for the Afflicted		370
	8.5.2 Arsenic Mitigation		371
8.6	Taiwan		373
	8.6.1 Southwestern Taiwan		373
	8.6.2 Northern Taiwan		373
8.7	Vietnam and Elsewhere in the East		374
	8.7.1 Vietnam		374
	8.7.2 Nepal		374
	8.7.3 China and Japan		374
8.8	South America		375
	8.8.1 Argentina		375
	8.8.2 Chile		375
8.9	Africa		376
8.10	Europe		378
8.11	North America		379
	8.11.1 Mexico		379
	8.11.2 The US Standard for Drinking Water		379
	8.11.3 Setting the US MCL		381
	8.11.4 Cost/Benefit Analysis		383
	8.11.5 The MCL Revisited		385
	8.11.6 Fallon, Churchill County, Nevada		385
8.12	Other Small Systems		387
8.13	The Canadian Maximum Acceptable Concentration (MAC)		387
8.14	The Opportunists Knock		387
8.15	Epilogue		388
	References		389

Subject Index 398

CHAPTER 1
Medicinal Arsenic: Toxic Arsenic

I am an evil, poisonous smoke
But when from poison I am freed,
Through art and sleight of hand,
Then I can cure both man and beast,
From dire disease oft times direct them;
But prepare me correctly, and take great care
That you faithfully keep watchful guard over me;
For else I am poison, and poison remain,
That pierces the heart of many a one.

– **Valentini, 1694**

1.1 The Element

Arsenic is classified in Group 5 of the Periodic Table of the Elements. The elements in Group 5, in order of increasing atomic weight, are nitrogen, phosphorus, arsenic, antimony and bismuth. The first two are nonmetals and are essential for life, the last two are metals and not essential for life. Arsenic is a metalloid, neither a metal nor a nonmetal and is best known as a terminator of life.

The word arsenic is derived from the Greek *arsenikon* meaning valiant, bold or potent – an allusion to the ease with which it combines with many other elements. Some texts state that the element was first prepared around 1250 by Albert Bollstädt (aka Albert Magnus) by heating arsenic trioxide (Section 1.2) with soap, but soot seems more likely to have been used in the reaction.

Ancient civilisations were familiar with a number of minerals that we now know contain arsenic, but they didn't think that they were related in any way. Fire provided the only available means of causing chemical transformations, and this practice, which became known as smelting, eventually led to the

---

Is Arsenic an Aphrodisiac? The Sociochemistry of an Element
By William R. Cullen
© William R. Cullen 2008

discovery of metals and alloys that could be used for tools and weapons. Any arsenic in the minerals would usually reveal its presence in the form of a white smoke released from the heated mass, as in the verse above. The toxic nature of this smoke would have become quickly obvious to these early metallurgists; however, they also used the smoke for medicinal purposes (Section 1.4).

The earliest metal objects were made mainly from native copper (copper found in a chemically uncombined state), but once the move was made to copper ores, useful impurities, such as arsenic, allowed easier smelting and casting. It was found that the copper/arsenic and copper/tin alloys, known as bronzes, were harder than copper alone: they ideally contained 5 to 15 per cent arsenic or tin. Cretan and Western Mediterranean smiths worked with the readily available arsenic ores around 3000 BCE. Eventually the copper/tin bronzes dominated in all regions, possibly because of the health problems associated with working with the arsenic-rich ores. Tin became a strategic metal. Fuelling the industry's furnaces required an enormous amount of wood, so many areas of the Mediterranean became denuded of trees.[1] Europe was well into the Iron Age by Roman times and although arsenic was only a minor constituent of wrought iron, the smelting process was equally hard on trees.

The thawing of a glacier in the Italian Alps, near the Austrian border, resulted in the 1991 discovery of the well-preserved body of a man who became known as Oetzi. He was a hunter who had been dead for some 5000 years, making him older than the Egyptian mummies. There was considerable media interest in establishing whether he had died of arsenic poisoning as a result of being involved in mineral smelting: he had a high concentration of arsenic in his hair (20 micrograms of arsenic per gram of hair, 20 µg/g. We will abbreviate this to 20 parts per million, 20 ppm) (Section 1.9).[2] Speculation along these lines ceased once it was discovered he had an arrowhead in his back, close to a lung.[3]

Soon after the discovery of Oetzi there was also speculation that the deaths of six individuals, who were loosely connected to the mummy, were the result of the "Curse of the Iceman."[4]

1.2 Mineral Medicine

The ancient Greeks were able to identify two arsenic-based minerals: orpiment and realgar. The first is yellow and was known as *arrhenikon*, or *arsenikon* (sometimes *auripigmentum*), and the second is red and was known as *sandarach*. Centuries later these were both identified as arsenic sulfides of formula As_2S_3 and As_4S_4, respectively. Pliny and his contemporaries believed they were different substances and that they were made up of different proportions of the four so-called Aristotelian elements: air, earth, fire and water. These four were associated in turn with the four humours: blood, yellow bile, black bile and phlegm. Realgar was mined by condemned prisoners who died in the noxious atmosphere underground.[5] Greek physicians such as Hippocrates (460–357 BCE) administered a paste of the sulfides for treatment of skin ulcers.[6] In his play *Caesar and Cleopatra* George Bernard Shaw has Cleopatra recommending the use of realgar for baldness: "For bald patches, powder red sulfuret of arsenic and take it up with oak gum, as much as it will bear. Put on a rag and apply, having soaped the place well first."[7] But no matter how much you respect the author, caution is advised. The Greeks actually mixed arsenic sulfides with lime to give a product that *removed* hair, skin, and flesh, and was a deadly poison (This mixture was still being used in animal slaughter houses some 2000 years later.) A yellow-white powder, is obtained if either sulfide is heated. This product was used externally and internally to cure cancer, fistulas, asthma and other breathing problems. By the 11th century, Arabian writers recognised three forms of arsenic: white, yellow and red, which we now know as arsenic trioxide and the two sulfides. They were aware of the poisonous nature of white arsenic that seems to have been prepared in pure form around the 8th century by an Arabian alchemist who distilled the white form from the yellow.[8,9] In general, when the word arsenic, the "A-word", is used without qualification the reference is usually to this oxide, white arsenic.

There were parallel developments in the eastern world. The Chinese began to practice herbal medicine around 2800 BCE. This experience was eventually gathered into Shen Nong's *Materia Medica* (200 BCE) around the time of the unification of the nation and the beginning of the Han dynasty.[10,11] Shen, who became known as the Divine Farmer, was familiar with the two arsenic sulfides, and also arsenolite the mineral form of the oxide. Like the Greeks, the Chinese thought of matter in terms of combinations of the five elements water, fire, wood, metal and earth. These five are associated with the five planets visible to the naked eye; and these in turn are associated with five minerals. Thus, we have in one group the element water, the planet Mercury, and the mineral magnetite (Fe_3O_4). In other groups we have fire, Mars and cinnabar (HgS); wood, Jupiter and malachite ($Cu_2CO_3(OH)_2$); metal, Venus and arsenolite (As_2O_3); and earth, Saturn and realgar (As_4S_4). It is curious that two arsenic minerals, one of them very toxic, were selected for this honour. It is also curious that according to the book *Huai Nan Tzu* (*ca.* 125 BCE) a mineral elixir made from the "essences of the five planets" was claimed to give a man perpetual life, exempt from death forever.[12] There is the possibility that perpetual life was achieved when the frequent user entered the afterlife. The corpses were said to become stiff and metallic-looking and did not decay. The connection between arsenic and the preservation of corpses is taken up later (Section 2.8).

The Chinese called realgar "male yellow" and, despite its name, the best quality was said to be the colour of cockscomb (a livid orange-red). As early as 222 BCE, realgar was recommended for use against skin diseases such as carbuncles, abscesses and tuberculosis, but it also had magical properties. It was purported to kill ghosts and cure people possessed by demons. It was said to regenerate lost teeth. When worn as an amulet, Realgar warded off dangerous animals and venomous snakes, because snakes were said to have an aversion to the mineral and this belief was the basis for many myths.[13,14] In one of these, a monk suspects that the wife of a villager is really a white snake because of the marks on the husband's face. The monk persuades the husband to give realgar wine to his wife, who then falls ill. She goes to bed, covering herself with a blanket and asks the husband to leave her alone. The husband is curious, so he lifts the blanket to reveal a snake. A long story follows with an ambiguous ending suggesting that the husband should have left things as they were.

Drinking realgar wine was once an important part of the Chinese dragon boat festival that, in addition to boat racing, involves eating rice dumplings wrapped in bamboo leaves. (Children were not expected to drink the wine: they got a dab of red powder on their foreheads.) The benefits of the drink: "all illnesses are banished; scars disappear; grey hair turns black; lost teeth are regenerated; after a thousand days fairies will come to serve you."[13] This custom was based on the following legend that dates back from about 300 BCE: Through no fault of his own a respected government official named Chu Yun fell out of favour with the Emperor. In despair, he committed suicide by jumping into the Mi Lo River on the fifth day of the fifth month of the lunar calendar. His grieving friends rowed a boat out on the river making a lot of noise to scare any river creatures away from the body. Others threw rice dumplings into the river in the hope that river creatures would eat only the dumplings. An old doctor poured a bottle of realgar wine into the river to ward off river creatures. Moments later a dead dragon floated to the surface.[15]

The Dragon Boat festival remains a major event in Asian communities especially in Hong Kong. There are boats, noise, rice dumplings, and huge crowds, but, these days, no obvious signs of realgar wine. When shopkeepers and dispensers in Hong Kong are asked for realgar wine they offer alternatives such as snakes in wine.[16]

The Chinese have been making wine for more than 9000 years and probably discovered realgar wine as a result of using realgar in making drinking vessels.[17] Traditional Chinese herbal medicine balls that are dissolved in warm wine (or water) and drunk as tea are a modern source of realgar wine. These balls are readily available in the USA and many are rich in arsenic and mercury, posing a potentially serious health risk to consumers.[18]

Orpiment was known as "female yellow." It was used to cure some skin conditions but not much otherwise, apart from its use in correcting mistakes in written texts because the colour matched that of the paper. The rare oxide mineral, arsenolite, was known to be toxic and was used against some skin conditions and as a rat poison.[14] The rat catchers were said to have a short life span.

Around 973 CE, during the Song dynasty, the Chinese began to describe a new drug, *pishuang*, the sublimate (white smoke) obtained when some minerals were heated to decomposition. Unlike the Greeks, the Chinese eventually realised they were preparing a drug, arsenic trioxide, that they already knew from natural sources. This drug was prescribed for a number of "women's complaints" and for all intermittent fevers including malaria, which was on the rise because population pressure was forcing the settlement of marshy regions. One treatise of the time lists 139 formulations for treating intermittent fever, 39 of which contain arsenic oxide. The rationale for its use was that the heat of the drug was necessary to combat the heat of the fever. Some pills, such as the Pill of Marvellous Application, contained up to 100 mg of arsenic trioxide; very close to the lethal dose (in the range of 100 mg to 200 mg). There must have been many accidents as a result of compounding errors; nonetheless, the use of arsenic oxide continued in China until about the 16th century. The *Materia Medica* of 1590 lists about 220 kinds of preparation from minerals including arsenic sulfides and the oxide. In the 17th century, Chinese medicine began using arsenic to treat venereal disease.[19]

The Pharmacopoeia of the Peoples Republic of China (English 1997) lists 48 products that contain metals and metalloids; 23 of these include realgar and 41 include cinnabar (red mercuric sulfide).[20] Realgar is believed to be particularly effective when mixed with bezoars (stones found in the intestines of ruminant animals) (Section 5.10). In the Pill of Six Miraculous Drugs, realgar disperses the "stagnated substance" and another ingredient relieves swelling and alleviates pain. The prescription is applicable to carbuncles, breast abscesses, nasopharyngeal carcinoma, lung cancer and other complaints attributable to phlegm, fire and toxic material. Usually, prescriptions containing realgar carry warnings against roasting by fire because of the risk of forming arsenic trioxide.[21]

In February, 2007, Reuters used the headline "Chinese Police Fish for Arsenic and Old Carp," for a story about a 3 kilogram carp that had been soaked in an arsenic solution and hung outside to dry, prior to being used in traditional medicine. The alarm was raised in the city of Xiangfan when the fish disappeared. The police and citizens were mobilised and the media went into high gear, fearing someone would eat it. "Anyone who discovers suspicious headless preserved fish should call the police." And: "Anyone who finds any suspicious dead cats or dogs should call the police." There was no follow-up story.[22]

Yellow pills of the patent medicine known as Niu Huang Jie Du Pian (bezoar detoxifying pills) are readily available over the counter in many countries. Self-medication with this product, which is almost the Chinese equivalent of aspirin, has resulted in several reported cases of arsenic poisoning and many more likely go undetected and unreported. The amount of arsenic per pill in the form of realgar depends on the manufacturer, but can be as high as 28 mg with the daily dose of up to six pills. Many have argued that because realgar is insoluble in water there should be nothing to worry about because it will simply pass through the human gut without change. However, about 4 per cent of the realgar is bioaccessible, *i.e.* available to be taken up within the human body (Section 7.4), as is demonstrated by a substantial increase in the arsenic content in the urine of the consumer.[23,24] The presence of realgar in the antiasthmatic herbal preparation known as Sin Lak has also caused problems.[25] There is no question that these pills can pack an arsenic punch.

Arsenic trioxide was used as an aphrodisiac in India, particularly the Punjab region.[26] Arsenic is claimed to enhance the aphrodisiac properties of opium leading to toxicity in addicts.[27] Pills known as Kushtay that contain herbs compounded with arsenic and lead are described as an Indian aphrodisiac by the World Health Organization, although they are not recommended. The US Agency for Toxic Substances and Disease Registry (ATSDR) refers to the same product as "Asiatic Pills," also as "yellow root," a tonic used in Asian cultures to cure various sexual disorders.[28,29] Tayyeb Dawakhana Pvt Ltd. sells Kushtay formulated in different ways to produce specific effects: one variation known as "kushta sang jarahat" is used to treat gonorrhea: another "kishta sang yashb" treats insanity, loss of sexual power, premature ejaculation, and other problems. Pregnant women in India take arsenic-containing medicine during their first trimester in order to ensure the birth of a male child. Another herbal supplement popular as an aphrodisiac in North Korea contains high levels of arsenic, but the active ingredient is the chemical used in Pfizer's impotence drug, Viagra.[30]

In 1992 "Deep Red Powder" was among the offerings available from Indian ethnic practitioners: it contained 90 mg of arsenic trioxide and 550 mg of mercuric sulfide per dose. Consumption resulted in chronic poisoning.[31a] Related Ayurvedic medicines, some described as aphrodisiacs[31b] which have been produced in South Asia for thousands of years and are now consumed by people all over the world, are also plagued with toxicity problems. Many contain potentially harmful levels of lead and mercury, as well as arsenic.[32] One 125 mg dose of *Suvarna sameepannaga* contains approximately 5 mg of gold; 24 mg of mercury; 24 mg of sulfur; 24 mg of orpiment; 24 mg of arsenic trioxide; and 24 mg of realgar.[31b] Health warnings have been issued in the Western World,[33] but the problem has been exacerbated by the availability of many known toxic concoctions for sale over the Internet.

There was an outcry in 2005 from the leaders of the Ayurvedic Medicine Manufactures Organization of India (AMMOI) when the Indian government notified them that exports should be free of compounds of lead, arsenic and mercury, among other things (the current permissible levels are 10 ppm for arsenic and lead, and 1 ppm for mercury). AMMOI argued that the chemicals are used in combination and testing individual components is meaningless.[34]

Nevertheless, the concerns of health professionals and governments are justified. More people turn to these alternative therapies each year: 48 per cent of adults in the United States used at least one alternative or complementary therapy in 2004; and 23 per cent of Canadians aged between 25 and 54 have now turned to alternative care.[35,36]

Realgar is still with us. It is an ingredient in a recently patented Chinese medicine for treating psoriasis, ichthyosis and eczema.[37] But perhaps the biggest surprise came around 1999, when a patent was filed claiming that realgar was effective against acute promyelocytic leukemia, chronic myeloid leukemia, and non-Hodgkin's lymphoma.[38] We will come back to this topic later in the chapter (Section 1.5).

1.2.1 Theophrastus Philippus Aureolus Bombastus von Hohenheim, aka Paracelsus

During the time of the alchemists, when creating gold was seen to be much more important than creating cures, arsenic was well enough known in Europe for Chaucer, who died in 1400, to write:

I wol yow telle, as was me taught also,
The foure spirites and the bodies sevene,
By ordre, as afte I herde my lord hem nevene.

The firste spirit quik-silver called is,
The second orpiment, the thridde, y-wis,
Sal armoniak, and the ferthe brimstoon:
The bodies sevene eek, Lo! hem heer anoon:
Sol gold is, and Luna silver we threpe,
Mars yren, Mercurie quik-silver we clepe,
Saturnus leed, and Jupiter is tin,
And Venus coper, by my fader kin!

Skipping forward a century, we meet Paracelsus, who was born Theophrastus Philippus Aureolus Bombastus von Hohenheim, in Switzerland in 1493. He studied a little medicine, eventually claiming the title doctor for himself, but he espoused very unorthodox views for the time: "Nothing is a poison that benefits the patient; only that is to be considered a poison which injures him. . . . The dose makes either a poison or a remedy." He inspired others to think about the application of chemical substances in medicine (iatrochemistry) and wrote about arsenic as follows: "It is to be noted that there are those which flow forth from their proper mineral or metal, and are called native arsenics. Then there are those made by Art through transmutation. White of crystalline arsenic is the best for medicine. Yellow and red arsenic are utilised by the chemist for investigating the transmutation of metals, in which arsenic has a special affinity. . . . The virtues of arsenic are for ulcers, wounds, and other openings. The virtues of orpiment obtain in fistulas, cancers and eating ulcers." His recipe for a cancer cure: "Orpiment, fuligo [possibly mercury] and sal ammoniac. Reduce by the fourth grade of reverberation [*i.e.* heat in a furnace] a day and a night. Reduce into alkali. Take this prepared orpiment and mix with cinders of pigeon's dung and oil of yolk of eggs."[39]

Paracelsus also treated syphilis with arsenic and mercury, although only the mercury treatment continued in favour with his successors. The mercury treatment was almost worse than the disease: a cure took five to six years and patients experienced ulceration of the tongue, jaws and palate, swelling gums, dripping saliva and fetid breath. Many died. This was all summed up years later in the saying: "One night with Venus – an eternity with Mercury."[10]

1.3 Arsenic Eaters of Styria

In the 16th and 17th centuries, red and white arsenic were put into amulets that were worn around the neck and close to the heart to ward off the plague; but at least one population took a more proactive approach.

The arsenic eaters of Styria, now a region of Austria near Graz, first came to public attention in the United Kingdom in the middle of the 19th century. Two popular publications of the time, Chambers' Journal and Blackwood's Magazine, chose to comment on the revelations of Dr. T. von Tschudi of Vienna.[40,41] Thus, we have in Chambers Journal, December, 1851, "The Poison Eaters;" February, 1856, "Poison Eaters;" and July, 1856, "The Arsenic-Eating

Question." Blackwood's Magazine ran an article in December, 1853, "The Narcotics We Indulge In."

At the time, von Tschudi's seemingly preposterous thesis that people would voluntarily eat a known poison and build up a resistance to it over time, was the subject of wonder and ridicule. It was also the fuel that would drive some wonderful crime stories. In one of the greatest of these, "Strong Poison" by Dorothy Sayers, Lord Peter Wimsey remarks: "As you know it's not good for you in a general way, but there are people – those tiresome peasants in Styria one hears so much about – who are supposed to eat it for fun. It improves their wind, so they say, and clears their complexions, and makes their hair sleek, and they give it to their horses for the same reason; bar the complexion, that is, because a horse hasn't got much complexion, but you know what I mean."[42]

Hittrich is the German term for the white smoke, which is mainly arsenic trioxide, that accompanies the smelting of arsenic-rich minerals. In Styria, the mineral was arsenopyrite (FeAsS), also known as mispickel (Section 7.5), and the smelting was usually performed in small huts. The poisonous properties of the fumes that collected as a white or yellow deposit in the chimney or escaped to the outside air were well known. Industrial production developed in the 15th century and the product was transported to Venice overland for glass manufacture and sold to the Levant for medicine and rat poison.[43] The early supplies of arsenic oxide from Austria were impure and production was limited, resulting in 60 tonnes per year in the 16th century, and 40 tonnes per year in the 17th.[44]

Sometime during the 16th and 17th centuries, inhabitants of the region developed the practice of eating this deposit, earning the nickname *Hittrichfeitl* or *Hittrichhansl*. They did so for the following benefits:[8,45]

- arsenic eating puts a bloom in the cheeks of women and enhances their beauty;
- it improves the wind and thus increases the ability to undertake strenuous tasks such as hiking up mountains;
- it acts as an aid to digestion;
- it acts as a prophylactic against infectious diseases;
- it increases courage;
- it increases sexual potency; and
- it acts as a contraceptive.

All this was put into a romantic setting by J.F. Johnston:[46] "By the use of hidrach [*hittrich*] the Styrian Peasant-girl adds to the natural graces of her filling and rounding form, paints with brighter hues her blushing cheeks and tempting lips, and imparts a new and winning lustre to her sparkling eyes. Everyone sees and admires the reality of her growing beauty: the young men sound her praises, and become supplicants for her favour. She triumphs over the affections of all, and compels the chosen one to her feet. Thus even cruel arsenic, so often the minister of crime and the parent of sorrow, bears a blessed jewel in its forehead and, as a love awakener, becomes at times the harbinger of happiness, the soother of ardent longings, the bestower of contentment and peace."

About the same time that arsenic eating surfaced, arsenic was being introduced to the region as an ingredient in magic concoctions used against bubonic plague and other infections. The potions that contained arsenic were typical of the time: they contained something hot or spicy such as pepper or chili; something revolting such as wood lice, mashed spiders or snake feces; and some arsenic compound. The whole concoction was mixed into schnapps resulting in a "magic" drink. Similar ingredients were part of the conventional pharmacopoeias of the time and reflect the concept that the demons causing the disease needed to be frightened off by something really disgusting.[10] (A very successful modern advertising campaign for Buckley's cough syrup in North America, is based on a similar concept: "It tastes awful. And it works".)

Plagues were considered to be of demonic origin, so cures were in the realm of the church. In Styria, some chose to use the magic drink for self-medication because they could not afford a physician. Self-medication was regarded as a sin, so the arsenic eaters of this time did so in private to avoid punishment by both the church and the law.[8] During the Reformation era there was a witch craze that peaked between 1550 and 1650 and, if caught, arsenic eaters were tried as witches. In Germany, the centre of religious revolt, about 26 000 men and women were killed during this time following a trial in local secular courts.[8] Arsenic purchases were illegal in most countries in central Europe so an underground economy in *hittrich* developed. It was smuggled in from Hungary, or stolen from local arsenic works and glass works.[47] Some suggest that the need for secrecy resulted in the coverup of some inevitable casualties. Roscoe reports (1862) that there were an excessive number of poisoning cases involving arsenic in Styria.[47] He mentions the use of the usual remedy for overdoses: "hydrated peroxide of iron," implying that this knowledge was part of the culture. He was undoubtedly referring to the remedy "hydrated oxide of iron" discovered by Robert Bunsen in Germany in the same century! (Section 1.6.1).[48] Many earlier overdoses must have gone untreated.

It is likely that the practice of eating arsenic, as opposed to using arsenic in "magical" medicines, began with feeding arsenic to horses to make them healthy and increase their strength. This practice, possibly initiated by gypsies, was still said to be common in Vienna in the mid-1800s;[46] "Grooms and especially coachmen that are employed by persons of high rank feed their horses with arsenic. One method to treat a horse is to scatter arsenic powder into the oats. Another practice of arsenic feeding is to wrap the pea-sized arsenic dose in a cloth. This arsenic piece can be fixed at the bit of the bridle. If the groom puts the bridle on the horse its saliva will slowly dissolve the arsenic. Arsenic feeding is the cause of the shining and beautiful appearance of most refined coach horses."[8] Local grooms then started to eat arsenic in the belief that they had to eat the same medicine in order for the treatment of the horse to be successful. The combination of a belief in arsenic therapy for animals and in magic medicine for humans was peculiar to Styria and resulted in the localised practice of arsenic eating by the inhabitants.

In another version of the story, perhaps less credible, arsenic was thought to have power over evil, so the shepherds in the Carpathian Mountains used to eat very small quantities and feed it to their animals as a protection against vampires. Likewise, alchemists in Prague and other cities in Moravia and Bohemia used to burn arsenic to drive away the powers of evil. The toxic fumes have the same smell as garlic and the thrifty peasants noticed the connection. So expensive exorcisms conducted by alchemists were replaced by the inexpensive practice of eating garlic; and that is how garlic became associated with defence against vampires.[49]

Whatever the true story, the practice of feeding arsenic to horses spread over time to other countries. The jockeys in the National Horseracing Museum in Newmarket, England, were familiar with the use of arsenic in horse feed, even though the modern-day curator would not comment on the practice.[50] It was not uncommon for the urine of winning race horses to be checked for elevated arsenic concentrations.[51] Prominent Australian veterinarian Percy Sykes recently said that arsenic was widely used as a tonic in the 1930s, and 90 per cent of horses would have had it in their systems.[52] He was speaking in response to the report that analysis of hairs from the racehorse Phar Lap indicated that the horse had been given a massive dose of arsenic about 35 hours before his death on April 5th, 1932 (Box 1.1).[53] When Sykes was asked about the legality of the tonics he replied that nobody was testing in those days. Fowler's solution and Donovan's solution, prepared from arsenic trioxide and arsenic triiodide, respectively (Section 1.4), were considered the best for horses.

Box 1.1 Phar Lap

Phar Lap is the most famous horse in Australian racing history. He won 37 of his 51 starts and was a national hero. Phar Lap was shipped to Mexico in 1932 where he won the world's richest race, but two weeks later he collapsed and died a painful death. One very popular theory was that the horse was killed on orders from US gangsters who feared the champion would inflict big losses on their illegal gambling operations. Phar Lap was so much a part of the era that his hide was treated by a New York taxidermist, stuffed and put on show at the Melbourne Museum in Australia. This was the source of the hairs that were analyzed using X-ray spectroscopy (Section 7.3) and found to contain arsenic in the "skin end" of the sample "consistent with a single large dose between one to two days prior to death."[54] The investigators think that they found another arsenic compound, lead arsenate, which may have been used as a preservative, but this is not a common application.[54] Other explanations being offered are: the horse strayed from his stable and ate vegetation that had been treated with an arsenical pesticide; or the horse was given an accidental overdose of a tonic that contained

> **Box 1.1   Continued.**
>
> arsenic and strychnine intended "to enhance his natural abilities." There appear to be reliable witnesses to support these alternatives.[55]
>
> A post-mortem examination of some internal organs conducted at the time of Phar Lap's death did not point to arsenic poisoning as the cause.[56] There is little doubt that arsenic could have been used to help preserve the skin (Section 2.8.5).

Most "successful" human arsenic eaters in Styria consumed arsenic trioxide in doses of 300–400 mg, over periods of 30 or more years. Others consumed a mixture that they called yellow orpiment, because it looked like the natural yellow arsenic sulfide mineral. This "yellow orpiment" was also collected from chimneys and was mainly (90 per cent) arsenic trioxide contaminated with sulfur. Because the composition of this product could vary, it was not considered to be as safe for ingestion as white arsenic. However, yellow orpiment was easily available and was used in spite of the increased risk. Some arsenic eaters preferred a yellow mixture known as "artificial orpiment," made by melting the oxide with sulfur. They preferred to eat solid samples starting at about 10 mg, which they slowly increased every two or three days up to 300 mg to 400 mg. They cut portions from a bigger piece and learned to guess the correct amount for themselves. Some also increased and decreased the dose in relation to the phase of the moon. Most arsenic eaters ate arsenic with bread and bacon, because the combination was claimed to reduce the rate of absorption and eliminate heartburn. The available liquid preparations of arsenic did not have known reproducible concentrations, and were not reliable.[8]

The belief in the increased safety of arsenic plus fat was not confined to Styria. "One of the nine modes of trial by ordeal described in a *History of Indostan* consists of compelling the accused to eat, from the hand of a Brahmin, a preparation composed of clarified butter and arsenic trioxide. If the poison produces no visible effect, he is absolved; otherwise condemned."[57]

Many scientists expressed scepticism about the existence of arsenic eaters. This was especially true at the time of von Tschudi's publications in the 1850s. "Lastly let me urge upon all who adopt the Styrian system, to make some written memorandum that they have done so, lest, in case of accident, some of their friends may be hanged by mistake."[58] These sentiments underlie the legal strategy known as "the Styrian defence" that we will discuss later (Section 5.4). In modern times, reports of arsenic eating have been equated to sightings of flying saucers and the Loch Ness monster.[59]

The most popular counterargument was that eating arsenic trioxide conflicted with both toxicological knowledge and common sense. The supposed connection between arsenic eaters and magical practice gave rational scientists another excuse for disbelief in their existence. Some of the scientific opposition did accept that Styrian peasants ate a white powder – zinc oxide was suggested – as some kind of secret tonic, but maintained that it could not be arsenic trioxide because otherwise it would be toxic. Others suggested that the Styrians ate insoluble forms, or chunks, of arsenic oxides and sulfides. In support of this thesis, one dog supposedly habituated to a daily dose of 25 mg/kg of undissolved arsenic trioxide, was reported to succumb to a single dose of 15 mg/kg of dissolved arsenic trioxide.[60]

Because the number of arsenic eaters in Styria was relatively small, and because they were very secretive about their habit, it was difficult to unequivocally prove their existence. Nevertheless, there is a considerable body of scientific evidence that Styrian peasants did deliberately ingest poisonous arsenic trioxide.[8,47] In 1860, two arsenic eaters were introduced at the 48th Meeting of the German Society of Natural Scientists and Physicians in Graz, where they ingested arsenic in front of the scientific audience. One volunteer consumed 400 mg of arsenic trioxide and the other 300 mg of "yellow orpiment." Samples of their urine were analyzed by using Marsh's test (Section 5.4) and the results, which clearly showed the presence of arsenic in the urine of both volunteers, were presented to the conference audience. One of the volunteers was a very healthy 66-year-old man named Flecker who had been eating enormous amounts of arsenic for 36 years without any negative side effects. His father was also an arsenic eater who had lived until he was 77. Flecker started to take arsenic, orpiment mainly, when he was 30 years old. Generally, he ate arsenic once a week; more if he felt sick or when he was working at his job as a travelling tailor and needed to go into houses where someone was ill. Flecker was never ill except once for two days.[8]

The superintendent of an arsenic factory in Austria was advised to eat arsenic to avoid being poisoned by the factory fumes. He started by eating 0.2 g of arsenic trioxide per day and over 45 years worked up to a dose of 1.5 g per day. Schroeder and Balassa reported that, around 1930, they both had seen mountaineers eat material that was said to be "arsen." This blackish-gray powder was scraped from rocks and eaten with bread and butter, but was neither the oxide nor a sulfide. They suggest it might have been native arsenic.[61]

Professor Kurt Irgolic of Graz, Austria, an expert on arsenic, said in 1995 that arsenic eaters were common in Styria one generation ago and a medical professor used to have one appear at his lecture. The old man would butter his bread, sprinkle white arsenic on it, eat it in front of the class and depart unharmed with his fee. Rumour has it that he went immediately to the local whorehouse. Irgolic also told of an 86-year-old woman who was eating arsenic because her husband was also an arsenic-eater and she needed to accommodate him. And there was an elderly gentleman who claimed that if he stopped eating arsenic "a number of women in the village would know." Irgolic also said that some arsenic-eaters used a saturated solution of the oxide in alcohol to provide a known dose for man and horse.[62]

The Sryrians mixed arsenic with the tobacco they smoked in their pipes; a custom said to imitate Chinese use.[63] Roscoe describes how he saw a solution of white arsenic being added to cheese curds during manufacture.[47] "I ate some bread and drank some beer with the cheese and experienced nothing but a slight burning in the throat as from food containing much spices, and afterward, a pleasant warmth in the stomach and good appetite."

The Styrians may not have been unique. One source claims that as children Aztecs began to eat arsenic regularly and build up some immunity, and their skin colour is said to be associated with an otherwise unrecorded interaction between the sun and the arsenic under the skin.[64]

Although there were few reports of misadventure from Styria (one successful female eater, after she had her man, decided to increase the dose to achieve even greater effect and "fell a victim to her own vanity"[65]) the same cannot be said for the rest of the world. In 1865 a male patient of Dr. Parker of Halifax, Nova Scotia, was being treated for a syphilitic throat with medicines containing arsenic and mercury. To complicate matters, he, the patient, had for a number of years been taking about 65 mg of white arsenic per day on the side, and was disappointed that he did not get all the expected benefits: it "did not improve the wind," but he did have the impression that his genital organs were stimulated by the arsenic. His friends said that his complexion had improved and that he was notorious for his "amorous propensities." However, following ingestion of a larger than-usual dose of arsenic he became ill and died a protracted and painful death, which may have been assisted by the new medication he was given. This included, internally, trisnitrate of bismuth and opium, prussic acid and tincture of opium.[66]

In addition to mystery novels and crime reports, accounts of arsenic eating are also featured in more conventional fiction. In *Rose* by Martin Cruz Smith set in the coal mines of Lancashire in the late 19th century, two characters, Blair and Rowland, eat arsenic because they caught malaria while in Africa.[67]

The mining engineer Blair "opened the envelopes, poured lines of arsenic and quinine across his palms and tossed them into his mouth." Then Lord Rowland, an upper-class adventurer and hunter "emptied white powder into his palms, twice as much arsenic as Blair had ever seen in one hand before, and ate it in a single swallow." "Rowland put out his left hand. White streaks lined his nails, and the heel of his hand was a horny callus, trademarks of arsenic addiction."

There is one hypothesis that suggest that eating arsenic may have led to at least one benefit. According to Vladimir Bencko, Head of the Epidemiology Department at Charles University of Prague, and Walter Cosmos, a professor in the Chemistry Department of the University of Graz, eating high doses of arsenic enabled the eaters' bodies to switch from aerobic metabolism to anaerobic. "This switch enabled the eater to carry enormous loads at high altitude where the air is thinner." This metabolic switch, if it happened, would be particularly useful if the eater were a smuggler trying to bypass the innumerable tolls charged for using more conventional routes.[68]

1.4 Fowler's Solution

Another form of arsenic therapy, developed in Europe during the time of the bubonic plague, was based on the belief that bad air was the root of the problem. To minimise exposure to air, all rooms in houses were closed up during epidemics and at nightfall, fires were lit to produce smoke in the rooms to take away the poisonous air. Standalone smoke houses provided further protection and arsenic therapy was an added bonus: the oxide was vapourised on the hot coals. This cure possibly evolved from smelting practice and is very old. Pliny (23–79 CE) reported that asthma suffers benefited from exposure to arsenic smoke.[8] Much later, in Victorian times, arsenic was added to the tobacco smoked by asthmatics.[69]

Around the time that large doses of arsenic compounds were being claimed to be beneficial, Dr. Samuel Hahnemann (1755–1843) of Germany was siding with the view that the society of the time "was bingeing on harmful sedatives, tonics and narcotics, washed down with tea, brandy and other stimulants."[10] He consolidated some old Greek theories into what he called the Law of Similars, which became one of the cornerstones of homeopathy. The law states: a substance taken in small amounts will cure the same symptoms it causes in large amounts. He experimented with quinine, arsenic and belladonna. Arsenic is still a part of modern homeopathic practice. Known coyly as "Arsenicum Album prepared from arsenopyrite", it is recommended for fear and anxiety linked to insecurity and oversensitivity, and also used for food poisoning, flu, insomnia and skin problems.[70] Dr. David Spence, who works within the British National Health System, claims to be successfully treating patients for migraine and premenstrual tension with Arsenicum Album.[71] The product is available for purchase on the internet at $4.30 (US) per 30 ml dose and it is recommended for a whole page of problems of the mind, eyes, ears, face, throat, stomach, rectum, *etc.*[72] The actual dose of arsenic received in taking some preparations, such as Hyland's Homeopathic Arsenicum Album, is far from minimal.[73] In a surprising development, Arsenicum Album was investigated in a study in India to alleviate the effects of arsenicosis resulting from drinking arsenic-contaminated well water. "The results are highly encouraging and suggest that the drug can alleviate arsenic poisoning in humans."[74] Others urge caution.

In general, the 18th century was the age of quackery and there was a growing demand for medication-centred healing.[10] Arsenic was present in alteratives, antiseptics, antispasmodics, caustics, sedatives, tonics and antiperiodics and about 60 preparations, mainly solid, were in vogue. A fever cure available in Vienna consisted of white arsenic, myrrh, pepper, red bole, sulfur and bezoar.

Eat arsenic? Yes, all you can get
Consenting, he did speak up;
'Tis better you should eat it, pet,
Than put it in my teacup.

Joel Huck (1842–1914)

Peruvian bark, also known as Jesuit's bark and the first effective remedy for malaria, was introduced to Europe from Peru in about 1633, but it took some time to be accepted into standard practice. The church had a virtual monopoly on supplies,[10] so there was a lot of interest in a patent application made by Thomas Wilson on June 11th, 1771. He invented "a medicinal composition which, after much experience hath been found to be an infallible remedy for agues and intermitting fevers, even in the most obstinate cases where the bark and every other medicine hath proven ineffectual." Ague is a fever such as malaria that is marked by paroxysms of chills and sweating, recurring at regular intervals. He described its preparation as follows: "Take of centaurium minus, [a species of flowering plant], or common centaury, any quantity; burn to ashes; take these ashes, boil them in water for three hours, evaporate the liquid to dryness; take this mass, calcine it for four or five hours, keeping it constantly stirring. Take cobalt, powder it fine, put in a crucible; sublime the flowers; take these flowers, add them to the above, melt them together, and boil them for two hours in water; then take the santalum rubrum, or red sanders, boil it in water for four hours; mix all together." Cobalt was an arsenic-containing mineral related to pyrites, and santalum rubrum was colouring matter.[75]

In October, 1783, Mr. Hughes, Apothecary to the Infirmary of the County of Stafford, England, tried to imitate this patent medicine that was known as "Ta*ftelefs* Ague and Fever Drops," (Tasteless Ague and Fever Drops). Ague was endemic in many parts of England and the drops were immediately adopted into hospital practice and found to be efficacious. Hughes discovered that arsenic was the active ingredient. His colleague, Thomas Fowler, who was a physician, was inspired to dissolve arsenic trioxide in vegetable alkali, and his medical trials with this solution are reported in his book *Medical Report of the Effect of Arsenic in the Cure of Agues, Remitting Fevers, and Periodic Headaches*.[76] Fowler was very much influenced by the ideas of Paracelsus and wrote "Indeed it must be obvious to everyone that the Effects of a Scruple [1.3 g] of the Mineral as a Poison, and of the thirteenth Part of a Grain as a Medicine [0.005 g], will not admit of a Comparison." He was also a realist: "In the meantime, as the idea of a poison seems to be strongly connected with that of arsenic, it will be found very difficult to separate them in the mind, whenever that Term is named, and therefore to avoid as much as possible, such a disagreeable association of ideas in the practice of the healing art, the medicine now about to be introduced to notice of the public, will be distinguished by the name of the Mineral Solution."[76]

Fowler's original recipe was written out in Latin as follows: "64 grains arsenic oxide, 64 grains purest vegetable alkali, distilled water half pound. Heat until clear. Cool. Add half pound spirit of lavender and make up to 15 oz with water." The lavender was added to give a more medicinal appearance and the dose, taken in water, ranged from 6 mg of the oxide per day for a child to around 12 mg per day for an adult. It should be noted that Fowler's doses are well below the 100 mg/day used in Chinese medicine during the Song dynasty.[14] A daily dose of 0.12 mg to 5 mg is suggested for Fowler's solution in the British Pharmacopoeia of 1948. European pharmacopoeias in the early 1900's suggest 5 mg as the maximum single dose and 15 mg to 20 mg as the maximum daily dose.[75]

Fowler presented data for 271 cases of ague: 171 patients were cured by the solution, 45 were cured by the Peruvian bark after failure by the solution,

24 treatments proved unsuccessful because of irregular intake of the medication, and seven patients remained under treatment.

"William Blewer, of Walton, aged 14, admitted an out-patient, March 3rd 1784, with a quotidian Ague, of near a month's continuance: Stools regular. Ordered 20 drops of the solution, three times a day, for three days.
March 6th. The solution excited nausea, and purged him three or four times a day, with griping pains; and every dose, except the evening ones, proved emetic. He has had no paroxysms. After two days intermission, ordered to repeat the solution, twice a day, for three days.
March 13th. It has neither griped nor purged, and only proved emetic with nausea, at the second dose. He has remained quite free from the ague.
March 31st. Continues quite well."

Fowler concluded that the Mineral Solution cured agues, remitting fevers, and periodic headaches. He directed that the solution could be used with laudanum if necessary when bowel problems are encountered. The solution was tasteless and could be given to patients, especially children, who otherwise might object to the bitter taste of the Peruvian bark.

"There is also a good economic argument for using arsenic. An ounce of the best Peruvian bark costs two shillings and six pence and will seldom cure more than one person; whereas, the same money will buy enough arsenic to cure ten thousand persons of the same disease. Charitable institutions will experience a certain saving of considerable sums of money, annually expended in that costly foreign drug, the Peruvian bark.[76]

The story from the United States was much the same. Many cures on the market contained arsenic and were used by doctors and itinerants. Dr. Potter writes about an itinerant practitioner Lafferti in the state of Maryland, who in 1783 secretly used arsenic on ulcers with unparalleled success (Some patients were long deemed beyond the reach of surgical art.): "I have just procured a small parcel of his medicine; at first I thought it looked like corrosive sublimate but upon trial found myself mistaken. I put some of it on to the fire, which soon perfused the room with the smell of garlic, from which it must be arsenic."[57]

Thus inspired, Dr. Potter became the Fowler of the New World. He cured cases of syphilis, ulcers and cancer (but not all types), children of worms, and he witnessed a cure of leprosy. He does report some deleterious properties of arsenic, include blackening of teeth. Some dogs he was studying lost their teeth. Potter began with smaller doses than Fowler, to diminish the side effects, and increased the dose if all went well. If there was pain, he added laudanum.

Fowler's solution first appeared in the London Pharmacopoeia of 1809, in the United States Pharmacopoeia of 1822, and in numerous European compilations around the same time. The actual concentration of the arsenic trioxide in the original formulation was 0.84 per cent weight per volume. Changes in units (*e.g.* the wine pint became the Imperial pint in 1851) and common sense led to the eventual adoption of a 1 per cent solution of the oxide as the standard.[75]

The solution was quickly adopted into medical practice for the treatment of epilepsy, hysteria, melancholy, dropsy, syphilis, ulcers, cancer, worms and dyspepsia. Most physicians were conservative with their doses and were advised to look for "symptoms of nausea, pain, thirst . . . strong feeling all over the body . . . and . . . contraction of the stomach," signs that the mineral had reached its toxic level."[63] However, there were others who disagreed. One of these was James Begbie, Physician to the Queen of Scotland.

Begbie claimed that arsenic must be regarded as one of the most useful and available therapeutic agents. He believed in the "need to push the medicine to the full development of the phenomena which indicates its peculiar action." Often the treatment is abandoned too early because it looks like the patient is being poisoned.[77] Begbie reports cures for chronic rheumatism, chorea sancti viti, headache, psoriasis and eczema, among others. "I have seen an infant at the breast, oppressed and feverish, suffering from a diffused eczematous rash, in due time relieved through the same agent [arsenic] administered to the mother, at the time in perfect health – her eye and tongue proclaiming the power of the remedy which had passed through her system, her milk the vehicle for conveying it to her child – the mother alone demonstrating its physiological action, the child, through her, deriving the therapeutic effect."[77]

Doctors were equally divided on other aspects of arsenic delivery. Some administered it alone but others used it in combination with remedies such as laudanum, quinine, antimony, mercury and sarsaparilla. One wrote "in ague, of whatever type, quinine will rarely fail alone, and arsenic alone will rarely fail; but when one fails the other will often succeed; and a combination of both remedies in full doses, has never been known to fail."[63]

A number of official preparations of arsenic are listed in the 1885 first edition of Remington's *Practice of Pharmacy*.[78] Liquor Potassii arsenitis, Fowler's solution, has the original composition and was used as an alterative (a tonic) in doses of up to 5 mg. Acidum arseniosum, prepared from the oxide and bicarbonate dissolved in boiling water, is used as a tonic and was used externally as an escharotic (corrosive salve); in paste form was applied to cancers and ulcers. Liquor arsenii et hydrargyri iodidi consists of the iodide of arsenic and red iodide of mercury, in distilled water. It is used as an alterative. Harle's solution is the same as Fowler's, but with cinnamon water. Clemens' solution has the oxide plus bromine in water. Biette's solution is ammonium arsenite in water. Pearson's solution is the oxide in water – the addition of mercury chloride and morphine results in Esmarch's solution. In the event of an overdose the suggested antidote is hydrated oxide of iron, as was used by the Styrians.

Professor A. S. Taylor of Guy's Hospital, London (Section 5.4.6) wrote of a man who applied a mixture of arsenic and soap to his scrotum and armpits for the purpose of killing lice. In 12 hours he began to feel all the well-marked effects of arsenic poisoning: thirst, headache, irritability of the stomach, vomiting, pain, stiffness in the neck and difficulty in swallowing. He felt as if his bowels were on fire and his hair was being pulled out by the roots. The cuticle of his scrotum peeled off leaving an inflamed and bleeding surface.[79]

The British Pharmaceutical Codex of 1907[80] had more than 120 entries for mercury, one of the more interesting being a treatment for syphilis that involved the direct deposition of the vapour from heated mercury salts onto the skin. There are about 60 references to arsenic which when taken as a tonic: "has an action on the nutrition of the skin – the subcutaneous fat is increased and the complexion is improved. . . . The coats of horses and other animals are thicker and more glossy." Arsenic was claimed to be effective against skin diseases, malaria, pernicious anemia, neuralgia, rheumatism, epilepsy and syphilis. The oxide was prescribed directly in the form of pills or in solution. Fowler's solution was given in doses of 10–50 ml, 100–500 mg of oxide – well into the lethal range – with instruction that this may be exceeded. Donovan's solution was new (arsenic triiodide 1%, mercuric iodide 1% in water, dose 30–120 ml) and was in use to treat rheumatism, epilepsy, syphilis, and cutaneous disorders. The combination of arsenic therapy with iron was recommended for treating malaria and pernicious anemia; pills were compounded from ferrous sulfate and arsenic trioxide with the addition of sugar. The oxide was also prescribed mixed with haemoglobin preferably obtained from dogs, rats, or guinea pigs."

The use of arsenic to treat anemia encouraged new applications in the beauty industry; thus we find "bust pills" that were manufactured on the premise that subcutaneous fat would be increased. "Today if you see a really busty demi-monde people say 'she eats rat poison'."[81] Because of their over-enthusiastic quest for perfection, 149 females were admitted to a London hospital in 1906 suffering from arsenic exposure.

By 1912, physicians in Europe were recommending a combination of arsenic and strychnine for all forms of chronic disease, and arsenic became the "therapeutic mule," performing "the hardest kind of work under the most adverse conditions, with the least amount of exertion."[63]

Around this time, strong arsenical pastes were being applied to cancerous ulcers, resulting in dry gangrene that later separated as a slough. Some poisoning did occur. Similar treatments are still available in Canada today. One recipe, said to have come to a self-proclaimed healer via his mother and the local Indian band, also came with many authenticated cures. One of the healer's patients had been diagnosed in 2003 with cancer in his right eye and was treated by applying a patch of the oxide mixed up with plant material to his abdomen. Very soon after the application the area became red and a blister formed that slowly cleared. The patient said he had a feeling that the treatment was working because he could feel some heating of the regions of infection around the eyes and the groin. His cancer did regress.

In the 1930s inorganic arsenic was used to treat leukemia, either as pills or in solution, but by 1936 Fowler's solution had lost the tincture of lavender because the colour precipitated (there was concern that the deletion could cause confusion). The dose was 0.1 ml, or 1 mg of oxide, highly diluted for use in syphilis and chronic skin disease. Arsenic and mercury iodides in bicarbonate were prescribed for the same purpose in similar doses.[82] The Asiatic version of the pill made its appearance: it contained 5 mg of arsenic trioxide and black pepper. Baud's Pills contained iron carbonate, arsenic trioxide and strychnine. Pasta Arsenicalis was compounded from arsenic trioxide, morphine hydrochloride and creosote.[83]

Rather ironically, around 1936 arsenite of copper, Scheele's green, the pigment that was claimed to be the possible cause of many problems when it was used to colour wallpaper in the 19th century (Section 3.3), was being used to treat diarrhoea, cholera and anemia; the dose being up to 3 mg per day.[82]

Skipping the war years, the Codex of 1954 has more than 50 entries for mercury and about 30 for arsenic. Arsenic trioxide was taken either in Fowler's solution, or in 1–5 mg doses of the solid in pills mixed with ferrous carbonate and sucrose. The treatment was said to have a depressive action on bone marrow and was used for the alleviation of leukemia. In the 1979 Codex, arsenic warranted three entries.[84,85] The eleventh edition 1989 of the Merck Index[86] describes Fowler's solution (10 g arsenic trioxide, 7.6 g potassium bicarbonate, 30 ml alcohol, distilled water to one litre) as an antineoplastic for man [acting to prevent the growth of tumours] and a tonic for animals. Gay's solution, which was Fowler's solution plus digitalis, potassium iodide, and sodium phenobarbitol, was used by Dr. Gay to treat 1128 asthma patients over the two years prior to 1955. He reported that 80 per cent of his patients were greatly improved. A later study of the treatment led to the conclusion that "Gay's solution is effective in the treatment of severe asthma and that this effectiveness is dependent on the presence of arsenic in the mixture. Because of the high incidence of side effects we recommend that it be tried only in refractory cases and that they be followed carefully."[87]

Ancient Chinese practitioners *ca.* 500 BCE and Arabian doctors in the middle ages made use of pastes of arsenic sulfides to pack around diseased teeth, encouraging necrosis of the tissue and spontaneous loss of the tooth. Around 1836, New York dentist Shearjashub Spooner developed the technique of using arsenic trioxide to kill the pulp in infected teeth: sometimes opium was added to dull the pain. Proprietary dental products containing arsenic were available in Britain until the mid-1960s and they were still in use in the Third World in 1994.[88] An "official" preparation used by dentists in the 1950s for killing nerves in teeth prior to filling contained the oxide, plus cocaine, plus creosote.[84] This concoction was used when performing root canals and the odour emanating from the mouth during the procedure was very unpleasant.[89]

In the 1980s Bayer was marketing a tonic in Pakistan claiming it fought stress, improved appetite and served as a nerve food and nerve stimulant. The label on the bottle screamed "Bayer's Tonic fights against Stress with STRYCHNINE NITRATE." There was no mention that the tonic also contained arsenic. The recommended daily dose of the tonic provided 2.6 mg of arsenic, about the same as Fowler's solution. The tonic was neither licensed by nor marketed in West Germany, the country of origin. Australia's Medical Lobby for Appropriate Marketing took up this issue in 1986 successfully asking for the withdrawal of the product.[90]

1.5 Acute Promyelocytic Leukemia (APL)

There was no reference to arsenic in the Codex of 1994,[91] implying that arsenic had dropped below the medical horizon. But this was soon to change.

In the 1970s Chairman Mao Zedong, who wanted to modernise traditional Chinese medicine, heard about a doctor in the Democracy Commune in northeast China who was curing patients, some with cancer, with the aid of a concoction made from two types of rock and the venom of a local toad – apparently the need to add something really nasty continued. The mixture was effective following either internal or external application. The search for the magic ingredient revealed it to be arsenic trioxide. This rediscovery was put to good use in the hospital at Harbin where they found the treatment to be effective against a particularly unpleasant form of cancer known as APL: acute promyelocytic leukemia. APL is characterised by a rapid accumulation of abnormal white blood cells in the bone marrow, resulting in anemia, susceptibility to infections, bleeding and haemorrhage.

After many trials and some deaths, good results were consistently obtained when patients were treated with the arsenic solution intravenously once a day for 30 days. This course was repeated after a week if the leukemia was still present. The treatment was mild and did not overly distress the patients. The arsenic trioxide could even be replaced by the arsenic sulfide, realgar, delivered in the form of the Niu Huang Jie du Pian pills described earlier (Section 1.2).[24,92,93] The Chinese waited until the 1990s to inform the western world of their success, prompting a major study that was published under the title: "Complete Remission After Treatment of Acute Promyelocytic Leukemia With Arsenic Trioxide."[94]

The protocol consists of using 0.05 mg of arsenic trioxide per kilogram of body weight in 5 per cent dextrose, infused intravenously over a period of two to four hours, once per day. The drug is given daily until tests show the disease to be under control.[94] The adverse effects are variable but mild. In a way this was a case of *déjà vu* all over again. Fowler's solution had been used to treat leukemia in 1865 and showed an initial positive response, but patients soon succumbed to the disease. The solution was rediscovered in 1931 to treat myeloid leukemia in combination with irradiation, until it was supplanted by another drug, busulfan, in 1953.[95,96]

APL accounts for 10 per cent of adult acute myeloid leukemia in the United States. An estimated 1500 new cases of APL are diagnosed each year, and about 400 of these will not respond to or will relapse from first-line therapy. This is when arsenic therapy becomes a consideration.

Box 1.2 How Does Arsenic Cure APL?

APL is caused by the creation of a cancer-causing gene (oncogene) that is formed when two normal genes, let's call them A and B, get together.

> **Box 1.2   Continued.**
>
> (Oncogenes are the templates for the production of proteins that overcome the usual cellular control mechanisms, allowing cells to grow without restraint. They can also be distributed from the primary site to grow in other secondary sites such as lungs, brain, and liver in a process known as metastases.) Retinoic acid can bind to the A part of AB and is used successfully in therapy. However, resistance does occur, at which point salvage therapy is needed. Thus the discovery that arsenic treatment is effective in such cases was very welcome. It originally appeared that the arsenic interacted with the B part of AB, inducing cell death – otherwise known as apoptosis.[97] However, other mechanisms may have to be invoked.[98] The most recent suggestion being that arsenite selectively destroys APL cells by causing enzymes that are usually sequestered within the cancer cell to be released into the cell at large.[99]

We know that chronic exposure to arsenic compounds can have toxic effects and the evidence suggests that the ability of arsenic to "kill or cure" comes down to the same biochemical interactions. "Some types of neoplastic cells are particularly sensitive to arsenic and at low doses the therapeutic benefits outweigh the toxicities."[100]

Arsenic trioxide was approved by the US Food and Drug Administration on September 25th, 2000, as a drug for the treatment of rare diseases or conditions. Approval came only three years after trials were initiated. The drug is marketed under the trade name Trisenox and is manufactured by Cell Therapeutics Inc. of Seattle, WA. The US National Institute of Health recently announced a successful clinical trial involving 582 patients: 77 per cent of these were alive and in remission three years after diagnosis. The figure was 50 per cent for those on the conventional non-arsenic treatment.[101] In 2006 there were reports that arsenic trioxide was effective as a single-agent therapy for newly diagnosed APL and could be used in combination with other drugs against multiple myeloma.[102,103] These inexpensive treatments could be very attractive for countries with limited resources.

However, one major snag has recently surfaced. As David Cyranoski reports in Nature.com, Trisenox is inaccessible to all but the richest of people because a US company holds the patent. The cost is about $50 000 for a full course of treatment.

In mid-2007 there were 25 ongoing clinical trials listed by the US National Institutes of Health that involved arsenic compounds and a range of cancers. One of the most recent developments in the arsenic *vs.* cancer battle is the introduction of the arsenical S-dimethylarsinoglutathione (($CH_3)_2$AsSGlut, where HSGlut is the amino acid glutathione), known as ZIO-101 and said to be one of a family of related anticancer agents. The drug is claimed to be superior to arsenic trioxide because it is less toxic and higher doses can be given that are more likely to enter the cell. The new drug is active against solid tumours, unlike arsenic trioxide that can promote the growth of such cancers. According to the manufacturers, "ZIO-101 increases the reactive oxygen species within the cell leading to cell death." The drug is currently in trials for treating advanced myeloma and can be delivered in an oral form.[104–106]

ZIO-101 is an example of an organoarsenical: it contains an arsenic-to-carbon bond. Organoarsenicals have been used in medicine for many years, so we will need to examine their beginnings.

1.6 The Organoarsenicals

1.6.1 Robert Bunsen

Around the same time that Fowler was investigating his solution of arsenic trioxide, L. C. Cadet de Glassicourt – who, in 1760, was a pharmacist in the French army – heated arsenic trioxide with potassium acetate and obtained a vile-smelling, flammable liquid that became known as "Cadet's fuming arsenical liquid." Some 80 years later Robert Bunsen, who was a professor of chemistry at the University of Marburg, Germany, began to study the two constituents of Cadet's liquid. They became known as cacodyl and cacodyl oxide: the very appropriate names, which were suggested by the Swedish chemist J.J. Berzelius, come from the Greek word *kakodos*, meaning stench.

It soon became clear that these compounds contained a dimethylarsenic group, $(CH_3)_2As$, so cacodyl oxide had the formula $((CH_3)_2As)_2O$ (think of H_2O). Bunsen was able to prepare many related arsenicals such as cacodyl chloride $((CH_3)_2AsCl$, chlorodimethylarsine) and cacodyl iodide. He found that cacodyl oxide was easily oxidised to cacodylic acid, which is now known as dimethylarsinic acid $(CH_3)_2AsO(OH)$. Cacodyl itself remained a puzzle: it seemed that the formula was simply "dimethylarsenic," but this would have been chemically unreasonable. The problem was solved in 1858 by formulating the compound with an As–As bond, and cacodyl became tetramethyldiarsine $(CH_3)_2As$-$As(CH_3)_2$.

Bunsen showed remarkable experimental skills and courage in working with these noxious compounds. He was unable to breathe the air in the laboratory, but since neither respirators nor fume hoods were available, he rigged up a face mask, fitted with a breathing tube, that allowed him to access clean air outside the laboratory. He lost the use of his right eye, narrowly escaping death, when an experiment using cacodyl cyanide exploded unexpectedly. There was a time when all chemistry students knew Bunsen's name because the Bunsen burner was a standard piece of equipment found in all laboratories. It was designed to give a clean flame from burning the right mixture of air and coal gas. He required such a flame for his studies on photochemistry, which he carried out later at the University of Heidelberg. One generally unacknowledged contribution Bunsen made to medicinal chemistry was the discovery that the hydrate of iron oxide was an antidote to arsenic poisoning.[48,107] This is still in use today and the underlying chemistry is the basis for many of the technologies developed to remove arsenic from drinking water (Section 8.4). Bunsen's work on the methyl arsenic compounds initiated a vast field of organoarsenic chemistry that we will touch on only briefly.

Organoarsenicals come in two classes: aliphatic or aromatic, sometimes described respectively, as alkyl or aryl. In the aromatic compounds the arsenic atom is joined to a carbon atom that is part of a ring of the sort we find in benzene. Examples are shown in Figure 1.1. All other organoarsenic compounds belong to the aliphatic class, and these include Bunsen's cacodyl derivatives (see Figure 7.1).

The first known aryl arsenical was prepared by Béchamp in 1863 by heating aniline with arsenate. The product of the reaction was eventually given the name arsanilic acid and the structure of the sodium salt known as atoxyl, is shown in Figure 1.1.[108,109a] At the time, aniline was an essential chemical for the newly established and rapidly expanding dye industry.

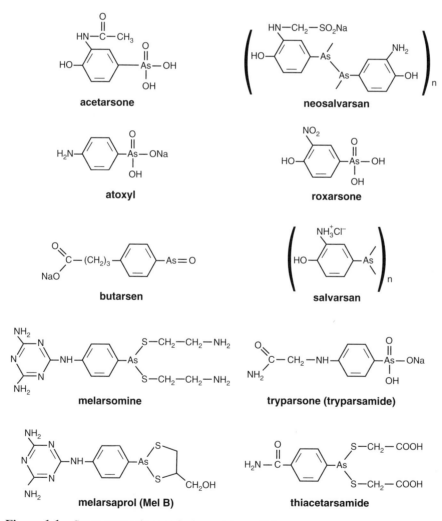

Figure 1.1 Some aromatic arsenicals used in medicine.

By the beginning of the 20th century, some organic arsenicals had been introduced into medicine, but the belief seems to have been that the arsenic becomes active only after the organic group separates from the inorganic part inside the body. The methylarsenicals were prescribed to treat anemia, malaria, TB, syphilis and skin conditions.[80] Salts of dimethylarsinic acid were given as pills that usually contained a strongly flavoured agent such as peppermint to disguise the "garlic breath" that resulted from the treatment.[109b] Salts of monomethylarsonic acid (sold as Arrhenal) were more popular because there were no odoriferous side effects. The sodium salt of arsanilic acid was used for anemia and skin diseases and could be injected intravenously. The salt became known as atoxyl (not toxic), but that was something of an exaggeration.

1.6.2 Paul Ehrlich

Paul Ehrlich was born March 14th, 1854, in Strehlen, Silesia. He was the younger cousin of the microbiologist Carl Weigert who had achieved considerable fame by discovering that some dyes were able to stain germs and tissues so as to render them more visible. Weigert introduced Ehrlich to this art and Ehrlich experimented with various combinations of dyes and tissues trying to understand the source of the interaction. In the 1890s Ehrlich worked with two Nobel Prize winning immunologists, Robert Koch and Emil von Behring, eventually earning his own Nobel Prize for medicine and physiology in 1908.[110]

Ehrlich moved from Berlin to Frankfurt-am-Main in 1899 as head of the Royal Prussian Institute for Experimental Therapy, and developed a very close working relationship with the Hoechst Dye Works. This was not an unusual situation as the pharmaceutical industry was developing in Europe and the US, but the ties were closest in Germany.[10,110] Ehrlich was supplied with money and resources and new dyes as they were developed. His objective was to find the *therapia magna sterilisans*, the "one-shot" cure, or "magic bullet."

Dr. David Livingstone, the same person who was greeted by journalist Henry Stanley on the eastern shore of Lake Tanganyika in 1871, made the important observation that arsenious acid had some effect on the tsetse fly disease of domestic animals in Africa, but a cure was seldom obtained even though the "animal's coat became so smooth and glossy that I imagined I had cured the complaint."[111]

In 1904 "sodium arsenic long employed in the treatment of the disease caused by the trypanosome organism" was found to have a direct action, on the parasites in rats, mice and rabbits.[112] Trypanosomes were recognised as the cause of African sleeping sickness in humans and a major success was achieved in 1905 when Dr. H. Wolferstan Thomas working at the Liverpool School of Tropical Medicine, discovered that atoxyl, Figure 1.1, could cure experimental trypanosomiasis in animals. Thomas demonstrated the low toxicity of the compound by giving himself high intravenous doses. Later it was found that the arsenical could cause vision problems and even blindness; nonetheless, it was adopted for human use in Africa. Around the same time, the spirochete

Treponema pallidum was identified as the causative agent of syphilis, and August von Wasserman developed the test that bears his name to establish whether or not a person is infected.[110,113]

In the belief that spirochetes and trypanosomes were biologically related, Ehrlich set out to test a wide range of arsenic compounds and dyes against both organisms. He began by studying the effect of changing the number and chemical nature of the groups attached to the aromatic ring of the arylarsenicals. Ehrlich developed the hypothesis that to be effective the arsenic needed to be in a reduced oxidation state and prepared many arsenic compounds meeting this requirement. Some 606 compounds later, in 1907, he happened upon the arsenical that was to become known as "606" salvarsan (Figure 1.1), later to be known as arsphenamine. However, one of his assistants erroneously reported that it was inactive against experimental trypanosomiasis. Ehrlich lost interest in 606 until 1909, when his Japanese coworker Sahachiro Hata used it to successfully treat animals infected with spirochetes. Ehrlich quickly arranged for trials against human syphilis, with very encouraging results. Two human volunteers agreed to act as Guinea pigs to establish a standard dose that would have no toxic effect.[110,113]

Ehrlich was very cautious about releasing 606 for general use. In the early days he personally supervised all the arsenical production, and he tested and retested. Finally he was satisfied and news of the success with 606 was released to the world at a conference in 1910. In one year more than 65 000 doses were distributed free of charge for trials and 606, prepared in the Hoechst factory, was put on the market to become "the most potent weapon that had ever been devised against any infectious disease."[113] However, 606 was not quite the ultimate magic bullet: it could kill, it was not well tolerated by the weak and it required more than one dose. However, it was effective, although not against the tertiary stage of syphilis.[112] The treatment usually resulted in nausea and vomiting, headache, fever and skin eruptions. Salvarsan was the name chosen for 606, "the healing arsenic," because the name would not "immediately betray to the patient or his friends the nature of the diagnosis."[114] During WWI, 606 also became known as arsphenamine, the name given to the drug as manufactured in the United States.

The press trumpeted that 606 was a radical cure for syphilis. Doctors travelled to Germany to secure samples for their own patients. They were invariably unsuccessful. The Hoechst Dye Works was forced to surround their production facilities with a high wall, fitted with iron spikes, to keep out the wave of humanity seeking the cure. This was one side of the picture. The other side was the public fear that the availability of a cure was "throwing down the barrier against venereal infection by removing the fear of syphilis."[114]

Box 1.3 Salvarsan

Salvarsan is easily prepared from 3-nitro-4-hydroxyphenyarsonic acid, which later became known as roxarsone (Figure 1.1) (Section 2.1). Salvarsan

> **Box 1.3 Continued.**
>
> it is a light yellow powder that is not very soluble in water. The formula is usually written with arsenic-to-arsenic double bonds, as in RAs=AsR, because this is the structure that would be expected based on the chemistry of nitrogen, arsenic's Group 15 companion. It took some time to recognise that salvarsan was probably a mixture of ring compounds with As–As single bonds with the trimer (three As–As bonds in a ring, Figure 1.1, $n = 3$) and the pentamer (five As–As bonds in a ring, $n = 5$) predominating.[115] One crystalline form of the parent compound, known as arsenobenzene and usually written $C_6H_6As=AsC_6H_5$, has six arsenic atoms in a ring.[116] This structure and a picture of Ehrlich adorned the 200 Deutschmark banknote in circulation in the late 1900s.
>
> Salvarsan was not an easy drug to use. Its purity was always suspect; it had to be stored sealed in glass in a vacuum or in an oxygen-fee atmosphere, and it was difficult to prepare in a form that could be safely delivered to the patient. Because of this problem Ehrlich introduced neosalvarsan (Figure 1.1), preparation number 914, in 1912. The new arsenical was easily prepared from salvarsan and was soluble in water. Its action and uses were those of salvarsan, although it was less stable. It was also less effective, so more treatments were needed. Ehrlich recommended that the drug be given intravenously, which, at the time, required surgery and a hospital stay.[110] This was a problem because syphilitic patients were often excluded from hospitals, so special venereal disease clinics were opened.[114]
>
> Neosalvarsan is shown as a polymer in Figure 1.1 by analogy with salvarsan, to avoid the As=As formulation. There is no hard evidence for this structure.[117]

It is useful to look at military medical records to give some indication of the extent of the problem of venereal disease at the turn of the 19th century. In the year 1898, 24 286 of the 65 397 British troops serving in India had been admitted to hospital with some form of VD. The army responded by reopening the supervised brothels that had been closed for some years because of protests from "home." The disease rate dropped.[118] Infected prostitutes were sent to "lock hospitals" and not released until they were "cured." (It is not at all obvious how this was accomplished.) Similar lockup facilities had been made available in Britain for the innocent wives of straying husbands since 1746.[119]

Venereal disease became a huge problem in WWI. The number of cases from the British forces treated in the period August 4th, 1914, to November 11th, 1918, was about 400 000, of which 24 per cent were for syphilis and 68 per cent were for gonorrhea. The infection rate was higher for Dominion forces, probably because the British went home for leave, but other troops went to major population centres such as London. On the whole, the flow of infection was from Britain to France rather than the other way. The army command initially turned a blind eye to the French Maisons de Tolérance because of the

Medicinal Arsenic: Toxic Arsenic 27

Figure 1.2 The army fights back. (With permission from the National Museum of Health and Medicine, Walter Reed Army Medical Centre, Washington, DC.)

belief that this would lower the infection rate and keep the troops fit to fight. They also engaged in an extensive antipromiscuity propaganda war (Figure 1.2).

Robert Graves writes in his autobiography: "I was now back in Bethune. The Red Lamp, the army brothel, was around the corner in the main street. I had seen a queue of 159 men waiting outside the door, each to have his short turn with one of the three women in the house. My servant, who had stood in the queue, told me that the charge was 10 francs a man – about eight shillings at that time. Each woman served nearly a battalion of men every week for as long as she lasted. According to the Assistant Provost-Marshal three weeks was the usual limit, after which she retired on her earnings, pale but proud."

"Blue lamps were for officers. There were no restraints in France; these boys had money to spend and knew they stood a good chance of being killed within a few weeks anyhow. They did not want to die virgins. The Drapeau Bleu saved the lives of scores by incapacitating them for future trench service."[120]

At the beginning of the war, Germany shut down the supply of drugs such as salvarsan, aspirin and novocain to the Allies, prompting speculation that the monopoly position achieved by the German chemical industry was part of a deliberate campaign to weaken her enemies: "When the war broke out our comfortable commercial contact with the I.G. (Interessen Gemeinschaft) (Section 6.3) became a stranglehold."[121] The British blockade in the Atlantic prevented medical supplies from getting to the US, and German patents prevented the sales of a locally manufactured product: the price for a dose of salvarsan rose from $7 to $35. Medical and financial pressures and entry of the United States into the war led to the abrogation of the German patents in 1917, opening the way to the local products that were sold under the names arsphenamine and neoarsphenamine. Britain also produced its own versions, sold as kharsivan and neokharsivan.[110]

The first cases of syphilis in British troops during WWI were treated only with mercury, but supplies of salvarsan and neosalvarsan became available early in 1915. The arsenical was given intravenously or more successfully, intramuscularly, and often other medications such as mercury, iron, iodine and antimony were given in tandem. In a typical course of treatment, a total of 2.8 g of salvarsan was given over 50 days, accompanied by 8 g of mercury. Blood tests every three months were used to check progress. The relapse rate was about 1.1 per cent after this treatment.

In the field, separate tents and huts were set up for salvarsan treatment, where injections were given to four patients at a time. The arsenical was delivered to each patient via flexible tubing from a single bag suspended above the beds; like a modern-day drip feed. When the deaths of eight patients at one treatment centre were investigated, it was found that arsenic was not to blame: the victims were full of malarial parasites even though only one of the victims was known to have served in the tropics. The modern equivalent is the spread of disease by needle-sharing among intravenous drug users.[122–124]

There can be no doubt that treatment with an arsenical allowed thousands of soldiers to return to active duty in the minimum time.[125] However, we will see in Chapter 6 that other arsenicals were being used as chemical-warfare agents at the same time, with the obverse aim of removing soldiers from active duty for the maximum time.

Ehrlich died August 20th, 1915, aged 61, greatly upset by the outbreak of WWI, and only a few months after Germany first used chlorine gas during the second battle of Ypres. He is universally regarded as the founder of modern chemotherapy.[112,126,127]

1.6.3 The Golden Age of Organoarsenicals

By 1922, US physicians were administering 6 million doses of organoarsenicals annually.[114] By 1932 more than 12 500 organic derivatives of arsenic had been tested for biological activity.[113] The British Pharmaceutical Codex of 1934 had about 130 references to mercury and half this number for arsenic, mostly for organoarsenicals.[83] Both salvarsan and neosalvarsan were in wide use to combat syphilis, with the latter being preferred because of its water solubility and lower toxicity. Mercury and bismuth were usually included in the treatment, which could involve at least 20 injections of salvarsan and 30 injections of bismuth over a period of 18 months.

But treatment was not without problems: "One should not be too much alarmed in a fresh case of syphilis by reactions such as rise in temperature, headache, nausea and malaise, seen after the first injection of arsphenamine. Concern is justified if the patient develops dermatitis, jaundice, or encephalitis following subsequent doses. The variability of the purity of the compound and the reaction of the patient is a problem, so care is necessary in administering the drug which should be started with a small dose."[82] Other arsenicals, such as tryparsone, tryparsamide and acetarsone, were introduced for treating various stages of syphilis, Figure 1.1.[82,83] Tryparsamide was also available for sleeping sickness. However, this drug had some serious side effects, particularly on the eyes and could lead to blindness.

The methylarsenicals were also being use to treat tuberculosis and syphilis in doses of up to 200 mg.[83] A daily dose of cacodylic acid for two weeks was prescribed by the Romanian army to build bodies and minds.[128]

The movie *Dr. Ehrlich's Magic Bullet* was released in 1940, featuring Edward G. Robinson – well known for his gangster roles – as Ehrlich. This film was incorporated into the Public Health Service's national campaign against syphilis that began in 1936. At the time no mainstream Hollywood producer had ever succeeded in overcoming the strictures of the Motion Picture Code that explicitly prohibited screen depictions of sexually transmitted disease. In the film Ehrlich is allowed to utter the word syphilis to the stunned guests at a dinner party. He was explaining his research to potential patrons: "It is caused by a microbe, just like any other infectious disease."[129]

One sad experiment that began during this period, ultimately resulted in the admission by the US Government that they had failed for four decades (1932–1972) to tell 399 mostly black, illiterate sharecroppers in Tuskegee, Macon County, AL, that they had syphilis. The researchers were studying the progress of the disease instead of treating it. The rationale for the non-treatment may have been that subjects were unlikely to have pursued the long course of arsphenamine therapy considered essential in 1932. Participants were given compensation in 1974: $37 500 (US) to each man involved. President Clinton apologised to the survivors in 1997.[130]

Venereal disease was again rampant in the armed forces during WWII. In spite of heavy fighting, the sick-to-wounded ratio was three to one in the Allied troops who liberated Italy. The so-called "Pro Kit" available to US forces around this time, which was effective against both gonorrhea and syphilis, consisted of an ointment containing calomel (mercurous chloride) and a sulfur drug, sulfathiazole.[131] Sulfanilamide (Prontosil), which was first patented and marketed as a dye, was discovered in the late 1939s to be effective against bacteria. Its development can be regarded as an outcome of Ehrlich's studies.[112] Penicillin, the first antibiotic, became available around 1944 and the wastage of men was greatly reduced. Penicillin had a great advantage in that it could be given over a short period of time, when the patient was most apprehensive and anxious for a cure.[96]

In the Codex of 1954[84] neoarsphenamine, although largely replaced by penicillin, was still listed as a consolidating treatment, post-penicillin, because the antibiotic did not completely eliminate *T pallidum* from the human host. Acetarsone was introduced to treat amoebiasis, vaginitis and congenital syphilis in children and infants. Pessaries were used for vaginitis, with a dose of 60–250 mg of the arsenical mixed with chalk and bicarbonate. British Anti-Lewisite (BAL), also known as dimercaprol, Box 6.2, was then in use for arsenic poisoning although it is toxic at higher doses.

The British Pharmacopoeia of 1973 has few references to mercury and arsenic compounds. Arsphenamine is absent, but acetarsone is still available for treating amoebiosis and vaginitis (250 mg twice a day for 10 days either as a pessary or powder). A new trypanoside, melarsaprol (Figure 1.1), became available for treating advanced sleeping sickness in which the parasite has invaded the central nervous system. The regime requires that the patient stay in hospital, but if the disease reoccurs after treatment the infection is usually resistant to all other arsenical drugs.

Arsenic warranted only three entries in the Codex of 1979. Melarsaprol and related compounds are no longer of interest: "they are available only in certain overseas countries and are used in advanced sleeping sickness." However, there is a new entry for the use of arsanilic acid in veterinary medicine for the prevention of enteric infections in fowl and to promote growth (Section 2.1). "The preparation should be withdrawn from feed not less than 10 days before slaughter for human consumption."[85]

Before we leave this aspect of illness and its treatment, it has been noted that there are remarkable medical similarities between AIDS and syphilis and the societal response to the two epidemics. There was confusion about the mode of transmission and the origin of both diseases. It took a few centuries to come up with a cure for syphilis in the form of arsenicals. We can only hope that the timeline is shorter for AIDS, although some 25 years after the beginning of the pandemic there is no general cure in sight. In 2006 there were 38 million people infected with the disease that had already killed 2.8 million in the developing world. There is continuing debate about how to prevent the spread of AIDS but there is a great reluctance to fund the necessary support programs. However, there is some hope: "2007 may bring a one-pill-a-day treatment for AIDS patients."[132–135] U.N. officials continue to warn against complacency.

Box 1.4 Mechanism of Action of Arsenicals

Two early observations were very important in the evolution of our interpretation of how arsenicals act on living organisms. First, arsenicals can be active against trypanosomes *in vivo*, but not active *in vitro*. Second, when infected animals are treated with arsenicals there is sometimes a considerable delay before the spirochetes or trypanosomes are cleared from the body. These results were interpreted as in Figure 1.3. The arsenical drug must be converted to an active form such as the oxide "RAsO" before it can exert its action by binding to the chemoreceptors on the parasite.

The selective action of the arsenicals may be due to differences in the rates of conversion of these to active compounds and to differences in their ability to penetrate the cell wall of the target organism. Needless to say, there should be a big difference in the rates of penetration of cells of target and host; otherwise the host would be attacked.[113] Salvarsan is converted to a compound that has low toxicity to humans and high toxicity to the target

Medicinal Arsenic: Toxic Arsenic 31

> **Box 1.4 Continued.**
>
> organism: it has a wide therapeutic window. Ehrlich thought that synthetic arsenoxides would be too toxic to be used directly as drugs, so he ended up inhibiting progress in the field that he had initiated[136] and as a result oxides such as butarsen, and melarsaprol, Figure 1.1, were latecomers on the therapeutic scene. 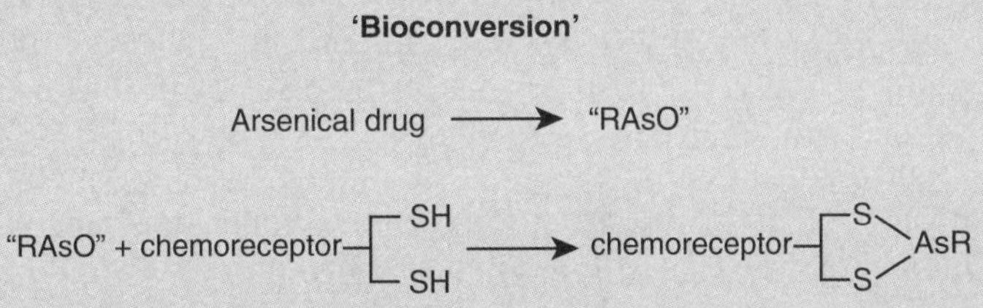
>
> 'Bioconversion'
>
> Arsenical drug ⟶ "RAsO"
>
> "RAsO" + chemoreceptor−[SH, SH] ⟶ chemoreceptor−[S, S]AsR
>
> **Figure 1.3** Ehrlich's view of how an organoarsenical such as salvarsan binds to a parasite.
>
> The model represented in Figure 1.3 is widely accepted and led to the development of British Anti-Lewisite (BAL) during World War II to counteract the effects of exposure to the war gas Lewisite (Section 6.9). However, there is very little direct evidence that can be offered in support. The best that could be said for many years is that most of the enzymes, or their cofactors, that are inhibited by arsenic compounds, contain thiol (SH) groups and these readily react with arsenic compounds.[137,138] However, some recent elegant studies have provided convincing direct evidence for the model in showing the binding of arsenic(III) species to SH groups in the protein metallothionein.[139]

1.6.4 African Sleeping Sickness

Tsetse flies are extremely important vectors of African trypanosomiasis, causing a disease known as nagana in livestock, which is fatal to horses and cattle; and sleeping sickness in humans, which is also fatal. After a person is bitten by an infected fly, the parasite enters the blood stream, resulting in fever, headache, sweating and enlargement of the lymph nodes. The central nervous system is then attacked, followed by the brain. The victim experiences violent mood swings, headaches and weakness, and eventually becomes comatose. Sleep becomes uncontrollable. Death follows quickly without treatment. The disease is caused by *Trypanosoma gambiense* in West Africa and *Trypanosoma rhodesiense* in East Africa. The colonisation of Africa triggered epidemics of the disease. Opening of the Congo to commerce flushed the disease into the central area. Missionaries conveying the sick to mission stations, inadvertently triggered infection in previously uninfected areas. Between 1896 and 1906, nearly a million Africans died of sleeping sickness.[10]

Another trypanosome, *T. cruzi* causes American trypanosomiasis, known as Chagas disease, resulting in severe debilitation rather than death. The parasite is spread by a small green beetle and infects most tissues, producing intermittent fever, enlargement of the liver and spleen and sometimes encephalitis. In some parts of Brazil about 20 per cent of the population is infected.[140]

We have seen that in 1904 H. W. Thomas discovered that atoxyl, Figure 1.1, was active against trypanosomiasis. The drug was massively used in treatment of sleeping sickness for many years, in spite of its problematic toxicity. The essential development of the US chemical industry during WWI resulted not only in the production of known arsenicals such as arsphenamine, but also in the discovery of new compounds such as the trypanocide, tryparsamide, Figure 1.1. This arsenical, a simple derivative of atoxyl, was found to be very effective against Gambian human sleeping sickness although it was of no value against the Rhodesian variation of the disease.[113,141] The intravenous treatment was seldom fatal but did cause problems with vision and sometimes resulted in blindness.

Tryparsamide was used for about 20 years but by 1947 there was an almost 100 per cent incidence of resistant strains of trypanosomiasis in certain areas of the Belgian Congo. In 1946 in French West Africa, 80 per cent of all second-stage cases of Gambian sleeping sickness were tryparsamide-resistant. The gravity of these figures was emphasised by the fact that of 50 000 new cases discovered in 1947, 60 per cent were already in the second stage of infection. Fortunately a new arsenical melarsoprol, Mel B, Figure 1.1, proved to be effective against both the Gambian and Rhodesian strains of the disease.[141] This compound, based on melamine, was developed as a result of the empirical observation: "all arsenicals with significant trypanocidal activity contain nitrogen in one form or another."[142] So much for rational drug development: the reader is invited to think about the immense number of compounds that would fit this requirement.

African sleeping sickness had almost been eradicated by the mid-1900s but in recent years it has staged a comeback. In 2004 the disease affected 36 countries in sub-Saharan Africa. The outbreak has reached epidemic proportions, particularly in the Sudan because of war, political unrest and the collapse of health care. The disease is a threat to about 66 million people, but only 7 per cent of these have access to treatment. Sleeping sickness attacks mainly the poorest and most isolated African communities. In 2004 it was estimated that 300 000 people were infected and that 60 000 would die from the disease every year.[143,144]

In spite of this grave situation, the one non-arsenical drug that was effective – eflornithine, known as DFMO and originally developed for treating cancer – was taken out of production in 1999 because it was not profitable. The drug, introduced in 1990, was so successful at bringing patients back from a sleeping sickness coma that it was called the "resurrection drug." At the time of its withdrawal industry did not deny there was more profit in drugs for western dogs than for African infants, but said sorry, profit is what the shareholders demanded. Industry also suggested that the countries most affected could do much more for their own people.

In the absence of DFMO the only other treatment available was the much cheaper – at $50 (US) per course – but far less desirable melarsoprol. DMFO costs $210 (US) per course. Little is known about the mode of action of the arsenical drug, so various empirically designed treatment schedules have evolved. One under investigation requires the patient to be hospitalised and comprises 10 daily injections of 2.2 mg/kg.[143,145] The physicians have no love for the treatment; they speak of injecting "arsenic in antifreeze" because the drug is dissolved in propylene glycol, which is a base for antifreeze. The injection is very painful, and encephalopathic syndromes cause the death of 3 to 10 per cent of patients. Other adverse effects include polyneuropathy, dermatitis, fever, headache, diarrhoea, and abdominal and chest pains. In addition, some forms of the disease are now becoming resistant to melarsoprol. Up to 25 per cent of the patients in some areas demonstrate this resistance. Melarsoprol is not used in the developed world – it would not meet the necessary medical ethics requirements.

This unfortunate situation was significantly improved in 2001 when, following public outcry, the manufacturer of DFMO Aventis now Sanofi-Aventis group, signed a deal with the WHO to supply a five-year global need for the drug. The company also agreed to supply melarsoprol and one other drug used for early-stage treatment of sleeping sickness, and to donate $25 million (US) to the WHO. The cynics are saying that this burst of magnanimity was triggered by the discovery that DFMO could be used as an ingredient in hair-removing cream.[146]

DFMO is neither the ideal nor the complete cure. It is difficult to administer, it causes adverse reactions and is effective only against *T. gambiense*. One hope is that research on the animal form of the disease, a separate but major economic concern, will lead to something better for human use.[147] Another hope lies in the Bill and Melinda Gates Foundation which is now funding the work of a consortium of research groups led by the University of North Carolina.[144]

The region in which sleeping sickness is found also has a high incidence of HIV/AIDS, but the scale of the AIDS problem is very much greater.[134] Tuberculosis and malaria also compete for scarce health dollars.

1.7 The Darker Side: Toxicity

Until now we have glossed over the toxicity of arsenic compounds, so it is time to make amends. The general public is well aware of this darker side, they associate the A-word with poison even if they do not recognise that arsenic is the name of an element that is present everywhere in the environment in many forms, many of which are nontoxic (Section 7.3). But in spite of its reputation arsenic is not the most toxic of the elements, being outclassed by far by mercury, beryllium, cadmium, thallium and many more. However, this negative image is not likely to change soon because arsenic continues to make the headlines as a homicidal agent (Chapter 5).

Arsenic trioxide is the usual arsenic compound implied in connection with toxicity and a dose of a few milligrams can cause serious symptoms in humans. A fatal dose is 100 to 200 mg, but recovery from 10 g (10 000 mg) has occurred.[79]

The effects of the ingestion of an acute dose of arsenic trioxide appear any time from half an hour to several hours after the event; the throat becomes constricted, there is violent abdominal pain, vomiting, bloody or watery diarrhoea and excessive salivation. There is also skin rash and anemia. Death usually results within 24 hours, but can take two to four days depending on the dose. Such a death is very unpleasant, as Madame Emma Bovary experienced (Section 5.4.9).[148]

Gradually her moaning became louder. A hollow scream was forced from her; she insisted that she was feeling better and that she would get up soon. But she was seized with convulsions; she cried out: "Oh, my God, this is agony!"

The cause of such total devastation is usually attributed to two processes. First, the binding of arsenic(III) species to bodily enzymes that contain SH groups (and at least 200 do) disrupts their ability to function. Second, arsenate can mimic phosphate and inhibit the production of ATP, which is the body's method of storing energy. The diagnosis of acute arsenic poisoning is difficult because the symptoms simulate bacterial dysentery, particularly cholera, and depend on the route of exposure, the dose and time elapsed since exposure. It is only really possible to come to a firm conclusion regarding the cause from knowing the history of exposure and/or identifying arsenic in biomarkers such as hair, urine and finger nails (Section 1.9).

Peripheral neuropathy, a problem with the nerves that carry information to and from the brain and spinal cord resulting in pain, loss of sensation, and inability to control muscles, can result after a week or so of exposure to chronic doses of arsenic. This is usually followed by skin changes such as hyperkeratosis, hyperpigmentation and skin cancer starting about six months after exposure, although symptoms could take much longer to appear – up to three years. When the dose is lower, about 0.5 mg per day, the skin changes, signalling the onset of arsenicosis can take anywhere from five to 15 years to appear (Box 1.3).

The acute toxicity of arsenic depends very much on its species and a relative scale can be determined by using experimental animals. The term frequently quoted is the LD_{50} value: the amount of the substance that will cause the death of half the exposed animal population under the conditions of the experiment. Thus the LD_{50} for arsenic trioxide is 35 mg per kilogram of body weight for mice. Some others for mice: sodium monomethylarsonate, 1800 mg per kg; sodium dimethylarsinate, 1200 mg per kg, arsenobetaine (Section 7.2) greater than 10 000 mg per kg.

These results suggest that organoarsenicals are less toxic than inorganic arsenic species and led to the widely held belief that methylation (the attachment of a methyl group to arsenic), a process that takes place within the human body and many other living organisms (Section 1.9.1), is a detoxification process. This mantra held up progress in the field for many years. It is now apparent that while dimethylarsinic acid is not so toxic, the reduced species dimethylarsinous acid is extremely toxic, and the same relationship is true for monomethylarsonic acid and its reduced species monomethylarsonous acid.[149–151] So a crude toxicity sequence looks like the following, where the gas arsine (AsH$_3$) is the most toxic and the methylarsenic species in oxidation state (V) the least:

AsH$_3$ > methylarsenic(III) > inorganic As(III) > inorganic As(V) > methylarsenic(V).

Chronic arsenic poisoning is also multisystemic and includes malaise, weakness, weight loss and peripheral neuropathy. Some of these symptoms can easily be attributed to constitutional causes, but the most characteristic indicators are the skin changes already mentioned. Perforation of the nasal septum is a common finding in workers in industries that liberate arsenic gases and powders during the production process – many of the affected don't know that they have been afflicted.[6] Chronic exposure can also lead to cardiovascular problems (high blood pressure, heart disease and stroke) and reproductive outcomes such as birth defects, although these are so far not quantifiable. Diabetes is sometimes listed as a consequence of chronic arsenic exposure but the current evidence is inadequate to establish a causal role for arsenic.[152]

Chronic exposure is also associated with hematomegaly, an abnormal enlargement of the liver[153] and, in addition to lung cancer, chronic exposure causes bronchiectasis – a disease that results in localised, irreversible dilatation of part of the bronchial tree.[154]

By far the most troubling outcome of chronic arsenic exposure is cancer.

1.8 Arsenicosis and Cancer

In 2002 there were 228 entries on the list of substances that are "known" or "reasonably anticipated" to pose a cancer risk to humans. Inorganic arsenic compounds belong to the "known" category and first made the list in 1980.[155] The document states that a listing in the report does not establish that such substances present a risk to persons in their daily lives. Such formal risk assessments are the responsibility of federal, state and local health regulatory agencies.

During the 1800s it was generally appreciated that Fowler's solution was toxic but some believed that the best results were achieved if the dose was sufficient to bring on the initial signs of acute arsenic poisoning. But what was the consequence of chronic exposure via this medium? Early reports of cancer outcomes were made by London physician Jonathan Hutchinson in 1887.[156] He described five cases of cancer preceded by skin keratosis – Box 1.5 – following prolonged internal administration of arsenical preparation. One patient, a clerk aged 34, had taken arsenic for a long time for psoriasis: "The palms of his hands and the soles of his feet were speckled over with corns when he applied at the skin hospital; finally epithelial cancer of the scrotum appeared, and was excised; the patient was then lost sight of."

Some time later 1450 patients who had been treated with arsenic were identified and invited to come in for a followup free medical examination.[157] They had all been treated for diseases such as psoriasis, neurodermatitis and chronic eczema. Twenty-one cases of skin cancer, some with multiple carcinomas, were found in the 262 who turned up for examination. The latency period for carcinomas appeared to be six to 14 years. Hyperkeratosis was the most frequent sign of arsenic toxicity occurring in 40 per cent of the patients. Melanoic hyperpigmentation was found in 5 per cent of the population and a few patients reported they had looked "stained" shortly after taking arsenic, however, this had regressed over the years. This study is very revealing, but from an epidemiological point of view it has problems. There was no control population and there could be bias in the population selected.[157] Nevertheless, the results of other studies, particularly from Taiwan, were persuasive enough for inorganic arsenic to be declared a Group A human carcinogen by 1988, particularly for skin cancer.[157] At the time this was thought to be a manageable risk so the conclusion was not particularly alarming. However, in the 1990s, evidence of cancer in other organs accumulated with the result that by 1999 there was little doubt that arsenic exposure could lead to cancer of the lungs and bladder, as well as the skin.[138,158] One estimate of the risk of combined cancer from drinking water containing 50 ppb arsenic was of the order of 1 in 100.[138]

The difficulty of detecting arsenic-related cancer cases in the US population (or any other population that does not have to contend with a high concentration of arsenic in its drinking water) is outlined in the following passage:[158] "A lifetime excess risk of bladder cancer incidence in males of 45 per 10 000 would represent only 13 per cent of the total risk for male bladder cancer in the United States from all causes. Epidemiological detection of such a risk would require study of a large population of individuals who consume drinking water containing arsenic at a concentration of 20 ppb over an extended period of time. Because background lung cancer mortality in the United States is almost tenfold greater than bladder cancer mortality, it would be even more difficult to demonstrate an association of arsenic in drinking water with lung cancer."

In this connection, extensive studies of 4058 residents of Millard County UT, whose drinking water ranges from 4 ppb arsenic to 629 ppb failed to demonstrate an association between arsenic intake and internal cancers. However, this study was deemed to be of limited value for quantitative risk assessment because, for example, the study population was "composed of individuals with a religious prohibition against smoking, and the unexposed comparison group was the overall population of Utah, where such religious prohibitions are not practiced by all residents."[158]

This topic will be taken up again in Chapter 8.

Box 1.5 Arsenicosis

Arsenicosis is a chronic condition caused by prolonged exposure to arsenic above the safe level, usually manifested by characteristic skin lesions with or without involvement of internal organs and malignancies. Based on the

Box 1.5 Continued.

experience in Bangladesh (Section 8.2) arsenicosis patients can be categorised into three stages:[159]

1st stage: melanosis, keratosis, conjunctivitis, bronchitis, gastro-enteritis, diabetes mellitus.

2nd stage: depigmentation (leukomelanosis) hyperkeratosis, edema (nonpitting, swelling of the leg), peripheral neuropathy (distal numbness, tingling, cold, weakness, impaired sensations), hepatopathy and nephropathy (early stage kidney disease).

3rd stage: hepatopathy and nephropathy (late stage), gangrene of the limbs, precancerous skin lesions and cancer.

Melanosis (hyperpigmentation). *Early melanosis*: diffuse or spotted blackening of palm, trunk, and/or mucous membrane (gum, tongue, *etc.*). *Late melanosis:* extensive, diffuse or spotted dense pigmentation affecting the trunk and other body parts.

Leukomelanosis. Depigmentation in hyperpigmented areas characterised by whitish patches, commonly referred to as raindrop pigmentation.

Hyperkeratosis. (Figure 1.4) Abnormally rough and dry thickening of palms and soles. It can be mild, just palpable thickening of palms and soles; moderate, palpable and visible multiple spotted diffuse thickening of palms and soles; or severe, multiple wart or plaque-like elevations on palms and soles, and may be also present on other parts of the body.

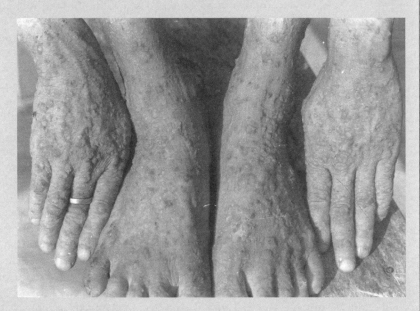

Figure 1.4 An arsenic victim.
(Photograph courtesy of Professor Dipankar Chakraborti.)

1.9 Biomarkers

1.9.1 Urine

The examination of urine for the purpose of diagnosing disease, known as uroscopy, has been practiced for thousands of years. Around 4000 BCE the Babylonians recorded colour and consistency. Around 400 BCE the Greeks, in particular Hippocrates, noted colour, odour and sediment. Urology blossomed in the Middle Ages, attracting both serious practitioners and quacks. Taste was included in the examination by the 17th century and in 1674 Thomas Willis noted the sweet taste of the urine of some diabetics, which proved to be a major discovery, although earlier Chinese and Hindu civilisations had made similar observations.[160]

The analysis of urine for arsenic was made possibly by the development of a test by the British chemist James Marsh in 1836 (Section 5.4), but large volumes were required and even in 1948 a text suggested that the method was capable of detecting arsenic only in the urine of individuals "who lived in an area in which arsenic-rich soft coal was being burnt."[161] Nowadays, it is possible not only to measure the total arsenic concentration but also the concentration of particular arsenic species in small samples (<1 ml). Some of the analytical methods are described in Chapter 7 (Section 7.2), and it is very unusual to find an individual following a normal diet whose urine contains no detectable arsenic.

The concentration of arsenic in urine has often been used as an indicator of recent exposure to arsenic because most ingested arsenic is excreted in the urine. If the intake is only inorganic arsenic the half-life in the body is about four days, and the total arsenic concentration in normal urine is well below 50 ppb. Some foods, especially seafood, mushrooms and seaweed, contain high concentrations of organoarsenic species and their consumption can result in temporarily elevated urinary arsenic concentrations, up to 1000 ppb, which could be misinterpreted as indicating excessive exposure to inorganic arsenic. So care must be taken when interpreting such results. Because of dietary dependence, we find mean total arsenic concentrations for adults from Europe of 17.2 ppm; Taiwan 20.7 ppb; Japan 121 ppb; and Argentina 274 ppb. The high value in Japan reflects high seafood consumption, while in Argentina it is the result of arsenic in the drinking water. These numbers show why individuals are requested to refrain from eating food that could be high in arsenic for a few days prior to collecting a urine sample for analysis.[138]

Most of any ingested inorganic arsenic is metabolised in the mammalian body to methylarsenic species, so the urine contains arsenate, arsenite, monomethylarsonic acid and dimethylarsinic acid. These species can be quantified in urine by using a variety of methods (Section 7.2). If the diet is rich in seaweed and other foods that contain arsenosugars (Section 7.1) the concentration of dimethylarsinic acid in the urine is increased over background, because this is the main arsenical metabolite. If the diet is rich in shrimp or other food that contains arsenobetaine, the urinary arsenic concentration increases because arsenobetaine is not metabolised and is excreted unchanged, adding to the background concentration. In the Japanese study mentioned above, the average inorganic arsenic concentration was 11.4 ppb, monomethylarsonic acid 3.6 ppb, dimethylarsinic acid 35.0 ppb, and arsenobetaine 71.0 ppb. The last two numbers are elevated because of diet.

As mentioned above (Section 1.7) there was a time when the methylation of arsenic was believed to be a detoxification based on the LD_{50} results. A lot of effort has gone into the measurement of the relative amounts of these methylarsenic species in urine because of the belief that some of the toxicity of arsenic might be associated with a failure to methylate and hence detoxify the ingested inorganic arsenic (*e.g.* Hopenhayer-Rich *et al.*).[162] These methylarsenic species are derivatives of arsenic(V) and their formation is accounted for by the Challenger pathway that we will meet in Chapter 3 (Section 3.9). But over the past 10 years the toxicity of the reduced arsenic(III) species, monomethylarsonous acid and dimethylarsinous acid, has become apparent. These are also part of the Challenger pathway, but their presence in urine went unnoticed for many years because the analytical methods used for arsenic speciation were not capable of detecting them. This is not an uncommon situation in science: we can see only where the light shines.

Now, with the help of improved analytical methods, these arsenic(III) species can be detected in the urine of some populations exposed to large amounts arsenic, giving rise to speculation that their presence could be an indicator of the metabolic processes that result in toxicity and perhaps be used as a surrogate for cancer risk.[163] The same species are found in the urine of patients undergoing treatment for APL with arsenic trioxide.[164] It has been suggested that the distribution of methylarsenic species in urine changes during fasting so the month-long observance of Ramadan could well have an unforeseen influence on the health of millions of Islamic people living in regions with arsenic-rich drinking water (Section 8.1).[165]

1.9.2 Hair

The Kelvin Royal Commission of 1901 delved into a number of areas of interest to modern chemists.[166] They noted that "Examination of hair might be of much value in cases where it is important to obtain indications of the past history of a patient in regard of arsenic." The hair from controls ranged from undetectable to 0.96 ppm, whereas the hair from patients known to be taking arsenical medicine was higher – up to 14 ppm. "Nearly always, in those cases, when portions of the hair were examined which approximately corresponded to the period during which arsenic had been taken, a relatively large amount of arsenic was detected." The idea of using hair as an indicator of the chemicals in blood was resurrected around 1945.

It is believed that growing hair is bathed in sebum from glands in the mammalian skin. Sebum is a fatty substance that protects, lubricates and waterproofs the skin: it is also a carrier of trace metals and metalloids such as arsenic that become attached to the hair, probably by binding to the sulfur-rich proteins (keratin). Scalp hair grows for about 900 days; then it dies off and eventually drops out. The whole cycle takes about 1000 days. The growth rate of hair is not constant and varies from about 0.2 mm per day to 0.5 mm per day, just over 1 cm per month. The idea of using hair as a recording filament for identifying metal and metalloid poisoning was put into practice in the 1970s (Section 4.7).[167] We will meet this topic again in later chapters, but in the meantime here are some guidelines as to the concentration of arsenic in hair to be expected from exposure to a source of contamination:[168a]

Normal: <1 ppm but <3 ppm in polluted areas (but can get up to 8 ppm)
Chronic poisoning: greater than 10 ppm (but can be lower)
Acute poisoning: greater than 45 ppm but can be lower (higher levels have been reported in subjects who survived)
External contamination: up to 800 ppm

The arsenic in the hair is essentially all inorganic with arsenic(III) species predominating. Concentrations of arsenic correlate with exposure *via* ingestion.[168b] A study in Thailand attempted to establish whether arsenic exposure, as reflected in hair concentrations, influenced the IQ of children. The results are ambiguous but suggest there is an undesirable relationship.[169]

1.9.3 Finger and Toe Nails

Like hair, nails are also primarily made of keratin and, as a consequence, like hair, they store arsenic. Nails grow about 3 mm per month. A bright opaque band across the finger nails sometimes appears about two months following acute arsenic poisoning. The bands, known as Mees' Lines, are about 1 mm wide, and the concentration of arsenic in the bands is up to 10 times as much as in the rest of the nail. The arsenic concentration in toe nails is being used to estimate people's exposure in an area of New Hampshire. The region has a high incidence of cancer that has been linked to drinking water from private wells. There, a nail concentration of 0.5 ppm arsenic was claimed to correspond to a drinking water supply of 100–150 ppb. But like the Utah study mentioned above (Section 1.8) the interpretation of these results is problematic.[158]

1.9.4 Saliva

The arsenic speciation in human saliva can now be determined because sensitive and selective analytical techniques are available. Sampling is noninvasive, and it is much easier to sample children's saliva than it is to take urine or nail samples. The limited results available indicate there are differences in speciation in exposed and nonexposed populations. For example, the major species in the saliva of subjects drinking Edmonton AB tap water containing less than 5 ppb arsenic, were dimethylarsinic acid and arsenite, with total arsenic in the range of 0.7 to 2.7 ppb. The arsenic concentration in saliva samples from Inner

Mongolia, where the drinking water contains up to 836 ppb, are in the range of 0.4 to 140 ppb, with arsenate and arsenite as the major species.[170]

1.10 Animal Models

Most of what we know about the toxicity and mode of action of chemical species is derived from studies on animals other than ourselves. In the case of arsenic, we have had to rely on studies of human populations because only a limited number of animal models have any relevance. One study conducted by Sam Cohen and his coworkers at the University of Nebraska Medical Center, in Lincoln NE, showed that dimethylarsinic acid given to rats in high doses (>25 ppm in water) produced bladder cancer. He suggests that the acid is reduced to the very toxic arsenic(III) species dimethylarsinous acid (Section 1.7) that, in the urine, attacks and kills bladder cells. The remaining cells then proliferate within the tissue beyond that which is ordinarily seen, resulting in cancer. (Rats accumulate arsenic in their red blood cells: it binds to the sulfur-containing cysteine in their haemoglobin as dimethylarsinous acid.[171])

Another animal model was selected on the premise that humans are much more sensitive to arsenic than other animals, so animals should be studied when they are most sensitive. To this end, Michael Waalkes and his colleagues at the US National Cancer Institute chose to work with C3H mice, a strain that is especially sensitive to chemical carcinogens. They also elected to dose the mice during a particularly active time of growth: *in utero*, transplacentally, via the mother's amniotic fluid. The results clearly showed that sodium arsenite was able to induce a range of cancers of particular relevance to humans, including bladder, skin, lung, liver and kidney.[172] The development of the model proved to be prescient: a closely related human study was published around the same time, finding "Increased mortality from lung cancer and bronchiectasis in young adults following exposure to arsenic *in utero* and in early childhood."[173] A study from Bangladesh reported an association, albeit weak, between arsenic exposure and birth defects although not between arsenic and stillbirth and childhood stunting or underweight children. Another study from the same region found an increased risk of miscarriage or losing a child in its first year. The latter effects were apparent at relatively low concentrations of arsenic in drinking water, 50 ppb.[174,175] Data from the world-renowned International Centre for Diarrhoeal Disease Research in Dhaka were used in both these studies.

1.11 Chelate Compounds and Chelating Agents

In Chapter 6 we will meet the compound known as British Anti-Lewisite (BAL, dimercaprol) that has the ability to bind strongly to arsenic species as shown in Box 6.2. The binding results in the formation of compounds containing five atoms in a ring. These, and ones with six atoms in a ring, have a special stability. The rings are known as chelate rings and the reagents such as BAL that form the rings are known as chelating agents or chelators. The chelate compounds are so stable that most of the arsenic that is already present in a living organism as a result of exposure, and probably bound up to proteins as shown in Equation 2 of Box 6.2, will switch to become preferentially bound to the chelating agent, if it is supplied. The stable chelate compound once formed is considered to be nontoxic and is easily eliminated via the urine.

BAL was difficult to administer and it is toxic, so water-soluble analogues, also with two sulfur atoms were developed. These are known under names such as DMPS (dimercaptopropanesulfonic acid) and DMSA (dimercaptosuccinic acid) and can be administered either orally or intravenously.[176] Experiments with animals demonstrate that all three chelating agents increased survival when administered soon after an acute dose of an inorganic or organic arsenical, but the effectiveness dropped off rapidly with time.[177] DMSA has advantages over BAL for treating human poisoning.[178] Ole Anderson[176] describes an incident in Argentina around 1990 in which 718 individuals were poisoned by sodium arsenite added to meat by "vandals". Because the supplies of BAL were very low only the worst afflicted, as judged by their urinary arsenic concentrations, were treated over ten days. No symptoms of toxicity were reported one month after the incident. BAL also aided in the complete recovery of a suicidal individual who consumed about 10 g of sodium arsenate.[176]

The current arsenic crisis in Bangladesh (Chapter 8) prompted the investigation of these chelate agents for reversing arsenicosis resulting from chronic exposure. The study found no differences in the responses of a group of adult men with chronic arsenicosis, including hyperpigmentation and hyperkeratosis, to DMSA and a placebo; leading to the conclusion that "DMSA is globally ineffective in the therapy of chronic arsenic toxicity in man."[179] Other Bangladeshi scientists suggest DMPS and BAL may be more effective against arsenicosis.[180]

The use of DMPS in a provocative or challenge test for content of metals in the body is well established, and the effect of the dose is seen in an almost immediate increase in the metal content in the urine. However, there is no way of establishing "whether a short-term increase in urinary excretion associated with chelation will result in a lower risk of long term adverse outcomes, such as cancer."[177] Vas Aposhian from the University of Arizona and his international team of coworkers, found that the administration of DMPS increased the arsenic content of the urine of the residents of two towns in Northeastern Chile by a factor of around 5. The towns were chosen because in the control town of Toconao, the arsenic content of the drinking water was 21 ppb, whilst in the other, San Pedro de Atacama, it was 593 ppb (the WHO standard is 10 ppb) (Section 8.1). They were able to show that arsenic was mobilised by DMPS; that arsenic was stored in the human body; that the use of DMPS results in a more accurate estimate of the body burden of arsenic; and that the subjects from the exposed town have a greater body burden of arsenic than those from the control town. In addition, the usual speciation pattern of the arsenic in the urine was changed in favour of an increase in the concentration of monomethylarsonic acid by a factor of about 10.[181]

1.11.1 Chelation Therapy

Chelation therapy is a controversial alternative treatment said to be beneficial for heart disease, stroke, diabetes, Alzheimer's disease, and many other ailments.[182] The chelating agent EDTA (ethylenediamine tetraacetic acid) is dosed intravenously: arsenic is not removed from the body by this treatment although metals such as lead, mercury, and calcium are. Some studies suggest that the treatment is a waste of money[183] although others indicate some benefit if DMSA is added following lead and arsenic exposure.[184]

1.12 Some Historical Connections
1.12.1 Charles Darwin

Charles Darwin, who achieved everlasting fame for his work on evolution and for his book *The Origin of the Species,* was never a healthy man. He appeared to be chronically ill, without energy and was possibly a hypochondriac. He kept a daily record of illnesses but not of any medication. He was known to be a victim of what the Victorians called "nervous dyspepsia," characterised by variable appetite, flatulence, diarrhoea, bad sleep, headaches, numbness of the extremities, nausea and vomiting. These symptoms were common to a number of other dyspeptics, including Thomas Huxley and Mary Anne Evans (aka George Eliot), and are akin to those of chronic poisoning. John Winslow claimed that 21 manifestations of Darwin's illness match those of arsenic poisoning (Section 1.7). He also noted two more pieces of evidence: a photograph taken when he was 51 reveals multiple corn-like growths on Darwin's left hand and his skin had a coppery colour. The anti-arsenic camp argues that the symptoms lack specificity and indicate nothing in particular, apart from flatulence. The warts and skin colour are ignored. Darwin's sometimes severe vomiting did not seem to affect his appetite.[185]

Darwin's father had a very successful medical practice. Without attending medical college, young Charles was allowed to treat patients and he was very proud of his successful treatment of one family with an antimony compound, tartar emetic.

So is there evidence that Darwin actually ingested arsenic? His school friend J.M. Herbert states that Darwin had taken arsenic when he was a student at Cambridge, to cure a lip problem. Use of arsenic was well established in Cambridge and the surrounding Fen country as a prophylactic against ague. Dr. Edmonds, a local physician, reports of the intake of "enormous" doses of arsenic for this purpose,[186] although the main treatment seems to have been opium, taken in massive and addictive quantities, prepared from locally grown poppies.[187,188]

Just before the voyage of the HMS Beagle, Darwin wrote to his father asking if he thought it would be advisable to take arsenic for a while because "my hands are not quite well. . . . I have always observed that if I get them well, and change my manner of living about the same time, they will generally remain well. What is the dose?" His father replied that the cure might be attended with worse consequences. This exchange is another indication that the son had used arsenic for this purpose before. At the time it is very likely that any doctor he consulted would have prescribed arsenic. As we have seen, doctors believed in very aggressive treatments and claimed that the medicine was not working unless symptoms of arsenic poisoning were evident.

The five-year voyage of the Beagle began when Darwin was 22 years old. He was very seasick and dosed himself with calomel (mercurous chloride) as an emetic. This was not an unusual treatment at the time, although it cannot have been very good for the patient. He did have access to arsenic because it was used for the preservation of the biological samples he collected (Section 2.8), although there is no record of him preparing his own equivalent of Fowler's solution. Some claim that during the expedition Darwin contracted Chagas' disease, American trypanosomiasis (Section 1.6.4), although the autopsy showed no evidence of it.[189]

He returned home tired and easily disturbed and withdrew from society to a country residence where he was able to enjoy regular exercise. His symptoms were brought on by encounters with people. By 1845 he was regularly taking a bismuth preparation and also took mineral acids, probably as a mixture of hydrochloric and nitric acids. He used snuff, a potential source of trouble in the form of lead[190] and possibly arsenic. Around the middle of the century, he investigated hydrotherapy (Box 1.6) the then-fashionable "water cure," and wrote that he put on a "compress prepared from broad wet folded linen covered with a Macintosh and which is refreshed with cold water every two hours." Darwin wore this most of the day.

Darwin's illnesses seemed to be milder and less frequent later on in life, perhaps because his doctors by then, around 1880, had become a little more careful in their use of arsenic.

These medical problems have been attributed to many diverse causes, ranging from disease to psychosomatic disorders brought on by feelings of guilt and sublimated hatreds. His father was said to have been tyrannical. Winslow[185] believes Darwin was just a normal man who was ill and whose arsenical-based medicine, which he regularly imbibed, contributed to the continuation of the illness diagnosed by others as "an anxiety state with obsessive features and psychosomatic manifestations."[190]

Box 1.6 Balneotherapy (Hydrotherapy)

Balneotherapy involves the treatment of disease by bathing, but the term has gradually come to be applied to all spa treatments, including drinking mineral waters and using hot baths and natural vapour baths. The principal constituents found in mineral waters are salts of sodium, magnesium, calcium and iron. The presence of other substances such as arsenic, lithium, bromine and iodine, is claimed to be therapeutic – particularly arsenic, and particularly by the French. In Germany, water from the spa at Baden Baden was once classified as arsenical water. The concentration of the element is

Box 1.6 Continued.

about 0.2 ppm, but this is not featured in modern advertisements. The novelist D. H. Lawrence, who suffered from tuberculosis, went to Baden Baden in the 1930s to "take the cure" and consult with doctors who prescribed arsenic and phosphorus.[191] On the other side of the Atlantic at least one resort is not so coy. Ojo Caliente Mineral Springs NM, included in its advertising package in 1998 offerings such as the Ojo Rejuvenator that consisted of a "private arsenic tub and milagro ["miracle"] wrap, 45 minutes. Traditional massage, 50 minutes. Radiant salt glow, 25 minutes. Approximately three hours, $116.00." The arsenic-rich spring flows at a temperature of 45 °C. The water is claimed to be beneficial in relieving arthritis, rheumatism, stomach ulcers and helps promote the healing of burns, eczema and contusions. One area of the spa has four water outlets labelled iron, soda, sodium sulfate and arsenic that visitors are invited to drink, Figure 1.5. The hype was undiminished in 2007: arsenic was claimed to aid arthritis, rheumatism, and skin problems.[192] The Native Americans knew of the curative powers of the springs long before the 1535 visit by the Spaniard Cabeza de Vaca, who described them as "the greatest treasure in the world; the fountain of youth" and named them Ojo Caliente [Hot Eye]. Evidently he was particularly impressed when the water cured two of his men of the "most dreadful form of syphilis."[193]

Figure 1.5 The author at work in Ojo Caliente.

1.12.2 Karin Blixen aka Isak Dinesen

Karen Dinesen was born in Rungstedlund, Denmark, in 1885. She also died there of malnutrition in 1962. In between these events at age 28 she married Bror Blixen to become the Baroness Karen von Blixen-Finecke and set out to make a life for herself growing coffee in Kenya. In July 1931, after a series of setbacks her farming venture finally failed and she returned to Denmark to live with her family. She wrote about her time in Africa in a romanticised style under the name Isak Dinesen because she believed that a male voice would receive more attention. Her most successful work, *Out of Africa*, was featured by the Book-of-the-Month club and received even wider attention as a film of the same name. She was nominated for the Nobel Prize for literature.

Very soon after arriving in Africa Karen Blixen contracted syphilis from her husband and was treated with mercury tablets by a doctor in Nairobi in 1914. She consulted a specialist in Paris who did not have much time for her because he had soldiers to deal with, but he said he doubted she would recover. She went back to Copenhagen in June 1915, under the care of Dr. Carl Rasch, who was familiar with Ehrlich's work. His approach was fearless: weekly injections of salvarsan and bismuth until Wasserman tests showed no continuing infection.

Karen Blixen suffered many periods of ill health throughout her life, which she attributed to syphilis in spite of the assurances and test results of Dr. Rasch. Much of this was the result of panic attacks. However, she was known to never miss a dramatic opportunity. It seems that she dosed herself with arsenic fairly frequently when she was in Africa, where it was readily available. It was used as rat poison and in the cattle dip which was a necessity every three to seven days to rid cattle of tics carrying East Coast fever. Biographer Linda Donelson writes that in September 1922, "Karin had grown plump over the previous months. She felt anxious about maintaining her allure; just at this time her face was too full to be beautiful. Several months before when she increased her doses of arsenic due to illness, her hair had begun to fall out." And when she began the journey back to Denmark permanently in July 1931: "She was suffering from jaundice, anemia, and the effects of chronic doses of arsenic. To cover her hair which had begun to fall out in clumps, she wore a close-fitting hat that looked like a turban and thick make-up to mask the discolouration in her skin caused by years of using arsenic. She was so thin and her eyes were so sunken that her face resembled a bare skull and her nose protruded like that of a witch."[194,195]

1.12.3 Alexander Borodin, Professional Organic Chemist and Amateur Composer

Borodin is world renowned for his achievements in classical music; however, his composing was only a part-time hobby and for most of his life he worked as a teacher of chemistry.

Alexander Porfir'yevich Borodin was born in St. Petersburg on November 12th, 1833, the illegitimate son of an elderly Russian prince. Alexander was well educated, and at an early age began to show an interest in music and musical instruments. He also enjoyed chemistry, especially making fireworks.

In 1850, Borodin entered the Medico-Surgical Academy in St. Petersburg, finishing his studies to high praise on April 6th, 1856. On May 15th, 1858, he presented a dissertation: "On the Analogy of Arsenic and Phosphoric Acid with Respect to Chemical and Toxicology." Although this resulted in him receiving the degree of Doctor of Medicine, he never practiced the profession.

The work described in his thesis consisted of observing the effects of administering known amounts of arsenic acid and phosphoric acid to horses, cats and dogs. As an example: "A healthy horse was given four grams of arsenic acid in 20 g of water. No special events were observed initially; however, after several hours tremors were observed in the extremities; the animal did not feed through the day, and spent most of the time prone in the pen. The horse died after 18 hours. Dissection revealed no visible signs of inflammation. . . . Arsenic was detected by the Marsh test in the urine and liver." Some of his conclusions: (1) Phosphorus and arsenic acids represent total analogy [are identical] not only in the chemical sense, but also in the toxicological one. (2) Their toxic effect is characterised by the absence of local action. (3) The two acids differ in the magnitudes of the dose required to achieve the same degree of activity; it is considerably higher for phosphoric acid than for arsenic. (4) Other acids, such as sulfuric and nitric, administered internally are poisonous only by virtue of their local effect on the tissues of the alimentary tract.[196] Modern toxicologists are probably glad he became a composer.[197]

After receiving his doctorate, he spent the next few years travelling and working in a number of chemical laboratories and attended lectures by Robert Bunsen in Heidelberg. He returned to St. Petersburg and was appointed to the post of adjunct-professor in the Academy of Physicians on December 8th, 1862, and lectured students in organic chemistry. One of his colleagues was Dmitri Mendeleev, a professor of chemistry whose great achievement was creating the Periodic Table of the elements.

Borodin was promoted to professor in 1864 and championed the cause of women in education. Toward the end of the 1860s he began to spend less time working on his chemical research and started to compose in earnest as a member of a group of musicians known as the "mighty handful" or the Russian Five: Balikirev, Borodin, Mussorgsky, Rimsky-Korsakov and Cui. Borodin died suddenly on February 15th, 1887 at a fancy dress party. His friends Nikolai Rimsky-Korsakov and Alexander Glazunov finished and published many of his compositions, including the opera *Prince Igor*.[197]

The 1953 musical *Kismet* is based on Borodin's music, and in 1954 he was posthumously awarded a Tony for this show. The most memorable song, *Stranger in Paradise*, has a lot to do with chemistry, but nothing to do with arsenic.

References

1. U. C. Davis, *The bronze age*, 2006, www.geology.ucdavis.edu.
2. W. Goessler, 1995, personal communication.

3. BBC, in *Science and Nature*, 2002, Feb. 7.
4. Agence France-Press, in *Vancouver Province*, 2005, p. A39.
5. R. Shepherd, *Ancient mining*, Elsevier Applied Science, New York, 1993.
6. W. D. Buchanan, *Toxicity of arsenic compounds*, Elsevier, Amsterdam, 1962.
7. G. B. Shaw, *Caesar and Cleopatra*, 1906.
8. K.-H. Most, PhD thesis, University of Graz, 1939.
9. S. Kind and M. Overman, *Science against crime*, Aldus Books Limited, London, 1972.
10. R. Porter, *The greatest benefit to mankind. A medical history of humanity*, Harper Collins, London, 1997.
11. K. Chan, in *The way forward for Chinese medicine*, ed. K. Chan and H. Lee, Taylor and Francis, London, 2002, p. 71.
12. P. Marshall, *The philosophers stone. A quest for the secrets of alchemy*, Pan Macmillan, 2001.
13. P. Kelly, in *Chem. Brit.*, 1997, April, p. 25.
14. F. Obringer and J. Lloyd, in *Innovations in Chinese medicine (Trans.)*, ed. E. Hsu, Cambridge University Press, Cambridge, 2000, p. 192.
15. *Traditional chinese festivals*, www.china.org.cn/english/features/Festivals/78316.htm.
16. W. R. Cullen and V. W.-M. Lai, 2003, personal experiences.
17. C. M. Henry, in *Chem. Eng. News*, 2005, Jan. 3, p. 32.
18. E. O. Espinoza, P. H. M. -J. Mann and B. Bleasdell, *New England J. Med.*, 1995, **333**, 803.
19. A. S. Lyons and R. J. Petruceli, *Medicine. An illustrated history*, H. N. Albans Inc., New York, 1987.
20. E. Ernst and J. T. Coon, *Clin. Pharmacol. Therapeut.*, 2001, **70**, 497.
21. M. Ou, ed. *Chinese–English manual of common-used prescriptions in TCM*, Joint Publishing (HK), Hong Kong, 1989.
22. Reuters, in *Reuters Foundation*, 2007, Feb. 6.
23. D. Bensky, A. Gamble and T. Kaptchuk, ed. *Chinese herbal medicine-materia medica*, Eastland Press Inc., Seattle, 1993.
24. I. Koch, M. Serran, S. Sylvester, V. W.-M. Lai, A. Owen, K. J. Reimer and W. R. Cullen, *Toxicol. Appl. Pharmacol.*, 2007, **222**, 357.
25. C.-H. Tay and C.-S. Seah, *Med. J. Aust.*, 1975, Sept. 15, 424.
26. J. W. Mellor, *A comprehensive treatise on inorganic and theoretical chemistry*, vol. IX, Longmans, Green and Co, London, 1929.
27. K. A. Winship, *Adv. Drug React. Ac. Pois. Rev.*, 1984, **3**, 129.
28. WHO, *Management of poisoning: A handbook for health care workers*, 2007.
29. ATSDR, *Case Studies in Environmental Medicine Arsenic Toxicity Exposure Pathways*, www.atsdr.cdc.gov/HEC/CSEM/arsenic/exposure_pathways.html.
30. Reuters, Oct. 23, 2007.
31. (a) J. Kew, C. Morris, A. Aihie, R. Fysh, S. Jones and D. Brooks, *Brit. Med.*, 1993, **306**, 507.

(b) P. Paranjpe, *Ayurvedic medicine. The living tradition*. ed. Chaukhamba Sanskrit Pratishthan, Delhi, 2003.
32. R. B. Saper, S. N. Kales, J. Paquin, M. J. Burns, D. M. Eisenberg, R. B. Davis and R. S. Phillips, *Am. Med. Ass.*, 2004, **292**, 2868.
33. CBC News, 2005, March 4.
34. T. Ramavarman, in *The Hindu*, 2005, Nov. 29.
35. L. R. Ember, in *Chem. Eng. News*, 1998, Dec. 7, p. 14.
36. A. Gill, in *Globe and Mail*, 2003, July 12, p. F6.
37. J. Hiayan, *China Patent*, CH1332011 2002-01-23, 2002.
38. D. Lu, *China Patent*, WO 9955344 Ai 19991104, 1999.
39. A. E. Waite, *The hermetic and alchemical writings of Paracelsus*, James Elliott and Co, London, 1984.
40. J. J. von Tschudi, *Weiner Medizinische Wochenschrift.*, 1851, Oct. 11, 455.
41. W. B. Kesteven, *Ass. Med. J.*, 1856, 721.
42. D. Sayers, *Strong poison*, Victor Gollancz Ltd., London, 1930.
43. K. Konkola, *J. Hist. Med. Allied Sci.*, 1992, **47**, 186.
44. W. Goessler, 2007, personal communication.
45. G. Przygoda, J. Feldmann and W. R. Cullen, *Appl. Organomet. Chem.*, 2001, **15**, 457.
46. J. F. Johnston, *Chemistry of common life*, D. Appleton and Company, New York, 1857.
47. H. E. Roscoe, *Memoirs of the Literary and Philosophical Society of Manchester*, 1862, vol. 1, 3rd series, p. 208.
48. T. Curtin, in *Great chemists*, ed. E. Faber, Interscience, New York, 1961.
49. R. Brautigam, *Garlic against vampires*, www.shroudeater.com/agarlic.htm.
50. W. R. Cullen, 1995, personal experience.
51. E. A. Reyes, *Rev. quim. farm. (Santiago, Chile)*, 1947, **4**, 2.
52. ABC News, 2006, Oct. 25.
53. P. Mickelburough, in *The Advertiser*, 2006, Oct. 23.
54. Ibid.
55. R. Chesterton, in *The Daily Telegraph*, 2006, Oct. 26.
56. S. Rintoul, in *The Australian*, 2006, Oct. 24.
57. N. Potter, MD thesis, University of Pennsylvania, 1796.
58. Anon, in *Chambers' Journal*, 1856, Feb.
59. J. Lenihan, *The crumbs of creation. Trace elements in history, medicine, industry, crime and folklore*, Adam Hilger, Bristol and Philadelphia, 1988.
60. E. W. Schwartze, *J. Pharm. Exp. Therapeut.*, 1922, 181.
61. H. A. Schroeder and J. J. Balassa, *J. Chronic Diseases*, 1966, **19**, 85.
62. K. Irgolic, 2005, personal communication.
63. J. S. Haller, *Pharm. Hist.*, 1975, **17**, 87.
64. W. Tichy, *Poisons. Antidotes and anecdotes*, Stirling Publishing Co. Inc., New York, 1977.
65. Anon, in *Chambers Journal*, 1851, Dec.
66. D. M. N. Parker, *Edinburgh Med. J.* 1865, **X**, 116.
67. M. C. Smith, *Rose*, Random House, New York, 1996.
68. K. Christen, *Environ. Sci. Technol.*, 2001, July 1, p. 291A.

69. P. W. J. Bartrip, *Eng. Hist. Rev.*, 1994, **109**, 891.
70. J. S. Segen, *Dictionary of alternative medicine*, Appleton and Lange, Stamford, Connecticut, 1998.
71. E. Grice, *Keep taking the arsenic*, www.telegraph.co.uk/health/main.jhtml?view=DETAILS&gri.
72. Homeopathystore, *Arsenicum album-30*, usa.homeopathystore.in/arsenicum_album_3030_ml_liquid.htm.
73. H. D. Kerr and L. A. Saryan, *Clin. Toxicol.*, 1986, **24**, 451.
74. A. R. Khuda-Bukhsh, S. Pathak, B. Guha, S. R. Karmakar, J. K. Das, P. Banerjee, S. J. Biswas, P. Mukherjee, N. Battacharjee, S. C. Choudhury, A. Banerjee, S. Bhadra, P. Mallick, J. Chakrabarti and B. Mandal, *Evidence-based Compl. Alt. Med.*, 2005, **2**, 537.
75. M. A. Langenhan, *Trans. Wisconsin Acad. Sci. Arts Lett.*, 1918, **XX**, 1.
76. T. Fowler, *Medical report of the effect of arsenic in the cure of agues, remitting fevers, and periodic headaches*, J. Johnson and W. Brown, London, 1786.
77. J. Begbie, in *Contributions to practical medicine*, ed. J. Begbie, Adam and Charles Black, Edinburgh, 1862.
78. J. P. Remington, *Practice of pharmacy. A treatise*, 1st edn., J. B. Lippincott Company, London, 1885.
79. A. S. Taylor and T. Stevenson, *The principles and practice of medical jurisprudence*, J. & A. Churchill, London, 1883.
80. *British Pharmaceutical Codex of 1907*, Pharmaceutical Society London, London, 1907.
81. A. Abels, *Arch. Kim.*, 1913, 320.
82. E. F. Cook and C. A. LaWall, *Remington's practice of pharmacy. A treatise*, 8th edn., J. B. Lippincott Company, London, 1936.
83. *British Pharmaceutical Codex of 1934*, Pharmaceutical Society London, London, 1934.
84. *British Pharmaceutical Codex of 1954*, Pharmaceutical Society London, London, 1954.
85. *British Pharmaceutical Codex of 1979*, Pharmaceutical Society London, London, 1979.
86. *The Merck Index*, 11th edn., Merck and Co. Inc., Rahway NJ, 1989.
87. J. G. Harter and A. M. Novitch, *J. Allergy*, 1967, **40**, 327.
88. P. Kertesz, *J. Royal Soc. Med.*, 1994, **87**, 186.
89. F. Katz, 1990, personal communication.
90. New Internationalist, *Arsenic with a straight face*, www.newint.org/issue169/briefly.htm.
91. *British Pharmaceutical Codex of 1994*, Pharmaceutical Society London, London, 1994.
92. E. Rosenthal, in *The New York Times Magazine*, 2001, May 6, p. 70.
93. J. H. K. Chu, *Acute promyelocytic leukemia (APL)*, alternative-healing.org/acute_promyelocytic_leukemia.htm, 2004.
94. S. L. Soignet, P. Maslak, Z.-G. Wang, S. Jhanwar, E. Calleja, L. J. Dardashti, D. Corso, A. DeBlasio, J. Gabrilove, D. A. Scheinberg,

P. P. Pandolfi and R. P. Warrell, *New England J. Med.*, 1998, **339**, 1341.
95. D. A. Sears, *Am. Med. Sci.*, 1988, **296**, 85.
96. D. M. Jolliffe, *J. Royal Soc. Med.*, 1993, **86**, 287.
97. G. Q. Chen, X. G. Shi, W. Tang, S. M. Xiong, J. Zhu, X. Cai, Z. G. Han, J. H. Ni, G. Y. Shi, P. M. Jia, M. M. Liu, K. L. He, C. Niu, J. Ma, P. Zhang, T. D. Zhang, P. Paul, T. Naoe, K. Kitamura, W. H. Miller, S. Waxman, Z. Y. Wang, H. D. The, S. J. Chen and Z. Chen, *Blood*, 1997, **89**, 3345.
98. W.-C. Chou, A. L. Hawkins, J. F. Barrett, C. A. Griffin and C. V. Dang, *J. Clin. Invest.*, 2001, **108**, 1541.
99. S. Waxman, in *Business Wire*, 2007, April 16.
100. W. H. Miller, H. M. Schipper, J. S. Lee, J. Singer and S. Waxman, *Cancer Res.*, 2002, **62**, 3893.
101. National Cancer Institute, *NIH News*, 2007, Jan. 24.
102. V. Mathews, B. George, K. M. Lakshami, A. Viswabandya, A. Bajel, P. Balasubramanian, R. V. Shaji, V. M. Srivastava, A. Srivastava and M. Chandy, *Blood*, 2006, **107**, 2627.
103. C. Berrie, in *Doctors Guide*, 2006, June 16.
104. T. Manshouri, *Blood*, 2005, Nov. 21, abstract 4446.
105. *ZIOPHARM oncology accomplishes key clinical development milestone; receives FDA clearance to initiate ZIO-101 phase I/II myeloma trial*, www.genengnews.com/news/bnitem.aspx?name=1135636XS.
106. *ZIOPHARM presents data highlighting oral ZIO-101 at AACR*, biz.yahoo.com/bw/070417/20070417005164.html.
107. R. Bunsen and A. A. Berthold, 1834. Published as a pamphlet reported by D. Maclagan, *Edinburgh Med. Surg. J.*, article IX, March 10, 1840.
108. M. A. Béchamp, *Compt. Rend.*, 1863, **56**, 1172.
109. (a) P. Ehrlich and A. Bertheim, *Ber*, 1907, **40**, 3292.
(b) V. Puntoni, *Ann.d'ig. Rama*, 1917, **27**, 293.
110. E. Bäumler, *In search of the magic bullet*, Thames and Hudson, London, 1965.
111. D. Livingstone, *Brit. Med. J.*, 1858, 360.
112. I. Galdston, *Behind the sulfa drugs*, D. Appleton-Century Company Inc, New York, 1943.
113. G. O. Doak and L. D. Freedman, in *Medicinal chemistry*, ed. A. Burger, Interscience Publishers Inc, New York, 1960.
114. P. S. Ward, *J. Hist. Med. Allied Sci.*, 1981, **36**, 47.
115. N. C. Lloyd, H. W. Morgan, B. K. Nicholson and R. S. Ronimus, *Angew. Chem. Int. Ed.*, 2005, **44**, 941.
116. W. R. Cullen, in *Advances in organometallic chemistry*, ed. F. G. A. Stone and R. West, Academic Press, London, 1966 vol. 4.
117. B. Nicholson, 2007, personal communication.
118. B. Farwell, *Armies of the Raj. From the mutiny to independence, 1858–1947*, W. W. Norton and Company, New York, 1989.
119. M. Harper, B.Sc. thesis, Wellcome Institute for the History of Medicine, London, 1989.

120. R. Graves, *Goodbye to all that*, Anchor Books Doubleday, London, 1957.
121. V. Lefebure, *The riddle of the Rhine. Chemical strategy in peace*, E. P. Dutton and Company, New York, 1923.
122. A. M. B. Golding, *J. Royal Soc. Med.*, 1993, **86**, 282.
123. T. Appleby, in *Globe and Mail*, 1996, Sept. 14.
124. R. Mickleburgh, in *Globe and Mail*, 1996, Sept. 21.
125. W. G. Macpherson, W. P. Herring, T. R. Elliott and A. Balfour, *History of the great war medical services. Diseases of the war*, His Majesty's Stationary office, London, 1922.
126. J. R. Aronson, in *Chem. Eng. News*, 1998, Sept. 7, p. 7.
127. J. P. Swann, *Medical Heritage*, 1985, **1**, 137.
128. I. Haiduc, 1998, personal communication.
129. S. E. Lederer and J. Parascandola, *J. Hist. Med.*, 1998, **53**, 345.
130. J. H. Jones, *Bad blood. The Tuskegee syphilis experiment*, Free Press, New York, 1993.
131. T. H. Sternberg, E. B. Howard, L. A. Dewey and P. Padget, in *Preventive medicine in World War II. Communicable diseases. ed. J. B. Coates*, Office of the Surgeon General, Department of the Army, Washington, 1960, vol. V.
132. S. McCarthy, in *Globe and Mail*, 2006, June 2.
133. N. Read, in *Vancouver Sun*, 2006, June 17, p. C7.
134. D. Saunders, in *Globe and Mail*, 2006, Dec. 30, p. F3.
135. A. Zuger, in *New York Times*, 2006, June 6.
136. D. V. Frost, *Fed. Proc.*, 1970, **26**, 194.
137. O. M. N. Dhubhghaill and P. J. Sadler, *in Structure and bonding*, Springer Verlag, Berlin, 1971, vol. 78, p. 130.
138. *Arsenic in drinking water*, National Research Council, Washington, 1999.
139. G. Jiang, Z. Gong, X.-F. Li, W. R. Cullen and X. C. Le, *Chem. Res. Toxicol.*, 2003, **16**, 873.
140. J. Mann, *Murder, magic and medicine*, Oxford University Press, Oxford, 1994.
141. H. W. Mulligan and W. H. Potts, *Trypanosomiases*, George Allen and Unwin Ltd., London, 1970.
142. B. J. Berger and A. H. Fairlamb, *Trans. Royal Soc. Trop. Med. Hygiene*, 1994, **88**, 357.
143. C. Burri, S. Nkunku, A. Merolle, T. Smith, J. Blum and R. Brun, *Lancet*, 2000, **355**, 1419.
144. MSF, *Sleeping sickness or human African trypanosomiasis*, MSF Fact sheet, 2004, May.
145. A. Picard, in *Globe and Mail*, 2000, Sept. 23.
146. S. Boseley, in *The Guardian*, 2001, May 7, p. 13.
147. D. G. McNeil, in *New York Times*, 2000, May 18.
148. G. Flaubert, *Madame Bovary*, International Collectors Library, New York, 1857.

149. M. J. Mass, A. Tennant, B. C. Roop, W. R. Cullen, M. Styblo, D. J. Thomas and A. D. Kligerman, *Chem. Res. Toxicol.*, 2001, **14**, 355.
150. S. Nesnow, B. Roop, G. Lambert, M. Kadiiska, R. P. Mason, W. R. Cullen and M. J. Mass, *Chem. Res. Toxicol.*, 2002, **15**, 1627.
151. L. Vega, M. Styblo, R. Patterson, W. R. Cullen, C. Wang and D. Germolec, *Toxicol. Appl. Pharmacol.*, 2001, **172**, 225.
152. A. Navas-Acien, E. K. Silbergeld, R. A. Streeter, J. M. Clark, T. A. Burke and E. Guallar, *Environ. Health. Perspect.*, 2006, **114**, 641.
153. D. N. G. Mazunder, *Tox. Appl. Pharmacol.*, 2005, **206**, 169.
154. D. N. G. Mazunder, C. Steinmaus, P. Bhattacharya, O. S. v. Ehrenstein, N. Ghosh, M. Gotway, A. Sil, J. R. Balmes, R. Haque, M. Hira-Smith and A. H. Smith, *Epidemiology*, 2005, **16**, 760.
155. *11th Report on carcinogens*, US Department of Health and Human Services, 2002, ntp.niehs.nih.gov/ntp/roc/toc11.html.
156. J. Hutchinson, *Brit. Med. J.*, 1887, 1820.
157. *Special report on ingested inorganic arsenic. Skin cancer: nutritional essentiality*, US Environmental Protection Agency, Washington DC, 1988.
158. *Arsenic in drinking water. 2001 update*, National Research Council, Washington DC, 2001.
159. M. A. K. Barbhuiya, M. H. S. U. Sayed, M. H. Kahn, M. A. Jalil and S. A. Ahmed, *Arsenic contamination and its consequences on human health*, Local Government Division, Ministry of Local Government, Rural Development and Cooperative Government of Bangladesh, Dhaka, 2002.
160. C. E. Dukes, *Urine examination and clinical interpretation*, Oxford University Press, London, 1939.
161. L. Gershengeld, *Urine and urinanalysis*, Romaine Pierson Publishers Inc., New York, 1948.
162. C. Hopenhayn-Rich, M. L. Biggs, A. H. Smith, D. A. Kalman and L. E. Moore, *Environ. Health Perspect.*, 1996, **104**, 620.
163. O. L. Valenzuela, V. H. Borja-Aburto, G. G. Garcia-Vargas, M. B. Cruz-Gonzalez, E. A. Garcia-Montalvo, E. S. Calderon-Aranda and L. M. DelRazo, *Environ. Health Perspect.*, 2005, **113**, 250.
164. Z. Wang, J. Zhou, X. Lu, Z. Gong and X. C. Le, *Chem. Res. Toxicol.*, 2004, **17**, 95.
165. N. Burton, in *Chemical Science*, 2007, Jan. 8.
166. Lord Kelvin, W. H. Dyke, W. S. Church, T. E. Thorpe, H. C. Bonsor and B. A. Whitelegge, *First report of the royal commission appointed to inquire into arsenical poisoning from the consumption of beer and other articles of food or drink*, Houses of Parliament, London, 1901.
167. A. Chatt and S. A. Katz, *Hair analysis. Application in the biomedical and environmental sciences*, VCH Publishers Inc., New York, 1988.
168. (a) T. J. Hindmarsh, D. Dekerkhove, G. Grime and J. Powell, in *Arsenic Exposure and Health Effects*, ed. W. R. Chappell, C. O. Abernathy and R. L. Calderon, Elsevier Scientific, Amsterdam, 1999, p. 41.

(b) J. Yanez, V. Fierro, H. Mansilla, L. Figueroa, L. Cornejo and R. M. Barnes, *J. Environ. Monit*, 2005, **7**, 1335.
169. U. Siripitayakunkit, P. Visudhiphan, M. Pradipasen and T. Vorapongsathron, in *Arsenic exposure and health effects*, ed. W. R. Chappell, C. O. Abenathy and R. L. Calderon, Elsevier Science, Amsterdam, 1999, 141.
170. C. Yuan, X. Lu, N. Oro, Z. Wang, Y. Xia, J. Mumford and X. C. Le, *Clinical Chemistry*, 2007, in press.
171. M. Lu, H. Wang, X.-F. Li, L. A. Arnold, S. M. Cohen and X. C. Le, *Chem Res Toxicol.*, 2007, **20**, 27.
172. M. P. Waalkes, J. Liu and B. A. Diwan, *Toxicol. Appl. Pharmacol*, 2007, **222**, 271.
173. A. H. Smith, G. Marshall, Y. Yuan, C. Ferreccio, J. Liaw, O. v. Ehrenstein, C. Steinmaus, M. N. Bates and S. Selvin, *Environ. Health Perspect.*, 2006, **114**, 1293.
174. R. K. Kwok, R. B. Kaufmann and M. Jakariya, *J. Health Popul. Nutr.*, 2006, **24**, 190.
175. A. Rahman, M. Vahter, E.-C. Ekstrom, M. Rahman, A. Haider, M. G. Mustafa, M. A. Wahed, M. Yunus and L.-A. Persson, *Am. J. Epidemiology*, 2007, **165**, 1389.
176. O. Andersen, *Chem. Rev.*, 1999, **99**, 2683.
177. M. J. Kosnett, in *Arsenic exposure and health effects*, ed. W. R. Chappell, C. O. Abernathy and R. L. Calderon, Elsevier Science, Amsterdam, 1999.
178. M. J. Kosnett and C. E. Becker, *Vet. Human. Toxicol.*, 1988, **30**, 369.
179. D. N. G. Muzumder, U. C. Ghoshal, J. Saha, A. Santra, B. K. De, A. Chatterjee, S. Dutta, C. R. Angle and J. A. Centino, *Clin Toxicol.*, 1998, **36**, 683.
180. M. M. Rahman, U. K. Chowdhury, S. C. Mukherjee, B. K. Mondal, K. Paul, D. Lodh, B. K. Biswas, C. R. Chanda, G. K. Basu, K. C. Saha, S. Roy, R. Das, S. K. Palit, Q. Quamruzzaman and D. Chakraborti, *J. Toxicol. Clin. Toxicol.*, 2001, **39**, 683.
181. H. V. Aposhian, A. Arroyo, M. E. Cebrian, L. MD Razo, K. M. Hurlbut, R. C. Dart, D. Gonzalez-Ramirez, H. Kreppel, H. Speisky, A. Smith, M. E. Gonsebatt, P. Ostrosky-Wegman and M. M. Aposhian, *J. Pharm. Exp. Therapeut.*, 1997, **282**, 192.
182. T. Murcott, *The whole story. Alternative medicine on trial*, Macmillan, 2005.
183. L. Priest, in *Globe and Mail*, 2001, March 22.
184. S. J. Flora, G. Flora, G. Saxena and M. Mishra, *Cell Mol. Biol.*, 2007, **53**, 27.
185. J. H. Winslow, *Darwin's Victorian malady. Evidence for its medically induced origin*, American Philosophical Society, Philadelphia, 1971.
186. D. Edmunds, *Brit. Med. J.*, 1887, 1820.
187. A. Nicholls, B.Sc thesis, Wellcome Institute for the History of Medicine, 1997.

188. K. Warren, in *Fenland notes and queries*, ed. W. D. Sweeting, Geo. C. Caster, Peterborough, 1987, vol. III.
189. A. W. Woodruff, *Darwin's health in relation to his voyage to South America*, Julian Friedman Publishers Ltd., 1974.
190. R. Colp, *To be an invalid, the illness of Charles Darwin*, University of Chicago Press, Chicago, 1977.
191. J. Worthen, *Biography of D.H. Lawrence*, www.nottingham.ac.uk/mss/online/dhlawrence/biog-full/chap8.phtml.
192. M. Goodfleisch, in *American-Statesman*, 2007, May 27.
193. M. J. Mauro, *History of Ojo Caliente*, 1998.
194. L. Donelson, *Out of Isak Dinesen in Africa: the untold story*, Coulsong List, Iowa, City Iowa, 1995.
195. K. Blixen a.k.a. Isak Denesen, *Letters from Africa 1914–1931*, University of Chicago Press, Chicago, 1981.
196. A. Borodin, Doctor of Medicine thesis, St. Petersburg, 1858.
197. G. B. Kaufmann and K. Bumpass, *Leonardo*, 1988, **21**, 429.

CHAPTER 2
Arsenic Where You Least Expect It

In Chapter 1 we dealt with arsenic compounds as they related to human health. In this chapter we will look at some of the many other ways in which we have used and abused arsenic compounds, beginning with two animal medications – one of which is under considerable scrutiny from a public-health angle.

> **Box 2.1   Arsenic Production and Consumption**
>
> Arsenic trioxide, now the main source of arsenic for the world's needs, is a byproduct of the mineral industry, although there was a time when it was produced for its own sake (Section 7.7).
>
> From 1941–1944 the annual global use of arsenic was 42 000 tons of white arsenic distributed as follows: 27 000 tons for insecticides, 7700 tons for weed killers, 3000 tons for glass manufacture, 570 tons for dyestuffs and 5000 tons for all other uses. By 1951 consumption had dropped considerably reflecting the shift to other pesticides and herbicides.[1,2]
>
> The USA is the world's biggest consumer of arsenic and peak domestic production of around 25 000 tons was recorded in 1944. At the time, the world was producing around 50 000 tons per year. US domestic production dropped off rapidly until it ceased in 1985 when the American Smelting and Refining Company, Asarco, closed its smelter in Tacoma partly because of the costs of bringing the plant into compliance with new emission restrictions (Section 7.7). World arsenic production was about 40 000 tons at that time. The average US import figure was about 20 000 tons per year until 2003, and 86 to 90 per cent of total consumption, was used for the production of the wood preservative chromated copper arsenate CCA, Figure 2.1.[3,4] "Arsenic

---

Is Arsenic an Aphrodisiac? The Sociochemistry of an Element
By William R. Cullen
© William R. Cullen 2008

Box 2.1 Continued.

compounds were also used in fertilisers, fireworks, herbicides, and insecticides. Elemental arsenic was used as a hardener in lead storage batteries. Addition of less one per cent arsenic also hardens small-arms ammunition used by the US Military. High-purity arsenic was used for gallium arsenide semiconductors and for germanium-arsenide-selenide optical materials."[3]

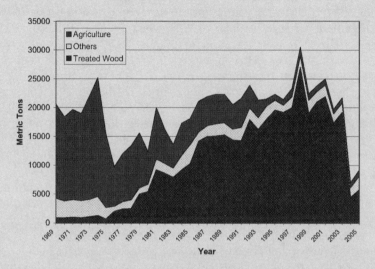

Figure 2.1 US demand for arsenic (1969–2005) compiled from published US Geological Survey data.

By the end of 2004 arsenic trioxide imports had dropped to about 6100 tons from the approximately 20 000 tons imported the year before. This 70 per cent decline was the result of a voluntary decision by the wood-preserving industry to reduce its production of CCA-treated wood for the domestic market. However, during 2005 US domestic imports of arsenic trioxide increased approximately 35 per cent over the previous year to around 8300 tons.

The US remains the world's leading consumer of arsenic, primarily as a component of CCA for commercial applications. China was the main source (51 per cent) of the oxide imported into the US. Morocco was the source of 30 per cent, and Hong Kong 10 per cent. Neither Morocco nor Hong Kong is on the US Geological Survey's list of arsenic producers so the actual source of this arsenic is undeclared. The main producers are China, 30 000 tons per year; Chile, 11 500 tons; Peru, 3600 tons; Russia, 1500 tons; and Kazakhstan, 1000 tons.[3]

Imports of elemental arsenic have generally declined since 2001, and in 2005 the US imported 812 tons, down from 872 in 2004. China was the leading source, accounting for 718 tons, down from 739 tons in 2004."[3]

2.1 Animal Feed Additives

We have already noted the custom, established in the 1700s, of feeding arsenic trioxide to animals to bring out desirable characteristics such as a sleek coat (Section 1.3). The Kelvin Royal Commission of 1901 was informed that arsenic was being given to poultry to fatten it up for market and was told about experiments that showed that the flesh did not become contaminated, but arsenic did get into the feathers, just as it did into human hair.[5,6] In 1975 the oxide was recommended for fattening young cattle at doses of up to 10 g per day.[7] But the story of modern usage really dates to 1924, when Dr. Charles Salsbury arrived in Charles City ID, from England to practise as a veterinarian. Over the years he established a sideline business of treating diseases in poultry, and eventually devoted all his efforts to developing pharmaceuticals for poultry under the banner of Dr. Salsbury's Laboratories. He died in 1967. His company looked into the use of arsenicals as animal food additives and its first patents, describing the beneficial effect of organic arsenicals given as animal food additives, appeared in the late 1940s. Arsanilic acid and 3-nitro-4-hydroxyphenylarsonic acid (roxarsone), a precursor to salvarsan (Figure 1.1), were effective in controlling coccidosis and related infections in chickens.[8,9]

Roxarsone is also known as 3-Nitro, a trade name used by Alpharma Inc. The initial results for chickens showed a weight increase of 4.1 per cent that could be attributed to the arsenical in the food. Similar results were observed for turkeys and swine. "There were a few instances reported in which no improvements in growth and feed efficiency were obtained, but the high frequency of positive effects from arsenicals gives the producer that small edge that may well be the difference between a profit and loss in modern swine and poultry feeding enterprise."[9] The arsenic compounds are approved in the US and Canada for feeding, alone and in combination with antibiotics such as chlortetracycline, the combination having a synergistic effect on growth. Roxarsone is claimed to be superior to arsanilic acid,[10] and the most popular poultry starter and grower diets contains 44–47 ppm of the arsenical.[11]

One relatively recent estimate, disputed by the pharmaceutical industry as being too high, finds that 25 million of the 28 million pounds of antibiotics produced in North America annually are fed to animals to improve their growth.[12] They are added to chicken feed to control infection. Roxarsone, although not classified as an antibiotic, probably promotes growth via a similar mechanism. The regulations governing the use of roxarsone require that the arsenical not be fed for the last five days before slaughter. This is to ensure that any residual arsenic in the marketed product does not exceed the maximum level allowed by the US FDA of 0.5 ppm in the uncooked muscle, and 2.0 ppm in the fresh uncooked edible byproducts.

There was some initial concern about using arsenicals in this way, but any fears were allayed in North America.[13] The European Union has not approved the use of roxarsone for animal feeds[14] and does not import poultry raised on feeds containing arsenicals. In 1959 France banned the use of arsenicals in animal feeds but did allow food supplemented with related antimony compounds in the production of *foie gras*, which certainly is associated with objectionable practices.[15] China is currently a major source of roxarsone.

In the US poultry farmers raise chicks, delivered by poultry-producing companies such as Simmons Foods and Tyson Foods, in houses that hold about 20 000 birds. It takes five to seven weeks for the chickens to grow big enough for slaughter and the cycle is repeated in each house about six times each year. To maintain semisanitary conditions, the floors of the houses are covered with bedding material such as corn husks and wood chips. The companies supply the necessary food, but they are not responsible for getting rid of the litter which is removed periodically from the houses. Standard practice is to spread it on the land, resulting in concern about contaminated water and elevated levels of phosphorus and arsenic in the soil.[16] The 600 million or so chickens raised each year on the Delmarva Peninsula in the eastern US are estimated to produce 1.5 billion kg of chicken litter that contains 20 000–50 000 kg of arsenic.[17]

Some regions of the US are running out of land on which to spread the litter, so alternative methods of disposal are being developed. One of these is to use the material as fuel for energy production and a Philadelphia, PA-based company. Fibrowatt, hopes to burn 477 000 tons of litter each year in Arkansas. If approval is given for the plant, the litter will be hauled away at no cost to the producers.[18] Another use is to transform it into pellets for sale to home gardeners under such catchy names as Cock-a-Doodle-Doo. The litter was once used as cattle feed.[19]

The initial studies showed that roxarsone was partially converted to 3-amino-4-hydroxyarsonic acid as it passed through the chicken; that is, the nitro group was reduced, but essentially all the arsenic was excreted without being converted to inorganic arsenic.[9] Modern analytical results confirm this finding, but there is much confusion about the ultimate fate of the arsenicals once they exit the chicken and enter the environment via the litter. There is no standard method for storing or treating litter, so each sample has a different history, especially with regard to microbial action. Composting, which provides optimal microbial conditions, will promote transformation of the arsenical. Under some conditions inorganic arsenic species – soluble and insoluble – are produced, but we know little about how they are formed: volatilisation of arsenic, although conceivable (Chapter 3), is a very minor biological pathway.[20–22] In rural Delaware, air arsenic levels close to chicken farms are no different from the low numbers recorded in other sampling locations in the state.[23]

The application rate to soils is normally based on the nitrogen content;[24] if this litter is spread at 10 000 pounds per acre, about 0.35 pounds of arsenic are also delivered per acre. This added amount is within acceptable experimental and sampling error ranges for soil analysis so it is not surprising to find reports that state "the use of poultry litter containing arsenical feed additives does not appear to affect the arsenic content of soil and crops."[19,25] The US Geological Survey shares the view of the National Chicken Council that this spreading activity is yet to pose a problem, in spite of the fact that the Department of Agriculture predicted "that in 2006, 9.1 billion broiler chickens will be produced in the US. About 70 per cent of these are fed roxarsone to control infections and increase weight."[26]

Claims have been made that arsenic in chicken litter spread on fields has contaminated homes and caused a variety of illnesses. Litigation is ongoing but the first case to be tried was decided for the defendants Alpharma and Alpharma Animal Health Ltd. by a jury who deliberated for 21 minutes.[27]

This story helped sensitise the US public to the amount of arsenic that is found in the chickens they are eating, prompting statements such as the following: "As the average American's chicken consumption rose from 28 pounds a year in 1960 to 87 pounds a year in 2005, the government's tolerance level for arsenic in chicken has remained at 500 parts per billion."[28,29] One 2004 study that contributed to the hype actually used the arsenic concentrations reported in chicken livers – 0.33 to 0.43 ppm during the years 1994 to 2000 – to estimate the amounts of inorganic and organic arsenic in chicken meat. None of the calculated results exceeded the government standard, but the authors confidently stated that "the higher arsenic concentrations observed in chickens compared with other poultry and meat products is consistent with the use of chicken feed containing additives including arsenic compounds."[30] Given this news, consumers are saying that they will deliberately seek out poultry producers that do not use arsenic. They are not aware that their chickens will contain some arsenic no matter what they do (Chapter 7).

Tyson, the world's largest protein processor, stopped using arsenicals in feed in 2004, but sales have plummeted because of the double whammy of publicity about arsenic and avian flu.[31]

This is not a new problem, nutritionally or politically. As early as 1977 Canadian newspapers reported that a Broiler Marketing Board seized 4000 pounds of the 60 000 pounds of fresh Missouri chicken imported by Super Valu Stores, which was challenging the board's British Columbia provincial monopoly. The board found arsenic in the chickens' livers and claimed that it had come from roxarsone. The board then said that this was an indication that local chickens, while more expensive, were of higher quality.[32]

The Minnesota-based Institute for Agriculture and Trade Policy an advocacy group for organic food is using the arsenic-in-chicken issue to advance its own agenda. One commentator writes: "if you are pushing organic farming, you won't change a lot of consumer behaviour by touting mixed results about possible contamination. You get further telling them there's arsenic in their chicken."[33]

2.2 Heartworm

Heartworm is a disease that afflicts dogs and cats and is carried by the parasite *Dirofilaria immitis*. There are three main stages of the disease, which is transmitted by mosquitoes: infectious larvae, adult worms and microfilariae. Mosquitoes deposit larvae on an animal's skin; the larvae penetrate the skin and migrate to the heart where they grow to mature worms, 9 to 14 inches long, in 60 days. These worms can clog the heart, leading to congestive heart failure. The symptoms are not usually obvious until too late. The worms reproduce sexually to spawn microfilariae that circulate in the blood and get transferred to feeding mosquitoes. The microfilariae grow to larvae in the mosquitoes' kidneys, and the cycle continues. The disease was initially confined to the southern US but is spreading north. About 240 000 dogs and 3000 cats were infected in the US in 2005.[34,35]

The arsenical thiacetarsamide (Figure 1.1) was used to kill adult worms in infected dogs, followed by arsenic-free dithiazanine to kill the microfilariae. Another oral drug, ivermectin, approved in 1986, kills larvae so it can prevent infection in uninfected dogs.[36] Current canine guidelines suggest that it is beneficial to administer a prophylactic dose of ivermectin for one to six months prior to the administration of another arsenical, melarsomine (Figure 1.1). This regimen allows older larvae and immature cells to develop to an age at which they can be killed by the arsenical.[35]

2.3 Pesticides and Herbicides

Ancient attempts to keep pests away from crops involved animal sacrifices, a lot of praying and strategies such as hanging a crayfish in the middle of a garden to ward off caterpillars. Evidently, the presence of a nubile woman with bare breasts and unbound hair also helped these efforts.[37] More effective treatments evolved that were based on applications of vegetable or mineral concoctions. A mixture of realgar and ashes was smeared on grapes to prevent rot. Applications of realgar mixed with olive oil lees kept vines free from beetles and ants. Tar and realgar had the same effect. Hellebore, a known toxic flowering plant believed to have magical medicinal properties, was mixed with arsenic, presumably a sulfide, to kill flies.[37] The Greeks developed mixtures of lime and sulfur, which were both readily available, and their use extends to the present day as Bordeaux mixture.

Medieval Europeans associated pest outbreaks with superstitions and divine punishment for sins, and pursued everything from magical rituals to legal and religious proceedings. They even tested the effect of formally excommunicating and banishing the troublesome invaders. It was not until 1668 that flies were shown to have life cycles involving eggs and maggots. In a major breakthrough, around 1867, Paris green – copper acetoarsenite (Sections 2.8 and 3.3) – was found to be an efficient pesticide against the Colorado potato beetle which was creating havoc in the eastern United States. It was also effective in orchards against the codling moth in apples, and quickly grew into international use against mosquitoes. These applications, coupled with the discovery in 1882 that lead arsenate was effective against the gypsy moth, marked the beginning of the modern era of pest control.

2.3.1 Lead and Calcium Arsenates

Lead arsenate proved to be better than Paris green; it was less toxic and more effective. Farmers initially mixed their own from readily available lead and arsenic salts.

A 1936 text aimed at the US market encouraged pharmacists to be resourceful and look for additional sources of revenue. "With a general knowledge of insects and recognition of the chemical insecticides, the pharmacist can be of real service to his community and at the same time build up a profitable clientele and avoid stocking his shelves with worthless and unsaleable trade-named materials."[38] The same text describes how lead arsenate is a stomach poison that kills underground grubs and worms and hinders the growth of common weeds such as crab grass and dandelion. "It can be applied with soil as a top dressing at 700 kg [sic] per acre." Soon a commercial product became available and its relatively benign effect on vegetation resulted in its use in many countries, including the US, Canada, Australia, the UK and North Africa, particularly against the codling moth, but also against mosquitoes. It was sprayed as a slurry, by hand, directly on the infestation: no moth, no spray. Countries, especially the USA, still have trouble today with land that was contaminated by frequent applications of lead arsenate (Section 7.11).[39]

During World War II (WWII) the price of lead escalated so other arsenates were introduced for pest control, particularly calcium arsenate; however, only the lead salt was effective against the codling moth. Even though residual contamination was high, no substitute was found until 1945 when DDT and 2,4-D became available, although this proved not to be such a benefit in the long run.[40] Nonetheless, something had to change because the moth was developing an immunity to the lead salt and annual application rates had to be increased to about 125 kg of lead and 45 kg of arsenic per acre in 1947.[41] The amount of lead arsenate used in Washington state orchards in the first half of the 20th century resulted in an average annual application of 59 kg of lead and 23 kg of arsenic per acre.

Global lead arsenate use peaked at 90 million pounds in 1944: peak calcium arsenate use amounted to about 80 million pounds per year around the same time. The arsenicals were applied by using air-blast sprayers and as aerial dust. Bait boxes containing arsenicals were used inside homes.[42a] Lead arsenate use continued after WWII, sometimes mixed with DDT, but faded from the scene in the 1980s, not because of any prohibition, but because a new air exposure limit for arsenic in the workplace of 0.01 mg/m^3 forced manufacturers to close up shop. Other countries had also stopped the practice by this time. In a bizarre turn of events, lead arsenate gained the acceptance of some "organic gardeners" in the 1970s. Presumably because its use was already well established and it was not a product of the agrochemical industry.[42b]

In the 1980s lead arsenate was being used as a growth regulator on 17 per cent of the US grapefruit crop. In order to get grapefruit to market early, the trees were sprayed with the arsenical after blooming to reduce the acidity of the fruit. This meant that the grapefruit reached legal maturity in September or October, when it would naturally be ready in November. "The application of lead arsenate spray was highly satisfactory in the production of early passing legal maturity ratios in the juice. Even after legal maturity the grapefruit could be "stored" on the tree for months, merely increasing in size, and extending the marketing season."[43]

In 1986, the US Environmental Protection Agency, US EPA, determined that the toxicological risks from the use of lead arsenate outweighed the limited benefit, but it made an exception for the grapefruit use. The urinary arsenic levels of workers in the industry in Florida were deemed at the time to be satisfactory, because a value of 200 ppb was used as being indicative of arsenic poisoning. The current cautionary level is 50 ppb (Section 1.9).[44] A later study of orchard workers in Washington State suggested that there was an elevated risk of death from coronary heart disease in those exposed to lead arsenate spray.[45] Lead arsenate was officially banned for use as a pesticide in the US on August 1st, 1988.

The boll weevil was a serious pest in US cotton fields. "The number one agricultural pest in America"[46] caused more than $15 billion (US) in losses from damages to crops and costs of pest control over the past century. It crossed from Mexico into Texas around 1892 and by 1922 the insect – which can produce eight generations per year – had infested more than 85 per cent of the eastern cotton-growing states. Early efforts to control the weevil included burning the crop and poisoning with a mixture of arsenic in molasses that could be sprayed onto the crops. A more effective treatment was to dust the crop with calcium arsenate. This treatment was costly but was the best available until synthetic organic pesticides became available after the end of WWII. The successful treatment started with organochlorine compounds and progressed to carefully timed treatment with organophosphates once these became available. Now, by using a combination of pheromone traps for detection, chemical treatment for control, release of sterile males, and improved agricultural practices such as destroying cotton stalks after harvest, it is possible to rid the pest from huge areas. The goal of total eradication from the US is now expected to be reached by 2009.[46]

Calcium arsenate was a constituent of "London purple," a byproduct of the dye industry that was used as a pesticide in the early 1900s. It was compounded into an insect repellent as in the following patented mixture: hog's lard, linseed oil, sulfur, lime, London purple, carbolic acid, and fish oil. When mixed this gives an "ointment which is applied to the hide of the animal by means of a brush or swab or in any convenient way."[47]

2.4 Arsenic Trioxide

2.4.1 The Black Death

Although bubonic plague, in the form of the Black Death, disappeared from Europe around 1722, it continued to break out in other parts of the world much later and still occurs in Asia. Various suggestions have been offered for Europe's mysterious immunity: improvements in sanitation, displacement of the black rat by the brown rat or development of immunity in either the rat or the human population, but none of these is entirely satisfactory. The following is a summary of an explanation that has been offered by the historian, Kari Konkola.[48]

Plague is a disease of rats that is transferred to humans via infected fleas. The normal rat population is determined by the amount of food available, and a critical population of rats is needed for the transfer of the disease to humans to become significant. Also, specific fleas are needed for the transmission, and to make matters even more difficult, these fleas need to develop an intestinal blockage to make them effective transmitters. Thus plague is an uncertain and chancy phenomenon and also intermittent. However, one thing is sure: no rats – no problem.

Cats and traps are not efficient exterminators and because rats are very prolific the overall population is not affected by low casualty rates. The best way of getting rid of rats is by using poison. However, rats are also very fussy eaters, so getting them to eat poison is not easy. Any poison must be tasteless, highly toxic and easily available in large quantities. Arsenic trioxide fits this profile. Only a small amount of the chemical is needed to kill a rat: about 0.02 g. It is tasteless and can be mixed with fat and flour, even baked, if desired, for presentation to the animal.

We have seen that impure arsenic trioxide was available from Austrian sources by the end of the 14th century, but soon the mining industry began to develop elsewhere in Europe. White arsenic was described in a book on glass making written in 1604, as a product of smelting the cobalt ores in Schneeberg, Germany. This process seems to have been initiated around 1570. Large-scale arsenic production began around 1650 and continued into the 1700s, resulting in a drop in the price of the oxide. After 1650, the Austrian arsenic industry declined because Germany could get arsenic to Venice for sale at a much lower price than could the Austrians. The arsenic was shipped to Venice as ballast, so there was a lot of it. Consequently, even though there were some restrictions on general access to the oxide, such as the French law of 1682 (Section 5.4), and even though sales were supposed to be restricted to medics, goldsmiths, colour makers and other officials, the oxide was readily available, at low cost, for use against rats by the late 1700s.

Konkola estimates that Europe needed about 2000 tons of arsenic trioxide per year to kill off the rats: a reasonable amount, given that the output from Schneeberg alone was about 1500 tons. He finishes his account of this very plausible hypothesis with the following: "It just might be that the beginning of industrial-scale toxic pollution brought about a vast improvement in the health conditions of Europe."[48]

Somewhat the same practice is involved in combating termites or "white ants" in Australia. These pests cause major structural damage to all types of wooden buildings. The worker ants digest timber by the use of symbiotic protozoa in their gut and feed their partly digested semiliquid food, either via regurgitation or through their anus, to the other termites. In extreme cases the ants can destroy all the timber in a home, walls and roofs, within three months of infestation. They do more damage to homes worldwide than any other cause combined, including fire, flood and storms.

To counter this threat, Furniss White-Ant Exterminator was widely used in Australia in the early 1900s and cans of the product are still occasionally discovered on farms. The label on one found in 1970 states: "POISON Contents arsenic trioxide, 80 per cent by weight. Eradicates white ants from houses, buildings, fences, fruit trees, vines, *etc.*" "The termites feed on their dead, so the poison is rapidly transferred through the whole colony or nest to the queen and a complete extermination is assured."

Current strategies for termite control use lethal, but arsenic-free, chemicals that stick to the insects' bodies and are carried back to the colony to continue the job.[49]

Other pests were introduced to Australia and New Zealand by settlers trying to recreate the familiar environment of Europe. These imports included rabbits whose population rapidly got out of control. Many methods were tried to reduce their number, including poison in the form of arsenic-laced carrots and icing sugar distributed around their warrens.[50] The deliberate introduction in 1950 of myxomatosis, a lethal disease of rabbits, rapidly eased the situation. However, the surviving rabbits acquired partial immunity so a second virus (rabbit calicivirus) was introduced into the rabbit population in 1996. Accidental release of myxomatosis wiped out most of the rabbits in France and Britain.[51]

Along with the rabbits came a few prickly-pear cactus plants that were imported to Australia by the first wave of European settlers in the early 1800s. Because the climate was favourable and there were no natural local predators, the growth rate of the "pears", like the rabbits, was also phenomenal, and by 1870 the prickly pear was a serious rural hazard. The spiky cactus formed dense stands that made the areas it infested nearly impossible to farm, driving many settlers from their land. In 1901 the government of Queensland was offering a reward of £5000 to the person who discovered a method to destroy the plant. Arsenic compounds provided some relief in the form of a mixture of arsenic trioxide, salt and caustic soda. This was replaced by arsenic pentoxide, sprayed or injected. In 1918 the introduction of "Roberts' improved pear poison," a mixture of 80 per cent sulfuric acid and 20 per cent arsenic pentoxide made by O.C. Roberts Ltd. of Willangarra on the New South Wales/Queensland border, proved to be superior to all other chemicals for plant control. The arsenic was originally imported from Japan, but a local product was soon available from a smelter on the Mole River near Willangarra.[52] The "pear poison" was either applied using hand-held sprayers, or by allowing fumes from the boiling mixture to drift across the pear's growing area, as shown in Figure 2.2.[53] Roberts also used arsenic trichloride, a dense fuming liquid, on the prickly pear and manufactured 4000 gallons for this purpose in 1920.

The eventual solution to the prickly-pear problem came in the form of biological warfare. The amazing spread of prickly pear in Australia was considered to be one of the botanical wonders of the world; its virtual destruction by cactoblastis (*Cactoblastis cactorum*) caterpillars is regarded as the world's most spectacular example of the successful biological control of a weed. The first liberations of the caterpillars in Australia occurred in 1926. Within six years, most of the original, thick stands of pear were gone. Properties previously abandoned were reclaimed and brought back into production.

Figure 2.2 Prickly-pear eradication using fumes from Roberts' improved pear poison (courtesy of Les Tanner[53]).

Back in the United States, arsenic acid was being widely used in agriculture in much the same way, particularly in cotton harvesting. Traditionally, cotton was harvested by hand, but when machines were brought in to do the job in the 1940s some chemical help was required. Arsenic acid was used as a desiccant to kill the plant, facilitating the harvest and making it possible earlier in the season. About two million acres of Texas and Oklahoma were treated annually. Spraying also helped control the boll weevil.[54]

Arsenic trioxide was widely used as an insecticide, a soil sterilant and termite eradicator. It was used in sheep dips, flypapers, and wallpaper glue. It was also used in ant bait, as in the following: "Mix granulated sugar 1.5 lb, water 1.25 pints, tartaric acid 15 grains, sodium benzoate 15 grains, add 1/8 oz sodium arsenite and 3 lb honey. Results from the use of this bait for ants should not be anticipated within a week or ten days because the action is slow. The dilution is deliberate so that the workers can carry the poison to the young in the nest and gradually kill out the entire colony."[38]

We will encounter the clandestine use of arsenic from flypapers in Chapter 5. In the meantime here is an excerpt from the Michigan Penal Code, Section 438:[55] "Any person who shall manufacture, compound, sell or offer for sale, or cause to be manufactured, compounded, sold or offered for sale, any flypaper or other form of fly killer which contains arsenic or other poison in sufficient quantity to be dangerous to the life or health of persons ... shall be guilty of a misdemeanour."

The city of Boston got rid of a surfeit of 20 000 pigeons in 1966 with the aid of arsenic-soaked corn.[56] In 1996 the state of Massachusetts tried something similar on gulls that were endangering the survival of two local bird species, the piping plover and the roseate tern. Nearly 2000 gulls were killed but they died a slow death and seemed to prefer to do so on roads, ponds and backyards rather than out of sight on the ocean. Bullets eventually proved to be a less distressing alternative. The program lasted for four years.[57,58]

Sodium arsenite began to be used for weed control around 1900, particularly by the US Army Corps of Engineers to deal with water hyacinth in Louisiana.[2] Mine dust, aka arsenic trioxide, from the Asarco smelter in Tacoma WA, was used all over the US and in parts of Canada during the 1940s to 1970s to cover the ground surface of electrical power substations. It was laid down to a depth of 2 inches by untrained individuals who had no idea what they were handling. No weeds grow, though.[59]

2.5 Wood Preservation

2.5.1 Chromated Copper Arsenate (CCA)

Although wood is one of the world's most valuable renewable resources there are two main disadvantages to its use. First, it is flammable, second, it deteriorates biologically because it is a source of food for fungi, insects and other organisms.

Charring with fire was one of the first methods used to preserve wooden stakes. Since then, many hundreds of chemicals and combinations of chemicals have been used as preservatives, but only a few proved to be both effective and devoid of objectionable properties from the viewpoint of mid-1900s society. These are commonly classified into three main groups: (1) preservatives derived from coal tar, petroleum or wood tar, *e.g.* creosote; (2) preservatives soluble in organic solvents, *e.g.* pentachlorophenol, and (3) waterborne chemicals, *e.g.* chromated copper arsenate (CCA).[60] We will be concerned mainly with the waterborne class.

Patents were issued in the early 1800s for the use of mercuric chloride and copper sulfate to preserve wood, but although the chemicals were effective they were easily washed out of the wood in spite of the use of vacuum and pressure to assist the impregnation. Then, in 1933, a mining engineer from India, Sonti Kamesam, discovered that the addition of chromate to copper sulfate resulted in improved fixation in the wood. The first US patents incorporating this finding date to 1938.[63] Oak posts were preserved successfully by filling holes drilled in the wood below the ground surface with arsenic trioxide.[61] This discovery led to Dr. Karl Wolman's patent of 1927 which described the use of mineral oil mixed with arsenic salts.[62]

Subsequently, arsenic-containing preservatives were formulated with mixtures such as: potassium dichromate, copper sulfate and arsenic pentoxide[60] and were marketed under names such as Greensalt, Tanalith-C, Chemonite and Wolman CCA. The name Wolman is so closely connected to arsenic-based wood preservation that CCA-treated wood is often sold as "Wolmanized." Material Safety Data Sheets refer to Wolmanized® Treated Wood and Lumber.[64]

In the 1970s, as the supplies of normally rot-resistant wood such as cedar and redwood decreased and the demand for inexpensive replacement material increased, the construction industry was forced to use less desirable wood species such as Ponderosa pine, Lodgepole pine and coastal Douglas fir that had to be treated. In the 1990s the use of CCA increased relative to other wood preservatives because of its ease of use in "do-it-yourself" construction projects, especially involving decking, landscaping and fencing, Figure 2.1.

CCA's advantages were seen to be that it: (1) provided excellent protection against most pests – although for marine use, where only pine wood was suitable, a combined heavy treatment of CCA and creosote offered the most protection; (2) produced no smell or vapours; (3) was suitable for use indoors; (4) was nontoxic to nearby growing plants; (5) allowed treated surfaces to be painted; (6) resulted in a permanent weight increase of only 1–2 per cent after treatment.[65] Its disadvantages were perceived to be (1) it did not protect wood from excessive weathering; (2) it provided poor protection against marine borers; (3) it was not heat stable above 60 °C, therefore not suitable for thermal treatment processes; and (4) the wood experienced a temporary weight increase of 20–90 per cent immediately after treatment and the associated swelling could produce defects when the wood was redried. But all in all, arsenical preservatives were viewed as being very good; and were only to be avoided in wood products that came in contact with food and in products such as railway crossties, where a long life in a harsh environment is expected.

CCA is usually applied to wood using the Bethel process, named after the inventor, in which the preservative solution is forced into the wood under pressure. Some types of wood are more easily treated than others and range from the easily penetrated Ponderosa pine to the very difficult Douglas fir.

About 70 per cent of single family homes in the US have decks or porches made of CCA-treated wood. Approximately three 45-year-old trees are needed to supply the wood to build one backyard deck. Working with treated lumber, the deck will last up to 50 years: with untreated lumber the deck will need to be replaced every few years. On this basis the industry claims that the use of pressure-treated wood saves trees.[65]

The most widely used composition for CCA contains 18.5 per cent copper oxide, 47.5 per cent chromium trioxide, and 34 per cent arsenic pentoxide. Depending on the usage, elemental metal concentrations in the wood may range from 1100 ppm to 11 600 ppm for copper; from 1400 ppm to 25 000 ppm for chromium; and from 800 ppm to 21 600 ppm for arsenic. In 1990 there were 540 wood-preserving plants in the US and they consumed about 70 per cent of the arsenic produced worldwide.[65]

At a rudimentary level the arsenic protects the wood against insects, the copper acts as a fungicide, and the chromium acts as the glue to fix the copper and arsenic in the wood. There is a good deal of evidence to support the view that most of the copper is associated with wood components, but before it is reduced by wood sugars the excess copper that is not fixed by the wood, reacts with the hexavalent chromium to form fixed copper chromate. The reduced chromium reacts with the arsenic acid, creating chromium(III) arsenate ($CrAsO_4$) and other byproducts. Some copper may also be fixed as copper arsenate ($Cu_3(AsO_4)_2$).[63]

Health and safety factors were an early consideration because the product contains both arsenic and chromium and trade organisations involved in promoting the use of concrete milked this aspect in campaigns against the use of CCA-treated wood.[66] In 1986 the US EPA finished a study that was initiated in 1978 and concluded that CCA-treated wood was safe to handle. This was received with more than a little skepticism by some members of the public because at the time the EPA was in the process of banning the use of many arsenic-containing pesticides (the decision to ban them was made in 1988). The US Consumer Product Safety Commission, persuaded by the industry – possibly with the help of flawed data, according to environmental activist Joe Prager[67,68] – determined that the benefits of using CCA-treated wood outweighed the risk and in consultation with the industry, which balked at the proposal that the wood carry a warning label, produced a set of guidelines to protect the treater, the end user and the environment. CCA was classified as a "restricted use" pesticide, meaning that the sale and use of the chemical was restricted to certified applicators. The guidelines included suggestions such as: CCA-treated wood should not be used for cutting boards or counter tops; it should not be burnt in open fires or stoves; and anyone sawing or machining treated wood should wear a dust mask and work outdoors if possible.[65] One major problem was that although the information may have reached the distributor, it was not passed on to the consumer, and accidents happened.

Jimmy Sipes worked for the US Forest Service in Indiana where he occasionally sawed wood to make picnic tables. In 1983 after 10 days of sawing he began to vomit blood. He recovered slowly but had a relapse after another stint with the saw. In hospital he was found to have around 100 times the normal concentration of arsenic in his hair and finger nails. A jury awarded him $100 000 (US).[69] His lawyer was David McCrea of Bloomington IN, who won two more jury verdicts and settled three similar cases.

Laurie Walker, of Salt Lake City UT, ran a few large slivers of CCA-treated Douglas fir through her middle and index finger, resulting in tissue narcosis that necessitated the amputation of part of a finger. Two adjacent fingers were subsequently amputated as well. She settled for $150 000, which was not enough to pay her medical expenses.[70,71]

Lynn Milam, of Hernando MS, was advised by the police to leave her husband because there was good evidence he was poisoning her with arsenic. She had been hospitalised six times with vomiting and diarrhoea. It was subsequently determined that she and her husband had been building a cabin from

CCA-treated wood and had done all the wrong things: they took no precautions, wore no protection, and they burned the treated wood indoors. In spite of the finding of high arsenic levels in her husband as well, the case went to trial because the lumber companies claimed that "this could not be." The jury found Mr. Milam not guilty.[72]

These settlements have encouraged lawyers and putative victims to initiate class action suits against producers such as Osmose Inc. and Arch Chemicals Inc., and distributors such as Home Depot Inc. and Lowe's Companies Inc. To date, class action status has not been granted, so CCA-treated wood has not yet turned out to be the "next asbestos" for the lawyers, although the hope is probably still alive.[69,73] David McCrea, who won the Sipes and Walker cases, claims that CCA-treated wood is unreasonably dangerous because no warnings were given and the product will generate liability many years after the sale.

In the 1990s, the American public became sensitised to new knowledge about the risks of chronic exposure to inorganic arsenic compounds (Section 8.11).[74,75] The characteristic green colour of CCA-treated wood is easily recognised (provided it is not too old) so the sight of children crawling all over playground equipment made from this material – found in 14 per cent of the public playgrounds in the US – increased the public's level of anxiety. One fear was that children, well known for their hand-to-mouth activity, could end up ingesting arsenic-rich particles from the surface of the wood. To add to the concern, elevated arsenic levels were found beneath back-yard decks in 1996. A 2001 article by Florida journalist Julie Hauserman, "The Poison in Your Backyard,"[76] really brought the issue to the fore and the media embraced the topic.[77,78]

In 2001, the EPA initiated a program requiring manufactures to tag treated wood with labels specifically saying the wood contained arsenic. This was not welcomed by the industry, even though most individuals who claimed to be injured by the product did not know this fact in advance. A manufacturer took Florida wood treater Pat Bischel to court for revealing this information and asking consumers, "Is it time for you to switch to a treated wood without arsenic?"[76,79] Finally, on February 12th, 2002, the EPA announced, rather clumsily, a "voluntary decision by industry to move consumer use of treated lumber products away from a variety of pressure-treated wood that contains arsenic by December 31st, 2003, in favour of new alternative wood preservatives. This action will result in a reduction of virtually all residential uses of CCA-treated wood within less than two years." Under the agreement Osmose, of Griffen GA; Arch Wood Protection Inc. of Smyrna GA; and Chemical Specialties Inc. of Charlotte NC were to gradually reduce their production of CCA to give the estimated 350 wood-treatment plants throughout the US time to change over to other arsenic-free preservatives. The decision affected virtually all residential uses of CCA and was claimed to remove about 80 per cent of the US CCA market.

The EPA was careful to point out that it believes that CCA-treated wood does not pose any unreasonable risk to the public and there is no reason to replace existing structures or surrounding soils. Osmose vice-president John Taylor said in 2002, "Basically we did it [switched away from CCA] for market reasons."[80] The EPA news release re-emphasised the need for common sense to minimise exposure to treated wood and wood dust.[81] These precautions are spelled out in information sheets available to the public. Frequent sealing with a penetrating stain is recommended to prevent leaching. Paint is not as effective.[82,83]

As an aside, the magazine Chemical and Engineering News published an article in 2002 on the problems associated with arsenic-treated wood. It was accompanied by a photograph of a man sawing a wood plank. A response from a reader contained the following comment about the photograph: "It shows a man sawing a wood plank with an electric saw. The man is holding the plank with his left hand. The plank is apparently not clamped; the electric cord is in front of him, near the saw blade, and he is not wearing safety goggles. This man's chance of coming to grief from arsenic poisoning are negligible compared to the likelihood of him cutting his leg off, electrocuting himself, and being blinded."[84]

CCA is still widely used for treating utility poles, marine timber and pilings, plywood flooring, shakes and shingles, highway construction, sound barriers and structural/agricultural timbers. A 1997 polemic, "Poison Poles," estimated that there were between 80 and 135 million wood utility poles in the United States or 28 per mile, with about 3 million replaced each year.[85] Forty-two per cent of existing wood utility poles have been treated with arsenic.

Around the same time, 2002, the European Union imposed a ban on the use of arsenic in wood preservatives rather than a voluntary phase-out, and classified CCA-treated wood waste as hazardous, requiring special treatment on disposal. Even so, permitted uses include telephone poles and railway ties.[86] In the US, efforts continue to have CCA banned.[68] The Environmental Working Group, a nonprofit organisation founded in 1993 and based in Washington DC (www.ewg.org), took up this challenge as part of its mandate "To protect the most vulnerable segments of the human population – children, babies, and infants in the womb – from health problems attributed to a wide array of toxic contaminants." The group conducted wipe tests on 263 decks, play sets and sandboxes across 45 states and found elevated arsenic levels on wood surfaces 20 years after installation.[87] The review of this information, which took one and a half years to finish and cost $1 million (US), concluded that children exposed to CCA-treated wood in playgrounds faced an increased risk of lung and bladder cancer, of between two in a million and one in 10 000. The calculated risk was based on estimates of exposure duration, the amount of residue on the hands, and the amount transferred to the mouth. The risk was deemed to be acceptable.

One petition to the Consumer Product Safety Commission, which was ultimately denied, was made in 2003 on behalf of special needs children.[88] Children born with Down's syndrome, roughly one out of every 500 births, experience the world through their mouths more than typical children. Consequently, there is an increased likelihood that these children will mouth deck railings and deck surfaces, possibly to their own detriment.[89]

In spite of repeated assurances by industry and governments that the structures are safe, many local authorities are replacing playground equipment with arsenic-free structures, usually made from metal. Studies continue in evaluating the safety of CCA-treated wooden play equipment. The material that can be dislodged has been fed to hamsters and a bioavailability of about 11 per cent has been determined (Section 7.4).[90] Other studies involved extracting the material with human sweat obtained from volunteers who cycled vigorously for an hour dressed in a Tyvek® suit with collection bags taped around their feet. All this information is fed into models that are used to predict effects on human health. But, as with all such studies, the results depend on the information fed into the model and vary from no risk to unacceptable risk.[91,92]

A more direct approach was taken in Canada by Professor Chris Le of the University of Alberta. The city of Edmonton AB, had 316 playgrounds in 2003, and 222 of these had CCA-treated equipment. His research group found that the arsenic concentration in the soil and sand of the Edmonton playgrounds was normal however, the chromium concentrations were slightly elevated. Children's hands were tested for arsenic and chromium after they had played on the equipment. The maximum amount of arsenic measured on the hands of a child was about 4 micrograms (μg), which was not enough to raise any alarm, even if it was all ingested, because the normal average daily ingestion of arsenic by a Canadian child is about 15 μg per day (the adult figure is 38 μg per day). Likewise, the chromium on their hands was insufficient to be of concern.[93,94]

2.5.2 Disposal of Treated Wood

It was estimated in 2005 that, worldwide, the preservation industry treats approximately 30 million cubic meters of wood each year, consuming some 500 000 tons of preservative chemicals, and approximately two thirds of this volume is treated with CCA.[95] The disposal of this wood is a growing problem as it reaches the end of its useful life. Because of its warm wet climate, Florida has a particularly high number of CCA-treated wood structures. Consequently, it is a centre for studies of the material and also the centre for litigation associated with its use.

Florida researchers estimate that, of the approximately 28 000 tons of arsenic imported into the state before the year 2000, 4600 tons have already leached into the environment. They predict that about 40 per cent of the arsenic in the wood will leach out over the products' life span, leaving 60 per cent for disposal.[96] In 1996, 110 000 m^3 of treated wood was disposed of in Florida: by 2006 the amount was expected to be around 400 000 m^3 and by 2016, 900 000 m^3. The amount of chemical released is predicted to increase from about 300 tons in 1996 to 2700 tons in 2016. Needless to say, the Wood Preservative Science Council is not happy with these projections.[97]

The available disposal options depend on geography. In some countries the waste is classified as nonhazardous, which effectively means it can be landfilled or incinerated, leaving any problems that might be created to be dealt with in the future. In the US, CCA-treated wood qualifies as hazardous waste; but it was given a special exemption – perhaps a sop to the industry – so that it could be disposed of in regular unlined construction/demolition waste sites.[17] In 2004 the EPA recommended that landfills for CCA-treated wood be lined so that the effluent could be collected for treatment, but as of 2007 it was not insisting. The arsenic content of the leachate from landfills is expected to increase with time, although the current impact from Florida landfills is not noteworthy.[98] Hurricane Katrina created 72 million m^3 of debris in 2005 when it wreaked havoc in Louisiana and Mississippi. This material was placed in unlined landfills and it is estimated to contain 1740 tons of arsenic, as CCA, that could eventually leach out.[99,100]

If and when the waste is declared a hazard, the available options are not all that attractive. The problems begin with the collection and storage of the wood, and deciding if the treated wood should be separated from the rest of the waste to enable subsequent processing. Visual sorting is not very effective because all dirty or weathered wood looks much the same, so other methods – some very sophisticated – are being tried. These include using X-rays (zapping the wood with X-rays and examining the radiation given off for an indication of arsenic – expensive), laser ablation (zapping the wood with a laser and examining the vapourised material for arsenic – very expensive), and chemical staining (a colour change indicates the presence of arsenic – inexpensive). The options available for the waste, once separated, are, in order of preference: reuse; refining and recycling; treatment and destruction; and disposal. Making mulch or compost from recycled wood is not an option as this product is classed as hazardous, but even so, mulch made from CCA-treated wood, often dyed black or red to disguise its origin, is found on the US market and has caused health problems.[101]

In practice, only thermal destruction and landfilling are practical, but there are difficulties with both of them. The arsenic concentration in CCA-treated wood is on average about 3000 ppm but some applications result in levels as high as 2.1 per cent or 21 000 ppm. When burned, this material results in ash containing on average 33 000 ppm of arsenic. Consequently, some ash samples from the combustion of wood piles containing as little as 5 per cent CCA-treated material have to be treated as hazardous waste; hence the urgency of developing reliable streaming methods for waste wood. If the wood is to be burned to supply energy, waste will have to be transported long distances to the thermal treatment plants to gain any economy of scale.[95,102–104]

Burning the wood at beach and campground fires has recently become a concern in the US;[105] however, Australian barbecue fanatics learned about the perils of "green wood" many years ago. (Even the ash from forest fires in California in 2007 contains unacceptably high arsenic concentrations.) In connection with this, in the 1990s two Native Americans incarcerated in a New Mexico prison wanted to have a sweat lodge ceremony. Under state law, all aboriginal customs are to be allowed in prison, so they were granted permission. The ceremony takes place in a sauna-like environment: fire is lit to heat stones, then water is poured on the stones to generate steam. The prison guards generously supplied the wood for the fire but the prisoners realised that the wood was pressure-treated with CCA. They assumed the worst and killed the guards. The court later recognised that the guards' actions amounted to a deliberate racist attack and the sentence was lenient – just a few more years on the prisoners' pre-existing life sentences.[59]

2.5.3 Alternatives to CCA

Alternatives to CCA are available. One involves pressure treating with ACQ, a 2:1 combination of a proprietary alkaline copper carbonate complex and a quaternary ammonium chloride that has the additional advantage of not containing chromium. Chemical Specialties Inc. of Charlotte NC received a Green Chemistry award for this development.[106] Another pressure treatment utilises copper boron azole (CBA) sold as Wolmanized Natural Select. However, many individuals and groups continue to be unhappy with these products, especially their use in the marine environment because of the extreme sensitivity of aquatic plants and animals to copper.[107] For them a possible alternative is TimberSil, wood injected with sodium silicate.[108,109]

Box 2.2   A Different Attitude in Hawaii?

A building material called canec, found only in Hawaii, was made from sugar-cane stalks. Arsenic was added to discourage mildew and insects. Canec was used in schools and private homes a few decades ago much like Gyprock® is used today. Some of the structures in Hawaii are still standing. Leslie Au, a toxicologist with the state Department of Health's Hazard Evaluation and Response Office, advises Hawaiians to leave it alone unless it is deteriorating. "If you leave it alone, then the arsenic won't come out," he said. If a home is 80 years old and its walls are soft and you can make a groove in the material just by scratching it with your finger nail, then Au says it is more than likely to made from canec. There might be some concern if the canec was deteriorating and falling to the floor. In that case, little children crawling along the floor might pick it up on their hands, then stick their fingers into their mouths. But even then, while it's possible for them to ingest some arsenic, it's "very unlikely," Au said.[110]

The Hawaiian state legislature does not seem to be too upset by reports of classrooms becoming littered with arsenic from deteriorating ceilings. One sample of the material taken from an Army building about to be demolished in 2001 contained 1700 ppm arsenic, yet the contractor was able to put the material in a commercial land fill.[111,112]

The Hawaiian Cane Products plant in Hilo manufactured canec from bagasse – the fibre that remains after sugar cane stalks are crushed for their juices – from 1932 to the early 1960s. The residues from the process were dumped into a pond that was connected to Hilo Bay via a stream, so both

Box 2.2 Continued.

became contaminated. The disastrous tsunami of 1960 that ended canec production also helped mix contaminants in the pond and bay. The pond was still contaminated in 2000 but there were no plans for a cleanup.[113]

However, the state is taking other arsenic issues very seriously. Soil that was used for sugar-cane plantations is badly contaminated because of the past use of arsenicals – 500 ppm arsenic is common – and residents are being monitored for exposure. The housing boom has increased pressure to carve new neighbourhoods out of old sugar fields and the state is trying to establish if this can be done safely.[114]

2.6 Monomethylarsonic Acid and Dimethylarsinic Acid

2.6.1 Use in the USA

These alkylarsenicals and their salts made the transition from human medicine to pesticide after World War II.[115] They displaced the more toxic inorganic arsenic species in use in agriculture and by the mid-1980s only four arsenicals remained in use: arsenic acid; monosodium methylarsonate (MSMA, the acid is also known as methanearsonic acid and monomethylarsonic acid); disodium methylarsonate (DSMA); and dimethylarsinic acid (also known as cacodylic acid, CA).

MSMA and CA are water-soluble contact herbicides that are applied as sprays to the foliage of weeds. They have no pre-emergent activity so they are not added to the soil to prevent weed growth. A few of the many trade names are Ansar 138, Ansar 170L, Silvisar 550 and Arsycodile. The dose is in the range of 2 kg to 5 kg per hectare.

MSMA saw much use as a weed-control agent for the maintenance of noncrop areas such as ditch banks and rights-of-way. Around 1979, 3.3 million pounds of MSMA were applied annually to US roadside rights-of-way.[116] Cacodylic acid is not selective and is used to clear ground of plants and bushes prior to reseeding.[117] This aspect is taken up further in Chapter 6, when we will meet Agent Blue (Section 6.11).

In the 1990s about 3000 tons MSMA and DSMA were applied per year to cotton fields in the US.[118] In Florida in 1992 about 41 580 pounds of MSMA (34 650 acres treated), and 9900 pounds of DSMA (8900 acres treated), were used for agricultural purposes particularly to control weevil infestations. The amounts increased with time until, for example, in 1997 149 000 pounds of MSMA were used on 74 803 acres of cotton.[4] Treatment was costly, but the best available until organic pesticides came along. In 1997, a genetically engineered species of cotton was introduced. These Roundup-ready® plants were not affected by the broad spectrum herbicide glyphosate which could be easily used in situations in which MSMA would kill the plants. The use of arsenicals on cotton rapidly decreased.[4]

There are about 1400 golf courses in Florida, and MSMA was routinely used on nearly 100 per cent of these for weed control. It selectively eliminated two big problems, Johnson grass and crab grass, from golf greens. Bermuda grass, a desirable species, is highly tolerant to MSMA. The Gainsville-based Florida Centre for Solid and Hazardous Waste Management, in its estimates of the "Quantity of arsenic within the State of Florida" used an annual application rate of 1.94 lb of arsenic per acre to calculate that MSMA used in this way contributed 116 tonnes of arsenic to the soil each year.[4]

Golfers can thank MSMA for flawless, weed-free greens and fairways, but experts question whether the arsenic-containing pesticide is safe for the environment and human health. Despite industry claims, MSMA, applied to golf courses degrades to inorganic arsenic, which can leach into groundwater.[119] Other studies, described below (Section 2.6.2) document that MSMA can move through wildlife food chains. This news came as Canada and the United States were re-evaluating registrations for MSMA.[120,121] Golf courses were fingered as the source of high arsenic concentrations in the groundwater of Naples FL,[122] although industry representatives again denied the charges, claiming that MSMA rapidly becomes bound to soil particles and is not easily released.

The final chapters on the use of these organoarsenicals are now being written. The US Federal Register of August 9th, 2006, carried the following notice: "The agency has determined that all products containing MSMA, DSMA, CAMA [calcium methylarsonate], and cacodylic acid are not eligible for registration. The agency's primary concern is the potential for applied organic arsenical products to transform into a more toxic inorganic form of arsenic in soil with subsequent transport to drinking water." Exhaustive animal tests indicated low mammalian toxicity.[123]

2.6.2 Canada

In Canada, MSMA is registered only for thinning conifers and controlling bark beetles.

The pine forests of British Columbia began to turn red and die in the 1980s, signalling an attack of the mountain pine beetle. The beetle is native to North America. Its habitat ranges from near sea level in British Columbia to 11 000 feet (3353 m) in southern California. The beetle is 0.3–0.8 cm long. It spends its life under the bark of the tree and emerges for a few weeks in midsummer to breed in new trees. Normally cold weather keeps its numbers low and it can't compete with other beetles, but in warm weather it thrives. Trees fight back by exuding toxic gum, but the beetles have another weapon: they carry the spores of the blue stain fungus in special compartments in their mouth. As they chew they inoculate the tree with the fungus which grows and shuts down the tree's defence system. The beetles mate and lay eggs in galleries under the tree's bark. The larvae soon hatch and feed on the tree, riddling it with holes as the tree dies.

Usually these infestations only last one summer because the hatched larvae are killed off by the winter's freezing temperatures. But in recent years, the weather has not been cold enough to kill the larvae and the infestation has spread. Until about 140 years ago, the aboriginal communities of British Columbia kept the forests healthy using controlled burns. But fire suppression is now a pillar of forest management, so this control method is limited to cutting down the infected trees and burning them. The only other option is to use MSMA. The arsenical is applied in solution to the base of a newly infected tree via a frill cut with an axe. Each tree gets about 22 g of the arsenical that travels through the tree's outer layers to kill the newly developing insects. The treatment is effective if the timing is right and it is claimed to be less expensive, at $25–30 (Cdn) per tree, than the "cut and burn" process, which costs $60–100 per tree and requires the use of helicopters. Spruce beetles, an equally costly pest, are also vulnerable to MSMA, however, in this instance trees are treated before the beetles attack, and are then cut down as bait trees.

The arsenic treatment program in British Columbia was expanded in the 1990s in an attempt to stop the beetle invasion. In one district near the town of Smithers, "68 000 trees were injected with 1568 kg of arsenic," but to no avail. The fact that the supply of the pesticide was drying up helped make the decision to abandon all further treatment a little easier. In 2005 the federal government was forced to admit the extent of the disaster and predict that by 2006 the beetles would infest more than 8.7 million hectares (215 million acres) of forest and destroy 90 million m^3 of timber. This is the largest insect epidemic recorded in North American history.[124,125]

As of 2007, the ravages were expected to continue for another 10 to 15 years, and a mortality rate of 95 per cent was seen to be not impossible. In all, the beetles will claim about 1 billion m^3 of timber in British Columbia.[126] The resulting loss of forestry jobs will have serious economic consequences for the province and it will be decades before new plantings become harvestable. The beetles have also recently invaded the neighbouring province of Alberta.

Another unfortunate aspect of the beetle attack is that the wood, now blue, is considerably reduced in value even if the tree is harvested immediately. Some entrepreneurs are attempting to develop a niche market for this "Denim Wood;"[127] others are salvaging the beetle-killed wood and converting it into inexpensive fuel for European wood-fired electrical generation.[128]

A number of organisations contested the use of MSMA, including the Canadian Association of Physicians for the Environment. Dr. Josette Weir was particularly concerned that the arsenic could be transferred to the bird population, especially to woodpeckers, via the beetles and their larvae. Indeed this seems to be the case. Woodpeckers are attracted to beetle infestations, and they eat 100–1000 beetle larvae per day. MSMA concentrations are high in the beetles and their larvae, and also in the blood of woodpeckers living in the area.[129]

2.7 OBPA

In 2007, only three inorganic arsenicals remained registered by the US EPA: arsenic acid (herbicide, insecticide, rodenticide); arsenic pentoxide (fungicide, insecticide, rodenticide, herbicide); and arsenic trioxide (insecticide and rodenticide). One organoarsenical, known as OBPA, also remained.[130]

OBPA, 10,10′-oxybisphenoxarsine, is an aromatic arsenical that is structurally related to the chemical-warfare agent Adamsite (Section 6.4). OBPA is currently approved in many countries for the control of bacteria and fungi, and is marketed under names such as Vinzene, Vinadine and Intercide. It is found in polyurethane for shoes, in vinyl for shower curtains, tents, mattress covers, vinyl flooring and wall coverings; in polyolefins for drainage pipes, trash bags and geotextiles; and in cotton fabric in tropical and subtropical regions. In 1970, 90 per cent of all flexible vinyl films were treated with OBPA. Some recent patent applications claim that the microbial growth inhibiting properties of OBPA can be used in water tanks for humidifiers, netting for protecting picked fruit, and polyurethane foams. The business is valued in the tens of millions of dollars.

Only a few years ago a Canadian outdoor goods company, Mountain Equipment Co-op, felt obliged to warn its customers regarding a product used for carrying water. "We have learned that trace amounts of OBPA, a chemical substance used in the laminate layer of the bag's fabric, leach through the NSF-grade laminate layer and into stored water. OBPA is an organic arsenical. There is no known human health effect associated with trace amounts of OBPA, other than short-term mouth and throat irritation. The arsenical is commonly used as a bacteriostat, disinfectant, and fungicide in plastic products. The manufacturer has stopped using OBPA in the bags and is offering to exchange bags with a new model that does not contain the substance. A large number of bags are involved and owners are asked to be patient, because it may take time to deal with all claims. Owners can also apply for a refund." The letter finishes with, "It is unfortunate that this problem has come up, but a company's action in such situations shows where its priorities lie." We will meet OBPA again in Chapter 4 (Section 4.2).

2.8 Arsenic in Other Products and Processes

2.8.1 Ironite

Ironite is a mine-waste-derived fertiliser used as a source of iron and nitrogen for agricultural crops, lawns and gardens. Most of the material, in the form of waste tailings, comes from the Iron King Mine in Yavapai County AZ. This is crushed and amended with nitrogen sources such as urea, and pelletised for sale in outlets such as Home Depot, Target and Lowe's. One web site claims, "A popular choice for home, lawn and garden for 45 years, Ironite is manufactured from naturally occurring rock containing essential minerals." Ironite's label, like those of nearly all fertilisers, does not disclose the presence of possibly harmful materials as this is not required by federal law, although individual states can make their own laws on this. Fertiliser makers need only list ingredients that are claimed to be beneficial. The Ironite label does not advise users to take any precautions during or after use, such as keeping children or pets away, or washing hands after handling the product.

Arsenic and lead are present in the commercial product in high concentrations. Reported values for arsenic range from 2500 ppm to 6020 ppm and it is sufficiently soluble for the product to be classified as hazardous waste.[131–133] But it isn't. This information became public knowledge around 1998 when company founder Heinz Brungs accused the Dallas Morning News of libel when it suggested that the fertiliser contained high levels of the two elements and could taint soil. Competitors sent copies of the article to sales outlets and one went so far as to include an accompanying note reading: "If you were exposed to Ironite, go home, gargle with Listerine and get ready for the end."[134] It seems that we have a similar situation to the one we encountered with CCA-treated wood: Ironite is exempt from classification as a hazardous material in the US because under the Bevill exemption, solid wastes from the extraction, beneficiation and processing of ores and mineral mine wastes are excluded from the definition of hazardous wastes.

The manufacturer claims that the source of arsenic in the fertiliser is the mineral arsenopyrite (FeAsS) and because this is a stable mineral the arsenic in it is not bioavailable (Section 7.4). However, recent studies show that this is incorrect;[135] the arsenic is bioavailable and it is weakly associated with iron oxy(hydroxyl) species, as well as with arsenopyrite. Moreover, the arsenopyrite is not as inert as claimed when released in the environment. The studies' authors urge caution and suggest all environmental costs should be considered before invoking the Bevill exemption.[133,136] Some states, such as Washington and Minnesota, have taken action against the product and issued a warning to consumers.[137] The US EPA has also put out a caution.[138]

Canadians can be thankful that for once their regulators did not follow the lead of the United States. Ironite was banned for sale in Canada in 1997 because the company was unable to prove the product was safe.

2.8.2 Gallium Arsenide

There are many applications for arsenic in the electronics industry. Elemental arsenic is added to lead/antimony grids in industrial batteries to improve their life span, and to selenium-based photoconductors used in photocopying machines. However, the most important product is the semiconductor gallium arsenide, GaAs. By the early 1980s, the silicon monopoly on the semiconductor industry was declining as producers began using new materials derived from elements in Groups 3 and 5 of the Periodic Table. One particular combination, gallium arsenide, proved to be an efficient emitter of photons – it converted electrical energy into light – and the wavelength can be tuned by the addition of controlled amounts of gallium phosphide. Elegant new technologies were developed for the growth of the necessary thin-film epitaxial structures, in which a crystal layer of one material is grown on the crystal base of another material in such a manner that its crystalline orientation is the same as that of the substrate. One of these technologies, vapour-phase epitaxy, uses organometallic compounds and/or metal hydrides as the source of the elements. This widely used industrial process is termed metalorganic chemical vapour deposition (MOCVD), or organometallic vapour phase epitaxy (OMVPE). An example is the formation of thin, single-crystal epitaxial films of gallium arsenide GaAs from trimethylgallium, $(CH_3)_3Ga$, and arsine:

$$(CH_3)_3Ga + AsH_3 \xrightarrow{H_2, 700\,°C} GaAs + 3CH_4$$

Only four decades ago it was confidently asserted that "arsine is unique among the industrially toxic arsenic compounds, being of no economic importance and seldom manufactured outside the laboratory."[1] This comment demonstrates how rapidly a technology such as MOCVD can develop. Organoarsines such as ethylarsine, in which one of the hydrogen atoms of arsine is replaced by an ethyl group (C_2H_5), are used in place of arsine when appropriate, because of their lower toxicity.[139]

GaAs has found use in light-emitting diodes, solar cells and lasers that operate continuously at room temperature.[140] Currently, magnetic ions such as manganese are being incorporated into ordinarily nonmagnetic GaAs to prepare semiconductors that exhibit ferromagnetism. The aim is to develop devices that can use electron-spin information in much the same way that charge information is now being used. The buzz-word is spintronics.[141]

2.8.3 Glass Making

We saw in Chapter 1 that arsenic trioxide was used in glass making in Venice during the Middle Ages. It seems that it was first added to Murano glass around 1450 and that one of the sources was Styria the home of the arsenic eaters (Section 1.3). In 1993 the glass industry was the second-largest consumer of arsenic in the US, using about 17 per cent of the total imported. However, it is difficult to obtain information on how it is used.[54] In 1983, Corning Glass Works (now Corning Inc.) was using 1000 tons of arsenic acid solution per year: this was deemed to be safer than the solid arsenic trioxide previously used. The amount of arsenic in the finished product ranges from 0.004 per cent to 4 per cent arsenic by weight. The function of the arsenic compound is to clear any bubbles from the molten glass by releasing oxygen, and to oxidise iron in the melt from blue-green Fe(II) to colourless Fe(III). It also aids crystal growth in the manufacture of glass ceramics such as CorningWare®, missile nose cones, and windows for fireplaces. The white stripe behind the mercury column in medical thermometers usually has a high arsenic content.[142] During WWII, arsenic became a "strategic element" in Canada and its use in glass making was deemed to be "nonessential" and therefore allowed only for certain types, such as fiberglass and optical glass. Similar restrictions were applied in the US.[143]

In 2006, Corning announced that after 11 years of research the company was able to produce arsenic-free glass for the thin screens used in liquid crystal display (LCD) televisions.[144] Schott North America Inc. has just introduced new arsenic- and lead-free glass for the manufacture of high-performance lenses found in camera phones, digital cameras and medical devices.[145] In 2007 Apple announced that it would be removing arsenic and mercury from its LCD displays.[146]

2.8.4 Embalming

The preservation of human corpses, in the absence of deliberate chemical treatment, usually requires extreme conditions such as very cold, very dry or very wet. Anaerobic (without oxygen) conditions with high levels of toxic minerals to inhibit the growth of micro-organisms are also effective.[147] Dehydration from diarrhoea in chronic arsenic poisoning probably aids preservation. Bodies sealed in lead coffins or buried within a building, rather than in the earth, survive particularly well. The state of preservation of the bodies of saints was taken as proof of their incorruptibility. For example, the corpse of St. Cuthbert of Lindisfarne, patron saint of Northumbria, UK, appeared to be unchanged when examined a few years after his death in 687.

The Atacama Desert of northern Chile is said to be the driest place in the world. The people living in the region, the Atacameño, used to bury their dead in shallow graves, which led to natural mummification. The limited water supply is highly contaminated with arsenic (up to 1000 ppb) from local mineralisation, so there was an interest in establishing whether the mummification was arsenic-assisted. Recently, the skin of one 1000-year-old mummy was examined using X-ray spectroscopy (Box 7.2), to reveal that its high arsenic content came from ingestion and not from external application.[148]

Self-mummification was practiced by a Buddhist sect in Japan until the ritual was outlawed toward the end of the 19th century. The monks deliberately starved themselves to death while drinking a toxic tea to induce dehydration and aid in the preservation of the body after death. The whole process took about 10 years. In this way they became one with the greater being and were not reborn into the world. The local spring was loaded with arsenic, which might have helped.[149]

Although the practice of embalming is ancient – there is evidence that embalming was used in the kingdom of Pharaonic Egypt (2660–2180 BCE) – we know little of the chemicals used because embalmers had to swear an oath of secrecy about their work. Modern developments that allow the investigation of very small samples find mainly plant materials and, to a lesser extent, animal oils. The Egyptians mainly used resins from pine trees, bees wax, palm wine, and camphor oil. The ingredients had drying and antibacterial properties.[150]

It was not until the early 1800s that arsenical solutions were used as preservation aids. One procedure employed a solution of one pound of arsenic in five pounds of wine. This was injected into the femoral artery without prior drainage of blood. The abdomen was also opened, the bowels removed, and the cavity moistened with the injecting solution. Jean-Nicholas Gannal (1791–1852) developed the use of a mixture that included acetic, arsenious, nitric and hydrochloric acids. Cinnabar was occasionally added to provide some colour. The mistress of a member of the French nobility who died rather suddenly, was accused of poisoning him with arsenic because some was found in the body. She was tried and almost convicted before Gannal admitted he had used some arsenic in the embalming process. The mistress was freed but the legal community was outraged and the first law prohibiting the use of arsenic for embalming was passed by the government of Louis-Philippe of France in 1846.[151,152]

Around the same time, John Meyer of Dublin was advocating the use of a mixture of salts that included potassium arsenate and alum.[152]

The modern period of embalming was initiated during the American Civil War because of the need to get the bodies home. Dr. Thomas Holmes, "the father of modern embalming" was called in by the War Department to help. He had an embalming practice in Washington and dealt with three bodies per day for the four years of the war. He charged $100 per body and used up to 6 kg of arsenic per body.[153] A sample of Holmes's civilian work is shown in Figure 2.3.

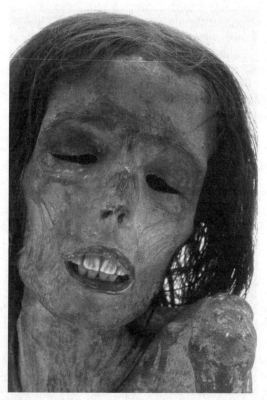

Figure 2.3 Embalmed head by Dr. Thomas Holmes *ca.* 1850 (with permission from the National Museum of Health and Medicine, Walter Reed Army Medical Centre, Washington DC).

The original dating from the 1850s is on display at the National Museum of Health and Medicine Walter Reed Army Medical Centre, Washington DC.

Other civilian embalmers set up business during the war alongside the medical tents.[153] F.A. Hatton patented a preservative mixture of alcohol, arsenic, mercuric chloride and zinc chloride. Benjamin Lyford used a similar solution but in addition filled the abdominal cavity with powdered arsenic trioxide. The body of President Lincoln, the last major casualty of the Civil War, was viewed, undoubtedly arsenic-stabilized, by hundreds of thousands of mourners as it was paraded around the nation by train. He was assassinated on April 15th, 1865, presumably too early to have been treated by Crane's Electrodynamic Mummifier, a powder patented in 1868 that contained arsenic and most of the salts mentioned already. In these applications drainage of the corpse was not necessary. Arsenic was also the primary ingredient in the first commercially available embalming fluids, and held this status from the 1860s through the 1910s.[154] The amount used per body was in the range of 110 g to 6 kg, and it very effectively killed the micro-organisms responsible for decomposition. The treated bodies were relatively supple and could be easily dressed and positioned.

A subplot in a crime novel *If I'd killed him when I met him* is based on the notion that a cemetery full of the bodies of Civil War soldiers preserved with arsenic could ultimately result in poisoned well water.[155] In fact, the US Geological Survey still includes cemeteries as a source of arsenic contamination in its annual surveys.[3] It became illegal to use arsenic for embalming in the United States around 1920. By then formalin was commercially available[156] which was commonly pressure injected for preservation.

There are a number of recorded instances in which the corpses of individuals exposed to arsenic, either through medication or homicide, seemed to be particularly well preserved when they were exhumed (Box 2.3). Napoleon (Figures 4.1 and 4.2) is a particular example we will meet again in Chapter 4, others in Chapter 5. One exception is the body of an arsenic eater who probably over indulged (Section 1.3). His doctor noted the "unusual rapidity with which putrefaction set in at a cold season of the year."[157]

Box 2.3 Mummies and Circuses

Mummies were an essential ingredient in all circus sideshows and freak shows until the demise of the business in the mid-1900s. The peak period for mummy viewing was in the 1930s and 1940s. Here are the stories of three:

Elmer McCurdy was a member of a gang who intended to steal a safe from a train on October 4th, 1911. Unfortunately they stopped the wrong train and their major prize was several gallons of whisky. McCurdy was cornered, drunk, a short time later, and was killed in the resulting gun battle. The body was eventually embalmed with arsenic and put on hold until it

> **Box 2.3 Continued.**
>
> could be claimed by a relative. None came. He was then dressed, given a rifle to hold, and began his career as "the bandit who wouldn't give up" or "the Oklahoma outlaw." Eventually after many public appearances in 40 states, he ended up on a film set as part of a tableau featuring his body hanging from a noose. The director asked that the dummy be moved. A crew member obliged and on grabbing the supposed dummy its arm came off, revealing the truth on December 7th, 1976. The now badly beaten-up mummy was examined and found to be otherwise in very good condition. "The preservation of the tissues was almost unbelievable. The heart was nearly perfectly preserved. Red and white blood cells were easily recognised."[156] Elmer was finally retired to a cemetery in his home state of Oklahoma in 1977, on a day close to his 100th birthday.
>
> Another mummy in circulation was known as Hazel Farris. She was said to have killed a few policemen, but rather than go peacefully to jail when cornered by the law in 1906, she got drunk and then swallowed enough arsenic to kill herself and mummify her corpse. Tests in 2002 revealed she had "astronomically high levels of arsenic in her hair and body, suggesting she had been dipped or washed in a solution of the preservative" There are no records of a Hazel Farris in any state documents, but she was finally cremated and the ashes were placed in a crypt in Madison WI, in 2002.[161]
>
> The mummy of Marie O'Day had a similar history: she is said to have died a violent death in 1925 and the body was preserved by being thrown into the Great Salt Lake of Utah. Recent examination showed the cause of death was probably tuberculosis, and that arsenic had been used to embalm the body.[162]

Johnston wrote about the graves of the Styrian arsenic eaters (Section 1.3) as follows: "In this part of the world when a graveyard is full, it is shut for about 12 years, when all the graves which are not private property by purchase are dug up, the bones are collected in the charnel-house, the ground is ploughed over, and burying begins again. On those occasions the bodies of arsenic eaters are found almost unchanged and recognisable by their friends." Many people surmise that the finding of these bodies lends weight to the ancient myth of the vampire and the undead of Bram Stoker.[158,159]

The related element antimony, the one below arsenic in the Periodic Table, is also associated with the preservation of bodies. In the records of the trial of George Chapman, who used tartar emetic (potassium antimony tartrate) to dispatch three women in London around 1900, we read that the bodies of two of his victims were found to be well preserved when they were exhumed after being interred for one and five years, respectively.[160a] Margery Allingham writes about the discovery of a well preserved body buried in antimony-rich earth. The clue for Inspector Campion was the antimony-tolerant cow parsley growing in the cemetery.[160b]

2.8.5 Taxidermy

The National Museum of Natural History at the Smithsonian Institution, Washington DC, has used many pest-control measures since the beginning of the 19th century. Arsenic and mercury compounds played a prominent role. Early collectors immersed wet specimens in "alcoholic spirits", while "dried specimens were preserved in salts, stuffed with herbs, heat treated or rubbed, painted, immersed, or brushed with arsenic or mercury compounds."[163] One conservationist went so far as to label his objects "Palmer poisoned" and evidently did not know that Dr. William Palmer, a particularly nasty individual who poisoned friends, a wife and their children, mostly for financial gain, used antimony and not arsenic (thirty thousand people turned out to watch Palmer hang outside Stafford Prison on June 14th, 1856).[164]

In 1887, Walter Hough, head curator of the anthropology department of the Smithsonian Institution recommended the following general insecticide for museum objects: 1 pint saturated solution of arsenic acid and alcohol, 25 drops strong carbolic acid, 20 grains strychnine, 1 quart strong alcohol and 1 pint naphtha. This solution is "most satisfactory for preserving nearly every kind of specimen."[163] More modern recipes of the 20th century suggest painting objects such as cloaks, robes and bird skins with the equivalent of Fowler's solution: soaking for a week is also recommended. Hough retired from the Smithsonian in 1935, so it is probable that arsenic compounds were in use at least until then (See Phar Lap, Box 1.1).

Arsenic trioxide was used for the preparation of dusting powders and supplies were taken into the field (Box 2.4). Felts that had been pretreated with sodium arsenate solution were added to wooden storage cabinets to create a pest-free seal. The use of arsenicals continued until the mid-1980s, when they were displaced by a succession of organic pesticides such as DDT, carbon disulfide, carbon tetrachloride and ethylene dibromide, sometimes using a "fumatorium" for the treatment. In more recent years sulfuryl fluoride has been employed, as has freezing. Modern management calls for frequent inspections and preventive conservation.

The Museum of New Mexico acknowledges the use of ethylene oxide (1930s–1950s) and a mothproofing spray, Sibur®, which consisted of 9 per cent sodium arsenite and 91 per cent water (1940s–1960s). The museum made a general comment advising caution when dealing with the spray.[165] Tests at the Provincial Museum of Alberta revealed that most of the bird collection is contaminated with arsenic compounds, but the mammal collection less so.[166]

The University of British Columbia's Museum of Anthropology was using lead arsenate against moths around 1958, and a solution of mercuric chloride, copper sulfate, sodium arsenite and copper naphthenate, in addition to pentachlorophenol, was recommended for the preservation of totem poles.[167] Many objects on display in the City of Vancouver Museum were recently found to have high amounts of arsenic compounds on their surface and some of these came from the part of the collection that was open to the public, especially to children, for hands-on examination. The exhibit was rapidly rearranged.[168,169]

So, when hunting trophies are in good condition many years after they were collected, be advised that the unfortunate animal may be able to have the last laugh through the medium of the arsenical preservative.

Analysis of the samples is often carried out using an X-ray fluorescence spectrometer. This is a portable instrument that beams X-rays onto a small area of an object and can detect the presence of elements such as arsenic and mercury. The method does not measure concentration in the traditional sense, as in grams of arsenic per total number of grams, but gives an indication of the relative amounts of the element of interest on the surface being studied. The method has the advantage of being rapid and nondestructive. An alternative approach is to wipe the object with a water-soaked pad and then analyse the pad using traditional chemical methods (Section 7.2). If a fragment of the object can be spared (usually it can't), this can be decomposed prior to chemical analysis to ensure that the results reflect the total composition.

Problems can be encountered when objects that have been treated with preservatives are to be returned to their original owners. Researchers and curators must disclose whether there has been any treatment and, if so, whether the integrity of the object has been affected. Aboriginal representatives should always be consulted prior to initiating an investigation: sampling must be done with great care and sensitivity, and may not be allowed at all for certain sacred objects. Seattle's Burke Museum offers a testing program for arsenic and other elements on cultural objects.[170]

Although no arsenic contamination was found on a number of culturally significant items that were repatriated to the Hoopa Tribal Museum in California, the mercury and pesticide residues were deemed to be too high for the objects to be worn in religious ceremonies.[171]

Box 2.4 Arsenic in the Arctic

Researchers from the Royal Military College of Canada were very surprised to find high concentrations of arsenic at the historic site of Fort Conger, Quttinirpaaq National Park. It is located on the northeast tip of Ellesmere Island NU, 500 miles from the North Pole, and was the staging area for a number of Arctic expeditions. Fort Conger was built in 1881 by an expedition led by American Adolphus Greely. This large fort built from wood brought up from the south, housed 25 men for several years during their explorations and scientific studies of the North Polar Region. In 1900 Robert Peary, with Matthew Henson and others, established a winter base at the site during one of their expeditions to the North Pole. Peary returned in 1905–06 and again in 1908–09. The expedition members built several huts out of the wood from Fort Conger. After the Peary era, the site provided shelter to American, Norwegian, Danish and British/Canadian expeditions in 1915, 1920, 1921 and 1935. Two of these huts are still standing, as are the

> **Box 2.4 Continued.**
>
> roofless remains of a third. There are also artifacts from the British Arctic Expedition of 1875–76, when HMS Discovery wintered at the site.[172–174]
>
> The arsenic at Fort Conger is localised in a number of patches in the ground and in or on some plants. The source was eventually traced to leaking cans of arsenic trioxide. Some of the white residue found on the ground proved to be essentially pure oxide. The only way to account for its presence in such large quantities is that it was brought in for sample preservation.
>
> This investigation is ongoing, but proceeding at a snail's pace because official permission has to be obtained before anything is done. To move a can on the ground to see if it contains anything becomes a formidable task.

2.8.6 Pigments

We have seen that the Chinese used yellow arsenic sulfide, orpiment, to cover writing blemishes because it was the same colour as the paper (Section 1.1). Its use as a pigment in its own right developed once civilisations began to decorate the printed page and to produce other works of art. There were not many good alternatives so orpiment, known as King's yellow, is found universally in paintings and in illuminated manuscripts. Until very recently it was believed that orpiment was the only yellow arsenic pigment used by artists, and that the main source of the pigment was the mineral itself. Another source appeared to be the yellow powder that forms on samples of realgar, the red arsenic sulfide, when it is exposed to sunlight. However, in 1980, a Canadian group discovered that this yellow was not orpiment but another form of realgar, now named pararealgar. It has the same chemical formula (As_4S_4) but a different structure that accounts for the different colour.

Differentiating between the two yellows is not an easy task because the pigments are finely ground before they are applied, so the analyst needs to be able to study the individual particles without destroying the usually fragile sample. These analyses require a microscope that can recognise different molecules. Such a tool became available in the form of Raman microscopy, which was soon used to identify pararealgar on a 13th-century manuscript.[175] The use of the pigment was deliberate for highlighting, so the artist had gone to the trouble of finding samples of realgar that had sprouted the yellow powder. Orpiment was used in the same object but for covering larger areas. Tintoretto used pararealgar in his masterpiece *The Dreams of Men*.[176]

The discovery that some fragments of stucco taken from the ancient city of Samarra, dating around the 9th century, contain orpiment and pararealgar has the curators at London's Victoria and Albert Museum in a tizzy. Jennifer Viegas of Discovery News reported: "The fragments are stored in a locked cabinet and only handled as little as possible by curators who wear sturdy gloves." It was suggested that "researchers also might wear face masks and work in a 'fume cupboard'." One outside expert is reported to say: "It was interesting to see the painters were poisoning themselves with arsenic."[177] More was written, but that should be enough to show the power of the A-word to stupefy even people who should know better.

In general, realgar saw fewer applications as a pigment probably because of the ready availability of red minerals such as cinnabar (HgS), litharge (PbO), and red lead (Pb_3O_4). Pre-10th-century Chinese manuscripts are generally coloured with cinnabar; the red in others probably came from madder root. However, the Egyptians appreciated realgar's bright colour, as did the Persians.[178,179]

Minerals are also used to colour pottery and this may be how realgar wine was discovered (Section 1.2). More recently, arsenic may have crept into the pottery manufacturing process in Oregon, where a group of nuns began to show symptoms of arsenicosis. Their hair samples averaged 165 ppm arsenic. Although the source of the element was never positively identified, some changes to their production methods resulted in improved health.[180]

Near the end of the 18th century a green pigment containing arsenic became available. One version of the 1775 preparation of what became known as Scheele's green has the Swedish pharmacist Karl W. Scheele dissolving arsenic sulfide (or oxide) and potash in water and heating it, then slowly adding the solution to a warm solution of copper sulfate. When the pigment precipitated, the liquid was poured off and the solid washed and dried on gentle heat. The material produced is copper arsenite. Very soon after this discovery, a much better green was prepared in 1814 by dissolving verdigris (copper carbonate) in vinegar and reacting this with a solution of arsenic trioxide in potash. This product was copper acetoarsenite and it became known by a variety of names, such as Schweinfurt green, Brunswick Green, Paris Green and emerald green. It did not become widely available until 1922 when Justus von Liebig published the synthesis that until then was a trade secret. Turner began to paint with the pigment in 1832.[179]

Here is a 19th-century recipe: Dissolve in a small quantity of hot water, 6 parts of sulfate of copper; in another part, boil 6 parts of oxide of arsenic with 8 parts of potash, until it throws out no more carbonic acid; mix by degrees this hot solution with the first, agitating continually until the effervescence has entirely ceased; these then form a precipitate of a dirty greenish yellow, very abundant; add to it about 3 parts of acetic acid, or such a quantity that there may be a slight excess perceptible to the smell after the mixture; by degrees the precipitate diminishes the bulk, and in a few hours there deposes spontaneously at the bottom of the liquor entirely discoloured, a powder of a contexture slightly crystalline, and of a very beautiful green; afterwards the floating liquor is separated.[181]

2.9 Some Historical Connections

2.9.1 Clare Boothe Luce

To quote from Stephen Shadegg's biography, "Everyone who has been privileged to come into contact with Mrs. Luce – friend or foe – is quick to confess that she is one of the most challenging, brilliant, intriguing, vexing, inconsistent, charitable, complex women of this century." She was born Clare Boothe in 1903. After a failed marriage and many affairs she married Henry Luce, the founder of Time magazine. She became a roving war correspondent for the magazine in 1939. After a spell as a Republican Congresswoman for Connecticut, she was sent to Rome by President Eisenhower in 1953 as "Ambassador Extraordinary and Plenipotentiary of the United States of America to Italy."

She elected to sleep and work in the study of the Ambassador's quarters in Rome because she worked late at night and slept late. Her husband, who was used to more regular hours, got the bedroom. The building itself, a palace, was built in the 16th century and, to prevent damage from earthquakes, the second and third floors had been suspended on chains to provide flexibility. A new American electric washing machine had been installed on the third floor above the study, and the noise and vibration from this made sleeping late rather difficult.

Mrs. Luce proved to be very effective in the job but her health began to deteriorate after a year in office. She became physically weak and she was losing her hair. She went for a checkup at the American Naval Hospital, where the doctor was astute enough to suggest that she might be suffering from lead poisoning and asked for a urine test. For reasons of secrecy the sample was submitted under the name John Paul Jones. The doctor also suggested that she return to the US for a rest and to remove herself from the vicinity of any poisoner.

The analysis showed the presence of arsenic, not lead, which made the likelihood of deliberate administration more likely. The CIA became involved and further tests suggested she had been ingesting small doses of arsenic for a long time. So who was the poisoner? The State Department sent a team of experts to Rome to investigate while she continued her recovery. The experts working undercover got nowhere until a film of dust was noticed on the Linguaphone record in the study (Mrs. Luce was becoming fluent in Italian). On asking the maid, they were informed that this was not the result of sloppy housekeeping – the dust accumulated as fast as it was removed. Analysis of the dust indicated that it was lead arsenate and the mystery was solved without any diplomatic crisis: vibration from the washing machine or other sources shook the ceiling and dislodged paint from the stucco roses. [Nobody seems to have mentioned that lead arsenate is colourless.]

The ceiling was repainted, the washing machine moved, and the Ambassador returned to duty and the ignominy of jibes about "arsenic and old Luce." She died in 1987.[182,183]

2.9.2 The Peale Family

Charles Willson Peale (1741–1827) was a well-respected American portrait painter and naturalist. He opened a museum in Philadelphia in 1782 and by 1786 the local newspaper was reporting: "We hear that Mr. C.W. Peale has acquired the means of preserving birds and animals in their natural form, and that he intends to place in his collection of curiosities every species of birds and animals that he is able to obtain, belonging to North and South America."[184]

The preservative used was a solution containing "three pounds of arsenic, to which was added mercuric chloride for larger animals, which was brewed in a cauldron over a flame. Thus, even the largest quadrupeds, such as grizzly bears or a five-legged cow, could be preserved."[184] Peale was aware that the process was dangerous and bad for his health and the health of his customers. He took an antidote containing iron from time to time when he felt indisposed and placed signs in the exhibits: "Do not touch the birds, they are covered with arsnic [sic] Poison." When that did not work, he arranged the exhibits in appropriate groups in glass cases. Some exhibits from the museum are still well preserved, and one curator commented that even 200 years later they make the fingers tingle when handled.

Raphaelle Peale (1774–1825) was the oldest of Charles Peale's 17 children, and was America's first important still-life painter. In spite of the known dangers his father insisted Raphaelle take on the task of chief taxidermist in the museum, a position he held for 23 years. He developed a whole range of health problems that doctors diagnosed under the catchall term "gout." He may really have had this affliction, but many of his symptoms are better described as the result of arsenic and mercury poisoning.

Father Charles claimed alcohol was the prime cause of Raphaelle's ill health, saying that the son was using it to escape the nagging of his wife – who had borne him 10 children in 12 years. But Raphaelle's nephew was to write: "How much more likely was Uncle Raphaelle's case one of mercury and all his gouty suffering also from the same cause. . . . Arsenic was all that was used for birds and small animals, but by its use the skins and hair of larger animals were not protected from the ravages of moths and other insects and bichloride of mercury, corrosive sublimate, was resorted to. The skins were soaked in an alkaline solution, and this was freely handled by Uncle Raphaelle for days together."

This opinion is supported by at least two modern historians,[184] but challenged by another who attributes the illness to lead poisoning resulting from drinking some form of alcohol stored in a lead-lined or lead-joined vessel. Raphaelle was a known alcoholic.[185]

There is also the suggestion that his health problems might have been exacerbated by his use of the arsenical pigments, Scheele's green and emerald green, in his paintings (Section 2.8). He used a lot of green in about 70 paintings, but other artists of the time were not affected.[184]

2.9.3 King George III

The following was announced on the BBC program Medical Mysteries:[186] "Last year a remarkable exhibit came to light. Hidden in the vault of a London museum was a scrap of paper containing a few strands of hair. The paper was crudely fashioned into an envelope, but the words on it immediately caused a stir: "Hair of His Late Majesty, King George 3rd." The undated hair was subsequently found to contain "300 times the toxic level" of arsenic: 17 ppm.[187] The account of the hair discovery given in the primary literature is much less exciting:[187] it suggests that there was nothing remarkable about the sample and its discovery. Measurement made along the length of the hair by using laser ablation to volatilise sections of hair (essentially heating small sections of the hair to a very high temperature) and a detector to measure the atoms in the vapour, showed a uniform but high arsenic concentration along its length. Readers may remember that King George III was generally believed to be insane. He had five bouts of long illnesses including a notable one from October 1788 to February 1789 when he was 50 years old.

Following a re-examination of the records and his symptoms and behaviour, two psychiatrists, a mother and son team of Ida Macalpine and Richard Hunter, came to the conclusion that the King suffered from a rare disease known as porphyria, later refined to variegated porphyria.[188] The source of the disease was traced back to Mary, Queen of Scots, and apparently it was passed on in recent times to William Prince of Gloucester who died in 1972.

Professor Warren and coworkers claim that although there is no established proof that arsenic alone provokes acute attacks of porphyria, if George III did have the disease he would be sensitised to the effects of arsenic and other metals such as lead and mercury whose concentrations were also elevated in the hair samples. "Indeed amounts of toxic metal insufficient to induce frank poisoning would, in all probability, exacerbate porphyric attacks in a susceptible individual."[187]

The records show two likely sources for the arsenic; a wig powder, and James' Fever Powder. The latter, the Georgian equivalent of the aspirin,[189] contains antimony (potassium antimony tartrate) and was given to the King, frequently by force, as medication. It is likely that the antimony was impure and contained arsenic. Warren and coworkers guesstimate that the antimony in the pills could contain up to 5% arsenic so the dose could have contained around 0.9 to 2.25 mg of arsenic. Apparently the King occasionally was given one dose every six hours amounting to 4–9 mg of arsenic per day, certainly a chronic dose. So they concluded that arsenic might have contributed to the Kings unusually severe and prolonged bouts of illness.[187]

Unfortunately, the thesis is not backed up by any measurements of antimony in the hair samples. If this were the source of the arsenic, the antimony concentrations should be off the scale. In addition, the researchers seem to have chosen to ignore the possibility that the arsenic could have come, as suggested, from the powders commonly used to protect the wigs. Another possible explanation is that arsenic was used to preserve the hair sample. The BBC mentioned that the human hair wig of a contemporary, Admiral Cornwallis, was found to be contaminated with arsenic, but the concentrations were much less than found in George III's hair; however, no data were provided. Rather surprisingly, the BBC release of July 13th, 2004 shows a picture with the caption "King George's wigs were laden with arsenic" supporting the contamination theory. We also have the problem of timing. It is claimed that the lock of hair was taken from the King's corpse. He died in 1820. Hair grows at about 1 cm per month so the sample provides a record of arsenic exposure for the last few months of his life so why did this arsenic fail to bring on an attack of his illness assuming he was still being medicated at the time? (In order to see the hair section corresponding to a dose received during an illness in the 1700s the strand would have to be over one meter long.)

The arsenic in the King's hair probably originated in life from the use of wig powder. A preservative applied to the sample after death would be expected to result in much higher arsenic concentrations (Section 4.9.4).

Finally, it seems that even more caution should be advised. The diagnosis of porphyria is questionable; indeed, the word fable has been used.[190,191] The BBC broadcast ended with the words: "For the professor it is the end of a very long trail,"[186] but it seems that there is still a long way to go to connect the death with arsenic.

References

1. W. D. Buchanan, *Toxicity of arsenic compounds*, Elsevier, Amsterdam, 1962.
2. H. J. Sanders, in *Chem. Eng. News*, 1981, Aug. 3, p. 20.
3. *2007 Mineral commodity summaries*, US Geological Survey, 2007. www.minerals.usgs.gov/minerals/pubs/commodity/arsenic.
4. H. Solo-Gabriele, D.-M. Sakura-Lemessy, T. Townsend, B. Dubey and J. Jambeck, *Quantities of arsenic within the State of Florida* 03-05, Florida Centre for Solid and Hazardous Waste Management, Gainesville Florida, 2003.
5. Lord Kelvin, W. H. Dyke, W. S. Church, T. E. Thorpe, H. C. Bonsor and B. A. Whitelegge, *First report of the royal commission appointed to inquire into arsenical poisoning from the consumption of beer and other articles of food or drink Vol. I*, Houses of Parliament, London, 1901.
6. Lord Kelvin, W. H. Dyke, W. S. Church, T. E. Thorpe, H. C. Bonsor and B. A. Whitelegge, *First report of the royal commission appointed to inquire into arsenical poisoning from the consumption of beer and other articles of food or drink. Vol II*, Houses of Parliament, London, 1903.
7. F. Lizal, *Zivocisna Vyroba*, 1975, **20**, 919.
8. N. F. Morehouse and O. J. Mayfield, US 2450866, 1948.
9. C. C. Calvert, in *Arsenical pesticides*, ed. R. F. Gould, American Chemical Society, Washington DC, 1975.
10. Alpharma, *3-Nitro vs arsanilic acid* Technical bulletin No 3-nitro QA/QC, 1999.
11. S. M. Shane, *Technical update on 3-Nitro (roxarsone)*, Alpharma animal health, 2004, July 16.
12. B. Hileman, in *Chem. Eng. News*, 2001, Feb. 14, p. 47.
13. G. O. Doak and L. D. Freedman, in *Medicinal chemistry*, ed. A. Burger, Interscience Publishers Inc., New York, 1960.
14. Opinion, *European Food Safety Authority Journal*, 2004, **121**, 1.
15. D. V. Frost, *Fed. Proc.*, 1970, **26**, 194.
16. J. C. Bowles, G. Huitinkanu, K. Van Devender, *Utilizing dry poultry litter: an overview*, Cooperative Extension Service, University of Arkansas, 1995.

17. D. A. Belluck, S. L. Benjamin, P. Baveye, J. Sampson and B. Johnson, *Int. J. Toxicol.*, 2003, **22**, 109.
18. K. Parker, in *The Daily Times*, 2007, July 5.
19. J. L. Morrison, *J. Agr. Food Chem.*, 1969, **17**, 1288.
20. J. R. Garbarino, A. J. Bednar, D. W. Rutherford, R. S. Beyer and R. L. Wershaw, *Environ. Sci. Technol.*, 2003, **37**, 1509.
21. I. Cortinas, J. M. Field, M. Kopplin, J. R. Garbarino, A. J. Gandolfi and R. Sierra-Alvarez, *Environ. Sci. Technol.*, 2006, **40**, 2951.
22. J. F. Stolz, E. Perera, B. Kilonzo, B. Kail, B. Crable, E. Fisher, M. Ranganathan, L. Wormer and P. Basu, *Environ. Sci. Technol.*, 2007, **41**, 818.
23. DATAS, *Delaware air toxics assessment study*, Division of Public Health Environmental Health Evaluation Branch, 2003.
24. B. C. Bellows, *Arsenic in poultry litter: Organic regulations*, ATTR National Sustainable Agriculture Information Service, 2005.
25. Y. Arai, A. Lanzirotti, S. Sutton, J. A. Davis and D. L. Sparks, *Environ. Sci. Technol.*, 2003, **27**, 4083.
26. K. Christen, *Environ. Sci. Technol.*, 2006, May 1, p. 2864.
27. NWAnews, in *Arkansas Democrat Gazette*, 2006, Sept. 26.
28. M. Burros, in *New York Times*, 2006, April 5.
29. B. Schwarz, in *The Charleston Gazette*, 2006, April 12.
30. T. Lasky, W. Sun, A. Kadry and M. K. Hoffman, *Environ. Health Perspect.*, 2004, **112**, 18.
31. A. Awbi, *Tyson's sales tumble amid safety fears* 2006, March 21, www.foodanddrinkeurope.com/news/printNewsBis.asp?id=67224.
32. Vancouver Sun, in *Vancouver Sun*, 1977, Feb. 15.
33. *Activists play chicken with arsenic*, Centre for Consumer Freedom, 2007.
34. Globe and Mail, in *Globe and Mail*, 1996, April 27.
35. J. W. McCall, *Vet. Parisitol.*, 2005, **133**, 179.
36. L. Raber, in *Chem. Eng. News*, 2005, June 20.
37. A. E. Smith and D. M. Secoy, *J. Agric. Food Chem.*, 1975, **23**, 1050.
38. E. Cook and C. H. LaWall, *Remington's practice of pharmacy*, Lippincott, Philadelphia, PA, 1936.
39. F. J. Peryea, *16th World Congress of Soil Science*, Montpellier, France, 1998.
40. R. Carson, *Silent spring*, Houghton Miffen Company, Boston, 1962.
41. F. J. Peryea, *Estimation of soil arsenic and lead concentrations resulting from use of arsenical pesticides in apple orchards.*, Tree Fruit Research and Extension Centre, Washington State University, Wenatchee, Washington, 1991.
42. (a) *EPA Pesticide fact sheet* Fact Sheer Number 112, US Environmental Protection Agency, Washington, 1986.
    (b) R. S. Adams, in *Chem. Eng. News*, 1976, Aug. 16, p. 3.
43. E. J. Deszyck and J. W. Sites, *Florida State Horticultural Society Proceedings* 1954, **67**. Quoted from *F.S.H.S. Newsletter*, 2005 **17**(1).
44. G. A. Wojeck, H. N. Nigg, R. S. Bramen, J. H. Stamper and R. L. Rouseff, *Arch. Environ. Contam. Toxicol.*, 1982, **11**, 661.

45. K. Tollestrup, *Arch. Environ. Health*, 1995, **50**, 221.
46. D. Hogan, in *Chemistry*, 2003, Spring, p. 22.
47. E. P. Reister, *USA Patent*, US 642368, 1900.
48. K. Konkola, *J. Hist. Med. Allied Sci.*, 1992, **47**, 186.
49. *White-ants control in Australia*, www.termite.com/white-ants.html.
50. A. Campbell, 1989, personal communication.
51. P. Bartrip, *Hare today-History of myxomatosis in the UK*, www.wellcome.ac.uk/doc_WTX024879.html.
52. I. D. Rae and J. H. Todd, *Mole River arsenic workings*, New South Wales Government Department of Planning, 1993.
53. L. Tanner, *Prickly pear history*, http://www.northwestweeds.nsw.gov.au/prickly_pear_history.html.
54. R. A. Zingaro, *Environ. Int.*, 1993, **19**, 167.
55. Michigan Penal Code, 2006, section 438.
56. W. Grady, in *Globe and Mail*, 1995, Nov. 4, p. D8.
57. Canadian press, in *Globe and Mail*, 1995, Jan. 15.
58. B. Melley, in *South Coast Today*, 1997.
59. G. Miller, 2001, personal communication.
60. W. P. K. Findlay, *Dry rot and other timber troubles*, Hutchinson's Scientific and Technical Publications, London, 1953.
61. N. Knight, *J. Ind. Eng. Chem.*, 1909, **1**, 261.
62. K. H. Wolman, F. Peters and H. Pflug, *US Patent*, US 1622751, 1927.
63. D. D. Nicholas, *Wood deterioration and its prevention by preservative treatments*, Syracuse University Press, 1973.
64. MSDS, *Wolmanized® treated wood and lumber, fiberglass coated*, Wood Preservers Inc, 2004.
65. I. Stalker, M. Applefield, B. R. Evans and F. T. Milton, *The preservation of wood. A self study manual for wood treaters*, Revised edition by F. T. Milton, University of Georgia, Athens GA, 1986.
66. J. N. R. Ruddick, 1976, personal communication.
67. J. Prager, *Treated wood's "smoking gun." 1977 memos from an industry insider reveal CCA wood toxicity*, BANCCA, 2003, Aug. 14.
68. J. S. Prager, *Assessing the assessment: Reviewing the consumer product safety commission's CCA briefing report. Great science or government boondoggle*, BANCCA, 2003, Feb. 26.
69. D. Hechler, *Nat. Law J.*, 2003, March 20.
70. N. Groom, in *Reuters News Service*, 2002, Oct. 8.
71. D. McCrea, 2003, personal communication.
72. A. Liptak, in *New York Times*, 2002, June 26.
73. B. A. Langer, *New York Law J.*, 2003, Dec. 3, p. 4.
74. *Arsenic in drinking water*, National Research Council, Washington DC, 1999.
75. *Arsenic in drinking water. 2001 update*, National Research Council, Washington DC, 2001.
76. J. Hauserman, in *St. Petersberg Times*, 2001, April 16.
77. J. Kluger, in *Time*, 2001, July 16, p. 44.

78. L. Priest, in *Globe and Mail*, 2001, March 31, p. A6.
79. International Bankruptcy Library (IBL) 2001, **3**, April 17, no. 75.
80. E. Pianin, in *Washingtom Post*, 2002, February 13, p. A2.
81. D. Ryan, *Whitman announces transition from consumer use of treated wood containing arsenic*, www.epa.gov/pesticides/citizens/1file.htm.
82. *Consumeraffairs*, 2005, May 12, www.consumeraffairs.com/news04/2005/epa_arsenic.html.
83. *Fact sheet on chromated copper arsenate (CCA) treated wood*, Health Canada Pesticide Management Regulatory Agency, Ottawa, 2005.
84. P. Haberfield, in *Chem. Eng. News*, 2002, March 18, p. 6.
85. J. Feldman and T. Shistar, *Poison poles*, National coalition against the misuse of pesticides, 1997.
86. K. Christen, in *Environ. Sci. Technol.*, 2003, March 1, p. 89A.
87. S. Gray and J. Houlihan, *All hands on deck*, Environmental Working Group, Washington DC, 2002, Aug. 28.
88. L. Janak, *ACCA ban petition. Petition HP 10-32*, Consumer Product Safety Commission, Washington, 2003.
89. E. Criss and K. Giles, *CPSC denies petition to ban CCA pressure-treated wood playground equipment*, Comsumnet Product Safety Commission, Washington, 2003, Nov. 4.
90. H. V. Aposhian, 2005, personal communication.
91. M. Mittelstaedt, in *Globe and Mail*, 2003, Jan. 15.
92. P. S. Nico, M. V. Ruby, Y. W. Lowney and S. E. Holm, *Environ. Sci. Technol.*, 2006, **40**, 402.
93. E. Kwon, H. Zang, Z. Wang, G. S. Jhangri, X. Lu, N. Fok, S. Gabos, X. -F. Li and X. C. Le, *Environ. Health Perspect.*, 2004, **112**, 1375.
94. C. Hamula, Z. Wang, H. Zhang, E. Kwon, X. -F. Li, S. Gabos and X. C. Le, *Environ. Health Perspect.*, 2006, **114**, 460.
95. L. Helsen and E. V. d. Bulck, *Environ. Pollut.*, 2005, **134**, 301.
96. K. Christen, in *Environ. Sci. Technol.*, 2006, Feb. 1, p. 634.
97. M. C. Kavanaugh, N. Kresic and A. P. Wright, *Environ. Sci. Technol.*, 2006, **40**, 4809.
98. B. I. Khan, J. Jambeck, H. M. Solo-Gabriele, T. G. Townsend and Y. Cai, *Environ. Sci. Technol.*, 2006, **40**, 994.
99. B. Dubey, H. M. Solo-Gabriel and T. G. Townsend, *Environ. Sci. Technol.*, 2007, **41**, 1533.
100. B. I. Khan, H. M. Solo-Gabriel, J. Jambeck, T. G. Townsend and Y. Cai, *Environ. Sci. Technol.*, 2007, **41**, 347.
101. J. Matarese, *Arsenic in mulch?*, www.wcpo.com/wcpo/localshows/dontwasteyourmoney/da5e8b.
102. H. Solo-Gabriele and T. Townsend, *Waste Management Res.*, 1999, **17**, 378.
103. H. M. Solo-Gabriele, T. G. Townsend, B. Messick and V. Calitu, *Haz. Mater.*, 2002, **B89**, 213.
104. K. Christen, *Environ. Sci. Technol.*, 2006, Feb. 1, p. 634.
105. KSBW, channel.com, 2004, July 2.
106. S. K. Ritter, in *Chem. Eng. News*, 2002, July 1, p. 26.

107. J. Pelley, *Environ. Sci. Technol.*, 2002, April 1, p. 126A.
108. M. K. Flynn, *Environ. Sci. Technol.*, 2006, **40**, 2871.
109. Newscripts, in *Chem. Eng. News*, 2006, May 8.
110. J. Watanabe, in *Star Bulletin*, 2006, Feb. 8.
111. US Army Corps of Engineers in *Headquarters Engineering and Construction News*, 2001, June, vol. III.
112. Department of Health, *Report to the twenty-third legislature State of Hawaii*, Honolulu, 2004.
113. I. Birnie, 2000, personal communication.
114. P. Natarajan, in *Pacific Business News*, 2006, April 2.
115. A. E. Hiltbold, in *Arsenical pesticides*, ed. E. A. Woolson, American Chemical Society, Washington, 1975, p. 53.
116. J. W. Mason, A. C. Anderson, P. M. Smith, A. A. Abdelghani and A. J. Englande, *Bull. Environ. Contam. Toxicol.*, 1979, **22**, 612.
117. J. R. Abernathy, in *Arsenic industrial, biomedical, environmental perspectives*, ed. W. H. Lederer and R. J. Feinstein, Van Nostrand Reinhold Inc., New York, 1983.
118. A. J. Bednar, J. R. Garbarino, J. F. Ranville and T. R. Wildeman, *J. Agric. Food Chem.*, 2002, **50**, 7340.
119. Y. Cai, American Chemical Society, Spring Meeting, Atlanta GA, 2006.
120. J. Pelley, *Environ. Sci. Technol.*, 2005, March 15, p. 122A.
121. Science News, *Environ. Sci. Technol.*, 2005, Feb. 9.
122. J. Cox, in *Naples Daily News*, 2005, Oct. 2.
123. EPA *Fed. Register*, 2006, **71**, 45554.
124. *Summary of forest health condition in British Columbia 2005*, Report 15, BC Ministry of Forests and Range, Victoria, 2005.
125. T. Glavin, in *Globe and Mail*, 2006, April 22, p. F7.
126. V. Palmer, in *Vancouver Sun*, 2005, June 15, p. A3.
127. G. Hamilton, in *Vancouver Sun*, 2002, Nov. 17.
128. V. Palmer, in *Vancouver Sun*, 2005, June 25, p. A3.
129. C. A. Morrissey, C. A. Albert, P. L. Dods, W. R. Cullen, V. W. -M. Lai and J. Elliott, *Environ. Sci. Technol.*, 2007, **41**, 1494.
130. *Arsenic*, www.pesticideinfo.org/Detail_Chemical.jsp?Rec_Id=PC35165.
131. R. Suwol, *Maine considers actions against chemicals in fertilizer*, Pesticide Action Network, 2007, www.beyondpesticides/org/daily_news_archive/2002/01_18_02.htm.
132. A. Pignataro, in *Orange County Weekly*, 2003, May 8.
133. A. G. B. Williams, K. G. Scheckel, T. Tolaymat and C. A. Impellitteri, *Environ. Sci. Technol.*, 2006, **40**, 4874.
134. K. Brown, in *The Business Journal*, 1998, April 27.
135. B. Dubey and T. Townsend, *Environ. Sci. Technol.*, 2004, **38**, 5400.
136. R. Renner, *Environ. Sci. Technol.*, 2004, Sept. 22, p. 382A.
137. *Risk assessment Facts about ironite*, www.health.state.min.us/div/sh/risk/studies/ironite.htm.
138. *Release of heavy metals from ironite*, www.epa.gov/nrmrl/lrpcd/wm/projects/135367.htm.

139. R. M. Lum and J. K. Klingert, *J. Cryst. Growth*, 1991, **107**, 290.
140. R. K. Willardson, in *Arsenic industrial biomedical environmental perspectives*, ed. W. H. Lederer and R. J. Fensterheim, Van Nostrand Reinhold, New York, 1983.
141. M. Jacoby, in *Chem. Eng. News*, 2006, Aug. 28, p. 30.
142. R. J. Bauer, in *Arsenic: Industrial biomedical, environmental perspectives*, ed. W. H. Lederer and R. J. Fensterheim, Van Nostrand Reinhold Company, New York, 1983.
143. R. C. Bateman and A. H. Williamson, *Wartime industries control board*, National Archives, Ottawa, 1942, Reel C-5003.
144. L. Blackwell, in *PC World*, 2006, March 23.
145. Schott, *North America Schott has introduced its new lead and arsenic free P-SF67 glass*, Schott North America Inc., 2006.
146. J. Galbraith, in *Macworld*, 2007, May 4.
147. A. T. Chamberlain and M. P. Pearson, *Earthly remains. The history, science of preserving bodies*, British Museum Press, London, 2001.
148. J. Charnock and J. Feldmann, in *Portal*, 2006, Autumn/Winter.
149. Oniko, *Buddhist mummies of Japan*, www.sonic.net/~anomaly/japan/dbuddha.htm.
150. H. Mayell, in *National Geographic News*, 2001, Oct. 30.
151. B. W. Richardson, in *The Asclepiad*, ed. B. W. Richardson, 1888, Vol. V.
152. R. G. Mayer, *Embalming history, theory, and practice*, 3rd edn., McGraw-Hill, New York, 2000.
153. R. Sussingham, in *Chemistry*, 2000, Spring, p. 17.
154. J. L. Konefes and M. K. McGee, in *Dangerous places, health, safety, and archaeology*, ed. D. A. Poirier and K. L. Feder, Bergin and Garvey, Westport, Connecticut, 2001.
155. S. McCrumb, *If I'd killed him when I met him,"* Random House, 1995.
156. C. Quigley, *Modern mummies: The preservation of the human body in the twentieth century*, McFarland and Company Inc., Jefferson, North Carolina, 1998.
157. D. M. N. Parker, *Edinburgh Med. J.*, 1865, X, 116.
158. J. F. Johnston, *Chemistry of common life*, D. Appleton and Company, New York, 1857.
159. B. Stoker, *Dracula*, Archibald Constable and Company, 1897.
160. (a) H. L. Adam, *The trial of George Chapman*, William Hodge and Company, London, 1930.
(b) M. Allingham, *The case of the late pig*, Hodder and Stronghton, London, 1937.
161. C. Ferguson, in *The Tennessean*, 2002, March 14.
162. J. Robinson, *Marie O'Day*, sidehowcentral.com.
163. L. Goldberg, *J. Am. Inst. Conserv.*, 1996, **35**, 23.
164. M. F. Farrell, *Poisons and poisoners*, Bantam Books, London, 1994.
165. *Health hazards and pesticides on museum objects*, Museum of New Mexico, Santa Fe, 1997.
166. C. Found and K. Helwig, *Collection Forum*, 1995, **11**, 6.

167. M. Clavir, 2001, personal communication.
168. C. Brynjolfson, 2001, personal communication.
169. W. R. Cullen and V. W.-M. Lai, 2001, unpublished results.
170. L. V. Mapes, in *Seattle Times*, 2005, Oct. 6.
171. P. T. Palmer, M. Martin, G. Wentworth, N. Caldararo, L. Davis, S. Kane and D. Hostler, *Environ. Sci. Technol.*, 2003, **37**, 1083.
172. R. A. Blanchette, *Research on the microbes attacking the historic woods at Fort Conger and the Peary huts on Ellesmere Island*, forestpathology. coafes.umn.edu/Fort%20Conger.htm.
173. Parks Canada, *Quttinirpaaq National Park of Canada*, http://www.pc.gc.ca/pn-np/nu/quttinirpaaq/natcul/natcul6_e.asp.
174. K. J. Reimer, 2007, personal communication.
175. R. J. H. Clark and P. J. Gibbs, *Anal. Chem.*, 1998, Feb. 1, 99A.
176. K. Trentelman, L. Stodulski and M. Pavlosky, *Anal. Chem.*, 1996, **68**, 1755.
177. J. Viegas, *Ancient Iraqi art determined poisonous*, Discovery Channel 2007, Jan. 22.
178. L. Stodulski, E. Farrell and R. Newman, *Studies in Conservation*, 1984, **29**, 143.
179. P. Ball, *Bright Earth. Art and the invention of colour*, Farrar, Straus and Giroux, 2001.
180. P. D. Whanger, P. H. Weswig and J. C. Stoner, *Environ. Health Perspect.*, 1977, **19**, 139.
181. Wikipedia, *Emerald green*, webexhibits.org/pigments/indiv/history/emerald.html.
182. M. Kesterton, in *Globe and Mail*, 2000, Oct. 14, p. A16.
183. S. C. Shadegg, *Clare Boothe Luce: A biography*, Simon and Schuster, New York, 1970.
184. P. Lloyd and G. Bendersky, *Perspect. Biol. Med.*, 1993, **36**, 654.
185. L. B. Miller, *Trans. Studies College of Physicians of Philadelphia*, 1994, **XVI**, 101.
186. BBC, in *Medical Mysteries*, 2004, July 14.
187. T. M. Cox, N. Jack, S. Lofthouse, J. Haines and M. J. Warren, *The Lancet*, 2005, **366**, 332.
188. I. Macalpine and R. Hunter, *Brit. Med. J.*, 1966, 65–71.
189. R. Porter, *The greatest benefit to mankind. A medical history of humanity*, Harper Collins, London, 1997.
190. J. T. Hindmarsh, *The Lancet*, 1997, **349**, 364.
191. J. T. Hindmarsh and P. F. Corso, *Eur. J. Lab. Med.*, 1999, **7**, 135.

CHAPTER 3
Arsine, Scheele's Green, Gosio Gas, and Beer

This chapter is concerned mainly with some of the problems that industrialised societies encountered when arsenic was introduced into their everyday lives through pigments/dyes and food. The extensive use of arsenic for criminal purposes during the same period is described in Chapter 5.

3.1 Arsine

In 1775, Karl Scheele (Section 2.8.6) discovered that a gaseous compound of arsenic and hydrogen, now known as arsine (AsH_3), is given off when a solution of arsenic acid reacts with zinc. He collected some of the gas in a bladder and found it to be flammable when exposed to a burning candle: "The flame took the direction towards his hand which was thereby coloured brown."[1] Nowadays, the gas would be collected in a glass or metal container, or in the scientific equivalent of a plastic bag. Arsine is extremely poisonous and has a slight garlicky odour. Many of its victims are unaware of its presence.

The toxic action of arsine is different from that of other arsenic compounds.[1,2] Poisoning is by inhalation and, in severe cases, becomes clinically evident within an hour or two. The initial symptoms are commonly a feeling of malaise and apprehension, soon followed by giddiness, headache, abdominal pain and vomiting. Diarrhoea is usual and may progress, as with ingested inorganic arsenic poisoning, to the stage of rice-water stools. However, the most characteristic reaction, hemoglobinurea, appears soon after the exposure: the urine becomes coloured to all degrees of red, up to resembling a port wine. On the second to third day jaundice sets in and rapidly spreads over the whole body surface developing into a deep copper-bronze hue. If death occurs, it is most common after the fourth day. Arsine is fast acting and little can be done for the victim. Recovery is slow for the lucky survivors. The mechanism

---

Is Arsenic an Aphrodisiac? The Sociochemistry of an Element
By William R. Cullen
© William R. Cullen 2008

of arsine toxicity is unknown: treatment with dimercaprol (BAL) and other chelating agents (Section 1.11) can provide some benefit.

Arsine is easily generated by the reaction of a metal, such as zinc or copper, on an acidic or basic solution of arsenic. We will encounter this reaction in the form of the Marsh test for arsenic, in Chapter 5. Accidentally generated arsine has been a common cause of poisoning in industry. In the early days of ballooning, the hydrogen gas used to fill the balloons was generated on site by reacting an acid with a metal. Arsenic impurities in the reagents resulted in the release of arsine along with the hydrogen. A related case was reported in 1901 to the Royal Commission Appointed to Inquire into Arsenical Poisoning from the Consumption of Beer and Other Articles of Food or Drink (Section 3.12)[3,4] when four vendors of rubber balloons for children became ill after filling the balloons with hydrogen obtained from commercial sulfuric acid and common zinc. One eventually died.

Mr. T. M. Legge, Chief Inspector of Factories for Britain's Home Office, reported to the same Royal Commission that in 1990 there were 14 cases of accidental exposure to arsenic compounds in chemical works and three deaths, but in the two succeeding years there was not a single death. All the cases occurred in two chemical works and they were all due to the evolution of arsine. In the UK there were 224 cases of industrial exposure to arsine between 1929 and 1974, resulting in 51 deaths.[5]

Here is an account of one of those earlier cases:[2] "He was a robust, powerful man, and he had been engaged on this particular work [preparing zinc chloride from zinc and hydrochloric acid] for 16 years. He had been at work there all morning and left his work at 2 pm. At 2:30 pm he felt sick, nauseated and depressed, with a hot burning pain from his throat to his stomach and with an intense thirst. This was soon followed by violent vomiting, at first of food, then of everything as soon as it was swallowed, even ice water. This again was followed by an equally severe diarrhoea, first of loose fecal matter, then of rice-water discharge and finally blood. Added to this, there was hemoglobinurea and a rapidly developed jaundice which, within 24 hours, assumed an intense coppery hue. As is usual in these cases, which without any knowledge of the surrounding circumstances and the obvious cause, a diagnosis of cholera would have been pardonable within the first 12 hours. The whole effect was that of an irritant poison taken by mouth, being evidence that the arsenic in the gaseous state was adsorbed by the blood direct from the lungs. The feeling of prostration deepened into extreme prostration, the features were sunk and cyanosed, the pulse thready and the voice lost. These severe symptoms lasted, with gradual diminishing severity, for several days. The green colour of the skin, which supervened on the disappearing jaundice, lasted for several weeks and it was only after the lapse of five weeks that he was able to return to work."

Similar accidents can occur while galvanising iron, which requires iron sheets to be cleaned by dipping them in dilute hydrochloric acid before they are dipped into molten zinc. Arsine can be released if either the iron or the acid is arsenic rich. Sulfuric acid tank cars contain arsenic-rich sludge, so cleaning them can become a problem. The law in the UK required "... a sufficient supply of nonmetal spades, scrapers and pails for cleaning out any chamber or other vessel which has contained sulfuric acid or hydrochloric acid and any other substance which may cause the evolution of arsine."[2]

Many industrial processes produce arsenic-rich material that can release arsine when exposed to water. One tragic accident involved the S.S. Vaderland on a voyage from Antwerp to New York in 1905. Fifty steerage passengers in accommodation over a hold containing a cargo of ferrosilicon, used in steel manufacture, became ill. Eleven died during the voyage and more died on arrival. The initial diagnosis was pneumonia and/or plague, so the ship was placed in quarantine. Later, the true cause was established: water had reacted with the cargo to produce arsine and phosphine from the arsenic and phosphorus impurities. Phosphine gas is similar to arsine although less toxic, and is now commonly used as a fumigant.[2] Cleaners consisting of aluminium and caustic soda can also liberate arsine from clogged drains[6] and it seems that sufficient arsine was generated during a cleaning of a cyclorama *The Battle of Atlanta*, housed in Atlanta GA, to incapacitate two workers in 1981. Arsenical pigments were used in the paint of this 80-ton circular artwork and the toxic gas was liberated by reaction with the mild alkaline cleaning solution.[7]

Car batteries generally contain small amounts of arsenic (0.5 per cent) and slightly more antimony (9 per cent) in the lead. Arsine and its antimony analogue, stibine, can be evolved during the charging process, which can be a problem in industrial-scale facilities. Arsine was encountered accidentally by submariners who became exposed to the gas when storage batteries were being charged. Vomiting and breathing difficulties were the commonest complaints among the crew, but there were unexplained instances in which crewmembers claimed to experience no ill effects in spite of the many problems evident in comrades.

Arsine can also be released during the electrochemical production of metals such as copper, and operators must take care to control the applied voltage.[8] A seemingly related problem was encountered in El Paso TX around 1980 at the Asarco plant when a worker, who had been called in to work for one day, reported that he had been poisoned by arsine. The only problem was that he was not "pissing blood," as an expert witness elegantly told the jury at the resulting trial. It transpired that his wife had added arsenic, presumably as the oxide, to his lunch that day, because she was aware of the possibility that her husband could encounter arsine gas. The wife took off and was never seen again. The lawyer sued the company anyway in the hope of making a dollar or two, but the case was thrown out.[9]

Box 3.1 On the Presence of Arsine in the Vapours of Bone Manure: A Contribution to Sanitary Science

This was the title of a pamphlet that James Adams, MD, wrote and printed in 1876 for private circulation,[10] about a study he made on behalf of the family of John Frazer. Frazer was exposed to offensive-smelling vapours from a bone-manure-manufacturing factory on his way to work and died a few days later.

Box 3.1 Continued.

The factory was attached to the railway station that the victim used, so he could not avoid it. The factory's business consisted of collecting animal bones, cleaning off the remaining flesh (usually by decay over time), crushing the residue, and treating this with sulfuric acid to make superphosphate fertiliser for agricultural purposes. This last operation was carried out in iron vats.

Dr. Adams was able to generate arsenic-containing gases by reproducing this process in his laboratory because the sulfuric acid was contaminated with arsenic. He writes: "The first discovery of arsenic in sulfuric acid in Britain was made by Dr. G. O. Rees, a lecturer in Medical Jurisprudence in London, who in 1841 found twenty two and a half grains of white arsenic in twenty fluid ounces of supposed pure acid sold to him at eight pence per pint."

At the time, Spanish pyrites, the source of the sulfur – and the arsenic (Section 7.5) – for the sulfuric acid, contained 1 ton of arsenic in 91 to 303 tons of mineral; Portuguese pyrites, 1 ton in 203 tons; Irish pyrites, 1 ton in 166 to 556 tons; and Cornish pyrites, 1 ton in 86 to 250 tons. Adams notes that iron in the vat could have been involved in the generation of arsine from the arsenic-rich acid but also adds that "fragments of scrap iron are bought in by dealers in bones. A manufacturer tells me that he has frequently seen several pounds weight of iron in a parcel of bones not exceeding a hundredweight [112 pounds], and that it is often added to increase the weight."

The pamphlet gives an account of some other contemporary poisonings by arsine, including what was probably the first: In 1815 Professor A. F. Gehlen of Munich was preparing hydrogen by reacting acid with a metal. In order to check for leaks in his apparatus he sniffed strongly at the joints, found a leak, but died some days later from the effects of arsine inhalation.

Clearly any legal action did not go well for the Frazer family because Adams ends with a tirade against management and local authorities that includes: "A woman employed in a manure work, on removing the cover of an apparatus encountered concentrated exhalations. Illness led to death in a few days." Several later cases in similar circumstances resulted in four deaths "These incidents occasioned a casual newspaper notice, soon neutralised by a quickly following paragraph, to the effect that, on inquiry, there was evidence of drinking habits in one or more of the cases, and that British cholera probably accounted satisfactorily for all."

The guidelines for exposure to arsine vary among jurisdictions, but as an example, the US workplace time-weighted average, the Federal Standard, is 0.05 ppm (0.2 mg/m^3) and the concentration that is judged immediately dangerous to life or health, (IDLH) is 3 ppm.[11] The toxic dose on a weight basis is about the same as for arsenic trioxide, the difference being that the dose is received very quickly. For comparison, the US Federal Standards for carbon monoxide, hydrogen cyanide and phosgene are much higher, being, respectively, 50 ppm, 10 ppm and 0.1 ppm. Arsine is found in low concentrations in the atmosphere above areas where large-scale microbial action is taking place, such as landfills and hot springs.[12]

3.2 Scheele's Green

Scheele, who discovered arsine, also discovered the green gas chlorine in 1774, which was identified as an element in 1810 by Humphry Davy. Scheele also discovered oxygen but the credit for this has been given to others.[13] Scheele was the first to describe a pigment known as Scheele's green, copper arsenite $CuHAsO_3$, which is the solid obtained on mixing solutions containing copper sulfate and arsenic trioxide. The development of the more brilliant Paris green (copper acetoarsenite), also known as Schweinfurt green among others, soon followed (Section 2.8).

These greens and a few other arsenic-based pigments provided colour for such a large number of commodities, particularly wallpaper and cloth, that during much of the 19th-century Europeans were living in a very green environment, albeit not the sort to which we currently aspire.[14] Some authorities became aware of possible problems with having so much arsenic around, and as early as 1815 the Prussian government directed that colour should be rubbed from green walls only when wet and never when dry. In 1839, distinguished chemist Leopold Gmelin suggested the possibility that the arsenic in the wallpapers could be dangerous because it might be volatilised. His article was published in a Sunday edition, November 24th, of the Newspaper of Karlsruhe: *Government Gazette of the Grand Duchy of Baden*. Gmelin wrote: "In newer times, for green wallpapers and room paintings there is usually used a colour material, which is named Green of Schweinfurt, of Vienna, *etc.*, and which impresses by the vividness of its colour, but which threatens the health due to its considerable content of arsenic... Moisture settling on the walls causes a slow decomposition process of the paper and the paste, in which the green colour is dragged in. The result of this is the development of an adverse, mouse-like odour, which is easily noticed by entering a room that was not aired for some time. There is no doubt that this smell is caused by traces of arsenic, which volatilise as a special compound, probably "alcorsine." Brief inhalation of such air is without danger: but longer daily stays in such rooms can cause harm; headache and undefined indisposition were noticed as a consequence of it, but even longer effect of this poisonous atmosphere can cause chronic poisoning by arsenic... These are the experiences which I have made over several years, especially numerous during the autumn, in this town, and about which I feel obliged to publish. The question obtrudes if this colour material should not be prohibited totally for wallpapers and paintings, except in oil." Gemelin was probably thinking that the arsenic could be volatilised to toxic gases related to the odiferous alkylarsines being studied at the time by Bunsen (Section 1.6).

This recently published translation[15] describes the odour as "mouse-like" although it has become usual to speak of a garlic odour. A public health official[16] described the odour of the rooms as leek-like and having some similarity to the smell of breath after eating horseradish; he did not mention garlic. We will come back to this important point later (Section 3.10).

It has been guesstimated that by 1860 there were 100 million square miles of arsenic-rich wallpaper in British homes – 700 tons of Scheele's green were made in England that year.[14] The use of wallpaper, at that time made from rags, was encouraged by a removal of some tax and the development of machine printing. (Paper made from trees was not universally available until the late 1800s.) The colour green was in favour with the fashion consultants of the mid-1800s, and the increasingly affluent population eagerly purchased these new products for their own use and to impress the neighbours. But arsenic was also present in other dyes (yellow, gold, red and blue). Clothes, books, kitchen utensils, glass, plaster, artificial flowers, packaging, confectionery (and wrappers) and foodstuff were all likely to contain the element in some form or other.[14]

Around 1850, a physician, W. Hinds, was one of the first to alert the British public to the possibility that living in rooms whose walls were covered with arsenic-rich wallpaper might be hazardous to their health. He had hung green wallpaper when he redecorated his study and reported that on reoccupation of the room he experienced symptoms of severe depression followed by feelings of nausea. He also experienced occasional severe pains in the abdomen and a feeling of faintness. This happened shortly after he entered the gas-lit study every evening: the door was usually closed. He found that the wallpaper was arsenical and his symptoms disappeared once he changed the paper.[14] Some years later the death of a child that was attributed to arsenic in wallpaper received much attention. This was a strange case, in that the "diagnosis" was made on the basis of "peculiar petechial spots" (tiny patches of burst blood vessels) on the stomach of the three-and-a-half-year-old child which are "so characteristic of arsenic poisoning [sic]." The flock paper was covered with Scheele's green that was easily brushed off by slight friction. Another two-year-old child in the same household had similar symptoms but her fate is not recorded. Dr. H. Letheby, who was involved in the case, wrote: "The French, who are our competitors, have long since abandoned the use of such pigments and have outstripped us in the brilliancy of tint. It is high time that our manufacturers should imitate their example."[17]

The death of a woman who ate an arsenical-green artificial grape, plus some well publicised criminal poisoning cases (Section 5.4) added fuel to the fire, as did a short story published in Chambers' Journal in 1962 entitled *Our Best Bedroom*. In it, an uncle attempts to murder his nephew, the heir to a fortune, by insisting that he sleep in the "Green Room."[14]

Illness and even death became attributable to the use of arsenical colours and the fear that arsenic could stealthily attack individuals in their own homes was at a zenith in the early 1860s. So we need to examine whether this fear was justified. How much arsenic was on the wallpapers? What was the extent of the illness and how good was the diagnosis? How was the arsenic transferred from the papers to the victim: by ingesting dust, or by breathing in some volatile arsenic compound, or both? How could arsenic be volatilised? And finally, if arsenic were volatilised, what was the product and was it toxic?

3.3 Arsenical Wallpaper

The *London Times* of October 20th, 1862, offered advice on the detection of arsenic on wreaths, dresses and paper. The suggestion was to put a drop of strong liquid ammonia on the sample: "If it turns blue copper is present; and copper is rarely present without arsenic being also present, the green compounds being arsenite of copper. It is therefore indirectly a very reliable test and if every lady would carry with her when she was shopping a small vial of liquid ammonia instead of the usual scent bottle, the mere catch of the wet stopper on the suspicious green would betray the arsenic poison and sever off business quickly." This advice, which was deemed important enough to be preserved in a recipe and memorandum book belonging to the family of Charles Darwin,[18] elicited a response in *The Times* on October 23rd, 1862. "I do not wish to disparate the ammonia test recommended by Dr. Letheby, which is only conclusive of the reaction for copper, but as it is quite possible to make greens without arsenic, Brunswick green for instance, it would be desirable in all cases to establish its presence. This is easily accomplished by steeping the article for a minute or so in a mixture of about one teaspoon fully each of water and liquid ammonia. Next invert a wine glass and place on it one or two drops of the deep blue liquid, then introduce the fragments, about the size of a mustard seed, of nitrate of silver, where, if arsenic is present, yellow spots will be obtained, which is arsenite of silver." No suggestion was made as to how ladies might discreetly conduct the test whilst shopping.

The Victoria and Albert Museum in London has a large collection of wallpapers, but only one sample on view, E 1243-1937 pressmark DW45, is actually noted as being arsenic-rich. Another example, recorded in their catalogue is shown with the caption, "This paper was analysed and was found to have been printed with Scheele's green, a copper-arsenic green of poisonous type the use of which was abandoned later in the century."[19] Many other sample papers in the collection from the same period, mid-1850s, have similar colours and presumably a similar chemical make-up, but the museum's expert Gill Saunders says the museum does not have the cash to do the analysis. (She also added that some of the museum surfaces were coated with an arsenic-containing paint.[20])

A visitor to the Historical Library at Yale University's Sterling Hall of Medicine would be rewarded with the sight of 100 or so wallpapers, described by the librarian as mostly hideous, that were investigated by W. M. Kenna around 1892.[21] He submitted a thesis to the Yale medical faculty on "The Analysis of the Arsenic Burden of Wallpaper." According to Kenna, it was common to find wallpaper with an arsenic burden of about $5\,g/m^2$ in the 1870s, but 20 years later Kenna found much lower amounts. His carefully conducted survey revealed that 90 per cent of the papers were either non-arsenical or contained less than $4\,mg/m^2$, 5 per cent contained arsenic in the range of 4 to $40\,mg/m^2$. Only one paper of the 100 selected had more than $40\,mg/m^2$, in spite of a selection process biased in favour of papers believed to be highly arsenical. This distribution of concentrations was much the same as that found by others around the same time, 1890. The price of the paper was not an indication of the arsenic content. The most expensive, which was said to be arsenic-free, was high in arsenic. In addition, although the arsenic was found in green wallpaper, it was also found in other colours, especially dark red.

Kenna had to take large samples, 20 cm by 20 cm, that were first destroyed with acid in order to get enough arsenic for analysis by the Marsh method (Section 5.4). In contrast, modern methods of analysis can be applied, non-destructively, to much smaller samples. Thus, X-ray methods (Section 2.8.5) revealed that a brown rosette (possibly once blue) on the wallpaper taken from Napoleon's residence on St. Helena in 1825 contained 1.5 g of arsenic/m^2. The beige background had a concentration of 0.04 g/m^2.[22] We will come back to this issue in Chapter 4. The background value could indicate that arsenic oxide had been mixed with the wallpaper paste to discourage vermin. Higher values – 6.5 g/m^2 – were obtained from an 1864 sample taken from a British National Trust property. Because these recent results were obtained from the analyses of small areas that had maximum concentration of colour, they probably represent the extremes of the concentrations in use at the time of manufacture.

Back in mid-19th-century London, some concerned individuals – including Lady Derby, the wife of the British Prime Minister – took samples for analysis to the Royal College of Science, a teaching/commercial laboratory in London. A Mrs. Bowman was one of the first customers requesting this service, on January 12th, 1870.[23,24] This laboratory was part of the foundation of what is now the Imperial College of the University of London.

3.3.1 Coal Tar Dyes and the Decline of Arsenical Colours

William Perkin was working at this same institution and in his home laboratory in 1856 when he accidentally prepared the synthetic mauve dye, mauveine, also known as aniline purple, from aniline. This event marked the birth of the chemical industry and the decline in the use of arsenicals as colouring agents, especially for wool, silk, cotton and linen. By 1960, mauve had become the most desirable shade in the fashion houses of Paris and London.[25] Shortly after Perkin's discovery, Professor August W. Hofmann, Perkin's employer, prepared aniline red (magenta) as well as aniline blue and aniline black. He used arsenic compounds in their production so the new dyes were arsenic-rich. A byproduct known as London purple, largely calcium arsenate, was widely used as a pesticide in the US but less so in Europe (Section 2.3).[26] According to author Susan Lanman, the association of the colour magenta with industry and arsenic led to its banishment from the thoughts and works of socially aware individuals such as William Morris (Section 3.13). The colour became labelled as "malignant magenta:" "The selective use or avoidance of magenta by gardeners reveals the aesthetic perspective informing their garden designs as well as their apprehension about technological and social change."[26]

3.4 Medical Problems

Details are available for five medical cases from 1859 in which arsenical gas/dust exposure was suspected; two from 1860; 21 from 1868; and about one per year from 1873 to 1881. According to this record, there were no deaths, and most patients recovered once they were removed from the source of the arsenic, assuming that this was the problem.[27]

In 1890 a physician named Malcolm Morris, who was secretary of a subcommittee appointed by the Medical Society of London to investigate arsenic poisoning, sent out 1500 circulars to medical practitioners and to fellows of the Royal Society of London, requesting a response to the question: have you had under your observation any cases clearly traceable to arsenical poison produced by arsenic in wallpaper, paint, *etc.*?[28] He got 224 replies; of these, 54 were positive and half of these described situations involving the families of friends of doctors. Diagnosis was generally admitted to be extremely difficult. The distribution of symptoms from 70 cases reported by Morris is as follows: 35 cases had diarrhoea, nausea and intestinal disorders, 16 had severe depression and 19 had conjunctivitis. Kenna lists similar vague symptoms: headache, sleeplessness, depression, convulsions and paralysis.[21] A modern-day physician suspecting subacute and chronic arsenic poisoning would look for something more specific (Section 1.7). At the time, Professor A. S. Taylor, a much-respected toxicologist (Section 5.4.6) wrote in his *Manual of Medical Jurisprudence*:[29] "there can be no doubt that the effects of arsenical wallpapers are to a great measure attributable to idiosyncrasy. The dust which escapes from these papers and the *arseniurated hydrogen* [emphasis added: arsine] which is now proved to be emitted by them are in sufficient quantity to affect a few persons; but the greater number who inhabit these rooms escape." The possibility of habituation was raised in order to account for some of the "survivors". Here are some typical examples of cases:[28]

A practitioner's wife had suffered for more than a year from repeated attacks of enteritis. She had been in the habit of sitting in a room that had a green paper on the wall. The paper was changed without reference to the wife's sickness and it was noticed that soon after, the symptoms disappeared. An examination of the paper showed arsenic present in large quantities. Another physician and his wife suffered from conjunctivitis. It was noted that these symptoms were apparent early in the morning and passed away somewhat during the daytime. The bedroom paper was found to contain arsenic and upon its removal the symptoms abated. A child showed symptoms of arsenic poisoning and was removed from one room in which it was sleeping to another, but without improvement. Finally, the child was removed to a neighbour's house in which the wallpaper had been tested and found free from arsenic. The child immediately improved in health, slept better and had an increased appetite. On being brought back home the child suffered a relapse and died. The paper in the bedroom contained traces of arsenic.

A number of attempts were made to look for arsenic in urine as an indicator of exposure because, during the 19th century, doctors noticed that abnormal intake of arsenic was usually associated with its presence in the urine

(Section 1.9). Sanger[27] did some very careful urine analysis on 15 individuals diagnosed as suffering from chronic arsenical poisoning. Most of the patients had been living in rooms covered with green arsenical wallpaper, although some wallpaper was also coloured red and blue. One child had been playing with a red flag that contained 336 mg/m^3 of arsenic and a Mr. F had been living with several stuffed birds and animals preserved by the free application of arsenic oxide, in addition to his arsenical wallpaper.

Here is a report of one case.[27] In the summer of 1883, Mr. A and wife took a house in Cambridge, of which four rooms, parlour, dining room, study, and bedroom had been recently papered. In the spring of 1885 the halls were covered, and later the other rooms with similar wallpaper. For several months no one experienced any trouble but toward the summer of 1884, Mr. A and his wife, together with a gentleman who occupied the house with them, began to feel some discomfort. This malaise disappeared in the summer during an absence of the family from the house, but began again soon after they returned in the autumn. The chief symptoms were trouble with the digestive organs and insomnia. Nausea was frequent and there was much languour and dizziness and inflamed eyelids. The summer holidays away from the house again brought relief, but the symptoms returned when they resumed occupancy in December, 1885.

The wallpaper in the parlour, the hall, the study and the bedroom was found to be rich in arsenic. Pending removal of the paper the family left the house and experienced immediate relief, especially in sleeping, but many of the other symptoms continued for some time afterward. A week after he left the house Mr. A's urine contained 0.008 mg/L of arsenic. The sample size was 1750 ml [modern tests use about 1 ml]. Four months later, the arsenic level was 0.005 mg/L, so it was bravely concluded that the elimination of the arsenic from the body appeared to be a slow process. Other amounts reported by Sanger range from 0.002 to 0.05 mg/L (2 ppb to 50 ppb).

Parallel studies conducted on "normal" urine around the same time revealed that arsenic could be detected in about 30 per cent of the samples. The results were interpreted as an indication of the widespread distribution of arsenic in articles of household use, and it was assumed that the arsenic got there, particularly in the chronic poisoning cases, by inhaling either arsenical gas or dust. Informed opinion of the time held that a distinction could not be made between the two pathways: the presence of any arsenic in the urine was taken to be the result of unnatural exposure. The possibility that arsenic would be found in most samples once the analytical difficulties had been overcome, was not a consideration prior to the early 1900s. The actual amounts detected in the supposed victims would now be considered as being in the normal range (<50 ppb).

The baleful reckoning does include a number of instances of poisonings that are better defined and involve some physical contact of victims with the arsenical paper or object, or with arsenical dust, or something similar. One such patient had no evident exposure to arsenic in wallpaper but had multiple neuritis. He was a shoemaker who had been in the habit of daily putting into his mouth a certain number of green labels which he used on the shoes. The labels were tested and found to contain arsenic in large quantities.[21] In a related instance the public records in the courthouse of Konigsburg had been piled against well-worn wallpaper that covered the green painted walls. Particles of green dust, which contained 8 per cent arsenic, were seen on the documents. Three clerks who had occasion to handle the documents suffered from undoubted symptoms of arsenic poisoning.[21]

A fatal case from the US was evidently caused by exposure to large amounts of arsenic. A whole family went to sleep in a house whose bedrooms had been cleaned with a solution of arsenic mixed with naphtha and turpentine, nine days earlier. The next day they were all seized with symptoms of acute arsenic poisoning.[21] This use of arsenic trioxide for household cleaning was not unique. The record shows that around 1872 arsenic was combined with soap to use in scrubbing British floors (Section 5.6.1).

3.5 Wallpaper Dust or Gas?

Early attempts to find arsenic in the air of suspect rooms were generally unsuccessful, although by about 1850 house dust was shown to contain arsenic just as it does today.[30] The manufacturers maintained that the arsenic in wallpaper was permanent and Alfred Fletcher of the East London Colour, Chymical, and Printing Works wrote in *The Times* of London: "Let the public be assured that it is not looking at cheerful walls, the fingering of brightly ornamented books, nor the wearing of tastefully coloured clothing that will hurt them, but the dwelling in ill-ventilated rooms and a continual dread of pure water."[31]

But in spite of these protestations from industry, the pro-dust lobby was convinced that if there was a toxic vector in the home, it was dust. They also believed that arsenic-rich dust could be generated from fabric, causing a London doctor to write in *The Times*: "The pallor and languor so commonly observed in those who pass through the labours of the London season are not to be altogether attributed to ill-ventilated crowded rooms and bad champagne, but are probably in great part owing to the inhalation of arsenical dust shaken from the clothing of a number of poisoners, who, though blameless are nonetheless pestilential." Some wearers of the dresses developed skin problems,[14] but the possibility of similar afflictions among the seamstresses seems to have been overlooked.

The proponents of the gas theory argued that volatilisation of arsenic must occur because people became ill in rooms that contained arsenical wallpaper even when the paper was covered by another arsenic-free layer. There was much speculation that the gas was arsine and unsuccessful attempts were made to trap this gas.[32] The British government became interested in the problem in 1859, around the time of maximum public concern, and commissioned a study of a room that was papered with a green unglazed paper that was said to contain 20 g arsenic/m^2. The room was closed for 36 hours and the air was then examined, but no arsenic was found. By 1884 analytical techniques were getting better but there was still no strong confirmation of gas evolution, even when an experiment was conducted in the room of an individual who showed symptoms attributable to chronic arsenical poisoning. No odour was obvious in the room.[32]

Sanger[32] also reports an experiment in which pieces of cardboard were coated with starch paste and then with green arsenic-rich wallpaper. This model room was placed for four years in a specially constructed box fitted with windows, during which time a copious growth of mould developed. When the air in the box was finally sampled, no arsenic was detected.

3.6 Gosio Gas

Throughout the latter half of the 20th century, many workers tried to show that arsenic compounds could be converted to a gas by some biological process. Krahmer was among the first to try, in 1852. He lived happily in an arsenic-rich room and was of the opinion that arsenic could not leave the walls even as dust. Nevertheless, he mixed up 4 g of an arsenical green with paste and lime taken from a damp part of a ground floor wall. After 19 days he observed no particular odour and none developed over the following five years. He tried trapping out any evolved gas with silver nitrate solution, but still no arsenic. In an attempt to get more biological activity, other workers mixed Schweinfurt green with meal and water, then adding this concoction to putrid cheese, putrid blood, or yeast. Their silver nitrate traps failed to collect arsenic. In 1875, Selmi provided some evidence for volatilisation. He worked with arsenic dust, horse dung, mouldy lemons and mouldy starch paste, and suspended silver-nitrate-treated test papers above the festering mess. After a day or so, the strips were found to contain arsenic.[33] Selmi considered that the results pointed to the formation of arsine, and Taylor refers to these experiments in the quotation above (Section 3.4). Another notable experiment extended over nine and a half years, finishing in 1886. In this the active concoction consisted of a mixture of glass, sand and human body parts (lungs, liver, kidneys and intestines) sprinkled with arsenic oxide. There was good evidence that arsenic was volatilised and also that the gas was probably not arsine. The investogator, Hamberg, concluded that a similar change takes place in the corpses of persons poisoned by arsenic: "In the course of years arsenic is given off as a gaseous compound: and this explains the disappearance of arsenic, which has been observed or conjectured by many toxicologists in the examination of parts of exhumed bodies."[34] Sanger, who did some very careful work with yeast and found no evidence for arsenic volatilisation, came to the conclusion that if arsenic was to be volatilised by biological activity, specific micro-organisms might be needed.[32]

Bartolomeo Gosio, an Italian physician working in the field of public health, finally achieved real success.[35–37] He exposed potato pulp containing 1 per cent arsenic trioxide to air and an intense garlic-like odour became apparent after the mixture became mouldy. He sucked the gas through a silver nitrate solution and then employed the Marsh test (Section 5.4) to show that the solution contained arsenic. Gosio isolated pure moulds from the pulp and found that three of them were mainly responsible for the gas production: *Penicilliun glaucum*, *Aspergillum glaucuus* and *Mucor mucedo*. Odorous arsenicals were not produced by the action of bacteria. Later on, Gosio isolated his champion gas producer *Penicillium brevicaule*, now known as *Scopulariopsis brevicaulis*, from a mouldy carrot. Gosio soon became convinced that the gas, which had a very characteristic garlic-like odour, was not arsine. He and his associates showed that the gas was probably an alkylarsenic derivative, as originally suggested as a possibility by Gemelin (Section 3.2). Gosio established that the gas could be volatilised from arsenic compounds such as the oxide, Scheele's green, and Schweinfurt green; so, in principle, arsenic could be volatilised from wallpaper. One of his experiments involved a box lined with Schweinfurt green-coloured wallpaper that had been infected with one of his gas-producing moulds, *Mucor mucedo*. The air was drawn through this model room for 39 days into a silver nitrate solution. The mould flourished and the trapping solution contained arsenic.

Gosio proposed the use of a smell test for the detection of traces of arsenic. The garlic smell of Gosio gas, as it soon was dubbed, is easily detected when an arsenic-containing sample is sprinkled onto a growing culture of *S. brevicaulis*. The detection limit is less than 1 microgram of arsenic in one gram of sample (1 ppm) and the test was at one time the most sensitive for detecting traces of the element.[38] The test was positive for Paris green, sodium dimethylarsinate, and even neosalvarsan, but no odour was detected from antimony compounds. In this connection, the action of brewer's yeast, *Penicillium glaucum* and *Aspergillus niger* on sodium cacodylate dissolved in beer wort was reported to produce an odour of "cacodyl gas" (Section 3.12).[4] In modern times, some of the unpleasant odours emanating from dead fish and crustaceans have been attributed to the formation of Gosio gas.[39] One patent[40] claims the odour is effective as a deer repellent.

Around 1925, Gosio also investigated a disease, known as Bay Illness, contracted by individuals who lived and worked in a lake inlet between Danzig (now Gdansk, Poland) and Königsberg (now Kaliningrad, Russia). There were 600 reported cases, including several deaths. The symptoms were severe pain in joints and muscles. Gosio found that arsenic-rich wastes were being discharged into the local river and he suspected that the problem was caused by the production of volatile or other arsenic compounds in the mud as a result of microbial action. He managed to isolate *S. brevicaulis* to give some credence to his hypothesis and once the river was cleaned up, the problem disappeared.[41a] It now seems likely that if arsenic caused the illness, it did so by leaching into the water, perhaps to an elevated concentration, as a result of microbial action on the factory wastes.

Around the time that Gosio was working, the use of arsenic for colouring wallpaper and other materials had effectively ceased;[14] so much so that when Lord Kelvin, aka physicist William Thompson, the chair of the Arsenic Commission of 1901 asked Mr. E. I. Pronk (a manufacturer and importer of colours of various kinds for food and for textiles)[4] whether he had ever come across injuries arising from wallpaper in the course of his work, the following exchange ensued:

Kelvin: You have not made, in the course of your trade, any special reservation in regard to selling colour which is to be used in wallpaper?"
Pronk: No.
Kelvin: Have you heard of injury arising from wallpaper?
Pronk: No, I do not think I have.

3.7 The Regulation of Arsenic, the "Verdant Assassin"

The Arsenic Act of 1851 was passed in response to the wide publicity given to the homicidal use of arsenic at the time. It regulated the retail trade in arsenic compounds: sales were to be to adults only, the substance was to be mixed with soot, and the purchaser was required to sign a "poison book" (Section 5.4). But no restrictions were placed on its use in industry, agriculture or medicine. The newly implanted fear of everything arsenical that peaked around 1860 saw members of the public calling for the wholesale regulation of arsenic and for calls for warnings to be placed on arsenical wallpaper prior to sale. According to Bartrip,[14] interest then waned until the publication of another book in 1869, "*The Green of the Period*" that had little literary merit and consisted of a series of stories about encounters with the "verdant assassin."[14] A periodical of the time published by the Natural Health Society *The Sanitary Record* kept the issue alive and in 1874 called for "the ladies, as controllers of the domestic environment, to eschew arsenically coloured goods, thereby discouraging their manufacture."

Around 1877, the use of an arsenic-adulterated baby powder caused the death of 13 children and the illness of many more. This "Violet Powder" was manufactured by Henry King who was accused of substituting arsenic trioxide for starch to save money. King claimed he was only a distributor of the product supplied by others and was acquitted of the charge of manslaughter. Although this case was not connected with wallpaper, the publicity was sufficient to prompt the Medical Society of London to investigate arsenic poisoning in general and the results of the medical survey by Malcolm Morris are described above. The Medical Society concluded that the arsenic trade should be restricted and that purchasers of arsenic-contaminated goods should be warned. The Society of Arts Manufacturers and Commerce joined the fray in 1880 but concluded that there was doubt as to whether domestic poisoning by arsenic was a significant problem. Bartrip notes that, "Owing to the virtual absence of hard evidence on morbidity and mortality rates associated with arsenic poisoning, production figures for arsenic compounds, and sales of arsenical goods, historical examination of the arsenical poisoning in the domestic environment is necessarily impressionistic and anecdotal."[14]

At the time, a survey of the practice in 22 European courts and the United States revealed that the arsenic trade was not regulated in the US, Belgium, Greece, Italy or the Netherlands. The German-speaking countries were the most regulated. Evidently, British law was not out of line and there was little mood for change in Parliament. The National Health Society of Great Britain concluded that wallpaper would be safe if it contained 6 mg/m^2 (Swedish law allowed one quarter of this loading), but attempts to legislate arsenic loadings were faced with the problem of apparent individual susceptibility, the size of the room, its ventilation and heating, and the kind of paper.[21]

Another problem at the time was the common knowledge that arsenicals were being consumed without problem by individuals with and without medical supervision (Section 1.4). By 1894 arsenic exposure ceased to be a domestic issue because there was by now general reluctance to buy goods that contained arsenic; moreover, the arsenic-based colours had become unfashionable. The "ladies of society" had triumphed without the necessity for government intervention. (Chemical contamination of domestic articles did not disappear. In 2007 a US Study revealed that hundreds of children's toys and clothing products contained potentially harmful levels of lead, arsenic and mercury, www.healthytoys.org. Even some tablets of the designer drug "ecstasy" are contaminated with mercury and arsenic.[41b])

Gosio gas was seen as the principal vector for arsenic transport from wallpaper to people, by those individuals who continued to believe in the power of the "Verdant Assassin". We will take up this topic again later (Section 3.10).

3.8 Other Assassins

Dr. Arthur Hassall wrote exhaustively on food adulteration in the mid-1800s, noting that confectionery products often were contaminated with a combination of red sulfuret (mercury sulfide), verdigris (copper acetate), blue vitriol (copper sulfate), sugar of lead (lead acetate), white lead (lead carbonate), and Scheele's green. He also reported on the wretched working conditions of the workers, mainly women and children, handling the arsenical colours. He described "general derangement of health:" loss of appetite and diarrhoea, accompanied by sore and runny noses, and sores on hands, feet, neck and other parts of the body.[42] One dyer of muslin sheets used for leaf-making had "greenish pimples on hands, hands stained greenish-yellow, eruptions on scrotum and groin." Hassall played a big part in persuading the British Parliament to pass the Adulteration of Food and Drink Act of 1860. This was the first general food law to be passed in any country.[43] He later wrote: "In the room in which I am now working there is a green Turkey carpet, a green velvet sofa, several green Morocco chairs, and three green table covers. Therefore I ought to be ill but I am not, and I ought to get rid of them. Also the yellow worsted and cotton goods contain chromate of lead so I should get rid of these too." His opinion was that the arsenic-containing colour that is used for paper hangings and other articles of furniture, dress and ornament, might be dislodged from unsized wall hangings; but not from sized or flock paper. There was no danger from volatilisation.[44] In 1872 a revised Adulteration of Foods Act incorporated Hassall's proposals and also made provision for the appointment of public analysts. In 1874 the Society of Public Analysts was founded with Hassall as its first president, and a select committee was set up to examine the working of the 1872 Act. Hassall again gave evidence and the report of this committee provided the basis for the Sale of Food and Drugs Act of 1875.[43]

Scheele's green is mentioned as a colouring agent for candies as early as 1820 but other less obvious sources of arsenic in food were in use later in the century.[43] Mr. Pronk told Kelvin's Royal Commission that the principal mineral colouring matter used in food was oxide of iron, often sold under the name "bole Armenia." The oxide, which has a great affinity for arsenic (Section 1.6.1), was used to colour sausages, anchovies, cocoa, and some sweets. A chocolate powder sold in London at a low price had about 7 ppm arsenic; sausages had about 1.5 ppm.

Mr. Pronk did not think that iron oxide should be used in sausages.[3] He also said that the magenta dyes were highly contaminated with arsenic when they first appeared but the synthetic method was later improved. Arsenic was present because of the use of impure sulfuric acid during dye manufacture. Magenta containing 6 per cent arsenic was used in colouring sweets and jam resulting in concentrations of up to 154 ppm arsenic in sweets. The use of the dye "apple green," supposedly arsenic-free, resulted in 10 ppm arsenic in sweets.

The arsenic did not actually have to be in the food to cause problems. Enamelled iron cooking vessels were capable of adding arsenic to the food during cooking. Mr. R. R. Tatlock reported to a meeting of the public analysts at Glasgow August 12th, 1876, that both lead and arsenic were released from the enamelled container during the cooking process. His analysis revealed that the coating in one sample (the worst offender) contained 1 per cent arsenic and 18 per cent lead, which was easily released because there was a deficit of silica in the enamelling mixture.[45] The arsenic was added to increase the "whiteness" of the enamel film.

By 1901, the Kelvin Commission was able to write in a footnote: "Coloured salts of arsenic such as Scheele's Green or Emerald green are so notoriously poisonous that it may be assumed that they are never used in food, and are avoided by manufacturers of toys or other articles which may be given to children or used about the house. The use of emerald green to colour wax tapers or candles appears to us objectionable and dangerous. Mr. William Thompson informed us last year that he had detected as much as 4 per cent arsenic in green tapers thus coloured."[3] A verdant assassin indeed (Section 5.7). Professor Andrew Meharg claims that these were nicknamed "corpse candles" and that there are reports of people being poisoned while reading books in bed by candlelight.[46]

We end this section with a tale of arsenic accidentally being substituted for an intended adulterant. In Bradford, England, in 1858, a sweet and confectionery seller named William Hardaker, known locally as "Humbug Billy," needed more stock for his stall in a local market. His supplier was in the habit of adulterating the sugar in his product with a substance known locally as daft, duck, or stuff. This could be anything cheap and bulky such as plaster of Paris or limestone that might be available from a local chemist and druggist. As a result of a series of mistakes, arsenic (12 pounds) was provided instead of the daft. The sweets were eventually made from 40 pounds of arsenic-adulterated sugar and four pounds of gum, and delivered to Hardaker, who sold five pounds of them on the evening of October 30th. Some 20 people, including young children, died and about 200 became severely ill. The first fatalities were diagnosed as cholera but the connection to the sweets was rapidly made. Humbug Billy was among the afflicted. No one was charged for the offence.[47]

3.9 Frederick Challenger

The chemical composition of Gosio gas was not established until 1932. Gosio and his colleagues believed it to be an organic derivative of arsenic rather than arsine because a garlic-smelling gas was produced from mould cultures containing sodium dimethylarsinate; hence, the gas probably had at least two methyl groups attached to arsenic. Sodium dimethylarsinate was also being used at the time as an internal medicine and the breath of the patients was notoriously foul (Section 1.6). Some bacteria isolated from the feces of these patients were found to produce a gas with a similar smell from cultures containing the arsenical.[48]

The successful investigation of the composition of Gosio gas was initiated following the unusual deaths of two children in the Forest of Dean, England, in December, 1931. Their parents and two other siblings survived. The inquest in Cinderford on January 19th, 1932, revealed that the walls of their house were covered with wallpaper and were mouldy. The County Analyst, Mr. R. H. Ellis, found traces of arsenic in four of the six subjects including the two dead children. The arsenic in the plaster, which had been prepared from cement and the ash from coke manufacture, amounted to 91 ppm, and the wallpaper, which was fixed to the plaster, contained up to 8.3 ppm. In the opinion of Mr. Ellis, moisture was able to come through the wall from the bank of soil outside the dwelling, dissolve arsenic from the plaster and deposit it on the wallpaper, where mould action converted it into a gas. This was maintained by Mr. Ellis in spite of the fact that he had difficulty proving the gas evolution. Tests on aspirated air samples were negative. He next exposed test papers on the walls of the house for up to 9 days: the Marsh test revealed some arsenic on the papers. The jury at the inquest returned a verdict that the death of the boy was from natural causes, but the death of the girl was from dysentery and exposure to arsenic generated in the house in a gaseous form. One of the chief factors that led to the verdict was the reported [but unlikely] finding that the amount of arsenic in the lungs was greater than in any other part of the body except for the large intestine. The diagnosis was not reliable and the possibility of direct transfer from the wall to the test paper does not seem to have been considered. The jury did comment that the house was not fit for human habitation in its present condition, and should be inspected by the Medical Officer of Health before the family was allowed to return to it.[49]

This case prompted a lot of discussion about the use of building materials that contained arsenic and the use of arsenic as a rodenticide,[50] because even though the arsenical greens were no longer in use, white arsenic was sometimes added to the horse-hoof sizing used for paper hanging in order to discourage rodents. This practice resulted in problems during renovations conducted many years later;[51] but the general public rightfully felt that they need no longer be concerned about Gosio gas.

The Forest of Dean verdict, based on the supposed toxicity of Gosio gas, prompted an investigation into the chemical structure of Gosio gas. In May 1932, Professor Fredrick Challenger of the University of Leeds and his coworkers identified the gas produced by the mould *Scopulariopsis brevicaulis*, Gosio gas, as trimethylarsine ((CH_3)$_3$As).[52,53] Other fungi afforded the same gas but not so efficiently. Challenger and his associates grew the mould on bread crumbs containing arsenic trioxide and sucked the gas that was produced through a solution of mercuric chloride in hydrochloric acid. The gas reacted with the solution to form a solid. The yield of the solid was never high even though some experiments were carried out for many weeks. The solid proved to be identical with that formed from an authentic sample of trimethylarsine and mercuric chloride, proving that Gosio gas was trimethylarsine. These classic studies provided a foundation for most of the recent, very pertinent work in this field (Section 1.7). Challenger proposed a widely acknowledged chemical pathway, now named for him, to account for the formation of Gosio gas from inorganic arsenic compounds (Box 3.2)[50] and showed that moulds could react in a similar way with compounds of other elements such as selenium and tellurium.[37]

Box 3.2 The Challenger Pathway

This pathway, Figure 3.1, is the one proposed by Challenger to account for the fungal production of trimethylarsine from inorganic arsenic species.[50] It consists of a series of steps that put the three methyl groups onto the arsenic, one at a time. These addition steps are preceded by a series of steps in which the arsenic is reduced – the "2e" notations are the two electrons that do the job – from the higher oxidation state (V) to the lower one (III) in order that the methyl group, as CH_3^+, can be added (the process is known as oxidative addition). Challenger believed that the whole sequence took place within the fungus: he was unable to find any of the intermediates along the path. We now know that some of the intermediates can be isolated and that the end product is not always the gas trimethylarsine; it is often the water-soluble precursor trimethylarsine oxide. Some bacteria also methylate arsenic and the process follows the same pathway but stops when two methyl groups are added. The arsenic is then eliminated either as water-soluble dimethylarsinic acid or as the gas dimethylarsine as indicated. Other bacteria stop the sequence after one methyl group has been added. There are few examples of the biological production of arsine.[54] We have seen that human metabolism of arsenic follows the same path stopping at the formation of dimethylarsinic acid that is eliminated in the urine (Section 1.9).

The subdiscipline of inorganic chemistry had a minor renaissance post-WWII, but by 1970 many practitioners were looking around for something

Box 3.2 Continued.

interesting that would make a connection with life processes and medicine, and possibly research grants. The situation did not look promising: inorganic does mean not organic – not related to life. However, the biochemistry of vitamin B_{12} was discovered to be based on compounds that had cobalt-to-carbon bonds – organometallic compounds. So the connection was made, and bioinorganic chemistry received a big boost.[55] The methylation of mercury to produce toxic methylmercury species became an important issue around the same time (Section 3.9), and cobalt compounds seemed to be involved, adding fuel to the flame. This notion then spread to arsenic and for quite a few years the accepted dogma was that B_{12} was involved directly in the methylation of arsenic.[56] The proposed chemistry was without foundation, but the situation eventually sorted itself out and S-adenosylmethionine (Section 7.2) was declared to be the methyl donor in the process, much as outlined by Challenger, with sulfur compounds supplying the reducing power.[57,58]

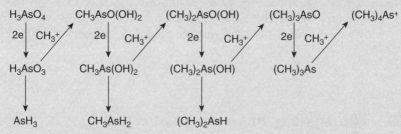

Figure 3.1 A modified Challenger pathway. The first two lines show how yeasts and fungi produce trimethylarsine ((CH_3)$_3$As) from inorganic arsenic species. The third line indicates how bacteria probably use the same pathway to produce arsine, methylarsine, and dimethylarsine.

Challenger became interested in microbiology in the course of his PhD studies in Europe and published some work on fungal metabolism during his tenure as a senior lecturer at the University of Manchester. His most important research was done after his appointment to the Chair of Organic Chemistry at the University of Leeds in 1930, which was where he remained for the next 23 years. He was a good administrator and a caring colleague and never strayed far from his roots in the north of England.[59] His biomethylation studies were novel, but at the time were well out of the mainstream of chemistry and biology and seemed to have little relevance to the real world. Consequently, he did not receive the recognition that he should have and was not elected to the Royal Society of London.

Two later happenings were to change this perception. The first was an accidental poisoning that took place in Minamata, a fishing village on Japan's southern island of Kyushu. This was described as the first major environmental disaster from human causes. In 1932, Chisso Corp, a major chemical producer began using Minamata Bay as a dumping ground for organic mercury waste. Dead fish began floating to the surface of the bay in the 1950s. Before long, residents who ate fish taken from the bay began to experience numbness, tunnel vision, slurred speech and spasms. Many suffered violent convulsions before going mad and dying. The children of those who had consumed the contaminated fish were born with Minamata disease, a severe mental retardation.[60] Over a period of 20 years thousands of people became afflicted, and hundreds died as a result of eating contaminated fish and shellfish from the bay.

In 1962 the toxic agent was found to be the positively charged methylmercury ion (CH_3Hg^+), linking the methylation of mercury by biological processes with that of arsenic. Chisso Corp, who suppressed scientific evidence from its own staff and continued to dump mercury wastes for another decade, reached a settlement with plaintiff groups after 30 years in Japan's courts.[61,62]

The second happening that directed the light back to Challenger was the discovery of nonvolatile methylarsenic compounds in many environments, including human urine, in the 1970s and the realisation that the methylation of arsenic was a biological response to ingested arsenic (Sections 1.9 and 7.2).[63]

Challenger remained active and interested in chemistry until his death in 1983, contributing to an American Chemical Society Symposium in 1978 at the age of 91.

3.10 The Toxicity of Gosio Gas

Gosio exposed rabbits to a chamber in which *S. brevicaulis* was growing in mass. The animals died, but probably from pneumonia from the fungal spores. For many years this evidence was the only basis for the statement that Gosio gas is toxic.[64] Nobody heeded Huss, who in 1914 was working in the Pharmaceutical Institute in Stockholm. He conducted tests with rabbits and mice and as a result declared that the arsenic-containing gas possesses little toxicity. However, there was one well-substantiated fatality in the early toxicity studies: one Guinea pig was bitten to death by another.[65]

Challenger started his work with arsenic in the belief that Gosio gas was toxic, despite the lack of evidence. Challenger wrote in 1944 that "it may be stated at once that the toxic compound has been shown to be trimethylarsine and that moulds growing on the damp wallpaper are responsible for its production."[50] He was undoubtedly influenced by Gemelin's newspaper article, which he quotes, but which refers to a gas with a mouse-like, not garlic, odour. "Since Challenger was fluent in German, writing his PhD thesis in that language, it seems that the only explanation for the error is that he had not actually seen the newspaper article. There is little doubt that Challenger's opinions were influential in promoting a rather general belief in the wallpaper-hazardous-to-health-scenario."[65] The toxic gas description held sway for more than 50 years following his identification of trimethylarsine.

Essentially all modern accounts of the story of Gosio gas suggest that many individuals, particularly children, were poisoned as a result of living in arsenic-rich rooms. Some put the number in the thousands.[22,37,66,67] Smith and Hattersley[68] wrote that "the fungal generation of arsines in conditions of mildew has been known for well over a hundred years. This killed thousands of children in Europe in the 1800s." There is just one dissenting voice.[65]

We have seen that arsine is one of the most toxic arsenic compounds (Section 3.1): in fact it was seriously considered for use as a chemical-warfare agent (Chapter 6). However, Gosio gas is not arsine, it is trimethylarsine, and there are now a number of reports that indicate that the toxicity of trimethylarsine is very much lower than was generally believed. Thus, although the LC_{50} of arsine (AsH_3) (the concentration of gas that is required to kill half an exposed population after a four-hour exposure) for rats is in the range 5 to 45 ppm; the LC_{50} for trimethylarsine for mice is $>20 000$ ppm.[69–71] The toxicity of trimethylarsine is "incomparably low" when trimethylarsine dissolved in olive oil is administered to mice by mouth.[72] Trimethylarsine is judged not to be a hazard for organo-metallic vapour phase epitaxy (OMVPE) applications, such as the manufacture of gallium arsenide (Section 2.8), whereas arsine is extremely undesirable from this point of view.[73] In this industry, trimethylarsine is viewed as being at least 6 500 times less toxic than arsine.[74] Experiments under way in Belgium show that rats survive well on long-term exposure to trimethylarsine.[75] However, it should be noted that replacing one methyl group of trimethylarsine with a hydrogen atom to give dimethylarsine (($CH_3)_2AsH$) results in a major increase in toxicity, to an LC_{50} of 130 ppm for rats after a four-hour exposure.

All the preceding discussion refers to acute toxicity; however, trimethylarsine has recently been shown to cause damage to DNA *in vitro*.[76] These studies showed it damaged isolated cellular DNA at a low gas concentration. However, the symptoms displayed by the supposed victims of Gosio gas were not those expected from exposure to a possible genotoxic agent.

We can rule out Gosio gas as the cause of any public health problems on at least four grounds. First, the garlic odour of Gosio gas does not correspond with the mouse-like odour reported to be present in some arsenic-rich rooms. The odour of trimethylarsine is unmistakable and easily detected. The only reports of garlic odours in rooms came from individuals who had some experience with the odour of trimethylarsine, but no experience with arsenic-rich dwellings. Second, there is no chemical evidence that Gosio gas has been produced in any room papered with arsenical wallpaper. Third, although Gosio gas can be produced by the action of certain moulds on arsenicals, it is unlikely that the gas would be produced by mould action on wallpaper because the gas production is inefficient and the gas-producing moulds are not all that common in the environment; furthermore, they do not grow well at high arsenic concentrations and are easily displaced by more robust micro-organisms. Fourth, and most conclusively, Gosio gas is not very toxic.[27,54,64,77–79]

The assumed association between human health and arsenic volatilisation by moulds in the form of Gosio gas achieved the status of fact in the early 1900s and justified the historical preservation of Gosio's name: the alliteration helped.

The possible association of gas production with the death of Napoleon[22] (Chapter 4) also contributed to the longevity of the myth. In another piece of irony, Gosio should have achieved international recognition following his discovery of the antibiotic action of a fungal metabolite now known as mycophenolic acid, but in this case others took the first prize for its rediscovery.[41a]

As for the arsenical pigments, the "verdant assassins" moved from the living room to the outdoors where their toxicity was put to use controlling such pests as the Colorado beetle (Section 2.3). Scheele's green was also used in doses up to 3 mg for the treatment of human diarrhoea, cholera and anemia (Section 1.4).

3.11 Sick-Building Syndrome?

Gosio's discovery of arsenical gas evolution gave a great boost to those who wanted to make a connection between a cluster of rather vague illnesses and the presence of arsenic-rich wallpaper-hangings and other articles, just because the arsenic was there. In a strange turn of events, a similar "sick-building syndrome" has become significant in the modern world. In some buildings there is no obvious reason for any of the medical problems, but in others the growth of fungi, particularly the notorious mould, *Stachybotrys chartarum*, is the probable cause.[80,81] *S. chartarum*, was actually first isolated from wallpaper in Eastern Europe and prefers to live in damp warm conditions on surfaces such as paper and wallboard.[82,83] There is considerable anecdotal evidence for human problems resulting from "sick" buildings, but the question remains controversial. "Exposure to various mould products, including volatile and semivolatile organic compounds and mycotoxins, and components of and substances produced by bacteria that grow in damp environments, has been implicated in a variety of biologic and health effects."[84] The symptoms reported include headaches, sore throats, general malaise, diarrhoea and fatigue, and are much the same as those of the supposed victims of Gosio gas.[85] The two sets of victims have another thing in common: their health improves when they move to a different environment.

3.12 The Manchester Beer Incident

In 19th century England, peripheral neuritis did not seem to be uncommon in alcoholics, but in the last gasps of the century the incidence of the disease reached epidemic proportions, particularly around Manchester. The presence of associated skin pigmentation led to the popular designation "khaki disease." The total number of victims was about 6000, with 2000 from Manchester, 1000 from Salford and 1000 from Liverpool. Almost all victims were poor and working class and drank beer; some consuming 10–30 pints per day. Even "infants at the breast" were not immune. Women seemed to be more affected than men, probably because they drank at home and got their supplies

Figure 3.2 *Talkative old lady (drinking a glass of milk, to enthusiastic Teetotaler, who is doing ditto)*. Yes SIR, SINCE THEY'RE BEGUN POISONING THE BEER, WE MUST DRINK *SOMETHING*, MUSTN'T WE! Punch 1901 with permission from the Bodleian Library, Oxford.

from one source, whereas the men spread their patronage around, Figure 3.2. Spirit drinkers were not affected.[3]

There were at least 70 fatal cases of arsenic poisoning but this does not represent the true picture because many of the initial deaths were attributed to chronic alcoholism. A Royal Commission chaired by Lord Kelvin was set up to "Inquire into Arsenic Poisoning from the Consumption of Beer and Other Articles of Food or Drink" (Section 3.1).[3,4]

The arsenic was eventually traced to the glucose and invert sugar that the breweries were using in the brewing process. The source of the arsenic was sulfuric acid that was rich in arsenic, having been made from arsenical pyrites (Box 3.1). The ore that was imported from Spain contained up to 1.1% arsenic that was released as the oxide along with the desired sulfur dioxide when the ore was heated. The arsenic stayed with the sulfur dioxide during its conversion to sulfuric acid. This impure sulfuric acid was then used to convert precursors such as cane sugar into the glucose/fructose mixture used in the brewing process, and the arsenic impurity continued on its journey. The acid was supplied by Messrs. Nicholson & Sons to Messrs. Bostock & Co., who then supplied the sugar to the breweries. The solid glucose from Bostock contained about 400 ppm arsenic; the invert sugar up to 500 ppm. The sulfuric acid contained about 1.5 per cent arsenic.

A Dr. Crawshaw made these notes about a middle-aged married woman living in Manchester in 1900:[86]

"For nearly 12 months the patient has suffered from morning sickness, loss of appetite, furred tongue due to gastric catarrh from alcoholism. About three months ago she first noticed a feeling of weakness in her legs, which gradually got worse. She became easily tired on walking short distances, and suffered from much pain in the feet and calves. A sense of numbness and tingling was felt in the hands. About the same time a rash appeared on the hands, arms and neck. It was red and burning and itched very much. After this subsided, the skin peeled off in various parts of the body, especially the hands and the feet. The hair came off freely at first but not much recently."

She also suffered from runny eyes, a cough, hoarseness, frequent vomiting, and constipation. "The patient presents a decidedly alcoholic appearance. The face is dusky, flushed, and somewhat puffy. The eyes are watery and inflamed. The skin over the forehead and at the sides of the face is slightly pigmented, the colour being a light brown. The superficial vessels over the cheeks are dilated. There is marked pigmentation of the skin in various regions of the body. ... The skin of the hands is thickened and is noticed to be peeling slightly. In the feet, especially the soles, the skin is much thickened, and is peeling off in thick flakes. No change is noticed in the nails. The hair is decidedly thin. There is a marked loss of the sense of heat; the husband of the patient says that she can easily place her feet on an oven-plate so hot that he cannot bear his hand to touch it. The reflexes, superficial and deep, are abolished. The appetite is very poor. No pain in the abdomen. Vomiting is very frequent, particularly in the morning. No diarrhea."[86]

"The patient drank two to four pints a day of stout or beer a day, and probably more at times. A sample of porter bought from the same shop from which she habitually procured her supply was found to contain 8/100 of a grain of arsenic per gallon [1.2 ppm]. A sample of beer contained 7/100 of a grain per gallon. The urine was examined on two occasions but no arsenic was detected. The patient steadily got worse and death occurred on December 17th, 1900."[86]

The arsenic concentrations found in the products of one brewery that made beer, pale ale and stout ranged from no detectable arsenic to 0.20 grain per gallon [2.9 ppm]. The highest concentration found was 1.40 grain per gallon [19 ppm], and this was said to have a very unusual effect on the person drinking it; one gallon of this would contain around 90 mg of arsenic, which is getting close to a lethal dose. Even at the lower end of 0.14 grain per gallon, Fowler's daily dose, about 5 mg (Section 1.4), would be achieved in three pints.

But not all the arsenic in the beer came from the sugar. Two years after the major outbreak there was a minor recurrence in the town of Halifax, where the local brew was found to contain about 0.3 to 0.9 ppm arsenic. In this case the source of the arsenic was the malt, which had been dried over local arsenic-rich gas coke. Some malt contained about 3 ppm arsenic.

One of the afflicted was 61-year-old George Shearing, a shoemaker who was admitted to hospital November 14th, 1901. His symptoms on admission were:

"marked pigmentation of the skin of feet, trunk and neck with clear patches; keratosis of the feet and scaly condition of the lower part of the legs. Loss of power in all four extremities and an inability to walk. Knee jerks were absent; hands were clenched; legs were flexed. His face was puffy; voice very husky; eyes running; he seemed muddled when spoken to and was generally very weak. This man was a heavy beer drinker, getting all his beer in a public house called The Cross Keys in Halifax."[86]

Medical witnesses dealing with the prevalence of alcoholic neuritis in Manchester and Liverpool at the time, stressed the relative frequency of the disease when compared with London, where the disease was rarely seen. They concluded that a considerable proportion of the beer brewed in some parts of the country before 1900 contained noteworthy amounts of arsenic and the amount of arsenic varied greatly in different parts of the land. Maltings in the south of England used arsenic-free anthracite coal, but in the north, coke was the most common fuel.[3] Coke can contain up to 50 ppm arsenic, mainly from pyrites, resulting in about 20 ppm in malt. Many claimed that the malting process added flavour to the final product (see also Section 8.3). We will meet a similar situation involving dried peppers in China in Chapter 7 (Section 7.10).

Following these revelations and the subsequent remedial actions by the breweries, another condition known as "alcoholic heart" ceased to be diagnosed.

As a result of the Royal commission, Messrs. Bostock & Co. were forced into liquidation. Lord Kelvin and his colleagues made it their business to ensure that the remaining stock of sugar, around 700 tons, was sold for nonfood purposes. England was lucky that the poisoning was restricted to beer. Bostock had just begun manufacturing table syrups mixed with fruit juices in two-pound tins (cans). "Fortunately at the date of the discovery of arsenic in their brewing sugars, scarcely any of these table syrups were on the market. A trial of tins (cans) had been sent out to retailers but in most instances the syrup in the tins was found to have become solidified through some accident of manufacture and in consequence the bulk of the tins had been returned to the factory."[4] Fourteen tons of the syrup, containing about 150 ppm arsenic were on the premises and were burnt on the orders of the liquidator.

The final act of the commission was to make a recommendation about "Proportions of Arsenic in Food Which Should Constitute an Offence Under the Sale of Food and Drugs Acts."

"Pending the establishment of official standards in respect of arsenic under the Sale of Food and Drugs Acts, the evidence we have received fully justifies us in pronouncing certain quantities of arsenic in beer and in other foods as liable to be deleterious, and at the same time capable of exclusion, with comparative ease, by the careful manufacturer. In our view it would be entirely proper that penalties should be imposed under the Sale of Food and Drugs Acts upon any vendor of beer or any other liquid food, or of any liquid entering into the composition of food if that liquid is shown by adequate test to contain one hundredth of a grain or more of arsenic in the gallon [0.14 ppm]; and with regard to solid food – no matter whether it is habitually consumed in large or small quantities, or whether it is taken by itself (like golden syrup) or mixed with water or other substances (like chicory or "carnos") – if the substance is shown by an adequate test to contain one hundredth of a grain of arsenic or more in the pound [1.43 ppm].

"All of which we humbly submit for your Majesty's gracious consideration *6th November 1903"*

These recommendations were put into law in the 1928 Food and Drugs (Adulteration) Act, Britain's first comprehensive act controlling the purity and quality of food.[87]

Modern beer drinkers do not have to worry; at least about arsenic: in a recent study only one of 33 European brews exceeded the 10 ppb World Health Organization limit.[88a]

In 1887, the Comte de Villeneuve, a wine producer in Hyères, France was sentenced to 20 days of jail for poisoning 500 people with the arsenic in his product: 20 died.[88b] The symptoms of the victims are almost identical with those seen in the beer drinkers.[88c]

3.13 An Historical Connection. William Morris

William Morris was born March 24th, 1834, in Walthamstow, Epping Forest, Northeast London. He was one of the principal founders of the British Arts and Crafts Movement and best known as a designer of wallpaper and patterned fabrics, a writer of poetry and fiction, and a pioneer of the socialist movement in Britain. His father was a discount broker who was involved in the financing of the Devonshire Great Consolidated Mining Company – Devon Great Consols for short – in 1844; Morris senior and his brother eventually became directors of the company. This provided considerable income and status for the family even after the father's death in 1847.

William Morris entered Oxford in 1853 where he "fell in love with medieval art and architecture and with medieval ideals of chivalry and of the communal life."[89] He started to write poetry. He inherited the first instalment of his annual income (£900) in 1856. In 1861 he helped found the firm Morris, Marshall, Faulkner and Company and in 1862 Morris designed "Trellis" – the first of the company's many influential wallpapers. These papers were hand printed from wood blocks, a very expensive process, by the subcontractor Jeffrey & Co, and many of them were coloured with arsenic-containing pigments as was the custom of the time. His last arsenic-containing design was made in 1872.[46,67,89]

The company was dissolved in 1875 and reconstituted as Morris and Co. with Morris as sole proprietor. During the 1870s Morris, who had previously made a strenuous effort to avoid political entanglements of any sort, became increasingly involved in leftist politics and the problems of poverty, the death of art [as he saw it], and the growing gap between the upper and lower classes resulting from the Industrial Revolution.

The Devon Great Consols, at the time under fire from the local authorities for its work practices and health record, went into the arsenic production business in 1867 (Section 7.7). "The arsenic works at Devon Great Consols were the largest in Devon and Cornwall, and indeed the world, during the 19th century, covering eight acres, with five ovens, three refineries, and 1200 metres of flues. This setup produced 2500 tons of arsenic a year in 1871, increasing to 3000 tons in 1884, and 3500 tons in 1891." The total output for the 60 years of operation was 700 000 tons of copper and 72 000 tons of refined arsenic.[90] It also produced a lot of local grief.[46] During this time, William served as company director from 1871–1875 and then sold off his shares; however, the family still kept a seat on the board.

Professor Andrew Meharg of the University of Aberdeen has written with considerable force about Morris and his business-driven relationship with the evils of arsenic on one hand, and his passion for the common cause on the other.[46,67] Meharg offers the following letters from Morris to his dyer, Thomas Wardle of Jeffrey & Co – his subcontractors for printing wallpaper – as proof of his indifference to the plight of others in general, and of Nicholson (perhaps a customer) in particular. Warner was an employee.

October 3 [1885]
"My Dear Wardle,

Thanks for your note about the papers. I cannot imagine it possible that the amount of lead which might be in a paper could give people lead poisoning. Still there should not be lead in them: especially by the way, in the red one. I can understand chromate of lead being in the green ones but only in small quantities. As to the arsenic scare a greater folly it is hardly possible to imagine: the doctors were bitten as people were bitten by the witch fever. I will see Warner next week to try to get to the bottom of the matter. My belief about it all is that the doctors find their patients ailing, don't know what's the matter with them, and in despair put it down to the wall papers when they probably ought to put it down to the water closet, which I believe to be the source of all illness. And by the by as Nicholson is a tea-totaller he probably imbibes more sewage than other people: though you mustn't tell him I said so.

Yours very truly,
William Morris."

Another letter was sent on October 6:

"My Dear Wardle,

Of course it is proving too much to prove that the Nicholsons were poisoned by wall-papers: for if they were a great many other people would be in the same plight & we should be sure to hear of it. I will get at Warner as soon as I can.

Yrs Truly,
William Morris."

Morris is saying that there may be some lead in the green wallpaper, probably as Brunswick green made from Prussian blue (ferric ferrocyanide), mixed with chromate of lead (chrome yellow), and then goes on to comment about the arsenic scare, which was well on the wane by 1885, so there is every reason to have some sympathy for what Morris is saying and little reason to condemn him for being an insensitive hypocrite. The discovery of Gosio gas, the vector needed by the pro-gas forces to "explain" the toxic effects of wallpaper, was announced in 1892, just a few years before Morris's death in 1894.

Meharg notes that at the time, the wallpapers of Morris & Co. were advertised as being arsenic-free, something that Jeffrey & Co. were doing 10 years earlier (rival companies stopped using arsenical pigments even earlier). The market was demanding arsenic-free papers. In fact, Morris ceased to use arsenic in his designs in 1872, just as he was beginning to feel the pull of personal political involvement. "The characteristic wall-paper and fabric designs for which he was, and is now, best known, date from the 70s and 80s."[89]

Meharg may be on firmer ground in his criticism of Morris for his apparent lack of sympathy for the people who were involved directly in the arsenic trade: the miners, the workers in the arsenic flues, the manufacturers and packers of sheep dip and Scheele's green (then being used as a pesticide). There is little doubt that by today's standards the working conditions were appalling, with heavy social costs. Although Morris was a director of the mining company for a short while, the bulk of his income did not come from company activities, and it could be argued that he used the money to better the lives of these individuals although they were not specifically named.

Morris may have had some success with his political activism because a few years after his death, when the Kelvin Royal Commission inquired into industrial problems, they noted the comparative rarity of severe poisoning among workmen engaged in recovering arsenic in arsenic-roasting works and in the manufacture of arsenical pigments and sheep dips. Peripheral neuritis and other symptoms of chronic arsenical poisoning, as were encountered in the 1900 beer-related epidemic, were almost entirely absent. Dr. Thomas M. Legge, the Chief Inspector of Factories, was able to state about the workers making emerald green: "Precautionary measures adopted in processes involving exposure to arsenical dust, in addition to the fans (which are of course the chief) are overall suits and head coverings, respirators, washing accommodation, and baths, alternation of employment (no man being allowed to work more than one day in seven). Periodical medical examination once a week [sic] has been instituted since 1899."[4]

There is an overwhelming sense of green in the William Morris Gallery located at Walthamstow, in what was Morris's family home from 1848 to 1856. The exhibits are dedicated to preserving his memory and the colour is omnipresent – in the wallpaper, the curtains, the clothing, and the upholstery. The same look is preserved, perhaps deliberately, 100 years later in many of the public houses of London, with green tiles, curtains, walls and upholstery.

References

1. W. T. Klimecki and D. E. Carter, *J. Toxicol. Environ. Health*, 1995, **46**, 399.
2. W. D. Buchanan, *Toxicity of arsenic compounds*, Elsevier Publishing Company, London, 1962.
3. Lord Kelvin, W. H. Dyke, W. S. Church, T. E. Thorpe, H. C. Bonsor and B. A. Whitelegge, *First report of the royal commission appointed to inquire into arsenical poisoning from the consumption of beer and other articles of food or drink, Vol. I*, Houses of Parliament, London, 1901.
4. Lord Kelvin, W. H. Dyke, W. S. Church, T. E. Thorpe, H. C. Bonsor and B. A. Whitelegge, *First report of the royal commission appointed to inquire into arsenical poisoning from the consumption of beer and other articles of food or drink. Vol. II*, Houses of Parliament, London, 1903.
5. B. A. Fowler and J. B. Weissberg, *New Eng. J. Med.*, 1974, **291**, 1171.
6. G. G. Parish, R. Glass and R. Kimbrough, *Arch. Environ Health*, 1979, **34**, 224.
7. P. L. Williams, W. H. Spain and M. Rubenstein, *Am. Indust. Hyg. Ass. J*, 1981, **42**, 911.
8. B. A. Fowler, in *Advances in modern toxicology*, Vol. 2, Ed. R. Goyer and M. A. Mehlman, Hemisphere Publishing Corporation, Washington, 1977.
9. K. Pannell, 1980, personal communication.
10. J. Adams, *On the presence of arsine in the vapours of bone manure: A contribution to sanitary science*, self-published, 1876.
11. M. Sittig, *Handbook of toxic and hazardous chemicals and carcinogens*, 3rd edn, Noyes Publications, New York, 1991.
12. A. V. Hirner, J. Feldmann, E. Krupp, R. Grimping and W. R. Cullen, *Org. Geochem.*, 1998, **29**, 1765.
13. H. M. Leicester and H. S. Klickstein, in *A source book in chemistry*, ed. E. H. Madden, Harvard University Press, Cambridge, 1968.
14. P. W. J. Bartrip, *Eng. Hist. Rev.*, 1994, **109**, 891.
15. T. G. Chasteen, M. Wiggli and R. Bentley, *Appl. Organomet. Chem.*, 2002, **16**, 281.
16. von Basedow, *Preuss. Med. Zeitung*, 1846, **X**, 43.
17. J. R. Metcalf, *The Lancet*, 1860, Dec. 1, 535.
18. R. Colp, *To be an invalid, the illness of Charles Darwin*, University of Chicago Press, Chicago, 1977.
19. C. C. Oman and J. Hamilton, *Wallpaper: a history and illustrated catalog of the collection of the Victoria and Albert Museum*, Sotheby, 1982.
20. G. Saunders, 1998, personal communication.
21. W. M. Kenna, M.D. thesis, Yale, 1892.
22. D. E. H. Jones and K. W. D. Ledingham, *Nature*, 1982, **299**, 626.
23. H. McLeod, *Herbert McLeod diary*, Imperial College London archives, 1870.
24. H. Gay, 2002, personal communication.
25. S. Garfield, *Mauve*, Faber and Faber Ltd., London, 2001.

26. S. W. Lanman, *Garden History*, 2000, **28**, 209.
27. C. R. Sanger, *Proc. Am. Acad. Arts. Sci.*, 1893, **29**, 148.
28. M. Morris, *Proc. Med. Soc.*, London, 1881, **V**, 70.
29. A. S. Taylor and T. Stevenson, *The principles and practice of medical jurisprudence*, J. & A. Churchill, London, 1883.
30. P. E. Rasmussen, K. S. Subramanian and B. J. Jessiman, *Sci. Total Environ.*, 2001, **267**, 125.
31. A. Fletcher, in *The Times*, 1858, Jan. 9.
32. C. R. Sanger, *Proc. Am. Acad. Arts. Sci*, 1893, **29**, 112.
33. Selmi, *Just Botan. Jahresber*, 1876, 116. Quoted from Sanger[32].
34. Hamberg, *Pharm. Zeitschr. f. Russland*, 1886, **XXV**, 779. Quoted from Sanger[32].
35. B. Gosio, *Science.*, 1892, **19**, 104.
36. B. Gosio, *Arch. Ital. Biol.*, 1901, **35**, 201.
37. R. Bentley and T. G. Chasteen, *Microbiol. Molecular Biol. Rev.*, 2002, **66**, 250.
38. H. R. Smith and E. J. Cameron, *Ind. Eng. Chem.*, 1933, **5**, 400.
39. F. B. Whitfield, *Water Sci. Technol.*, 1999, **40**, 265.
40. V. A. Baltrusaitis and M. Speyer, EP 200297, 1986.
41. (a) R. Bentley, *Adv. Appl. Microbiol.*, 2001, **48**, 229.
    (b) I. Fierro, L. Deban, R. Pardo, M. Tascon and D. Vazquez, *Forensic Toxicol.*, 2006, **24**, 70–74.
42. A. H. Hassall, *The Lancet*, 1860, 353.
43. www.rsc.org/Education/EiC/issues/2005Mar/Thefightagainstfood adulteration.asp.
44. A. Hassall, *The Lancet*, 1895, 95.
45. R. R. Tatlock, *Analyst*, 1876, **1**, 120.
46. A. Meharg, *Venomous earth*, McMillan, Basingstoke, UK, 2005.
47. I. F. Jones, *Pharmac. J.*, 2000, **265**, 938.
48. V. Puntoni, *Ann d'igiene*, 1917, **27**, 293.
49. Anon, *Analyst*, 1931, **56**, 163.
50. F. Challenger, *Chem. Rev.*, 1945, **36**, 315.
51. M. E. Gordon, *The Lancet*, 2000, **356**, 170.
52. F. Challenger, C. Higgenbottom and L. Ellis, *J. Chem. Soc.*, 1933, 95.
53. F. Challenger and C. Higginbottom, *J. Chem. Soc.*, 1935, 1757.
54. W. R. Cullen and K. J. Reimer, *Chem. Rev.*, 1989, **89**, 713.
55. A. W. Addison, W. R. Cullen, B. R. James and D. Dolphin, ed., *Biological aspects of inorganic chemistry*, Wiley-Interscience, New York, 1977.
56. J. M. Wood, *Science*, 1974, **183**, 1049.
57. W. R. Cullen, C. L. Froese, A. Lui, B. C. McBride, D. J. Patmore and M. Reimer, *J. Organomet. Chem.*, 1977, **139**, 61.
58. W. R. Cullen, H. Li, G. Hewitt, K. J. Reimer and N. Zalunardo, *Appl. Organomet. Chem.*, 1994, **8**, 303.
59. T. G. Chasteen and R. Bentley, *Appl. Organomet. Chem.*, 2003, **17**, 201.
60. *Minamata Disease The History and Measures*, Ministry of the Government, Government of Japan, www.env.go.jp/en/chemi/hs/minamata2002/.
61. J.-F. Tremblay, in *Chem. Eng. News*, 1996, June 3, p. 8.

62. J.-F. Tremblay, in *Chem. Eng. News*, 1996, March 25, p. 8.
63. R. S. Braman and C. C. Foreback, *Science*, 1973, **182**, 1247.
64. A. F. Lerrigo, *Analyst*, 1932, **57**, 155.
65. W. R. Cullen and R. Bentley, *J. Environ. Monit.*, 2005, **7**, 11.
66. T. J. Sprott, *The cot death cover-up?*, Penguin Books, Auckland NZ, 1996.
67. A. Meharg, *Nature*, 2003, **423**, 688.
68. L. H. Smith and J. Hattersley, *The infant survival guide: Protecting your baby from the dangers of crib death, vaccines and other environmental hazards*, Smart Publications, Petaluma CA, 2000.
69. R. M. Lum and J. K. Klingert, *J. Cryst. Growth*, 1991, **107**, 290.
70. M. Roychowdhury, *Am. Ind. Hyg. Assoc. J.*, 1993, **54**, 607.
71. MSDS, *Material safety data sheet: trimethylarsine*, MDL Information Systems Inc., Nashville TN, 2001.
72. H. Yamauchi, T. Kaise, K. Takahashi and Y. Yamamura, *Fund. Appl. Toxicol.*, 1990, **14**, 399.
73. G. B. Stringfellow, *J. Electron. Mater.*, 1988, **17**, 327.
74. L. D. Partain, R. E. Weiss and P. S. McLeod, American Inst. Physics Conf., Vol. 166, 1988, p. 29.
75. J.-P. Buchet and W. Goessler, 2004, personal communication.
76. P. Andrewes, K. T. Kitchen and K. Wallace, *Chem. Res. Toxicol*, 2003, **16**, 994.
77. C. R. Lehr, E. Polishchuk, M.-C. Delisle, C. Franz and W. R. Cullen, *Human Exp. Toxicol.*, 2003, **22**, 325.
78. C. Thom and K. B. Raper, *Science*, 1932, **76**, 548.
79. W. Merrill and D. W. French, *Proc. Minnesota Acad. Sci*, 1964, **31**, 105.
80. K. Wilkins, K. Larsen and M. Simkus, *Chemosphere*, 2000, **42**, 437.
81. K. Betts, *Environ. Sci. Technol.*, 2003, **37**, 407A.
82. J. D. Miller, in *Canadian Chemical News*, 2003, vol. 55, p. 19.
83. T. G. Rand and B. R. Jarvis, *Med. Mycol.*, 2003, **41**, 271.
84. *Damp Indoor spaces and health*, National Academies Washington, Washington DC, 2004.
85. E. Story, K. H. Dangman, P. Schenck, R. L. DeBernardo, C. S. Yang, A. Bracker and M. J. Hodgson, *Guidance for clinicians on the recognition and health management of health effects related to mould exposure and moisture indoors*, University of Connecticut Health Centre, Farmington CT, 2004.
86. T. N. Kelynack and W. Kirkby, *Arsenical poisoning in beer drinkers*, Ballière, Tindall and Cox, London, 1901.
87. F. A. Robinson and F. A. Amies, *Chemists and the law*, E. and F. N. Spon Ltd., London, 1967.
88. (a) W. Goessler, 2001, personal communication.
    (b) *New York Times*, 1988, July 29.
    (c) T. L. Brunton, *The Lancet*, 1901, May 4, p. 1257.
89. D. Cody, *William Morris: A Brief Biography*, www.victorianweb.org/authors/morris/wmbio.html.
90. *Morwellham Quay*, Dartington Amenity Research Trust publication number 78, 1983.

CHAPTER 4
Arsenophobia: A Connection between the Deaths of Infants and Napoleon I

4.1 Sudden Infant Death Syndrome

The general population is likely to be unaware of the chemical distinction between arsine and Gosio gas, or of any differences in their odours or toxicities. All that is required to generate fear and apprehension is to mention the word, arsenic. So the panic among parents in the United Kingdom was very understandable when they were informed on June 6th, 1989, via the news media, of a possible connection between an arsenic-containing gas and sudden infant death syndrome (SIDS); also known as cot death or crib death.

The concept of SIDS was first formulated in the late 1960s. It has recently been defined as "the sudden unexpected death of an infant less than one year of age, with onset of the fatal episode apparently occurring during sleep, which remains unexplained after a thorough investigation, including performance of a complete autopsy and review of the circumstances of death and the clinical history."[1] The incidence of SIDS ranges around the world from less than 0.1 per 1000 live births to 2.5 per 1000. Countries with low rates include, Portugal, Japan and Sweden, while New Zealand, Ireland, England have relatively high rates.[2]

SIDS is a diagnosis of exclusion. So how good can the diagnosis be if the cause is unknown? A recent analysis of 450 post-mortem reports, part of a UK government study entitled "A Confidential Enquiry into Stillbirths and Deaths in Infancy," found the misdiagnosis of SIDS to be a real possibility.[3]

Prior to 2002, in the UK, professor and paediatrician Roy Meadows was often called to act as an expert witness in cases of suspected child abuse. He developed Meadows' Law: "One sudden infant death is a tragedy, two is suspicious, and three is murder, unless proven otherwise." Meadows testified that the chance of two cot deaths happening in the family of Sally Clark, a solicitor, was one in

---

Is Arsenic an Aphrodisiac? The Sociochemistry of an Element
By William R. Cullen
© William R. Cullen 2008

Arsenophobia 131

73 million, and as a result, in 1999, she was convicted of murdering her two sons. Some three years later she was freed when it was established there was no statistical basis for Meadows' figures and that some evidence had been withheld from the jury at the trial.[4] In Britain the chances of having a SIDS baby are 1 in 8 500. That risk increases if the family has already experienced one SIDS death. The chances of having a SIDS baby are much greater, about 1 in 200, if the mother is under 27 years old and has had more than one previous child; and if both parents smoke and are unemployed.[5,6] The Sally Clark trial and others around the same time revealed problems with the collection and analysis of medical evidence, prompting the formulation of the definition of SIDS given above.[7,8]

There are difficulties in collecting worldwide statistics for SIDS because of differences in definition, record-keeping and medical and legal practice, but it does seem that the SIDS rate peaked in the western world around 1990 and dropped rapidly over the next five years or so to reach a plateau. For example, in the late 1950s the SIDS rate in England and Wales was estimated to be 1.6 deaths per 1000 live births. It increased to 2.5 per 1000 live births in the late 1980s, decreased to 1.4 in 1991 and then fell further to 0.65 in 1996. In the US the rate in 1991 was 3.4 per 1000 live births; falling to 1.4 by 1997. In New Zealand, the rate was 2.9 deaths per 1000 live births in 1990. This was split between the Maori population, with a particularly troubling 8.3 deaths per 1 000 and non-Maori (predominantly European) with 2.2 deaths per 1000. (There are about 600 000 Maori in New Zealand and three million Europeans.) By 1994, the NZ rates were about 5.5 per 1000 live births for Maoris and 1.5 for the rest of the population. The incidence of SIDS is generally greater among indigenous populations than among Europeans so, for example, in North America the native population is 3 times as likely to have a SIDS baby as the rest of the population. In Western Australia this factor is 5.

4.2 The Toxic-Gas Hypothesis

On January 11th, 1989, Mr. P. R. Mitchell, the managing director of Mitchell Marquees, Winchester, wrote the following letter to the UK Department of Trade and Industry:

"May I start by saying that I am not medically qualified but have accidentally stumbled on a possible cause of cot deaths.

My company hires out marquees [tents] and we have had problems with discolouration of the white material caused by fungal and biological attack. In order to overcome this we asked Mr. [Barry] Richardson who is a consulting scientist to look into this problem.

The simple solution is to add an arsenic-containing biocide to the PVC of the marquee. Mr. Richardson is however, not prepared to recommend this material because although there will only be about one teaspoon of arsenic in a marquee for 300 people, there is a risk of this arsenic being turned into arsene [sic] gas, which is extremely toxic and may affect the marquee erectors.

I have checked with the manufacturers of this biocide and it is approved for use in babies' plastic pants and also in the PVC covering on babies' cots and prams.

It would appear that if this arsenic is going to be affected by fungus to turn it into arsene gas this will happen to an old mattress.

There are several unexplained reasons why cot deaths occur. I list some of them below which give credibility to my theory:

(1) They have an older brother or sister. Most parents would use the same cot/pram for the second child, which would now be old.
(2) If another child in the family has died unexpectedly, in the same cot/pram?
(3) If the mother is a teenager: more likely to buy an old second-hand cot/pram.
(4) During the winter months. During the summer months the windows would be open giving better ventilation.
(5) There are less cot deaths in Scandinavia. There is another biocide available called 1BBC which is widely used in Scandinavia but not in the rest of Europe."

Barry Richardson, the consulting scientist mentioned in the letter, had written to the UK Health and Safety Executive the month before about the use of the biocide OBPA, 10,10'-oxybisphenoxyarsine (Section 2.7), in tents, because he believed that fungal action could generate toxic arsine gas from OBPA. It is clear that at the time Richardson was aware of Gosio's work on gas evolution but was under the misapprehension that the gas was arsine, when in fact, it is, as we have seen in Chapter 3, the less toxic organic derivative trimethylarsine. He also made an unjustified leap in stating that it could be generated from OBPA by the action of the fungus *Scopulariopsis brevicaulis*.[9,10]

At the time, OBPA was available from Akzo Chemie in the Netherlands as Intercide ABF, and from the Ventron division of Morton Thiokol Inc. in the US as Vinyzene BP-5-2. The manufacturers were licenced to use the arsenical in plastic materials intended for long-term performance in heavy-duty outdoor applications, such as roofing liners, tarpaulins and marquees, but not for use in babies' cots and pants. Nonetheless, the possibility of adverse publicity affecting sales of OBPA prompted wary discussions between Morton Thiokol and Richardson: the company had some doubt about Richardson's motives. In an attempt to deflect criticism, one company representative made the observation that "if there were to be any possibility of micro-organisms reducing OBPA into arsene [sic] gas, a possibility which can be checked easily, there would be an equal chance to reduce arsenic impurities from antimony trioxide, a widely used filler and flame retardant for soft PVC [polyvinyl chloride]. Even the antimony oxide could be reduced to stibine (SbH_3), also a poisonous gas. The oxide is added to PVC formulations from 2.0 to 10%."[11] Richardson seized on this unsubstantiated proposition and added stibine to his suite of toxic gases that could be emitted from cot mattresses.

He submitted a report entitled "Cot Mattress Biodeterioration and Sudden Infant Death" to the British Medical Journal for publication pending peer review. The report was never published; however, in a letter to the medical periodical *The Lancet*,[12] he stated the general hypothesis that "phosphine

Arsenophobia 133

(PH_3), arsine and stibine can be generated by biodeterioration of organic and inorganic phosphorus, arsenic and antimony compounds; alkyl compounds may also be produced, which are also exceedingly toxic gases."

Morton Thiokol scientists were in close contact with Richardson and informed him that they were unable to reproduce his results. They advised him that Gosio gas was trimethylarsine – the only gas they were able to identify from control experiments. In spite of this information, by June 6th, 1989, the media had the story. Here are some of the headlines, and not just from tabloid newspapers: **"TOXIC GAS THEORY ON COT DEATHS" "Cot deaths: Can this be the answer? HOPE FOR PARENTS" "FUNGUS POISON GAS LINK TO COT DEATH" "MOTHERS IN COT DEATH PANIC."** Richardson is quoted in one headline: "As a scientist I had to prove a link. As a father I had to warn the world." The media described Richardson as a factually precise, mild-mannered man, not given to emotional statements, but one of these reports quotes him as saying, "I had a terrible dilemma. Should I go on as a scientist and completely prove a link I was certain existed, or should I warn people." The British Broadcasting Corporation (BBC) had this to say: "A British company is claiming that cot deaths could be caused by lethal gasses given off by PVC mattresses. Barry Richardson, the technical director of the company, says that most of these mattresses produce gases that can kill. The main one is called stibine and it's extremely toxic." Hundreds of panic-stricken mothers jammed a help hotline after the broadcast.

Lady Sylvia Limerick, vice-chairman of the Foundation for the Study of Infant Deaths, said to the BBC the same day of the news release, June 6th, that there was no proof yet that this was a cause of sudden death in infants, so parents should carry on as normal. If parents were very anxious, the sensible thing would be to purchase a new mattress. This was Richardson's suggestion as well.

4.2.1 The Reaction

There was an immediate backlash against Richardson from mattress makers, doctors and other scientists. Only the manufacturers of OBPA had something to be happy about: their product was back below the horizon and not mentioned again. A fact sheet released by the US Environmental Protection Agency in 1993[13] indicates that OBPA was authorised for use in vinyl baby pants, even though Morton-Thiokol claims it never recommended the use of the product for crib mattresses. The chemical is still with us (Section 2.7), but the proposed connection with SIDS has been largely forgotten.

Richardson claimed that the British army issued crib mattresses containing OBPA to military families which he linked to a high incidence of SIDS: up to 5.8 deaths per 1000 live births. Richardson claims the mattresses were withdrawn from circulation following his disclosure.[14] The veracity of this story has been questioned,[15] with the British Ministry of Defence stating that arsenicals were never used in their mattresses. However, it seems that the abnormally high rates were recorded, even if the cause is disputed.

Richardson did acknowledge in 1989 that published reports refer only to the generation of arsenic-containing gasses but went on to say that phosphine and stibine can be similarly formed. He claimed that "the arsenic source in plasticised PVC is usually OBPA and the antimony source is usually antimony trioxide, which is added as a fire retardant. The phosphorus source is usually a phosphate plasticiser that is also used as a fire retardant." He described experiments in which he encouraged microbes "naturally" present on the PVC to grow, especially the fungus *S. brevicaulis,* and he examined any gas given off by using indicator papers to detect arsine (Sections 3.6 and 5.4).[16,17] One sample of his research notes reads as follows: "Forty used mattresses were examined. Thirty-eight were covered with plasticised PVC and all were naturally infected by *S. brevicaulis* in the area immediately beneath the baby, affected by warmth and perspiration. The infection was invisible but had caused plasticiser deterioration, resulting in splitting in some cases. Trihydride gases were generated when infected PVC samples were incubated. The most common gas was stibine, formed from the antimony trioxide used in the PVC as a fire retardant, although phosphine also occurred, formed from phosphate plasticiser, and sometimes traces of arsine from arsenical impurities in the antimony trioxide."[10]

There is very little information here about gas production in the real environment. Richardson comes to the conclusion that all mattresses are contaminated with the fungus *S. brevicaulis,* and that gas generation by the fungus is a possible cause of SIDS. He recommends the use of clean mattress coverings that do not contain phosphorus, arsenic or antimony, and that temporary protection against gas and spore generation can be achieved by wrapping a mattress in a clean polythene sheet, secured on the underside using adhesive tape. The test he was using relies on the production of a coloured spot on the test paper that changes from yellow, to orange, to black as the amount of arsine increases. Phosphine and stibine, if they are produced, interfere with the colour formation but, contrary to the claims of Richardson,[10,12] the spot test is not recognized for the analysis of phosphine and stibine. The colour test for arsenic as arsine is sensitive; however, it is useless for Gosio gas, trimethylarsine, which gives a colourless product. As mentioned above Richardson was also under the mistaken impression that OBPA would act as a source of arsenic for the production of Gosio gas. To cap it all, at the time there was no good evidence for the biological formation of any gas containing phosphorus or antimony, least of all phosphine or stibine.

4.2.2 The Turner Commission

Richardson was rather cautious in his public statements, but the publicity caused a mass panic that was advanced by the media even though there was little evidence that there was a real problem. The British government felt it had to respond, so in March 1990, a working group chaired by Professor Paul Turner of St Bartholomew's Hospital, London, was convened to consider the Toxic-Gas

Hypothesis. In the background hovered the insinuation that the government was worried because the fire retardants had been added to the bedding material as a result of government decree, consequently legal action against the government could be triggered if the hypothesis were found to be correct.

The Turner Commission was informed by the Laboratory of the Government Chemist that the evidence for the evolution of volatile antimony compounds, either in mattresses or in culture, was "hard to obtain: in fact we have none."[18] The colour changes that Richardson observed on his test strips were attributed to the evolution of sulfur compounds from common bacteria. *S. brevicaulis* was found on a few of the SIDS mattresses, although a later study[19] failed to detect this fungus. The commission concluded, at a cost of more than a quarter of a million pounds (about $500 000 US), that the Toxic-Gas Hypothesis was unfounded, but did recommend an investigation of the microbial population of cot mattresses and their covers to determine the significance of the presence of pathogenic micro-organisms, especially fungi.[20] In addition, they hedged their bets by stating that the need for chemical additives as fire retardants in cot furnishings should be carefully considered and only grades of antimony trioxide containing the lowest possible levels of arsenic should be used in the treatment of cot furnishings. The media lost interest, but the SIDS rate in the UK, which was about 2.3 per 1 000 live births at the time Richardson went public, began to drop, and reached 1.4 per 1 000 by the end of 1991. So something was having an impact – perhaps publicity about old mattresses or guidance on mattress wrapping. The proponents of the Toxic-Gas Hypothesis think that both of these had some effect and claim that the antimony compounds that were being added to mattresses in 1987, in response to a government requirement for fire retardants in mattress material (the 1988 Furniture and Furnishings Fire Safety regulations), were responsible for the SIDS peak around 1989. They believe that most of the drop in the SIDS rate after 1989 was because the manufacturers quietly removed the antimony from the mattresses following Richardson's revelations. But before we deal with these claims we need to consider what was happening elsewhere in SIDS research.

4.2.3 The Back-to-Sleep Campaign

A survey of New Zealand families, spearheaded by Professor Ed Mitchell of Auckland University, revealed that most infants who died of SIDS were discovered lying face-down, which at the time was the universally recommended prone sleeping position for babies. As a result the research team made the recommendation that infants be placed on their backs to sleep during the first year of life.[21] The medical profession slowly concurred, resulting in the "Back-to-Sleep" campaigns in New Zealand, England, and later North America. This change of practice lead to in dramatic improvements in SIDS numbers even though there appears to be no hard science providing evidence as to why the switch of sleeping positions was so effective. The SIDS rate dropped further in the UK, from 1.4 per 1000 at the beginning of the campaign in 1991 to 0.66 in 1993. Richardson accounted for the success of the Back-to-Sleep campaign in terms of statements such as: "Infants who are hyperthermic through overwrapping or fever are most at risk – especially those sleeping in the face-down position, who are most severely exposed to accumulations of heavy stibine.... Avoidance of the face-down position reduces the risk." He also argued as mentioned above that part of the SIDS decrease was attributable to the voluntary removal of fire retardants from mattresses by the manufacturers and his recommendations regarding the use of new and wrapped mattresses.[22]

It is true that a number of leading UK retailers, including Boots' Children's World and House of Fraser, later suspended sales of cot mattresses containing antimony or phosphorus.[23] A claim was made in the Limerick Report (Section 4.3) that antimony concentrations were actually being increased during the Back-to-Sleep period in continuing response to government regulations, but the data presented do not support this statement.[15] Nonetheless, the antimony concentration in new infant mattress covers was still high, although it did decrease from 1988 to 1995. UK imports of antimony trioxide peaked in the years 1987–88 and then dropped in 1989.[24] As for Richardson's mattress-wrapping claims, the Limerick Report states that the practice was not widely adopted and statistically had no effect on the SIDS rate.[15]

So we are left with the not-completely-satisfactory conclusion that the increase in SIDS prior to 1988 was mainly attributable to the adoption of the prone sleeping position; and the fall thereafter, to the adoption of the Back-to-Sleep recommendations, with the drop from 1988 to 1991 being the result of unsubstantiated dissemination of the results of the New Zealand study prior to the launch of the campaign.

Richardson's paper, "Sudden Infant Death Syndrome: A Possible Primary Cause," was finally published in 1994 having encountered some problems in peer review. The paper contained the first public report of the gas generation experiments described above and also some analytical results that were presented as evidence for elevated concentrations of antimony in some SIDS babies. One significant section reads: "it has also been reported that phosphines, arsines, and stibines cause depression of the central nervous system. This may indicate an anticholinesterase action."[25]

At the time of publication, there were no reports of the anticholinesterase action of any of the three gases or their alkyl derivatives, and none has appeared since. The nerve agents sarin and VX are examples of phosphorus compounds that have extreme anticholinesterase activity that has been exploited for military purposes (Box 6.3). Their mode of action is well established and depends very much on their chemical composition and structure. These compounds are chemically very distinct from the phosphines in Richardson's Hypothesis; nevertheless, the nerve gas description took on a life of its own in the hands of Richardson's disciples, as we will see later.

In the UK the emphasis shifted from arsenic to antimony and phosphorus because of the use of compounds of these elements as fire retardants and the absence of arsenic compounds in the mattress covers. There was a passionate public reaction to two episodes of the television program *The Cook Report* on

Arsenophobia

ITV, November 17th, and December 6th, 1994, which focused on the Richardson Hypothesis. The first generated 60 000 phone calls within 24 hours. These were fielded by individuals who gave advice on mattress wrapping, which the public heeded temporarily.[15] A well-known TV personality, Ann Diamond, who had lost a child to SIDS and who had become a SIDS activist, was featured prominently in one program saying: "The government has got to instigate a major scientific inquiry into this, and now." Richardson made an appearance and demonstrated his experiments on gas production. Some analytical data were presented, commissioned by the program, claiming to show elevated levels of antimony in SIDS babies.[26] The production of stibine was stated to be the transfer route of antimony to the baby from the mattress.

In the Plymouth Extra, March 16th, 1995, we find: "Recent tests on tissue samples taken from baby Adam at his post-mortem have revealed the presence of antimony poisoning in the lungs. . . . Baby Russell James, who died two years ago, also suffered the same antimony poisoning." The Extra, like other newspapers, called for a ban on the sale of second-hand cot mattresses and a total ban on the fire retardant. A free analytical service for antimony levels in infants' hair and tissues was made available to families who had lost a child to SIDS. The British Foundation for the Study of Infant Deaths, which had been opposed to the advice to wrap mattresses, reluctantly changed its stance in December 1994, to tell parents to wrap mattresses if they were worried. A BBC documentary on SIDS that aired on March 21st, 1995, attempted to calm the waters by emphasizing the importance of the Back-to-Sleep campaign.[27]

4.3 The Limerick Report

The public outcry and the threat of legal action (some independent research was funded by legal aid on behalf of bereaved parents – *e.g.* Pearce *et al.*[28] commissioned by Leigh Day and Co., London, UK) prompted another government-funded inquiry. This time it took the form of an Expert Group to Investigate Cot Death Theories chaired by Lady Sylvia Limerick. The Limerick expert group, whose members were seen by some as being too close to the problem to be completely objective, met first on December 16th, 1994, and on a further 23 occasions.[15]

The following is a formal statement of the Toxic-Gas Hypothesis as presented to the Limerick Committee by Richardson in 1994:[22]

"The hypothesis suggests that the primary cause of SIDS is poisoning by gaseous phosphines, arsines, and stibines generated by deterioration of cot mattress materials by micro-organisms, particularly Scopulariopsis brevicaulis, *an otherwise harmless fungus which is normally found in all domestic environments. All cot mattresses become naturally infected in use by micro-organisms, and these toxic gases are generated from all mattress materials that contain phosphorus, arsenic or antimony compounds. Whether an infant is unaffected or suffers irritability, illness or death depends on various contributory factors."*

Table 4.1 Candidate compounds for microbial metabolites.

Phosphines	Arsines	Stibines
PH_3	AsH_3	SbH_3
CH_3PH_2	CH_3AsH_2	CH_3SbH_2
$(CH_3)_2PH$	$(CH_3)_2AsH$	$(CH_3)_2SbH$
$(CH_3)_3P$	$(CH_3)_3As$	$(CH_3)_3Sb$

The hypothesis concerns the possibility of the evolution of gaseous compounds of three elements: phosphorus, arsenic and antimony. In principle we need to consider a huge number of compounds. Fortunately, we can narrow down the possibilities. When the hypothesis was first formulated in 1989, only arsenic compounds containing hydrogen and/or methyl groups were known to be produced by biological action, as set out in the middle column of Table 4.1. We now know that all these can be produced to a greater or lesser extent by the action of fungi and bacteria, although trimethylarsine, Gosio gas, is by far the most commonly encountered (Section 3.6). It is very unlikely that other groups would be found in any gaseous metabolites containing phosphorus or antimony so the total number of compounds to be considered is 12, as shown in Table 4.1.

It should be emphasised again that at the time the hypothesis was formulated in 1989, there was no compelling evidence for the production of phosphines or stibines (antimony derivatives) through microbial action; so the inclusion of these compounds was pure speculation. In the 1930s, Frederick Challenger and his coworkers had done some research on the interaction of antimony compounds with fungi without much success.[29-31]

The Richardson Hypothesis refers to micro-organisms in general and *S. brevicaulis* in particular. *S. brevicaulis* was singled out because, as outlined earlier (Section 3.6) it was associated with Gosio's classical studies of the biotransformation of inorganic arsenic compounds. The Limerick Committee focused its work on this organism and the possibility that it was involved in the volatilisation of antimony.

We should not forget that in the late 1900s trimethylarsine was generally believed to be highly toxic, and the same was assumed to be true for trimethylstibine. Statements to the contrary entered the debate only in 2005.[32]

4.3.1 Antimony Biomethylation

Since 1994, antimony compounds and particularly trimethylstibine ($(CH_3)_3Sb$) have been added to the list of gaseous compounds produced by microbial action.[31] Trimethylstibine was first identified in 1996 as being a product of the interaction of inorganic antimony compounds with unidentified soil micro-organisms growing in an oxygen-free environment. It was also identified in landfill and sewage gases and in the air above hot springs, presumably produced by similar bacteria.[33] More recently, the production of trimethylstibine, but not stibine, from antimony-containing cultures of *S. brevicaulis* growing in air has been established.[34,35] It turns out that the same biochemical path is followed during the biological production of trimethylarsine and trimethylstibine.[31] This pathway (Section 3.9) is not available for the production of the corresponding phosphorus compounds because the necessary two electron-reduction steps require too much energy.

Phosphine (PH_3) is found at low concentrations in the atmosphere but there is minimal evidence for its production by microbial action.[36–38] The other phosphorus compounds listed in Table 4.1 have not been detected in the environment and are unlikely products of biological action. Thus, trimethylarsine and trimethylstibine are the only two gases likely to be produced by microbial action on commonly encountered compounds of arsenic, antimony and phosphorus.

However, there are differences in the efficiencies of the two biological processes. First, trimethylarsine is produced from most soluble arsenic compounds whereas trimethylstibine is produced from a limited number, second, the production of trimethylarsine by *S. brevicaulis* is about one million times as efficient as the production of trimethylstibine so very sensitive analytical methods are needed to reliably and reproducibly detect the formation of trimethylstibine. But it isn't easy to volatilise arsenic. Even under ideal conditions a culture of *S. brevicaulis* releases less than 10 per cent of its arsenic as trimethylarsine over a growing period of a few months[39] and no trimethylarsine is detected if the initial concentration of arsenic species in the medium is low: less than 1 ppm. Under these conditions methylation results in non-volatile species.[40] Nonvolatile methylantimony species have been found in some cot mattress foams, suggesting that biological methylation has taken place at some time.[41,42]

4.3.2 Report Summary

Some of the information outlined above was available to the Limerick Committee and incorporated into its final report. Here is a summary:[15]

- Cot mattress contamination with *S. brevicaulis* is rare and no more common in SIDS mattresses than in other used mattresses. Richardson seems to have confused some bacteria with the fungus.
- There is no evidence for the biovolatilisation of phosphorus, arsenic and antimony from PVC cot mattresses by *S. brevicaulis* under conditions relevant to an infant's cot. Richardson's test paper experiments could not be reproduced. There was some suggestion that the observed colour changes might be caused by exposure to a sulfur compound [*but the ones identified seem to be unlikely candidates*].
- There is no evidence that SIDS is due to poisoning by phosphine, arsine or stibine or their methylated derivatives.
- The antimony concentrations in the tissues of SIDS infants are not exceptional and there are a number of sources other than fire retardants in

the mattress material to account for the presence of antimony. For example, antimony concentrations in household dust are naturally elevated,[43] and high concentrations of antimony, up to 4 per cent, are present in baby care products such as push-chair covers.
- Overall, we conclude that the Toxic-Gas Hypothesis is unsubstantiated.

All of this took great effort from many individuals and research teams. Richardson was invited to participate in some of the experiments to ensure that his experimental conditions could be replicated. He later claimed that, in spite of his input, there were some differences in the protocols that would account for the inability of the investigators to reproduce his results. The painstaking investigation, which focused almost entirely on antimony volatilisation by *S. brevicaulis,* took four years. The report that was published in May 1998, cost about £20 million ($40 million US).

There was some concern over the report's narrow focus. Some organisms identified on mattresses are capable of carrying out biomethylation[44] and the possible synergic interactions of microbial consortia were not considered. Furthermore, although antimony had been added to cot mattresses in the UK, it had not been used in other countries, such as New Zealand. Some critics of the Limerick report, including Dr. Jim Sprott of Auckland NZ, dismissed its findings as obsolete and irrelevant.[45]

4.4 Dr. T. J. Sprott

Dr. T. J. (Jim) Sprott of New Zealand is a very vocal advocate of Richardson's Hypothesis. He became a disciple and assisted in the preparation of the ITV *Cook Report* programs dealing with SIDS. In 1996 he published a book in New Zealand with the provocative title *The Cot Death Cover-up?*.[20] The book was released in the UK a year later. The first chapters were devoted to showing how, in 1986, Sprott had come to the conclusion that SIDS was caused by exposure to an extremely toxic nerve gas generated by microbial action on something in the baby's cot. He then read about the Toxic-Gas Hypothesis and, he says, it all fell into place. He is convinced that in 19th-century Europe thousands of children died because of exposure to Gosio gas. The remainder of the book is a presentation and defence of Richardson's Hypothesis, which he interprets as follows: there is one cause of cot death, and it is poisoning by one or more of the gases phosphine, arsine or stibine and/or their lower alkyl homologues.

Sprott is convinced that the medical profession has personal reasons for ignoring Richardson's hypothesis, particularly greed. An angry exchange on this subject between Sprott and a British professor in his office once ended in fisticuffs.[20]

Sprott has zero contact with the New Zealand Cot Death Association. He claims the organization has not replied to his many letters and has asked him to cease criticising its public campaign to collect money for research on SIDS. He took out full-page advertisements in New Zealand newspapers saying, in effect, that the cause of SIDS was known, so why waste more money on research.

He said he would back off if the Cot Death Association would endorse mattress wrapping, but no deal was reached.

In the meantime the cot death rate was dropping in New Zealand and sales of his personally endorsed slip-on cot mattress cover were brisk. The New Zealand Woman's Weekly described parents' panic over the conflicting advice, with the Cot Death Association dismissing the Toxic-Gas Hypothesis as bunkum.[46] One New Zealand mother is quoted as saying, "I've never met Jim Sprott, but he's saved my baby." Her son suffered from apnea attacks, during which he would stop breathing in his sleep. She installed a monitor but she and her husband still got no sleep; either reacting to the alarm or waiting for it to sound. In desperation they contacted Sprott, and on his advice she threw out the second-hand mattress and started again with a wrapped piece of foam rubber. "Blake's apnea attacks ceased immediately," she claimed. The Royal New Zealand Plunket Society, a government agency that monitors the health of infants and advises mothers on baby care, has incorporated mattress wrapping into its field staff training.

Both Richardson and Sprott were invited to speak at the SIDS International Conference held in Auckland in 2000, but both declined. There were about 300 delegates and approximately 100 papers were presented in oral and poster sessions. Sprott, however, was very evident outside the theatre at the opening ceremonies of the conference, handing out his "Cot Life 2000" pamphlet, which espouses his views. This action was clearly an embarrassment to the organisers, so although he was not a registered delegate, he was reluctantly invited to speak and presented a passionate defence of the Richardson Hypothesis. The audience response was rather muted, although there were some "did so" "did not" exchanges.[47,48]

Sprott criticised the Limerick report's focus on antimony and *S. brevicaulis*. He argued that the findings were not particularly relevant to New Zealand where PVC and antimony fire retardants were rarely encountered, yet the rate of SIDS was high – he claimed as a result of the evolution of arsines and phosphines. His use of the highly emotive phrase "nerve gas" continues unabated to this day.

Sprott is convinced that the big drop in SIDS numbers in New Zealand, especially after 1994, is directly attributable to mattress wrapping:[49] "Tens of thousands of New Zealand parents have wrapped their babies' mattresses in BabySafe mattress covers or in polythene sheeting." None of this is supported by hard evidence.[50] However, it is distressing that the almost universal drop in SIDS rates found during the 1990s did not extend to New Zealand's Maori population. The Maori rate, at 3.5 deaths per 1000 live births in 1997, was twice that of the rest of the population and has been attributed to a high prevalence of female smokers.

4.4.1 Sheep Skins

In 1996, Sprott wrote that arsenic in sheep skins was a major cause of cot death.[51] The use of sheep skins for bedding material is heavily promoted in New Zealand, Australia and Britain, and consequently a large number of babies sleep on them. Britain imports about 100 000 sheep skins for baby use every year, mainly from Australia and New Zealand. The skins are warm and comfortable for sleeping, but they can also be a sink for dirt and microbes if parents are not careful. Medical authorities write that it is unfortunate that sheep skins were ever recommended for infant bedding because at the very least they harbour dust mites, which can cause asthma.[52]

One New Zealand study published in 2000[50] found that there was neither a decreased risk of SIDS associated with mattress wrapping nor an increased risk of SIDS associated with sheepskin bedding. This study of 485 infant deaths revealed that 42 per cent had occurred on sheep skins.

Although no antimony or arsenic-containing additives are involved in the preparation of the skins for market, the proponents of the Toxic-Gas Hypothesis argue that, like humans, the skin and wool of the live animals accumulates arsenic, antimony and phosphorus. SIDS is then the result of inhaling toxic gas produced by microbial action on the natural burden of compounds of these elements.

One recent Canadian study was conducted to test this hypothesis.[44] Although the researchers did not find *S. brevicaulis* among the bacteria and fungi isolated from the sheep skin samples supplied by Sprott, a few of these isolates were able to methylate arsenic compounds, and one was able to produce small amounts of trimethylarsine gas in culture. The researchers concluded that, bearing in mind the newly appreciated low toxicity of trimethylarsine (Section 3.10) and trimethylstibine,[32] there was little chance that a baby sleeping on the skins could become exposed to a toxic gas generated as a result of the growth of bacteria or fungi on this bedding. By extension they concluded that, in general, the Toxic-Gas Hypothesis had little foundation. They did report one very important result in connection with SIDS, and recommended that health care professionals take note: pathogenic bacteria such as *Mycobacterium neoaurum* and *Acinetobacter* were isolated from the skins.[53,54]

The results from the sheep-skin study were published in 2003 after a major struggle with Sprott. He implied that the work was faulty and biased because researchers received funding from the sheepskin industry (there was none), and tried to block publication by threat of legal action directed at anyone connected with the publication, the authors, the publishers and editors of the journal, and university administrators from the president down. He wrote in a letter to the publishers: "It is quite clearly legal negligence for researchers/research journals to publish a known false conclusion in a research paper."[55] The publishers did not back off. Finally in April 2003, Sprott prepared a press release criticising the work even though it was not yet in the public domain. The resulting newspaper article managed to turn the story completely around to fit the headline "COT DEATH: A Canadian Study Gives New Weight to an NZ Campaigner's Theory." Letters of protest to the newspaper from bemused Canadian authors remain unanswered.

Other researchers have encountered similar problems.[28,50,56]

One finding that could be taken to support the Toxic-Gas Hypothesis originated in Scotland, where scientists found that the reuse of old mattresses resulted in a 2.5 times increase in the number of babies likely to die of SIDS.[57] The association was stronger if the mattresses came from a different home. The Scots considered the possibility that the increase was associated with the increased microbial population and the associated risk of disease, but there was not enough information to establish cause and effect.

4.5 Other Proponents of the Toxic-Gas Hypothesis

The rhetoric has not been muted in recent years, and SIDS is frequently discussed on a number of web sites. Anxious parents look at this material and write: "There is a lot of conflicting info around about how to put a baby to bed these days and it's a bit hard to sort out."[58] One weekly newsletter, posted by Dr. Joseph Mercola in November 2000,[59] summarises a book *The Infant Survival Guide: Protecting Your Baby from the Dangers of Crib Death, Vaccines and Other Environmental Hazards*.[60] In the book the Toxic-Gas Hypothesis is presented as the primary cause for SIDS, and SIDS can be prevented either by proper mattress wrapping or by using bedding materials that are free of phosphorus, arsenic and antimony. The hype is extraordinary: "The neurotoxic gases based on phosphine, arsine and stibine are about one thousand times more poisonous than carbon monoxide, which can kill a person in a closed garage with a running engine. They are about as toxic as sarin, used in the 1980s Iran–Iraq war and in a Tokyo terrorist subway poisoning in 1995. In probably the worst environmental disaster of the 20th century, these toxic gases have killed about one million victims of SIDS worldwide." The authors give a Biblical reference, Leviticus 14, to a pinkish stain (mildew) on house walls – first thought to be a mark of "leprous plague" – that they liken to the stain seen on babies' mattresses. The Biblical instruction are to abandon the house for seven days to see if the leprous mark gets bigger; if it has, the house is cleaned, the offending stones removed, and the wall replastered. If the stain returns, the house is demolished.[61] Sprott[14] explained the Biblical phenomenon by saying that the mortar used at that time was "pozzolana," a natural cement derived from volcanic regions where soils frequently contain arsenic. The growth of mildew could therefore have generated arsine (or trimethylarsine) with deadly results, as supposedly occurred in the Forest of Dean incident (Section 3.9).

With respect to the Limerick Commission, and its conclusion that the Toxic-Gas Hypothesis was unsubstantiated, Smith and Hattersley[60] say, "The commissioners, who were already biased against the theory, knew that revealing the truth could subject the British government to millions of pounds in liability lawsuits. Such lying for financial reasons, sadly, is the rule rather than the exception throughout medicine and science."

Less strident disciples of the Toxic-Gas Hypothesis, such as Dr. Denton Davis[62] repeat the arguments and recommend mattress wrapping. Dr. Davis believes that there is a connection between the high concentrations claimed for arsenic and antimony in children and the incidence of autism.[63]

4.6 Toxicity and Related Considerations

To assess a potential chemical risk, investigators need to have a source, a hazard and a receptor, and all three must be in the same place. In the case of trimethylarsine, we need to consider the risk associated with the microbial production of a small amount of a nontoxic gas (Section 3.10) within the environment of a sleeping infant. That risk is essentially zero. If you add to this the fact that, even in the best conditions, the production of the gas is inefficient, and that it is formed very slowly and in low concentration, and only when the amount of available arsenic is high, the risk of poisoning an infant becomes even lower. If the risk is exposure to trimethylantimony, there is every reason to believe that the toxicity of this compound would be much the same as trimethylarsine.[64,65] However, even if it were 100 times as toxic as trimethylarsine, this would be balanced by the fact that production of this gas is more than 100 times less efficient. In addition, only very small amounts of antimony can be leached from mattresses, under extreme conditions,[41] so there is no fuel for a fungus such as *S. brevicaulis* to work on. Again, there can be no risk to an infant.[40]

This leaves us with phosphorus and phosphine gas production. There is no evidence for the biological production of any of the phosphorus compounds shown in Table 4.1, so in this case the hazard does not exist. For the record, phosphine is about 25 times less toxic than arsine.[66]

What is known about SIDS today? The US National Institute of Child Health and Human Development advises that the Back-to-Sleep message still holds and that babies should be living in a smoke-free environment, be breast-fed and sleep on a firm mattress.[67] In addition, in 2005 the American Academy of Pediatrics advocated the use of pacifiers at nap times and at bedtime during the first year and was contemplating a ban on having babies sleep with their parents.[68]

The diagnosis of SIDS is difficult, and local cultural considerations, such as not permitting autopsies, can compound the problem. The future does not look rosy from an understanding or prevention point of view. One attractive model for the pathogenesis of SIDS, known as the triple-risk model, is based on the concept that the infants are not entirely "normal" prior to death, but possess vulnerabilities that put them at risk of sudden death (*e.g.* Filiano and Kinney[69]). The model requires a vulnerable infant, a critical development period in homeostatic control, and an external stressor. Thus, the external stress need not be the same for all deaths. Recent results at the genetic level suggest that environmental challenges and an abnormal gene carried by one in nine black children confer a 24-fold increase in SIDS in infants who receive a copy of the gene from each parent. Black infants are three times as likely to die of SIDS as white ones, and six times more likely than Latinos and Asians.[70]

Both Richardson and Sprott continue to publicise the Toxic-Gas Hypothesis. Sprott is taking full credit for the drop in New Zealand SIDS rates after 1994 because mattress-wrapping materials provide an impermeable membrane, free of arsenic, antimony and phosphorus. A big difficulty here is that the mattress

Arsenophobia 145

covers, at a cost of about $20 (US), are primarily used by the financially well-off who have fewer cases of SIDS in the first place. SIDS has evolved into a disease of the disadvantaged.[71] Sprott wrote to speakers at the 2006 SIDS International Conference claiming that the "success of his mattress wrapping campaign," no deaths from the 140 000 or so babies who have slept on them, proves the Toxic-Gas Hypothesis.[72] As for Richardson, in February 2002, he wrote: "My hypothesis and associated recommendations have become widely recognised around the world and SIDS rates have decreased as a consequence."[73] In May of the same year, Richardson restated his hypothesis somewhat ambiguously. "I cannot agree with the suggestion that the generation of stibines from cot mattress materials causes chronic poisoning in infants. I believe that deaths are caused by circumstances such as overwrapping, overheating through various reasons, and prone sleeping which results in acute poisoning. ... This same effect can be caused by phosphines and arsines, but some infants have a greater susceptibility to this anticholinesterase poisoning." He also had not given up his belief in the accuracy of his original results with test papers.[74]

In conclusion, a recent study established that *S. brevicaulis* does not break down OBPA so the original postulate on which the Toxic-Gas Hypothesis was built has no foundation.[34] Richardson's Hypothesis is an example of rampant arsenophobia: once it started, it took on a life of its own. Total Home Environment, a high-end houseware shop, opened in Vancouver BC, in 2003 and sells bedding for baby cribs made from organic materials, wool and cotton. Each mattress costs $418 (Cdn) plus tax, and they are flying out the door. Why? Because sales staff are telling customers that some research links ordinary mattresses with gas production and SIDS.[75] The Toxic-Gas Hypothesis was resurrected by some doctors in Germany in 2006[76] and in the USA the antimony and arsenic content of mattresses remains a concern that is exploited by the activist group "People for Clean Beds" which was founded by Mark Strobel whose company manufactures beds that can only be purchased in the US with a doctor's prescription.

4.7 The Death of Napoleon I of France

We have just seen that in an attempt to provide a badly needed explanation for the deaths of infants by SIDS, society was handed a rickety crutch based on arsenophobia. Some members of society also have a compelling need for an explanation of the death of their heroes, because heroes can't possibly die from natural causes such as old age or disease – someone has to have provided a helping hand. Arsenophobia provides all manner of imaginative death scenarios.

4.7.1 Was it the Arsenic in the Wallpaper?

Any mention of the death of Napoleon is frequently met by the response: Didn't he die of arsenic poisoning? It is true that Napoleon was so worried about being poisoned by his British captors during his exile on the island of St. Helena that he refused to eat anything that was not prepared by his own, presumably loyal, staff. It is also true that his final will contained the words, "I die prematurely, murdered by the British oligarchy and their hired assassin."

The wallpaper-as-arsenic-source of poison made the headlines in 1982[77,78] when analysis of a sample of wallpaper from the living room in Longwood, Napoleon's residence on St. Helena, revealed arsenic concentrations of about $0.12\,g/m^2$. The paper was hung in 1819, during Napoleon's tenure, but it is not known if previous wall coverings used at Longwood were arsenical. Although the measured arsenic concentration was low in comparison with other wallpapers of the period (Section 3.3) the hype was high when the BBC aired a program, *The Strange Story of Napoleon's Wallpaper*. David Jones from the University of Newcastle-upon-Tyne, UK, presented the evolution of Gosio gas from this wallpaper as a viable cause of illness. "Hundreds of luckless householders developed the various symptoms of arsenic poisoning, and quite a number died," Jones wrote of Napoleon's contemporaries in Britain and elsewhere. He concluded that the gas evolution could not have posed a serious threat to Napoleon's life, but it may have contributed to his illness, and provides a plausible explanation of why large amounts of arsenic were found in his hair (Section 4.8).[77] But as we saw in Chapter 3, the argument has little foundation, although the damp climate could well have resulted in mould growth and a case of sick-building syndrome. Illness was common at Longwood and bad health was the fate of many on the island. The mortality rate was high among the troops even though some were stationed there to recuperate from previous illness.[79]

4.7.2 The Autopsy

Napoleon was healthy when he was forced into exile on St. Helena in October 1815. St. Helena was actually a busy port, in spite of its isolation in the middle of the Atlantic Ocean. Once there, he got fat and lethargic, probably through lack of exercise. He was initially treated by a British naval surgeon, Barry Edward O'Meara, who lost his job and was dismissed from His Majesty's Service, because he said Napoleon was ill, possibly with hepatitis. Sir Hudson Lowe, the governor of the island, did not welcome this news because he insisted that his charge was living happily and healthily in a tropical paradise, whereas the French were claiming the reverse. Between July, 1818, and September, 1819, Napoleon reluctantly admitted another doctor, Stockoe, on five occasions. Stockoe was also dismissed by the governor because of his diagnosis of hepatitis. After this, Napoleon was treated by Francesco Antommarchi, a Corsican doctor trained in Italy. Antommarchi was chosen by Napoleon's family. He had little skill and accepted the hepatitis diagnosis, which again displeased Napoleon's captors who continued to insist that all was well. Around October, 1820, the patient began to decline rapidly and in the last five weeks of

Napoleon's life, Archibald Arnot, the senior army doctor on the island, was called in for consultation. Napoleon did not like doctors but he got on well with Arnot. However, Arnot did not agree that his patient was in danger; some claim for political reasons, others claim stupidity. Napoleon just got worse.[80]

The autopsy report following his death on May 5th, 1821, aged 51, stated that Napoleon had a chronic stomach ulcer but died of a cancer of the stomach. This conclusion would not have surprised Napoleon, as one of his parents had died this way. The official autopsy report on Napoleon, as amended by Sir Hudson Lowe, reads in part as follows:

"The body appeared very fat. . . . The heart was of the natural size but covered with fat . . . and exposing the stomach, that viscus was found the seat of extensive disease, strong adhesions connected the whole superior surface, particularly about the pyloric extremity to the concave surface of the left lobe of the liver, and on separating these, an ulcer which penetrated the coats of the stomach was discovered one inch from the pylorus sufficient to allow the passage of a little finger. The internal surface of the stomach to nearly the whole extent was a mass of cancerous disease or schirrous portions advancing to cancer, this was particularly noticed near the pylorus. . . . The stomach was found nearly filled with a large quantity of fluid resembling coffee grounds."

Doctor Antommarchi, who actually conducted the autopsy, wrote an independent report because the other five doctors, all British, did not want to write that the liver was enlarged, which would support Antommarchi's conclusion that the climate on St. Helena was unhealthy and contributed to Napoleon's premature demise. Such a diagnosis would place the blame on the British. There were similar problems with the earlier suggestion of hepatitis. The "cancer" verdict was acceptable to both the French and the English because it implied no blame on either side.[81]

Since 1821 in addition to cancer, Napoleon's demise has been attributed to everything from malaria, to syphilis, to schistosomiasis.[82]

In a book provocatively titled, *Assassination at St. Helena: The Poisoning of Napoleon Bonaparte*, Forshufvud and Weider claim that none of the six doctors officially present at Napoleon's autopsy diagnosed cancer. The insertion of that diagnosis was one of the main additions to the original report made by the governor of the Island, Sir Hudson Lowe. The authors suggest that the doctors were actually describing the massive corrosion of the lining of the stomach that would be expected if Napoleon had been poisoned.[81,83]

4.7.3 Arsenic Poisoning?

In 1955, Sten Forshufvud, a Swedish dentist who was a great admirer of Napoleon, came to the conclusion that the bouts of sickness Napoleon suffered toward the end of his life could be correlated with his dietary intake, as recorded in the then-newly published diaries of Napoleon's chief valet,

Louis Marchand. Forshufvud became convinced that arsenic had been administered to Napoleon in acute doses over a long period of time. In all, he noted 32 supposed symptoms of arsenic poisoning. "On those especially painful days he displayed symptoms of typical arsenical intoxication of an acute nature: palpitations of the heart, a weak and irregular pulse, very severe headache, an icy chill in his legs extending right up to his hips, pain in the shoulders and back, pain in the liver, a persistent dry cough, loosening teeth, a coated tongue, severe thirst, skin rash and pain in the legs, a yellow skin, yellowed whites of the eyes, shivering, deafness, sensitivity of the eyes to light, spasmodic muscle contractions, difficulty in breathing and nausea – they were all there, today's accepted, recognisable symptoms of arsenical poisoning. But still it was the damnable climate of the island – a place that British seamen were sent to regain health from sieges or illness – that was blamed for the Emperor's decline."[83]

Arsenic was available to the inhabitants of Longwood to kill rats, but was never used for that purpose, possibly because the smell of the dead rats would have been too much to bear.[83] But Forshufvud could not prove his initial poisoning thesis in 1955 without access to the body, which was a challenge because Napoleon's remains lie under the Dôme des Invalides in Paris within six coffins (tin, mahogany, two of lead, ebony and oak), enclosed in a dark red porphyry sarcophagus that rests on a green granite pedestal. However, he saw some support for arsenic poisoning in reports that the corpse of the Emperor showed little sign of decomposition when his tomb on St. Helena was opened 19 years after his interment, in order to identify the corpse before returning it to France as the Emperor desired (Section 2.8). Although the clothing on the figure was in tatters, Dr. Guillard, the supervising physician, said that the features were so little changed that the face was instantly recognisable. He merely looked like he was asleep (Figures 4.1 and 4.2).

Box 4.1   Neutron Activation Analysis (NAA)

In this analytical technique, induced radioactivity is used to analyse otherwise nonradioactive substances. The sample is first bombarded with neutrons when some atoms in the sample become radioactive. The source of the neutrons is usually a nuclear reactor because the neutron flux density has to be high, (10^{11}–10^{13} neutrons/cm^2). In the case of arsenic, a neutron is captured by an arsenic atom (^{75}As) to give a heavier but unstable atom, ^{76}As. The latter is radioactive and decays with emission of beta particles and gamma rays, which can be measured. The intensity of the radiation, number of counts per second, is related to the concentration of the radioactive arsenic atoms in the sample.

Around 1960, Dr. Hamilton Smith at the University of Glasgow, who had access to the atomic reactor at the Harwell Atomic Research Station near Oxford, England, was conducting research on the arsenic content of human hair. He established that the average arsenic content of 1000 random samples collected in Glasgow was 0.81 ppm (median 0.51 ppm) with a range of 0.03–10 ppm. Most of the samples, 99 per cent, had less than 4.5 ppm.

Box 4.1 Continued.

One industrially contaminated sample had 74 ppm.[84] The method that he used for the analysis was as follows:[85] The hair sample was weighed and irradiated with neutrons. In order to separate the radioactive arsenic atoms from any other radioactive elements, the hair was dissolved in acid, a small amount of nonradioactive arsenic was added as a carrier, and the resulting solution was treated with zinc and acid to release all the arsenic as the gas arsine, which was trapped in mercuric chloride solution. This is essentially a variation of the Marsh Test (Section 5.4). The activity of the solution was counted using a Geiger counter. A weighed sample of arsenic trioxide, which served as a standard, was also irradiated for the same length of time. The irradiated arsenic trioxide standard was then dissolved in alkali and the solution counted to provide a reference. The sensitivity of the method was about 10 times better that any other method available at the time.

Smith also studied the arsenic distribution along single hairs and found that broad peaks of activity could be mapped that seemed to correspond with growth during periods of exposure to arsenic. "The method can be used to determine areas of high arsenic concentration and approximately when the arsenic was taken" (Section 1.9).[85,86]

Figure 4.1 *The Translation of Napoleon's mortal remains: Exhumation at St. Helena* by André Claude Boissier (1760–1833) with permission from the Napoleon print collection of McGill University, Montreal, QC.

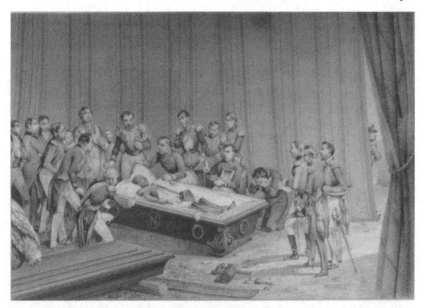

Figure 4.2 One of a set of eight lithographs by Victor Adam (1801–1866), *The Transfer of Napoleon's Remains to Paris*, with permission from the Napoleon print collection of McGill University, Montreal QC.

In life, Napoleon liked his hair cut short and earned the nickname "le Petit Tondu," the little crop-head. Napoleon's hair was shaved off the day after he died to facilitate the preparation of a death mask and to provide souvenirs of his passing. A very excited Forshufvud, who had heard about activation analysis (Box 4.1), was able to acquire one of Napoleon's hairs for analysis. The 2 mg sample was found to contain 10.38 ppm of arsenic. This result, "high by comparison with the normal mean arsenic content of about 0.8 ppm," was used to back up the claim that the Emperor had been poisoned by administration of arsenic.[82] No mention was made of the result from another hair sample dated 1805, which was analysed at the same time and contained essentially the same amount of arsenic, 10.53 ppm. Subsequently, another of Napoleon's hairs, with an average arsenic value of 4.91 ppm, when sectioned, was claimed to indicate that Napoleon was exposed to arsenic intermittently and that the timing of the exposure seemed to match the recorded history of Napoleon's periods of ill health; although no details were given.[87] The poisoning hypothesis was published in a book *Who Killed Napoleon*.[83,88]

The "definitive" paper on the poisoning hypothesis was published two years later in 1964.[89] In spite of their earlier reservations about the analytical method, authors concluded there was "no doubt Napoleon was exposed to arsenic on many occasions, at least 40, between the summer of 1820 and April 1821, because, on one hand, there is no record of clinical administration and, on the other, it is well documented that throughout his exile on St. Helena, Napoleon steadfastly

Arsenophobia 151

refused to take any internal drug prescribed by his surgeons." He distrusted medicine of that era and dubbed it *"la science des assassins."*[89]

The concentration of arsenic in the 1mm sections of the hair dated 1821 ranged from 4.4 ppm to 23.0 ppm; the 1818 hair, from 26 ppm to 18 ppm; the 1817 hair, from 1.8 ppm to 4.9 ppm; but the 1816 hair, cut in January, showed particularly high concentrations with sections ranging from 25.2 ppm to 60.0 ppm, indicating that Napoleon was exposed to "substantial amounts of arsenic on several occasions before being exiled to St. Helena."[89] But a few years later the conclusion was much more conservative following the analysis of other samples of Napoleon's hair that were also cut in 1816. This time we find: "The distribution of arsenic in the hair is similar to that found after the daily ingestion of excessive amounts of arsenic.... There is no evidence for any single massive dose during 1816."[90]

4.7.4 The "Real" Cause of Napoleon's Death

It would be reasonable to assume that the proponents of the poisoning hypothesis would have claimed that Napoleon actually died of arsenic poisoning, but this is not the case. The plot was much more devious. Forshufvud and Weider posit that arsenic was administered in what they refer to as the cosmetic phase of the crime, to keep Napoleon on his sick bed, to destroy his capacity to think and work, and to estrange him from his captors. This was followed by the lethal phase in which Napoleon actually died as a result of "acute intoxication by mercuric salts that had caused swelling of the pylorus muscle, large corrosion of the stomach leading to massive gastric haemorrhage, and toxic damage of body tissues, including the brain cells."[82,83,89]

Forshufvud and Weider credit the putative poisoners with looking into the literature for inspiration, where they would have discovered that the Marquise de Brinvilliers dispatched her father, who had been weakened by many low doses of arsenic, by finally administering tartar emetic (potassium antimony tartrate) (Section 5.3). To avoid suspicion she first ensured that the emetic was actually prescribed by a doctor; all she had to do was up the dose. According to the authors, Napoleon's assassins went one better. First the arsenic and tartar emetic routine was observed, with the latter being prescribed by the doctors and administered March 22nd, 1821. The dose was repeated the next day, and some more may have been given subsequently without Napoleon's approval. Next, the doctors were persuaded to administer a dose of the purgative, calomel (mercurous chloride), to the now heavily constipated and weakened victim who was also being given an almond-based drink, known as orgeat, as a tonic. The tartar emetic, calomel, and orgeat – although each said to be harmless in their own right – supposedly had the capacity to do great harm when given together. First, the natural expelling reaction of the stomach becomes inhibited by the orgeat, so that when calomel is administered with the orgeat, the mixture is not rejected by vomiting. In the stomach, the bitter almonds in the drink act as a source of cyanide.[83] According to Weider[93] the calomel (mercury chloride [sic]) combined with the cyanide from the orgeat to form mercury cyanide "which would then be expelled from a healthy stomach by vomiting. But Napoleon had been given several drinks containing a large quantity of tartar emetic and this would have inhibited the vomiting reflex. Consequently, the highly toxic mercury cyanide was retained." Death followed soon after.

4.7.5 Who Did It?

According to Forshufvud and Weider, the poisoning of Napoleon, orchestrated by the king of France, started long before St. Helena. The symptoms he showed at the battle of Waterloo were similar to those at his other critical battles of Borodino, Dresden and Leipzig: "He could not overcome Wellington and arsenic trioxide too."[83] Once Napoleon was on St. Helena, the Count d'Artois – who was the brother of the reigning king Louis XVIII and later became Charles X – is said to have directed the poisoning from France. He was able to install the Count de Montholon as Napoleon's field marshal on the island where, in this capacity, he was the only one with keys to the wine cellar. The authors propose that it was Montholon's task to gain Napoleon's complete trust while slowly poisoning him with arsenic, mainly via the wine. Montholon's wife became Napoleon's mistress and probably bore him two children on St. Helena: Napoleone-Marie, born in 1816 and Napoleone-Josephine, born in 1818. Montholon is said to have written to his wife after she left the island, "Calomel will soon end the Emperor's gardening activities." Montholon remained on St. Helena and assisted in drafting Napoleon's will to become a major beneficiary. He was said to have been involved in the decision to give the fatal dose of calomel over the objections of Doctor Antommarchi. Calomel was given without Napoleon's knowledge or approval on May 3rd. The dose of 10 grains is described as 40 times the normal dose[83] or five times the normal dose[91] depending on the particular bias of the authors. Shortly thereafter Napoleon passed out and died two days later, without regaining consciousness.

The conspiracy theory also requires Montholon to be the source of the many illnesses that plagued the members of Napoleon's court, as well as the murderer of the butler Cipriani in 1818 whose body strangely disappeared immediately after burial. It is also extended to account for the premature death of Napoleon's son, the Duke of Reichstadt, in 1832, also by poisoning.

The poisoning hypothesis has been the subject of four books[81,83,88,92] and innumerable presentations by Ben Weider (*e.g.* Weider[93]).

4.8 Some Analytical and Chemical Problems

4.8.1 The Preservation of the Corpse

This topic is discussed more fully in Chapter 2 (Section 2.8). Dr. Guillard, who supervised Napoleon's exhumation, reported that the clothing was in tatters but the corpse was well preserved: "The appearance was of one recently interred and the features so little changed that the face was instantly recognised by those who

Arsenophobia 153

had known Napoleon in life."[80] The stylised depictions of the disinterment (Figures 4.1 and 4.2) suggest that Napoleon was in better shape at that time than he was on the day he died. Clearly the painting of the event by Boissier shows more reverence than reality. Adam's representation (Figure 4.2) is a little more believable. Notions of incorruptibility probably played a part in the eyewitness accounts, but most of the credit should be given to the coffins. His body had been placed within four of them (solder-sealed tin, wood, solder-sealed lead, wood) without the use of any embalming agent.[83] These coffins are depicted in another version of the exhumation published by Weider[93] that shows the Emperor undefiled by death.

4.8.2 The Lethal Phase

There are considerable problems with the chemistry of the so-called lethal phase. Orgeat, the proposed source of cyanide, is made from crushed almonds that are separated from the oil, and only a few bitter almonds are included. A modern version of the recipe is available.[94] One web site run by reputable scientists suggests that several kilograms of bitter almonds would be needed to yield a lethal dose of cyanide.[95] However, even if cyanide was delivered as a consequence of drinking the orgeat, its toxicity would be reduced by its reaction with calomel (mercurous chloride). The expected product from this reaction, mercurous cyanide is unstable and decomposes to mercuric cyanide and mercury. Mercuric cyanide actually has a relatively low human toxicity: the lowest published lethal dose for human is 10 mg/kg.[96] Napoleon was given a "heroic dose"[83] of 0.6 g of calomel, which would yield at the most about 0.3 g of mercuric cyanide and 0.26 g of mercury. It is unlikely that this would have caused Napoleon much of a problem. The small amount of mercury would hardly be enough to account for the blackness of the enormous, copious, stools produced by the patient in response to the dose of calomel.

4.8.3 The Hair Analysis

It has been said that there is more Napoleonic hair in circulation than could ever have grown on his head, so before we start this section we need to bring up the thorny question of authentication. "My high value comes from a genuine hair, your low value comes from a fake", has been said more than once in this debate.[97] DNA samples would be required for unambiguous identification, but Napoleon's remains lie in that impregnable tomb in Paris. Comparison of the DNA from hair samples, to establish at least some common source, has yet to be performed. It seems that Madame Tussaud's Wax Museum in London once owned a mattress stained with Napoleon's blood that could have been a source of DNA; unfortunately, it, the mattress, was destroyed by fire in 1925.

Needless to say, Weider is crusading to open the tomb to get some DNA.[98] However, this might uncover yet another layer in the poisoning/conspiracy theory, because the claim has been made that it was the disappearing butler, Cipriani, who was buried a second time with military honours, while Napoleon's corpse was hidden in the crypt of Westminster Abbey. The body switch, the poisoning proponents claim, was made to conceal the fact that the British killed Napoleon, and the French went along with the plot for the good of diplomatic relations. This fantasy is based on supposed discrepancies in the records: Napoleon was buried in St. Helena inside three coffins. When exhumed, there were four. His teeth, notoriously yellow and rotting in life, were gleaming white. The urns containing his heart and stomach, which were said to be placed in the corners of his coffin, were found between his legs. Funeral masks, which multiplied after his death, look more like the butler than the Emperor.[99]

So much of the argument is based on the hair analysis that we should look at a study by Rita Cornelis from the University of Ghent, Belgium, "Neutron Activation Analysis of Hair: Failure of a Mission," which says quite a lot about the status of the analytical technique at the time of the first poisoning theories.[100] Her 1973 NAA results revealed that the distribution of arsenic over a population in Belgium was 0.052 ppm to 0.534 ppm, lower than that found by Smith in the early 1960s, and she finds a wide variation in the concentration of arsenic in hairs taken from the same head. She was uniquely fortunate to have access to an archive of hair samples collected from two brothers, identified as A. P and J. P, over the period 1898 to 1924. The variation of arsenic concentration over the years was dramatic and "spiky," in the range of 0.69 ppm to 9.32 ppm, with an average value of 1.49 ppm. Her conclusion: "The trace elements present in hair are not as constant, nor are their concentrations as specific, as was suggested by the enthusiasts who introduced neutron activation analysis so prematurely in court. It is not possible to relate a given sample of hair to a given individual."[100]

Lander and coworkers[101a] came to a similar conclusion following a study of hair samples taken from 254 people admitted to hospital in South Australia, between 1946 and 1963, suffering from acute arsenic poisoning. Evidently there was an unrestricted availability of highly toxic weed and pest control preparations (Section 2.3). One of their subjects was a child who died after an older brother put a shaving brush into his mouth – the brush had been dipped in a pool of weed killer on a path. The conclusion of the study: "The detection of arsenic in hair or nails in cases of poisoning may bear no relationship to the time or duration of ingestion. We conclude that the cherished concept that arsenic in hair and nails is a reliable pointer to chronic or remote ingestion is incorrect, and that it is impossible to separate these types of cases on the basis of this investigation. The reliance placed on the presence of arsenic in hairs and nails and on the distribution of arsenic along the length of these structures as being a reliable indication of the time of ingestion is evidently unjustified. Medical testimony concerning this point in past cases must be reviewed with suspicion." (To be fair, time course studies do seem to be more reliable when modern techniques are employed.[101b])

Forshufvud and Weider[83] assumed a growth rate of hair of 0.43 mm per day (13 mm per month). The rate of hair growth, however, is not constant and varies from about 0.2 to 0.5 mm per day. So when all this is put together the supposed correlation of precise dose with precise time of illness is probably fantasy.

The protagonists' publications generally assume that the background level of arsenic in hair in the early 1800s would be similar to that found today, at less than 1 ppm. Although the data are limited, the background at the time of Napoleon seems to have been higher, at about 3.8 ppm.[102]

Smith and coworkers were well aware that not all arsenic in hair was the result of ingestion: some women who had about 40 ppm of arsenic in their hair were washing it in a detergent that contained 74 ppm arsenic.[84] However, they claimed, not on very secure grounds, that none of their samples of Napoleon's hair had been subjected to external exposure and that the arsenic in their samples resulted from excessive intake.[103]

They were soon contradicted by results of 1.4 ppm arsenic from hair samples that were washed prior to analysis and published in a study entitled "Napoleon Bonaparte: No Evidence for Chronic Arsenic Poisoning."[104] Lewin and coworkers also found a high antimony concentration of 6.7 ppm in the same sample that may have reflected the doses of tartar emetic given to Napoleon in March prior to his death. Because the poison hypothesis requires the administration of arsenic right up to the end, the presence of antimony and the lack of arsenic are significant. Washing had no effect on the results. The authors suggest that earlier methods may have resulted in high arsenic numbers because of the small sample size and the possible inclusion of counts from radioactive antimony and bromine. This NAA study by Lewin and coworkers employed direct counting methods. The technique had been improved to the extent that it was no longer necessary to separate the arsenic from the sample after irradiation and before counting.

The US FBI entered the contest, with two results of 33.5 ppm and 16.8 ppm, from Napoleon's hair dated from 1816. They certified their results to be consistent with arsenic poisoning.[81]

Following the death of Forshufvud in the 1990s the controversy took on a distinctly Canadian aspect. Ben Weider became the spokesman for the pro-poisoning camp. Thomas Hindmarsh, a Professor of Pathology and Laboratory Medicine at the University of Ottawa, in collaboration with Philip Corso, of the Department of Plastic Surgery at Yale University, championed the death-by-cancer side.

Weider, a Montreal business man, runs his own physical fitness and sporting goods company. He lists interests in two disparate areas: body building[105] and Napoleonic history.[83,92,106] Weider is the founder and President of the International Napoleonic Society (www.napoleonicsociety.com) and in 1994 began an address as follows: "I am not here to defend the Thesis that Napoleon was poisoned on St. Helena because it is not a Thesis. It is a fact." Weider was honoured by the French government in 2000 for his Napoleonic work. There is also no doubt about the position of Hindmarsh and Corso: "Napoleon Bonaparte died on the island of St. Helena on May 5th, 1821, aged fifty one years, from carcinoma of the stomach, probably complicated by hematemesis. ... Possibly he died prematurely as a result of a dose of calomel."[91,97,107]

4.9 The Overall Picture

In 1998, Hindmarsh and Corso listed the results of 31 hair analyses. They averaged the results found for a particular hair if it had been sectioned for

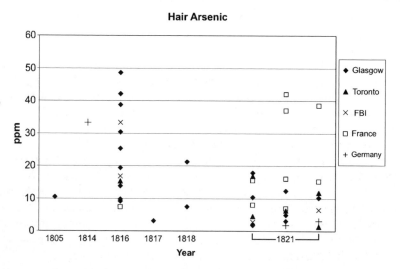

Figure 4.3 Arsenic content of samples of Napoleon's hair obtained during the years 1805 to 1821. He died in 1821. Results were obtained by Smith and coworkers, Glasgow; Hindmarsh and coworkers, Toronto; US FBI; Kintz and coworkers, France; Lin and coworkers, Germany.

analytical purposes.[91] These results are shown in Figure 4.3, together with data from ten more recent analyses.[108a,109,110]

Some of the newer analytical results generated fresh interest in June 2001, when the BBC and the Canadian Broadcasting Corporation carried the headline "Tests on Napoleon's Hair Back Poisoning Theory." The new evidence was released to a packed news conference in Paris. "The analysis showed there was major exposure, and I stress 'major,' to arsenic," said Pascal Kintz, a toxicologist who studied five samples of Napoleon's hair. The French rejected that the high levels could have come from other sources such as seafood or contaminated water. One sample from October 1816 contained 7.4 ppm, and four samples from May 1821, had values of 6.99 ppm, 15.20 ppm, 15.50 ppm and 38.53 ppm. These samples were washed to avoid the possibility of external contamination.[108a] Weider quotes these results as being definitive.[111] The same French group published the results for two more 1821 samples in 2007: 42 ppm and 37 ppm. A speciation study showed that this was essentially all inorganic arsenic.[108b]

Recent analyses performed in Germany,[109,110] employed NAA and the activity in the irradiated hair was counted directly, enabling the determination of a range of elements in the sample from one irradiation. (The decay curve is measured and fitted to the known decay rates of particular elements.)[112] The arsenic levels in Napoleon's hairs from 1821, 1.85 ppm and 3.05 ppm, are at the lower end of those reported for the St. Helena samples, Figure 4.3, and could be taken as an indication of normal exposure. Like Lewin and coworkers Lin *et al.* found elevated antimony concentrations in these samples. Hair from 1814, on the other hand, had high arsenic, 33.4 ppm; and mercury, 11.5 ppm; but low antimony, 0.37 ppm.

It is strange that, apart from the early reports from Glasgow, the neutron activation results from Toronto and Germany are significantly lower than those obtained by other workers who used more traditional analytical techniques. Unfortunately, the two methods have not been challenged by the same sample. Hindmarsh is of the opinion that " . . . hair arsenic is a useful test for confirming chronic arsenic poisoning in forensic cases, provided that external contamination has been excluded. However, its lack of clinical precision limits its usefulness in the evaluation of subjects exposed to arsenic from their water supply."[113] He quotes one Canadian study that revealed two individuals who were living in the same house and who were obviously suffering from arsenic poisoning: they had skin lesions and clinical neuropathy, but they had widely different hair arsenic levels, of 47 ppm and 4.2 ppm. He suggests as a guideline that chronic poisoning is indicated by hair concentrations greater than 10 ppm and that acute poisoning by concentrations greater than 45 ppm (Section 1.9.2).

Washing procedures developed by the International Atomic Energy Agency[112] are designed to remove material resulting from external exposure; however, Hindmarsh claims it is not possible to distinguish between external and internal application.[91,97,107]

Many values are above the 10 ppm line in Figure 4.3, and some are high enough to be in the chronic/acute range if the arsenic was acquired only by ingestion.

4.9.1 Sources of Arsenic

We should attempt to account for the fact that some hair samples contained elevated arsenic levels, some extremely so, while keeping in mind that some of the early work could be in error resulting from small sample size, especially the sectioned hair samples, and primitive methodology.

4.9.2 Wine and Water

One study of the arsenic levels in the hair of individuals exposed to high concentrations of the element in their drinking water found a range of 1 ppm to 32 ppm, with a mean of 6.75 ppm[114] (Section 8.2). The spring that was the source of water on St. Helena now contains little arsenic[91] and there is no reason to suspect that the concentration has altered over time. The water was used by the entire household, as was the food. Napoleon seems to have preferred meat over fish which would be a more likely source of arsenic, although of non-toxic species unlikely to result in elevated hair arsenic concentrations (Section 7.2).[115] Napoleon drank a South African wine called Constance that was specially imported from Capetown for his use on St. Helena. Apparently the shipments were organised by Count Emmanuel Las Cases, who originally accompanied Napoleon to exile on St. Helena but was forced to leave the island because of ill health. Perhaps the loyal Las Cases doctored the wine, but Montholon – who had the only key to the wine cellar – became the prime suspect in the assassination investigation. Apparently Napoleon occasionally made a gift of wine to his guests, some of whom later became ill, adding to the evidence against Montholon who is supposed to have confessed on his deathbed to poisoning Napoleon.[116] If true, the dose must have been carefully monitored to achieve the desired effect. It is said that Napoleon was not a big drinker, although records of his eating habits on Elba, on display at the Villa San Martino, suggest that he was well supplied. We have seen that beer drinkers in Manchester, England (Section 3.12) noticed that 19 ppm arsenic produced very noticeable immediate effects, and that one tenth of this produced nephritis, but these individuals were drinking a lot. Assuming Napoleon was drinking wine with an arsenic concentration of 10 ppm (probably noticeable), he would get about 3 mg of arsenic per day from a half bottle, about what he would get from a course of Fowler's solution. Assuming Napoleon's weight to be 85 kg and a lowest observed adverse effect level (LOAEL) of 0.05 mg arsenic/kg-day[117] Napoleon could probably tolerate 3 mg per day without problems for some time.

There is a good possibility that some of the arsenic in the 1814 and 1816 hair samples accumulated while he was on Elba. He resided there from May 5th, 1814 to February 26th, 1815, and had full sovereign rights. The geology of Elba is probably similar to that of Corsica to the west and the mainland of Italy to the east. Both terrains are rich in arsenic.[118,119] The Emperor insisted that drinking water from a mountain spring in Poggio be available for every meal. This could have been the source of the arsenic in some of the 1816 hairs (Figure 4.2). However, in 2006 the spring water was very low in arsenic, as was the bottled water from the same source labelled: Oligominerale Fonte NAPOLEONE, Poggio-Isola D'Elba.[120] The island's publicity booklets recommend the water for urinary and kidney problems.

4.9.3 Self-medication

Napoleon's use of arsenical preparations for medical reasons is generally ruled out by both sides of the poisoning controversy. They were unable to find reference to any such administration, self- or otherwise. However, there is a letter in the British Museum, dated July 16th, 1821 and written by Dr. James Watson Roberts who served as Deputy Inspector of Hospitals during the Napoleonic wars. He revealed that he had been informed that Napoleon had contracted leprosy (*Lepra Arabum*) in Egypt in 1803, for which he was taking large doses of arsenic.[121] This self-medication could account for the high 1805 values at least.

4.9.4 Arsenical Smoke and Preservatives

In addition to the presence of arsenic in the wallpaper pigment at Longwood (Section 4.7), the following have been suggested as external sources of exposure to the element: domestic and military smoke, cosmetics and preservatives for hair samples.

Arsenophobia 159

Smoke is a very remote possibility. Children living near a coal-burning power station in Romania had up to 10 ppm arsenic in their hair.[122] Residents of Changqing, China, who use arsenic-rich coal containing about 56 ppm arsenic for heating and cooking in unvented dwellings (Section 7.10), have on average 7.99 ppm arsenic in their hair.[123] But the British coal that was available on St. Helena is not high in arsenic and if Napoleon was exposed to arsenic-rich smoke, so was everyone else in the household. French scientists were struggling when they suggested that the arsenic in the pre-St. Helena samples could have come from exposure to wood smoke, or from guns "whose bullets contained the metal".[124] Napoleon was not often if ever in the heart of battle, and was occasionally out of it altogether because of illness.

So we are left with the real possibility, that some of the hair samples that were analysed had been preserved with arsenic (Section 2.8). A 1968 letter to *The Times* in London[125] stated: "It was the practice of craftsmen who made up locks of hair as mementos to add a little arsenic powder which acted as a preservative. On this simple basis a whole mythology regarding the death of Napoleon has been erected." This formed the basis for the claim from Hindmarsh and Corso: "We believe that arsenic must have been used to preserve the hair samples with As concentration greater than 30 ppm."[91,113]

Some of the hair samples were mounted professionally and hence, were more likely to have been treated with an arsenic preparation. Such samples would probably be proudly displayed as authentic and would be the first targets for analysis: they would also yield high arsenic levels. One hair sample from 1821 was donated to the protagonists by a Lady Holland.[108a] The strand was fixed to the centre of a medallion and had been stored in a golden box.[108a] The arsenic content was 38.53 ppm and would meet the suggested 30 ppm arsenic preservation threshold suggested by Hindmarsh, although this is probably unrealistically low.[47] The much-trumpeted study by Kintz and coworkers found only 6.99 ppm in another 1921 hair that came from the collection of the Museum of Arenenberg.[108a] Other museum samples have even lower arsenic concentrations.[109,110] The Capetown museum provided some of the hair analysed by Hindmarsh:[97] at 11.8 ppm, the concentration is not particularly high. Thus, it seems that the museum collectors did not routinely use arsenic to preserve the hair samples. Although others may have. The 1814 sample that is high in both arsenic and mercury could have been treated with a preservative.

The highest value reported from the hair samples is 42 ppm, so the evidence for aggressive use of arsenic preservatives is not particularly compelling. At the other end of the scale we have hairs that have a low arsenic content and should be taken as indicators of a lack of arsenic exposure. As mentioned above, the protagonists get out of this dilemma by claiming that these are not genuine samples. There is also the possibility that these hairs, or indeed any of the hairs, record exposure many months before the claimed growth period, because the hair is dead and is incapable of taking up freshly ingested arsenic.

It is possible that the results from the 1814 and 1816 hair samples are elevated because Napoleon drank contaminated water on Elba, and that self-medication might account for the 1805 result. However, we are left with a large number of samples that show somewhat elevated arsenic levels even if we accept the 3.8 ppm background level (Section 4.8.3). These concentrations seem to indicate that Napoleon was exposed to arsenic in some form while he was on St. Helena and Elba. So the answer to the question about whether he was being deliberately poisoned remains: maybe.

4.9.5 Arsenical Straws

Forshufvud and Weider believe that numerous attempts were made to poison Napoleon throughout his military career. Napoleon's health was never very good when he was on the battlefield and he was essentially *hors de combat* at Waterloo on June 18th, 1815. "Some of the symptoms could be taken as an indication of arsenic poisoning and the hair analysis indicated that he was possibly exposed of [sic] arsenic for many years." They offer this story to support their thesis. Following the Battle of Dresden on August 27th, 1813, Napoleon believed he had been poisoned by the garlic in his stew. In the normal course of events Napoleon would not have been served any garlic because his staff knew he did not like it. But if the garlic odour was actually an indication of the production of trimethylarsine by the action of the stew ingredients on added arsenic trioxide, the case is made, and "We [the authors] are forced to conclude almost without room for doubt that arsenic was added to the stew."[83]

And another: in December, 1820, vesicatories – blister-raising concoctions such as mustard plaster – applied to Napoleon's abdomen, did not act normally, and the flesh underneath turned cadaveric: "... Clearly a mortifying poison had been mixed into the substance of the vesicatory. It is easy to believe it was arsenic."[83]

4.10 Medical Evidence

Medical practitioners generally agree that because of the problems with the analysis of arsenic in hair – such as variation from individual to individual exposed to the same extent, variation in concentration over an individual's head, variations along hairs, and variations associated with exposure to external sources – it is unwise to make a diagnosis of arsenic poisoning based on hair-arsenic values alone. Medical evidence must also be considered.

Thirty-two symptoms of arsenic poisoning were noted in Napoleon by the pro-poisoning lobby in 1955 (Section 4.7). This list is certainly an indication that Napoleon was not in good health, but it does not include such primary indicators of chronic arsenic poisoning as hyperkeratosis and peripheral neuropathy. However, in support of arsenic exposure, Napoleon was balding and had no body hair when he died.

A lot has been written about the interpretation of the autopsy reports and how they were supposedly influenced by outsiders. The final verdict at the time was cancer but this was not a view shared by Antommarchi, Napoleon's doctor (Section 4.7). Medical doctors, real and amateur are still trying to interpret the written word and the background commentary.[79,91,106,107,126,127] The pro-poisoning camp interprets the autopsy report as an indication of corrosion by mercuric cyanide and mercuric chloride, but we have seen that this is somewhat far-fetched, given the chemistry supposedly involved. The pro-cancer school[91] writes with what seems to be unjustified conviction: "Microscopic diagnosis was not widely established until the end of that century, hence the uncertainty in some of the [autopsy] reports about whether the lesion was cancer or "scirrhous thickening fast advancing to cancer"; nevertheless, the reports clearly show that Napoleon had extensive scirrhous carcinoma of the stomach, probably complicated by partial gastric obstruction manifested clinically by intractable vomiting and hiccupping in the last few months of his life."

The one fact that was generally accepted is that Napoleon was fat when he died, and this is not usually associated with arsenic poisoning where weight loss is the norm.[91] However, it now appears that Napoleon actually lost a lot of weight just before he died. A study of the Emperor's pants, which had to be remade as he got slimmer, suggests that his weight dropped from 200 pounds to 165 pounds in the last five months.[128] This same international team of doctors received even more media coverage when they came to the conclusion that the initial diagnosis of death resulting from gastric cancer was probably correct. "Napoleon is likely to have had a long-standing *Helicobacter pylori* infection, which might have led to the development of a prepyloric ulcer that created the background for genesis of a gastric adenocarcinoma. A massive gastric haemorrhage that occurred in or around the advanced stage gastric tumour can be considered to be the immediate cause of his death."[129]

This takes us back to 1821 but it is probably not the last word.

References

1. H. F. Krous, J. B. Beckwith, R. W. Byard, T. O. Rognum, T. Bajanowski, T. Corey, E. Cutz, R. Hanzlick, T. G. Keens and E. A. Mitchell, *Pediatrics*, 2004, **114**, 234.
2. F. M. Sullivan and S. M. Barlow, *Paediat. Perinat. Epidem.*, 2001, **15**, 144.
3. P. J. Berry, E. Allibone, P. McKeever, I. Moore, C. Wright and P. Flemming, in *SUDI A confidential inquiry into stillbirths and deaths in infancy.*, 2003, vol. 17, p. 272.
4. P. J. Fleming, 2003, personal communication.
5. S. Knight, in *The Times*, 2005, June 21.
6. S. Freeman, in *The Sunday Times*, 2003, Dec. 14.
7. Q. Eastman, in *Science*, 2003, Vol. 300, p. 1858.
8. J. Willey, in *Daily Express*, 2003, Dec. 11, p. 6.
9. B. A. Richardson, *Wood preservation*, Construction Press and Longmans, London, 1978.
10. B. A. Richardson, *Cot mattress biodeterioration and sudden infant death*, Penarth Research International, Guernsey, 1989.

11. M. Robeers, 1989, personal communication.
12. B. A. Richardson, *The Lancet*, 1990, **335**, 670.
13. US EPA, *10,10'-oxybisphenoxarsine (OBPA)* EPA-718-F-93-003, Office of preservatives, pesticides, and toxic substances, 1993.
14. T. J. Sprott, *The cot death cover-up?* Penguin Books, Auckland NZ, 1996.
15. L. S. Limerick, *Expert group to investigate cot death theories: Toxic-Gas Hypothesis*, Final report, Department of Health, UK, London, 1998.
16. G. Charlot, *Colorimetric determination of the elements, principles and methods*, Elsevier Pub. Co., New York, 1964.
17. L. C. Thomas and G. J. Chamberlin, *Colorimetric chemical analytical methods*, Tintometer, Salisbury, UK, 1974.
18. P. Baker, *Volatile organometallics*, Laboratory of the Government Chemist, Teddington, 1995, Oct. 9.
19. D. W. Warnock, H. T. Delves, C. K. Campbell, I. W. Croudace, K. G. Davey, E. M. Johnson and C. Sieniawska, *The Lancet*, 1995, **346**, 1516.
20. T. J. Sprott, 1996, personal communication.
21. E. A. Mitchell and A. C. Engelberts, *The Lancet*, 1991, **338**, 192.
22. B. A. Richardson, in *Expert group to investigate cot death theories: Toxic-Gas Hypothesis Chair Lady Limerick*, London, 1994.
23. S. Cooper, in *The Independent*, 1995, Dec. 7.
24. B. McCutcheon, *Antimony*, www.nrcan.gc.ca/ms/cmy/content/1995/10.pdf.
25. B. A. Richardson, *J. Foren. Sci. Soc.*, 1994, **34**, 199.
26. A. Taylor and P. Hodges, in *Anal. Eur.*, 1995, Feb., p. 12.
27. P. J. Fleming, M. Cooke, S. M. Chandler and J. Golding, *Brit. Med. J.*, 1994, **309**, 1594.
28. R. B. Pearce, M. E. Callow and L. E. Macaskie, *FEMS Microbiol. Lett.*, 1998, **158**, 261.
29. F. Challenger and L. Ellis, *J. Chem. Soc.*, 1935, 396.
30. D. Barnard, Ph.D thesis, University of Leeds, UK, 1947.
31. P. Andrewes and W. R. Cullen, in *Organometallic compounds in the environment*, ed. P. J. Craig, John Wiley and Sons Ltd., Chichester, UK, 2003, p. 277.
32. W. R. Cullen and R. Bentley, *J. Environ. Monit.*, 2005, **7**, 11.
33. J. Feldmann and A. V. Hirner, *Int. J. Environ. Anal. Chem.*, 1995, **60**, 339.
34. P. Andrewes, W. R. Cullen, C. Q. Wang, E. Polishchuk and T. Liao, *Appl. Organom. Chem.*, 2000, **14**, 364.
35. R. O. Jenkins, P. J. Craig, D. P. Miller, L. Stoop, N. Ostah and T. A. Morris, *Appl. Organomet. Chem.*, 1998, **12**, 449.
36. R. O. Jenkins, T. A. Morris, P. J. Craig, A. W. Ritchie and N. Ostah, *Sci. Total Environ.*, 2000, **250**, 73.
37. J. Roels and W. Verstraete, *Sci. Total Environ.*, 2004, **327**, 185.
38. B. Schink, in *Metal ions in biological systems. Biogeochemical cycles of elements*, Marcel Dekker, 2005, vol. 43, p. 131.
39. W. R. Cullen and K. J. Reimer, *Chem. Rev.*, 1989, **89**, 713.

40. P. Andrewes, K. T. Kitchen and K. Wallace, *Chem. Res. Toxicol.*, 2003, **16**, 994.
41. R. O. Jenkins, P. J. Craig, W. Goessler and K. J. Irgolic, *Human Exp. Toxicol.*, 1998, **17**, 138.
42. R. O. Jenkins, T. A. Morris, P. J. Craig, W. Goessler, N. Ostah and K. M. Wills, *Human Exp. Toxicol.*, 2000, **19**, 693.
43. M. Thompson and I. Thornton, *Environ. Technol.*, 1997, **18**, 117.
44. C. R. Lehr, E. Polishchuk, M.-C. Delisle, C. Franz and W. R. Cullen, *Human Exp. Toxicol.*, 2003, **22**, 325.
45. T. J. Sprott, in *Cot Life*, 1998, July.
46. R. Wakefield, in *NZ Woman's Weekly*, 1996, Sept. 30, p. 26.
47. W. R. Cullen, 2000, personal experience.
48. *Weekend Herald*, 2000, Feb. 12, p. A8.
49. T. J. Sprott, in *Cot Life*, 1997, Dec.
50. A. P. K. Ford, P. J. Schluter and S. Cowan, *NZ Med. J.*, 2000, 8.
51. T. J. Sprott, 1996, personal communication.
52. R. Elfast, *NZ Med. J.*, 1995, **108**, 178.
53. A. J. d. Beaufort, A. T. Bernards, L. Dijkshoorn and C. P. v. Boven, *Acta Paediatricia*, 1999, **88**, 772.
54. A. Abbott, *Nature online* 2005, Aug. 10.
55. T. J. Sprott, 2003, personal communication.
56. C. Brett, in *North and South*, 1996, March, p. 50.
57. D. Tappin, H. Brooke, R. Ecob and A. Gibson, *Brit. Med. J.*, 2002, **325**, 1007.
58. C. v. Ginkel, 2001, personal communication.
59. J. Mercola, *Dr. Joseph Mercola's Health News Letter*, www.mercola.com.
60. L. H. Smith and J. G. Hattersley, *The infant survival guide: Protecting your baby from the dangers of crib death, vaccines and other environmental hazards*, Smart Publications, Petaluma, CA, 2000.
61. A. Kaplan, *The living torah*, Maznaim Publishing Corporation, New York, 1981.
62. D. Davis, *A simple explanation for SIDS*, www.mercola.com/1999/dec/19crib_death_explanation_treatment.htm.
63. D. Davis, 2001, personal communications.
64. J. Seifter, *J. Pharm. Exp. Therap.*, 1939, **66**, 306.
65. MSDS, *Trimethylantimony*, MDL Information Systems Inc., Nashville TN, 2001.
66. M. Sittig, *Handbook of toxic hazardous chemicals and carcinogens*, 3rd edn., Noyes Publications, Park Ridge NJ, 1991.
67. *Vancouver Sun*, 2004, Dec. 6, p. C4.
68. J. Grimaldi, in *Vancouver Sun*, 2005, March 12, p. E4.
69. J. J. Filiano and H. C. Kinney, *Biol. Neonate*, 1994, **65**, 194.
70. T. Maugh, in *Vancouver Sun*, 2006, Feb. 2, p. A15.
71. E. Mitchell, 2002, personal communication.
72. T. J. Sprott, 2006, April 20, personal communication.
73. B. A. Richardson, 2002, Feb. 21, personal communication.

74. B. A. Richardson, 2002, May 28, personal communication.
75. S. Reeder, in *Globe and Mail*, 2005, Oct. 25, p. L4.
76. A. V. Hirner, 2006, personal communication.
77. D. E. H. Jones, *New Scientist*, 1982, 101.
78. D. E. H. Jones and K. W. D. Ledingham, *Nature*, 1982, **299**, 626.
79. R. E. Gosselin, *Exhuming Bonaparte*, dartmed.dartmouth.edu/spring03/pdf/Exhuming_Bonaparte.pdf.
80. A. Chapman, in *Tenements of clay*, ed. A. Sorsby, Julian Friedman Publishers Ltd., 1974.
81. B. Weider and S. Forshufvud, *Assassination at St. Helena. Revisited*, John Wiley and Sons, Inc., New York, 1995.
82. S. Forshufvud, H. Smith and A. Wassén, *Nature*, 1961, **192**, 103.
83. S. Forshufvud and B. Weider, *Assassination at St. Helena. The poisoning of Napoleon Bonaparte*, Mitchell Press Limited, Vancouver, 1978.
84. H. Smith, *J. Forensic Sci. Soc.*, 1964, **4**, 192.
85. H. Smith, *Anal. Chem.*, 1959, **31**, 1861.
86. H. Smith, *J. Forensic Med.*, 1961, **8**, 165.
87. H. Smith, S. Forshufvud and A. Wassén, *Nature*, 1962, **194**, 725.
88. S. Forshufvud, *Who killed Napoleon?* Hutchinson, London, 1962.
89. S. Forshufvud, H. Smith and A. Wassén, *Archiv fur Toxikol.*, 1964, **20**, 210.
90. A. C. D. Leslie and H. Smith, *Arch. Toxicol.*, 1978, **41**, 163.
91. J. T. Hindmarsh and P. F. Corso, *J. Hist. Med.*, 1998, **53**, 201.
92. B. Weider and D. Hapgood, *The murder of Napoleon*, Congdon and Lattes, Inc., New York, 1982.
93. B. Weider, *The assassination of Napoleon*, members.tripod.com/amik78/Assassination-eng.htm.
94. Planet botanic, *Almond facts (Prunis dulcis)*, www.planetbotanic.ca/fact_sheets/almond_fs.htm.
95. N. Bunce and J. Hunt, *History of cyanide*, www.physics.uoguelph.ca/summer/scor/articles/scor176.htm.
96. NIOSH, *Mercury(II) cyanide*, National Institute for Occupational Safety and Health, Atlanta GA, 2000.
97. P. F. Corso and T. Hindmarsh, *Sci. Prog.*, 1996, **79**, 89.
98. Science, in *ABC Online News.*, 2001, June 6.
99. P. D. Broughton, in *Vancouver Sun*, 2002, Aug. 17, p. A22.
100. R. Cornelis, *J. Radioanal. Chem.*, 1973, **15**, 305.
101. (a) H. Lander, P. R. Hodge and C. S. Crisp, *J. Forensic Med.*, 1965, **12**, 52.
    (b) V. P. Guinn and R. Demiralp, *J. Radioanal. Nucl. Chem.*, 1993, **168**, 294.
102. I. M. Dale, J. M. A. Lenihan and H. Smith, *2nd International conference on nuclear methods in environmental research*, Columbia, 1974.
103. A. C. D. Leslie and H. Smith, *Med. Sci. Law*, 1978, **18**, 159.
104. P. K. Lewin, R. G. V. Hancock and P. Voynovich, *Nature*, 1982, **299**, 627.

105. B. Weider and J. Weider, *The edge. Ben and Jo Weider's guide to ultimate strength, speed, and stamina*, Avery Publishing, New York, 2002.
106. B. Weider and J. H. Fournier, *Am. J. Forensic Med. Pathol*, 1999, **20**, 378.
107. J. T. Hindmarsh and P. F. Corso, *Eur. J. Lab. Med.*, 1999, **7**, 135.
108. (a) P. Kintz, J.-P. Goullé, P. Fornes and B. Ludes, *J. Anal. Toxicol.*, 2002, **26**, 584.
    (b) P. Kintz, M. Ginet, N. Marques and U. Cirimele, *Forensic. Sci. Internat.*, 2007, **170**, 204.
109. X. Lin and R. Henkelmann, *J. Radioanal. Nucl. Chem.*, 2003, **257**, 615.
110. X. Lin, D. Alber and R. Henkelmann, *Anal. Bioanal. Chem.*, 2004, **379**, 218.
111. *National Geographic*, 2005, May, p. 10.
112. IAEA, *Trace element analysis in hair*. Report IAEA/RL/50, International Atomic Energy Agency, Vienna, 1978.
113. J. T. Hindmarsh, *Clin. Biochem.*, 2002, **35**, 1.
114. D. Das, A. Chatterjee, B. K. Mandal, G. Samanta, D. Chakraborti and B. Chanda, *Analyst*, 1995, **120**, 917.
115. N. Yamato, *Bull Environ. Contam. Toxicol.*, 1988, **40**, 633.
116. M. Keynes, *The Lancet*, 1994, **344**, 276.
117. T. A. Lewandowski and B. D. Beck, *Reg. Toxicol. Pharmacol.*, 2004, **40**, 372.
118. T. Guerin, N. Molenat, A. Astruc and R. Pinel, *Appl. Organomet. Chem.*, 2000, **14**, 401.
119. R. Barocci, *Maremma Avelenata (Poisoned Maremma) A report of an expected environmental catastrophe*, Margini series, Stampa Alternativa Publishing House, 2003.
120. W. R. Cullen and V. W.-M. Lai, 2006, unpublished results.
121. P. O'Leary, in *The Historian*, 1990, Spring.
122. V. Benko and K. Symond, *Environ. Res.*, 1977, **13**, 378.
123. A. Shraim, X. Cui, S. Li, J. C. Ng, J. Wang, Y. Jin, Y. Liu, L. Guo, D. Li, S. Wang, R. Zhang and S. Hirano, *Toxicol. Lett.*, 2004, **137**, 35.
124. *Daily Telegraph*, 2002, Oct. 30.
125. C. Farthing, in *The Times*, 1968, Aug. 12.
126. P. F. Corso, J. T. Hindmarsh and F. D. Stritto, *Am. J. Forensic Med. Path.*, 2000, **21**, 300.
127. B. Weider, *Am. J. Forensic Med. Path.*, 2000, **21**, 303.
128. A. Lugli, A. K. Lugli and M. Horcic, *Human Path.*, 2005, **36**, 320.
129. A. Lugli, I. Zlobec, G. Singer, A. K. Ligli, L. M. Terracciano and R. M. Genta, *Nature Clin. Pract. Gastroent. Hepat.*, 2007, **4**, 52.

CHAPTER 5
Arsenic and Crime: The Law of Intended Consequences

They put arsenic in his meat
And stared aghast to watch him eat;
They poured strychnine in his cup
And shook to see him drink it up;
They shook, they stared as white's their shirt:
Them it was their poison hurt.
– I tell the tale that I heard told.
Mithridates, he died old.

    A. E. Housman, 1896, *A Shropshire Lad*

5.1 Introduction

Mithridates VI, 133–63 BCE, was King of Pontus, a region south of the Black Sea and, like many in his time, lived in fear of being poisoned by his many enemies – or worse, by his friends. He experimented with poisons using his prisoners as subjects and became convinced that he had discovered a number of antidotes. He compounded them all into one universal antidote which became known as Mithridatium and contained about 41 ingredients. Before Mithridates was defeated by Roman invaders, he killed all his wives, daughters and concubines and tried to poison himself as well. But the poison failed to act because he had been taking his own antidote as a precaution. Instead, he elected to die by the sword of one of his obliging soldiers. After his death, his formula was taken to Rome, where an improved version was developed by one of Nero's personal physicians. This mixture included viper's flesh and became known as Theriac of Andromachus. These antidotes eventually became

---

Is Arsenic an Aphrodisiac? The Sociochemistry of an Element
By William R. Cullen
© William R. Cullen 2008

"universal cures" and were in medical use until the 18th century. The 1746 edition of the London Pharmacopoeia lists mithridatium and its 45 ingredients.[1]

5.2 Ancient Times

Toxic substances have been around a lot longer than mankind: many organisms use them for self-defence. It did not take long for humans to recognise that they could use these substances with nefarious intent. Some of the earliest records are found on Egyptian papyri written between 1900 and 1100 BCE. One of these, the Ebers Papyrus, lists all the drugs and poisons of the time. The Greeks of 500 BCE were familiar with arsenic, antimony, mercury, hemlock, aconite, and many more, and used poison for capital punishment. In a well known example, Socrates was sentenced to death in 402 BCE by drinking a concoction containing hemlock. (This practice of state poisonings continues today with death by lethal injection sanctioned in some US states – 35 of the 37 that allow capital punishment.)

The 2006 release of the movie *Alexander* prompted speculation about the cause of the hero's unexpected death at age 32 in Babylon, 323 BCE. Some have suggested that arsenic was the killer, although hellebore seems more likely.[2] Along the same lines, Attila the Hun may have been poisoned by his own men.[3]

Arsenic, aconite, opium and lead were well known as poisons in India and were used liberally. The fear of an untimely death supposedly led to the practice of *suttee* (also known as *sati*) – the burning of widows – instituted to discourage women from eliminating unwanted husbands. According to Indian mythology, Lord Brahma, one of the Holy Trinity of Indian Gods, is credited with the creation and distribution of poisons and their classification as vegetable, animal or mineral. This same classification was used by the Greek physician Dioscordes (49–90 CE).[4,5]

According to the Roman historian Suetonius, most of his contemporaries believed that the emperor Claudius was poisoned by his fourth wife Agrippina, possibly via a dish of mushrooms, his favourite food. Assuming that the story is true, Agrippina probably used the services of Locusta, a professional poisoner. Locusta opened a school in Rome to teach her art. The women of Rome became so adept at poisoning that the Senate was obliged to execute about 200 of them, including Locusta. Aconite (*Aconitum napellus*) seems to have been the tool of professional poisoners in ancient Rome; so much so, that growing the plant became a capital offence.[4,6,7]

5.3 European Excess: The Age of Arsenic

Acts of homicidal poisoning peaked in Europe in the 14th to 17th centuries. Practitioners, often professional, were seldom caught, and food tasters were essential dining companions for the wealthy when the emphasis shifted to the use of mineral poisons.

5.3.1 Italy of the Borgias and the Medicis

Powerful families throughout history used the papacy to further their ambitions. Popes were created and removed frequently. Poison accounted for the death of Pope John XIV (984); Clement II (1047); Benedict XI (1304); Leo X (Giovanni de' Medici) (1521); Adrian VI (1523); Clement VII (Giulio de' Medici) (1534), Alexander VI (Rodrigo Borgia) (1503).[8]

Two papal families, the Medicis and the Borgias, achieved special places in the poisoners' hall of fame; so much so that when Catherine de' Medici (daughter of Lorenzo de' Medici) married into the French royal family, the King of France insisted that a unicorn horn be part of the dowry so that he would be protected from his spouse (Section 5.10). Catherine is credited with introducing Italian poisoning methods to France and with contributing to the deaths of a number of French notables.[9]

Rodrigo Borgia, originally from Spain, was appointed Cardinal in the church by Pope Calixtus III (Alfonso de Borgia). Rodrigo flourished and his mistress bore him four children: Cesare, Juan, Joffre and Lucretia. Rodrigo bought his way into office and eventually became Pope Alexander VI in 1492. He was probably the worst of a bad lot of Renaissance Popes and died in 1503, after drinking from a poisoned goblet that had been intended for a visiting cardinal. The names of two of his children, Cesare and Lucretia, are particularly associated with cruelty. Lucretia seems to have been cast as an especially vicious woman as in the Donizetti opera *Lucretia Borgia* where in Act II Scene V, she sings: "You idly thought to pass unpunished; for your affront I have already full vengeance. Five funeral palls are prepared to cover your remains, since you are all poisoned."

However, the records of the time indicate that Lucretia was a pious woman who performed good works and died peacefully without a single murder on her conscience.[10] But there is little doubt that Cesare, a model for the Renaissance prince, and his father had fewer scruples. Cesare set up a special room to prepare poisons. One recipe, said to be from a Spanish monk, reads: "The abdominal viscera of a sow which has been poisoned by arsenic are covered over by arsenic powder and putrefaction is allowed to proceed until liquids flow from it. The liquid is concentrated by evaporation until it constitutes a white powder." This concoction was known as La Cantarella.

At the time certain ingredients were used for their own effect, and others in the belief that animal organs in an unhealthy state would produce indisposition in the corresponding organ of the victim. Intestines of humans were valued, especially if the original owner was syphilitic, cancerous or had been hanged. Leonardo da Vinci concocted his *fumo mortale* while he was in the service of Cesare. The ingredients: sulfur, realgar, arsenic [presumably in this case arsenic trioxide] toad and tarantula venom, and mad dog saliva.[6,8,11]

In 16th-century Italy, poisoning became a formal method of assassination. Thus the "Secret Circle", an arm of the government of Venice, put out contracts on the life of anyone who caused the government displeasure. The official records now on display in Venice display the word "Factum" to indicate the successful outcome. Contracts were carried out by professional poisoners whose fees depended on the rank of the victim. There were "circles" of poisoners in Venice and Rome that supplied the necessary knowledge and manpower. A contemporaneous book – *Poison Formulas* by Bapista Parta, 1589 – contained the following concoction which was described at the time as a "very strong poison": arsenic, aconite, *Taxus bacceta* (common yew), caustic lime, bitter almonds and powdered glass; all blended with honey and rolled into a ball.[5]

Giulia Toffana (La Toffana) attained particular fame as a poisoner in Naples. In 1709 she admitted, probably under torture, to murdering 600 people, including two Popes, Pius III and Clement XIV. She sold vials with labels such as "Manna di St Nicholas" and "Aqua Toffana" which contained juice that was produced by rubbing arsenic into the joints of freshly slaughtered swine. Women bought the products to use to improve their complexions (Section 1.3) but they also used them to remove unwanted spouses, lovers and other hindrances. La Toffana was executed in 1709. Another woman – Hieronyma Spara (La Spara) – sold an arsenic-based "Aquetta di Perugia" and operated on much the same scale but in Rome a few years earlier, in the mid-1600s. She founded a society to teach women how to rid themselves of husbands. Because of the spate of poisonings in various regions of Italy at the time, the contemporary British used the words "Italianated" or "Italianation" to describe poisoning and its end result.[9,12]

Box 5.1 Forensic Toxicology

Although the written word leaves little doubt that arsenic was in wide use in Europe during the time of the Borgias and the Medicis, there is always the desire to confirm this with the aid of modern science. Curiosity about their past is a big part of the Italian culture and the work of Pisa University's Centre for the Study of the Dead is closely followed by the public. Gino Fornaciari, a professor in Pisa, recently exhumed 49 sets of remains belonging to the Medici family. In 2004 he also exhumed the naturally mummified body of the "Lord of Verona" Cangrande della Scala (1291–1329). At the time of his death, it was widely believed that Cangrande had been poisoned with arsenic. The body was found to be exceptionally well preserved; so much so that some of Cangrande's internal organs could be examined. It was revealed that the actual cause of death was digitalis poisoning, probably administered under the guise of medical treatment.[13]

Francisco I de' Medici and his wife became ill in October 1587, during a visit from his brother Cardinal Fernando. They died 12 days later and their bodies were buried in a tomb in a nearby church. Their viscera, which had been removed during the autopsy ordered by Fernando, were placed in four terracotta jars and buried in the crypt of the same church. In 1857 the tomb was opened to move the bodies to the Medici chapel in San Lorenzo. At the time the bodies were described as being "fairly well preserved." Many years later this observation (Sections 2.8 and 4.7) and the contemporary medical records of their illness (vomiting, nausea, cold sweats, *etc.*) were taken as an

Box 5.1 Continued.

indication that the deaths were the result of arsenic poisoning and not malaria as originally maintained.[14a]

Hair samples obtained from the grave of Francisco I did not show an elevated arsenic concentration. This was rationalised by saying that the time between ingestion and death was too short for the arsenic to have been deposited in the hair. Samples collected from the crypt, identified as remnants of human tissue, did seem to show higher arsenic concentrations. The crypt samples were connected to Francisco via DNA profiling and "an extremely high degree of similarity with the DNA of a small skin fragment found attached to the beard hair of Francisco I." The scholars say they have definite proof of arsenic poisoning,[14a] but this is by no means the case. Well-preserved bodies are not necessarily associated with arsenic poisoning (Section 2.8), and arsenic poisoning is notoriously difficult to diagnose. The absence of arsenic in the hair is damaging to their argument, because there is evidence that deposition is actually very rapid.

Acute arsenic poisoning was claimed to be the cause of deaths of a Portuguese King in 1826 on the basis of the analysis of post-mortem soft tissues kept inside a Chinese porcelain container.[14b]

5.3.2 France: The Poisons Affair

Around the same time that La Toffana was in business in Naples, La marquise Marie de Brinvilliers was experimenting with arsenic on patients in a hospital in Paris where she volunteered her compassionate help. Her beds had a high turnover rate. She got her arsenic supplies from the King's pharmacist whom she visited in disguise, and without any apparent remorse, she poisoned her family including her father, two brothers and a daughter, as well as servants. Marie recorded it all in letters that were accidentally found on the death of her husband. Incidentally, he died of natural causes and is said to have protected himself with doses of Theriac of Andromachus. Marie was beheaded in 1676 and her convulsed face looking at the mob is preserved in a painting by Le Brun that hangs in the Louvre in Paris.[6,12]

Hints of poisoning on a far grander scale reached the authorities in 1673 when priests of Notre Dame revealed that, for some time, a great proportion of those who made confession accused themselves of having poisoned someone. "The Poisons Affair" came fully to light in 1679 when a policeman chanced to hear Marie Bosse, an arsenic supplier, boast: "What a marvelous job I have! What a superb clientele – only duchesses, Marquises, princes and lords. Three more poisonings and I retire, my fortune made."

The police found themselves dealing with allegations of witchcraft and poisoning on a massive scale. Subsequent investigation led to the arrest of many aristocrats and to a fashionable fortune teller known as La Voisin (Catherine Montvoisin) who held court in her house. She organised black masses with the help of real priests, and sacrificed children. She trafficked in poison and performed abortions – the police found themselves in the middle of what became known as "The Poisons Affair." About 400 women were eventually tried and burned as witches in Paris. Men experiencing domestic troubles were especially worried about poisoning whenever they experienced stomach upsets. There was one report of a shop that opened to provide poison to women who wanted to murder their husbands, "in order to give themselves over more freely to libertinage." The owner was allegedly discovered because a woman in the same business wished to get rid of the competition.[15–17]

A number of very important people from the then licentious court of Louis XIV were implicated in poisonings, including the King's powerful mistress Madame de Montespan, and a special commission was set up to preserve discretion and to ensure that justice was done. Torture was used to gain information: La Voisin, a major supplier of the poison, was burned alive, having confessed to murdering 2500 infants. The commission ended in 1682 by order of the King; most of the records were destroyed, and many not so innocent people were able to sleep a little easier in their beds. Judgment was made on 104 cases, and 34 of the accused were executed.[17] After the dust settled, the Sun King reformed his ways and became a moralist.

One outcome of The Poisons Affair was the establishment of the first register of poisons in 1682. "Among poisons should be counted not only those capable of causing rapid death but also those gradually undermining health or causing other maladies....With regard to arsenic, realgar and orpiment and sublimate, although they are wholly dangerous poisons, since they are used and employed in several necessary processes, we decree, in order to prevent in future their hitherto great and abused facility of use, that only registered merchants shall be allowed to sell and deliver them personally to doctors, apothecaries, surgeons: Which merchant, nevertheless, shall write what they have sold in a register kept for that purpose, containing the name and quantity of the buyer, and his domicile....No serpents, toads, vipers, or similar reptiles were to be used either as a part of a prescription or for the purpose of experiment."[15]

We can only assume that these regulations would put a mild damper on the availability of arsenic for homicidal purposes. Arsenic-based rat poison was still in use and without any reliable means of detection many deaths were probably attributed to cholera or malaria that were the result of administering the *poudre de succession* ("inheritance powder"), as arsenic was called at the time.

5.4 Forensic Science

Prior to the mid 1700's, poisoners usually escaped detection because the crime was difficult to prove. Whispers from friends, enemies and some of the clergy, who broke their vows of silence over what they heard in the confessional, lead to the arrest of suspects. Torture was then used to get confessions of multiple murders such as the 2500 infants attributed to La Voisin. However, even when forensic identification techniques improved, there was no guarantee that the results would be reliable or their interpretation correct, as we will see.

5.4.1 Mary Blandy

The trial of Mary Blandy was one of the first to capture the British public's attention. Her trial, which took place in Oxford on March 2nd and 3rd, 1752, was also noteworthy in that it was the first time that scientific evidence was presented at a criminal trial: certainly marking an advance over the use of torture. Blandy was accused of poisoning her father with the help of Captain the Honourable William Cranstoun, whom Blandy hoped to marry. Unfortunately, the Captain was already married, although he denied it. Blandy's father was obdurate and Cranstoun, with Blandy's knowledge, administered some powder to him to make him, the father, look more favourably on the proposed union. The powder (possibly sugar) did not work, although it did no harm, so the Captain, who had by now retreated to Scotland, sent Mary another "love potion" in the form of arsenic trioxide, which resulted in the painful death of her father after Mary mixed it with his gruel.

During the trial, Dr. Anthony Addington, appearing for the prosecution, was asked how he identified the "white arsenic." "For the following reasons: (1) This powder has a milky whiteness; so has white arsenic. (2) This is gritty and almost insipid; so is white arsenic. (3) Part of it swims on the surface of cold water, like a sulfurous film, but the greatest part sinks to the bottom, and remains there undissolved; the same is true for white arsenic. (4) This thrown on red-hot iron does not flame, but rises entirely in thick white fumes, which have the stench of garlic, and cover cold iron held just over them with white flowers; white arsenic does the same."

In her defence, Blandy said to the jury: "I really thought the powder an innocent, inoffensive thing, and I gave it to procure his love."

The jury took five minutes to come to a verdict of guilty and Blandy was sentenced to be hanged. She asked for a little time to settle her affairs, which was granted. The respite allowed time for a stream of letters and pamphlets with arguments for and against her guilt that included "A Letter from a Clergyman to Miss Mary Blandy, With Her Answer Thereto" printed by M. Cooper at the Globe in Paternoster Row, London; and in reply, there was "An Answer to Miss Blandy's Narrative," printed by W. Owen near Temple Bar, London. The pamphleteers kept the discussion alive a year longer than its subject.[18]

Had Blandy been found guilty of poisoning in earlier times, she could have been boiled to death as set out in an Act passed in 1532 when Henry VIII was king. At that time, poisoning was a new crime in England, whose inhabitants normally preferred to use a bludgeon or dagger to inflict serious injury. A harsh punishment was needed to dissuade potential poisoners because there was no real way of detecting that a crime had been committed. The Act was repealed in 1547 and hanging substituted, except when a woman poisoned her husband or child, in which case she could be burned.[6,19]

5.4.2 James Marsh

Around the end of the 1700s the identification of arsenic in any sample was a haphazard affair. Some of the recommended tests included:

(1) Burning the sample: white fumes and a garlic smell indicate arsenic. This test was applied to arsenic in water by evaporating the sample in an iron pot and throwing the residue on the fire.
(2) Infusing a small portion of the powder in a solution of vegetable alkali in water, letting it stand for an hour or two; then adding a solution of copper sulfate. Arsenic is present if the solution turns green and a green precipitate is produced. (This is the Scheele's green of Section 2.8.)
(3) Boil the sample with lime water. Arsenic gives a white precipitate that is soluble in acetic acid. Mix the precipitate with oil and throw onto the fire; a garlic smell confirms the identification.
(4) The most infallible method: add hydrogen sulfide to water containing arsenic. Arsenic is present if the liquid turns yellow or a precipitate of arsenic sulfide is obtained.[20]

In 1832, James Marsh was called as an expert in a murder trial. He was requested because he was the only chemist in the area. The case involved the suspicious death of a farmer in Plumstead, England, and the illness of his daughter and granddaughter. It was known that his dissolute grandson had purchased arsenic for poisoning rats. Marsh examined coffee pots and body parts, which duly produced the telltale yellow precipitate of arsenic sulfide that convinced the jury at the inquest to send the grandson for trial. The trial jury had other opinions – they did not want precipitates, they wanted to see the arsenic – so the grandson was acquitted. He admitted his guilt seven years later and was summarily transported to Australia.[21]

The verdict irked Marsh, and he set out to develop a means of providing what the jurors wanted: visible arsenic. He discovered that the gas, arsine, is easily and reproducibly generated from a sample, by using the method developed by Scheele (Section 3.2), and that it can be recognised because a film or mirror of arsenic is formed inside the apparatus when the gas passes through a heated section. The size of the mirror produced can be used to estimate the amount of arsenic in the sample. This procedure is the basis of what we now know as the Marsh test for arsenic.[22] One other important property of the gas that has been mentioned (Section 3.1) is that arsine reacts with test papers that have been treated with solutions of silver or mercury compounds. The intensity of the colour that develops can be used to estimate the arsenic concentrations in the sample. This variation is known as the Gutzeit method. By the mid-1900s, other more convenient and sensitive methods based on colour development were in use for arsine detection,[23] these were superseded in turn by even more sensitive methods to be described in Chapter 7 (Section 7.2).

Another important test for arsenic was developed by Hugo Reinsch in 1842. It involves dipping copper foil into a boiling solution of the sample. If arsenic is present a grey deposit of copper arsenide is obtained on the foil. Heating the deposit yields white crystals of arsenic trioxide that are easily identified with the aid of a magnifying glass.

These tests provided a sensitive and reliable means for detecting the presence of arsenic in places it should be and in places it shouldn't. Human bodies belong to the latter category and the availability of these tests meant that life suddenly became much more difficult for the homicidal poisoner.

Prior to the time of his discovery, Marsh had been employed for many years at the Royal Military Arsenal at Woolwich where he was working on such things as the development of a recoil break for naval guns. On Faraday's appointment to the Royal Military Academy in December 1829, Marsh became his assistant at a salary of 30 shillings a week. The Crown Prince of Sweden sent Marsh a small silver medal as a mark of appreciation for his services to science. But, despite his accomplishments, Marsh died penniless in 1846, leaving a wife and family without financial support.

The French were the first to hear evidence based on the Marsh test during the trial of Marie Lafarge – the British soon after. Even today some of the more sensitive analytical methods for arsenic are based on the evolution of arsine from a sample (Section 7.2). Reinsch's method remains in use in the 21st century.

5.4.3 Marie Lafarge

The population of France and much of the rest of the world was consumed with the trial of Marie Lafarge in 1840. Marie Fortunée Chapelle, an orphan, was brought up by relatives in Paris. She was left a small sum of 100 000 francs as her sole source of income. Marie's aunt found her a husband, Charles Lafarge, a widower, through a matrimonial agency. He was misrepresented to Marie as a rich merchant and, although she was not impressed, the wedding took place in 1838. She was obliged to move in with her mother-in-law in the country.

Marie Lafarge took care of the house, even though she felt trapped, and wrote to a chemist in the nearby town of Uzerche: "We are devoured by rats here. Will you, or can you, trust me with a little arsenic? You can count on my caution. It is to put in the linen cupboard."

Husband Charles went to off to Paris to raise money for his business and Marie asked her mother-in-law to make him some cakes. Marie later told the court that she had wrapped four or five of them in paper and mailed them to Charles in a parcel. In Paris, Charles collected the package but on opening it found, not the expected four or five small cakes, but one large round cake. He had a piece and went out for dinner. In the morning he began to vomit, had diarrhoea and a headache. Charles came home January 3rd, 1840, still sick, and went to bed, but his doctor assured him the problem was not serious. Charles died January 14th and the autopsy results suggested there was arsenic in the stomach although the test was botched because the test tube exploded.

Marie Lafarge was arrested and taken to prison in Brives where she was joined by her maid Clementine. Her cell became a second home for numerous friends and well-wishers. Many people, including her two lawyers, found her fascinating. Priests believed in her innocence. Her charm became legendary.

The courtroom was crowded at the time of the trial and temporary seating was arranged for the ladies who were dressed as for the opera. An alleged jewellery theft by Marie had become part of the case, providing further scandal. She was quickly found guilty of the theft. As for the murder charge, the defence produced an opinion from the famous Professor Orfila of Paris (see below) that there were problems with the analytical procedures and therefore the results. The president of the court then ruled that the analysis should be repeated by three chemists from Limoges. The next day they reported that they used the recently invented Marsh apparatus (Section 5.4.2) and found no arsenic in the stomach and none in the vomit, but they did find arsenic in some dregs of milk and eggs: "enough to kill ten persons."[24]

The Court requested that the body be exhumed and re-examined. No arsenic was found in the body. (Evidently the chemists had been heating the apparatus at the wrong place to produce a mirror.) This revelation caused more sensation and much sympathy for Lafarge, who was seen to be the victim of a family plot. The defence moved for an acquittal but the trial judge was not satisfied, so Orfila was sent for from Paris. He and two colleagues arrived and began to work on the body, which was close by, and the stench in the court was very unpleasant. Orfila found around half a milligram of arsenic in the body and the stomach liquids.

The defence was shocked by this turn of events: they had agreed to the reanalysis in the belief that no arsenic would be found. The defence then decided to appeal to Francois Raspail, another well known chemist and rival of Orfila. The message from Lafarge reads, "I am innocent and very unhappy, Monsieur! In my trouble I call both your science and your head to my aid. The experiments of chemists had given me back much of the reputation torn from me in the last eight months. But now M. Orfila has come and I have fallen back into the abyss. I now put my hope in you, Monsieur. Lend the aid of your science to this poor maligned innocent. Now that everyone is abandoning me, come and save me. Marie."

In the meantime the trial proceeded with the defence saying, "She is too good a person to kill. The amount of arsenic found by Orfila was not worth worrying about." Raspail did not arrive in time and the jury, when asked for a verdict, decided she was guilty but with extenuating circumstances. Marie Lafarge was condemned to hard labour for life and public exposure in Tulle.

Raspail, when he arrived, went to Lafarge, examined the evidence and found no arsenic. He, too, fell under Lafarge's spell and published his opinion in the newspapers making much of the small amount of arsenic found by Orfila. Raspail claimed that Orfila's reagents were contaminated with arsenic.[24] Lafarge's lawyers appealed her case to the Supreme Court without success. There was a general feeling after the trial that the Judge had continued the analysis process until the desired answer was obtained. The sentence was later commuted to life imprisonment and no public exposure.

Lafarge led a saintly life in prison and was set free after becoming ill early in 1852. She died in September the same year. Anyone who supported her cause was sent a small piece of tapestry. One recipient said: "Well, I'd rather have her tapisserie than her patisserie."[25]

Mateu Joseph Orfila (1787–1853) held the chairs of legal medicines and then medicinal chemistry at the University of Paris and is regarded as the father of toxicology. He wrote his encyclopedic "Trait de Poisons" in 1814, which systemised the subject. Orfila was chagrined by Marsh's achievement, but he did choose to use the Marsh test during the trial of Marie Lafarge – although he may have been overly confident with the results. Overconfidence was certainly the case a few years later when in 1859 forensic evidence was first admitted in a Canadian court. Dr. Henry Croft of Toronto testified that he found arsenic in the stomach of the deceased wife of Dr. William King. The jury was convinced and Dr. King was sentenced to hang. On his way to the gallows Dr. King admitted the crime but said he used morphine, not arsenic.[26]

Gustave Flaubert's *Madame Bovary* was first published in 1857, so he must have been aware of the trial of Marie Lafarge. In the book, Emma Bovary commits suicide by eating a handful of arsenic that she manages to get from the local apothecary. The description of her suffering is vivid and painful (Section 1.7). The book is based on the real experiences of a Madame Delamare, who had lived in Flaubert's neighbourhood. Flaubert said, "When I was writing the poisoning of Emma Bovary, I had such a strong taste of arsenic in my mouth, I had poisoned myself so badly that I suffered two attacks of indigestion in a row, two very real attacks, for I vomited up all my dinner."

A very unreal depiction of death following arsenic ingestion is portrayed in a painting that hangs in London's Tate Gallery: *Death of Chatterton* by Henry Wallis. Thomas Chatterton was a poet who committed suicide in 1770, at age 17. He tried to pass off his own work as that of a 15th-century monk and took arsenic because of rejection and disillusionment. He became a romantic cult figure – a genius destroyed by indifference.[27] The picture is very sentimental; set in a garret, there is a blue pallor to the body which is in a restful pose, the face is peaceful, and the bottle of poison lies on the floor.

5.4.4 The Arsenic Act of 1851

The following account is based on the article "A 'Pennurth of Arsenic for Rat Poison' The Arsenic Act, 1851 and the prevention of secret poisoning" by Peter Bartrip.[28]

In an 1848 address, the president of the British Pharmaceutical Society declaimed: "At present, as the law stands, any man, however, ignorant – an individual unable even to sign his own name – half of whose shop is stored with butter, bacon, cheese or tape, shall from the other half have the power of dispensing, to any person applying, preparations of mercury, arsenic, opium, *etc. etc.*" (Figure 5.1)

At the time, the medical and pharmaceutical professions had become interested in reform, and were exploiting public anxieties about accidental and intentional poisoning for their own purposes. For example, in 1842, there were 3839 violent deaths in one region of England but only 128 were attributed to poisons and of these only 15 were ascribed to arsenic. Nonetheless, those seeking change promulgated the notion of an epidemic of secret and premeditated poisoning. The British public, who had been following the Lafarge trial in France, was not convinced that Marsh had solved the problem of detecting arsenic so were sympathetic to the demands for regulation.

The pharmacists wanted the terms "chemist" and "druggist" to be defined by an Act of Parliament, and membership in the profession to be established by compulsory examination. The doctors wanted to look good and protect the public; they also wanted sole rights to prescribe medicine, particularly poisons such as arsenic where the effects of ingesting were likely to be misdiagnosed.

The first response from Parliament included the provisions that only chemists, druggists and apothecaries could sell arsenic; that only males could purchase arsenic; and that all sales be recorded in a "poison book." Parliament ultimately chose to include only the "record provision" in the Act (women strongly objected to the male-only sales suggestion). The Act was confined to sales of arsenic: "It was limited to one poison, partly because [while] it was feasible to debar the ignorant from using arsenic, it was less easy to prevent the knowledgeable from using alternatives." Grocers could still sell the poison.

The Act required the vendor to keep a registry of all sales of arsenic. "The entries were to be signed by the purchaser and a witness, unless such purchaser professes to be unable to write (in which case the person making the entries hereby required, shall add to the Particulars, to be entered in relation to such sale, the words "cannot write")." The Act requires the vendor to mix with the arsenic before the sale was effected "soot or indigo" in the proportion of one ounce of soot or half an ounce of indigo at the least, to one pound of arsenic. Strange to say, few of these sales records have been preserved.[29]

There were immediate prosecutions after passage of the act. *The Wisbech Advertiser* of September 15th, 1851, reports on a murder case in which arsenic was used and the supplier, a "grocer and general dealer", was charged with the sale contrary to the act. But by 1857, renowned toxicologist Professor Alfred Swaine Taylor, of Guy's Hospital[30] was reporting that: "The poison's great cheapness (one penny to twopence an ounce) places it within the reach of the poorest person. It is sold to any applicant on the most frivolous of pretenses. The better class of druggist does not sell arsenic by retail; the grocer, chandler, oilman and village shopkeeper are the principal vendors of this poison; and it is clear from the numerous deaths which take place from white arsenic, that they set the law at defiance and sell the poison in an uncoloured state."[28]

According to Bartrip[28] the Arsenic Act served little purpose. But he points out that the Act was important because Parliament moved away from its *laissez-faire* policies and established a precedent for later regulation of opiates, poisons, food and drugs. The act acknowledged the legitimacy of the medical association and also led to pharmaceutical reform in the 1868 Pharmacy Act, which confined the sale of a range of poisons to qualified pharmacists.

5.4.5 Madeleine Smith

Emile L'Angelier, who lived in Glasgow, was overcome with internal pain and vomiting once in February and twice in March, 1857. The third attack was fatal. A search of his possessions revealed numerous letters of a highly scandalous and passionate nature from a young woman named Madeleine Smith. At her subsequent trial, the Judge felt compelled to say about one of her letters: "Certainly such a sentence was probably never before penned by a female to a man." Smith came from a well-to-do and well-connected family living in Glasgow. L'Angelier was described as a profligate and "had an almost priggish standard for other people, particularly for any woman whom he might be debauching at the time."[31]

The cause of his death was determined to be arsenic poisoning and Smith was accused of administering the arsenic and ultimately his murder. The case attracted a huge audience for the trial in Edinburgh (it was moved from Glasgow because of the perceived difficulty of finding an unbiased jury) and was widely reported in the press. The autopsy report on the exhumed body presented in evidence was long and detailed. The contents of the stomach were described as follows: "This liquid measured eight and a half ounces. On being allowed to repose it deposited a white powder which was found on examination to possess the external characteristics and all the chemical properties peculiar to arsenious acid; that is the common white arsenic of the shops."

This chemical identification was much more certain than had been presented in earlier trials such as Mary Blandy's. The "Marsh process" gave the mirror of metallic arsenic. The Reinsch and Marsh tests were applied to the stomach fluid, and Dr. Frederick Penny – a professor of chemistry at Andersonian University in Glasgow – testified that the arsenic in the stomach amounted to "eighty two grains and seven-tenths of a grain" [5.5 g]. Solid arsenic trioxide was found in the small intestine and arsenic was detected in the liver, heart and brain.[31] Other details are given in Box 5.2.

Box 5.2 Some Analytical Results Reported at the Trial of Madeleine Smith[31]

"Sample No. 2 was a bottle containing prepared fluid from contents of stomach. This fluid was colourless and nearly transparent. (1) A stream of sulfuretted hydrogen threw down from it an abundant sulfur-yellow precipitate. (2) Hydrochloric acid being added to a portion of it, copper gauze was subjected to a boiling heat in the mixture, upon which, in a few seconds, the gauze became encrusted with a greyish-black coat. (3) This gauze, when washed, dried and heated in a glass tube, was restored to its original bright

Box 5.2 Continued.

copper-red appearance, and at the same time, a ring of sparkling crystals was obtained, the form of which was the regular octohedrae, or some form derived from it.

The fluid prepared from the contents of the stomach therefore contained oxide of arsenic, and in considerable quantity.

Sample No. 7 was a jar containing a portion of liver. The contents, being about four ounces of a liver, were subjected to a modification, proposed in 1852 by Dr. Penny, of the process of Reinsch for detecting arsenic in such matter. The liver having been cut into small pieces, and boiled in hydrochloric acid and distilled water in a glass flask, to which a distilling apparatus of glass was connected, the whole texture was gradually reduced to a fine pulp and a distilled liquor was obtained, which was collected in divided portions. These liquors were colourless, and nearly clear. The two first portions obtained did not contain any arsenic; the third gave faint traces of it; the fifth and sixth portions, when separately subjected to the action of copper-gauze, gave characteristically the usual dark-grey encrustation, and this, again, was driven off, as usual, by heat in a small glass tube, and yielded, in each case, a white sparking ring of crystals which were regular octohedrae, or forms derived from the octohedrae." [The fourth portion seems to have been overlooked.]

Box 5.3 The Styrian Defence

Because of the possibility that arsenic eaters really existed, judges and lawyers involved in criminal trials that centred on accusations of arsenic poisoning, were obliged to consider whether the victim might have accidently self-administered the fatal dose. Such an argument became known as the Styrian defence, after the revelation that peasants in Styria became immune to arsenic poisoning after ingesting large quantities of arsenic trioxide (Section 1.3).[33] The Styrian defence was used in a number of well-known criminal trials: notably Madeleine Smith in Edinburgh, 1857, and Florence Maybrick in Liverpool, 1889. The trial of Mrs. Maybrick for the murder of her husband offers considerable insight into the use and availability of arsenic in England at the time.

Mrs. Maybrick was accused of poisoning her husband with arsenic and claimed that, like Madeleine Smith, she was using the solution for cosmetic purposes, accounting for the fact that she had arsenic in her possession. The Styrian defence was based on the knowledge that Mr. Maybrick was in the habit of using arsenic that he obtained from a local druggist, as a tonic.

> **Box 5.3 Continued.**
>
> When asked about Maybrick's use of arsenic during the trial, Maybrick's servant Thomas Stannell said, "When I brought him the arsenic he told me to make him some beef tea. I went and filled a cup and brought it to him. He asked me to give him a spoon, and, taking the spoon, he opened the package and took a small bit out. This he put in the tea and stirred it up."
>
> During the same trial, James Bioletti, a hair dresser and perfumer in Liverpool, said: "Arsenic is used a good deal in the hair for some purposes and I have used it as a wash for the face on being asked for it by the ladies. It is used principally by the ladies for removing hair from the arms. I mix it with lime in powder. I generally use yellow arsenic, but I have used white arsenic."
>
> This evidence and more in the same vein certainly leaves the impression that a number of individuals in Victorian society used arsenic for much the same purposes as the Styrians: as a tonic, as an aid to digestion, as a cosmetic and as an aphrodisiac. It is also clear that some individuals increased the dose over time. The fact that Maybrick kept a mistress and fathered a number of children including two after his marriage could be taken as an endorsement of the practice.
>
> The belief that arsenic eating is good for the complexion seems to have resulted in an unfortunate death in more recent times. In Auckland, New Zealand, on August 31st, 1954, Mrs. Maud Wilson died of arsenic poisoning and her husband James Wilson was charged with murder. The defence argued there was no doubt that Mrs. Wilson was in the habit of self-medication with arsenic and that in a dulled state, because of bromide use she accidentally took an overdose of the poison (the Styrian defense). The prosecution claimed that James tampered with the medicine, came home knowing she would be ill from the overdose, and then "refrained from calling a doctor until such time as he knew his wife was beyond human aid." The jury deliberated for six hours but could not come to a verdict. Another trial was ordered, resulting in Wilson's acquittal.[34,35]

Smith testified: "I have bought arsenic on various occasions. The last I bought was a sixpence worth, which I bought in Currie, the apothecary's, in Sauchiehall Street, and prior to that, I bought other two quantities of arsenic, for which I paid sixpence each. ... I used it all as a cosmetic, and applied it to my face, neck and arms, diluted with water." She did this on the advice of a young lady and she had seen it recommended in the newspapers. But in order to make the purchase she had to claim, like many before her, she wanted to poison some rats: an unusual task for someone of her social station. "And [according to the apothecary], when she came back for a fresh supply, she simply reported that encouraging progress was being made against the rats but another dose was needed to complete the job."[32]

The question of how Smith managed her ablutions despite the soot and indigo was not asked (Section 5.4.4). What was a concern was the apparent absence of the colouring agents in the body of the victim. The prosecution argued, with medical evidence on its side, that the colouring matter would not be seen, yet it was evident in dogs that had been poisoned in this way. The size of the dose provoked the following from the defence council John Inglis: "We are asked to believe that he took from her hand a poisoned cup in which there lurked such a quantity of arsenic as was sufficient to leave in his stomach after his death 88 grains – such a dose, indicting the administration of a least double – ay, I think, as Dr. Christison said, indicating the administration of at least half an ounce, 240 grains [16 g] – and this he took that evening from the hand of the prisoner, with all his previous suspicions that she was practicing on his life. There is the greatest improbability of such a thing being done; it is a most difficult thing to conceive a vehicle in which it could be given." The prosecution argued that the dose was given in a cup of cocoa.

The incontrovertible fact is that the prosecution failed to prove that Smith met L'Angelier before any of the three occasions on which he was taken ill. There was no meeting after the arsenic was purchased although there was evidence, not allowed at trial, that this was probably not correct.

L'Angelier acknowledged that he was familiar with the use of arsenic to improve the wind of horses and had taken the substance himself for cosmetic reasons, to improve his breathing and to relieve the pain in his hands. During the trial, considerable reference was made to the articles on arsenic eating that had just been published in Chambers Journal and in Blackwood's Magazine (Section 1.3). Naturally, the defence made the most of these facts and invoked the Styrian defence for the first time in a British trial – both the victim and the accused had valid reasons for having arsenic in their possession (Box 5.3). The defence argued that it was impossible to suspend as much arsenic in any fluid. Smith freely admitted she had given him cocoa and he had been sick after one visit, but she had not purchased any arsenic at the time.

The verdict after nine days of trial was "not proven," as allowed in Scottish law.

The proceedings attracted huge crowds and was big news all over Britain. There were three points of view: Smith was innocent and L'Angelier had committed suicide; Smith had committed murder and she should pay the penalty; and the third school, "in which most students of the case have found themselves ever since, declared in effect 'probably she did it, but anyhow he deserved it'."[32]

Smith left Scotland after the trial and eventually married George Wardle, an artist who became a valued associate of the firm of William Morris (Section 3.13). The couple moved in fashionable circles in London. Smith eventually died impoverished in the United States, in 1928.

5.4.6 Thomas Smethurst

Dr. Thomas Smethurst, a married man, whose medical degree was probably purchased like many others in Victorian times, became enamoured of Isabella Banks. The two entered into a bigamous marriage setting up house in

Richmond, London, in 1858, around the time that Madeleine Smith was in the public eye. The new Mrs. Smethurst soon became ill and died.

One commentator wrote: "A delicate fragile woman who had suffered from uterine disease, and used vaginal injections, who was one of a family subject to bilious attacks, who was herself the frequent subject of nausea and liable to occasional vomiting, marries at the age of 42. Between three and four months after, she becomes pregnant, and about the same time, or about five to seven weeks before her death, she begins to suffer from vomiting and diarrhoea, which becomes exceedingly obstinate. Effervescing draughts with prussic acids, grey powder [mercury and chalk], Dover's powder, opiate enemata, bismuth, acetate of lead and opium, nitrate of silver, sulfate of copper, are tried successively, and all prove useless. Some of them are followed even by an increase of symptoms. This goes on from 28th March to 3rd May, when the patient dies, exhausted."[36]

Her doctors were suspicious and sent samples taken from the body and the residence to Professor A. S. Taylor who found arsenic in one bottle belonging to Smethurst. As a result, Dr. Smethurst was charged with murder. Taylor had been persuaded to work on a Sunday, which in the era of strict Lord's Day observance was certainly not his normal practice. He received his punishment: during the resulting trial that began on August 15th, 1859, when he had to admit he had made a mistake in the application of the Reinsch test. The arsenic he found was an impurity in the copper gauze he habitually used. No arsenic was found in any of Mrs. Smethurst's organs, although some antimony was detected. Nevertheless, the jury returned a guilty verdict and Dr. Smethurst was sentenced to hang.

There was an uproar from the public: a Dublin Medical Press Editorial contained the following remarkable sentence about Professor Taylor: "The man who, par excellence, was looked upon as the pillar of medical jurisprudence; the man who it was believed could clear up the most obscure case, involving medicolegal considerations, ever brought into a Court of Justice; the man without whose assistance no criminal suspected of poisoning could be found guilty in England; the man whose opinion was quoted as the highest of all authorities at every trial where analysis is required, is the same who has now admitted the use of impure copper in an arsenic test where a life hung upon his evidence, the same who has brought an amount of disrepute upon his branch of the profession that years will not remove, the ultimate effects of which it is impossible to calculate, which none can regret end, a lesson may be taught which will not be lost upon the medical jurists, and which may tend to keep the fountain of justice clear and unpolluted."[36]

The piece then went on to suggest that Professor Taylor should retire to private life in the country, not forgetting to take his arsenical copper gauze.

Thirty medical practitioners in London who were upset at the verdict, petitioned the Crown for mercy and wrote: "That, having considered the chemical evidence adduced at the trial, the manner with which the analysis was conducted, and the mistaken evidence given on oath by that eminent analyser, Dr. Taylor, on the investigation of this case, at Richmond, your memorialists are strongly of the opinion that there is no proof worthy of belief that arsenic was found in the evacuations or remains of the deceased, and still less that death was produced by poison."[36]

The pressure mounted and the Home secretary took the very unusual step of referring the case to a well-known surgeon who concluded: "there is no absolute and complete evidence of Smethurst's guilt." Smethurst was pardoned; however, he was immediately charged and convicted of bigamy. Taylor did not resign and went on to write more editions of his popular book on medical jurisprudence (the first edition was published in 1842 the tenth in 1897 well after his death in 1880).[37] In all, he spent 46 years as a professor at Guy's hospital in London.

As for the verdict, Leonard A. Perry, the editor of a book on the trial, writes: "The possibility that Mrs. Smethurst had been given drugs to produce an abortion or prevent a natural termination was not raised at the time."[36]

There was no Court of Criminal Appeal in the mid-1800s, so Smethurst was unable to appeal his conviction; hence the adjudication. Florence Maybrick encountered a similar situation in 1903.

5.4.7 Florence Maybrick

Florence Chandler was a pretty 18-year-old from Mobile AL, when she met James Maybrick, a 42-year-old Liverpool cotton merchant, on a transatlantic liner in 1880. They soon married and eventually settled in Liverpool, England. In May 1889 James Maybrick died of a mysterious illness and Florence Maybrick was charged with murder.

The jury heard that Florence had spent a few nights in London with another man in March 1889, but the tryst was far from idyllic. The fact that Florence knew that her husband, who was experiencing financial problems, had a longstanding mistress in the same town who had born him five children, was not allowed as evidence. Mrs. Maybrick bought 36 flypapers from two chemist shops even though flies were not in season. Each flypaper sheet contained about 100 mg of arsenic and she soaked them in water to remove the arsenic to make a cosmetic solution, just as Madeleine Smith had done some years earlier. The papers were displayed in prominent positions in the shop. Mrs. Maybrick also bought some arsenical cat poison. Mr. Maybrick died a few months after these purchases had been made. A chemist gave evidence that other ladies had purchased flypapers for cosmetic purposes, but the possibility of other suspicious deaths was not explored during the trial (Box 5.3). A lot of arsenic was found in various places around the house, more than was ever purchased by Mrs. Maybrick. She was seen handling some beef extract to be given to her husband and the prosecution claimed that this was how she administered the poison.

Only about 30 mg of arsenic were found in Mr. Maybrick's body and the cause of death was not clear. At one point in the trial an expert witness produced a tube containing an arsenic mirror supposedly formed from the arsenic in Mr. Maybrick's liver. The Judge had to be guided to the right place and was able to see it if he used a magnifying glass and a black background provided by a hat.

Mr. Maybrick was a hypochondriac. He first started taking arsenic when he developed malaria in the United States in 1884. He recovered with the help of pills made by the Taylor Brothers Pharmaceutical Chemists Norfolk VA, that contained "iron, quinine, arsenic, one capsule every three or four hours; to be taken after food." He dosed himself with everything available, including arsenic and strychnine. His doctors even prescribed arsenic for his terminal illness in Liverpool. As a result Mrs. Maybrick's defence relied heavily on the Styrian defence (Box 5.3).

Mr. Edwin Heaton the local druggist confirmed Maybrick's arsenic eating habit at the trial when Sir Charles Russell examined him for the defence. Heaton said Maybrick used to come into his shop and get a pick-me-up and drink it. He was taking one third of a grain of white arsenic per day [about 20 mg]. Other men came in for an arsenical pick-me-up as many as seven times per day.

A passage from the trial follows:

Sir Charles Russell: Do you know that liquor arsenicalis has aphrodisiacal qualities? Do you know that word?
Heaton: I do not.
Mr. Justice Stephen (The Judge): Did it excite passion?
Heaton: Yes, it had that effect.

Mrs. Maybrick was found guilty, mainly because of her adultery, and sentenced to death. There was much sympathy for her on the grounds that the amount of arsenic found in the exhumed body was not enough to cause death and that it probably resulted from self-administration. In addition, Maybrick's symptoms were not those of arsenic poisoning. The Judge was well past his prime and his handling of the trial was deemed to be incompetent and biased.

Almost half a million persons signed a petition to the Home Office urging Mrs. Maybrick's reprieve. All this pressure resulted in the sentence being reduced to life imprisonment, to be interpreted as saying: she did not kill him but probably tried to do so. Sir Charles Russell, her counsel during the trial, worked tirelessly for her release which was obtained in 1904 after 15 years in prison. She had to wait for the death of Queen Victoria, who had little sympathy for her. She returned to the United States and gave interviews and wrote her autobiography, *My Fifteen Lost Years*. Florence gradually faded from view living an eccentric life in squalour until her death in 1941.[32,38,39]

The Maybrick trial is over now, there's been a lot of jaw,
Of doctors contradictions, and expounding on the law;
She had Sir Charles Russell to defend her as we know,
But though he tried his very best it all turned out no go.

– **London street doggerel**

The arrival of Jack the Ripper in London in 1888 prompted much speculation about his identity, and there were many who suggested he was one of the known criminals of the time such as the poisoner George Chapman, who used antimony. One piece of fanciful speculation has it that James Maybrick, driven mad by his wife's infidelity, although heedless of his own, "in a haze of addiction to arsenic and strychnine took out his frustrations on the prostitutes in Whitechapel".[40]

5.4.8 Herbert Armstrong

Herbert Armstrong has the dubious honour of being the last lawyer to be executed in Britain. His wife died February 22nd, 1921, after coming home from a stay in a mental hospital. She was buried without suspicion but her body was exhumed on January 2nd, 1922, because there was evidence she might have been poisoned. Arsenic was found in her tissues and Mr. Armstrong was charged with murder.[41]

There was no dispute over the availability of arsenic in the Armstrong household. They had purchased a great quantity for killing weeds, particularly dandelions. The prosecution relied heavily on evidence given by a well-known pathologist, Dr. Bernard Henry Spilsbury, who claimed that he encountered arsenic cases at least monthly. He confidently stated that the amount of arsenic in the organs they recovered was more than enough to cause death. In addition, the body was exceptionally well preserved after 10 months' in her grave, prompting him to testify: "The condition of preservation was fully explained by the presence of the amount of arsenic that was found in those parts by analysis"[41] (Section 2.8).

The analytical results provided by John Webster, the Home Office Analyst, suggest there were at least 208 mg of arsenic in Mrs. Armstrong's body; enough to cause death. His results were given in amounts rather than concentrations but they can be converted to: 181 ppm (wet weight) in the liver, 100 ppm (ww) in the kidney, 1.7 ppm in the toe nails (10 sampled), 3.6 ppm in the finger nails (eight sampled), and 5 ppm in the hair.

These values were probably obtained by using the Marsh test and the liver and kidney results are appreciably higher than some reported more recently. For example:

Around 1963 a 9-year old child from Illinois was hospitalised twice for gastrointestinal distress and other dysfunctions and recovered. Some time later he was again hospitalised with similar symptoms. Once again he appeared to be getting well but suddenly he became very ill and died within 24 hours of his mother taking him some milk. His liver contained 4.7 ppm arsenic, his kidney 2.0 ppm. In Dallas, Nancy Lyon was poisoned by her husband Richard on or about January 14th, 1991. Her liver contained 30 ppm arsenic, her kidney 3.7 ppm. And in 1999 a 28-year-old male in Lyon France died three days after ingesting a massive dose of about 8 g of arsenic trioxide. His liver arsenic concentration was 147 ppm (dry weight), approximately 29 ppm (ww) his kidney concentration, 26.6 (dw) approximately 5.2 ppm (ww) (Section 7.2).[42–44a] 14 ppm is equivalent to about 20 mg in an average liver which is 70 to 80 per cent water. Higher liver and kidney results 62.8 ppm (ww) and 23.5 ppm (ww) respectively, were measured in a suicide victim who ingested 15 g of arsenic trioxide-he died eight hours later.[44b] Doses of 100 g of the oxide result in rapid death and liver and kidney concentrations of 1623 ppm (ww) and 771 ppm (ww).[44c] One forensic pathology text by Sankko and Knight (Arnold, 2004) gives average liver and kidney concentrations from 49 fatalities as 29.0 ppm (ww) and 15.0 ppm (ww).

The results reported for Mrs. Armstrong were described during the trial as the highest amounts Webster had ever encountered and are far greater than could be expected especially from a body burden of 208 mg. Surprisingly, the hair, finger nail and toe nail are not outside what could be expected from chronic exposure. It is possible that the results are reported as arsenic trioxide and not as arsenic; this would lower the numbers by 25 per cent but not change the conclusion. (Some analytical results for sea food obtained around the same time are also high by modern standards (Section 7.2)).

The average arsenic content of normal human liver is hard to pin down. Values quoted range from 0.46 ppm (ww) for British livers,[45a] and for Japanese livers, 0.29 ppm.[45a] A major study from the USA shows that the concentration is rarely above 0.02 ppm.[45b] So it seems that Mrs. Armstrong must have been getting some arsenic from somewhere.

The defence argued that Mrs. Armstrong herself had deliberately taken arsenic because she was of unsound mind and had suicidal tendencies; that Mr. Armstrong had no motive for murdering her; and that there was no evidence that he had ever administered arsenic. Medical evidence was given by Dr. F. S. Toogood stating that Mrs. Armstrong did not show any symptoms of arsenic poisoning until she took one dose on February 16th of about 400 mg, resulting in her death February 22nd. To account for the delayed response to the dose he put forward the theory that a large part of the arsenic must have become "encysted," or retained in a kind of capsule attached to the wall of the stomach and later released. In reply, Dr. Spilsbury agreed that a person might take a large dose of arsenic and live for a period of seven or eight days, but this would be exceptional. He did not agree with the proposition that the path to death could involve a sequence such as vomiting, sickness and diarrhoea for three days, remission for two days, and resumption leading to death a few days later.

One significant point was that Mrs. Armstrong, while in a mental institution shortly before her death, was routinely dosed with a tonic containing citrate of iron and ammonia, a hydrochloric acid solution of arsenic and *nux vomita*, three times a day. In this way she received a dose of 3/20 of a grain of arsenic (10 mg of the oxide) each day for 31 days – not an insignificant amount. Even so, her health improved during the time she was taking the tonic.

Mr. Armstrong was allowed to go into the witness box in his own defence (the law had been changed with the Civil Evidence Act of 1898, and the Court of Criminal Appeal Act of 1907[12]). He said the arsenic was for gardening purposes, but he had some difficulty explaining why he divided the arsenic up into little packets, one of which was found on his person when he was arrested. Evidently there were 20 dandelions, so 20 packets were needed. The betting on the street was in favour of an acquittal; but the jury saw it otherwise as did the appeal court. He was sentenced to death by hanging.[41]

In 1994, the trial was made into a BBC TV documentary entitled *Dandelion Dead*. According to a newspaper report, one of the Armstrong children, who had spent their lives trying to escape the memory of the events of 1921, "saw red" and gave permission for Martin Beales to re-examine the record. This resulted in a book *Dead, Not Buried* and the thesis that Mrs. Armstrong was slowly poisoned by an array of arsenic-based medicines and that Mr. Armstrong was innocent, framed by Oswald Martin, the other lawyer in the town, and Martin's father-in-law, Fred Davies, the local chemist. The supposed poisoning of Oswald Martin in October 1921, was the trigger that resulted in Mrs. Armstrong's exhumation in January 1922, and the enshrining of her husband in the Chamber of Horrors in Madame Tussaud's Wax Museum, London.[46,47] Certainly the analytical results, which were not questioned at the trial, were not of sufficient quality to back up the claim that Mrs. Armstrong died as a result of a fatal dose of arsenic administered within 24 hours of her death.

Modern lawyers are not immune from prosecution: A father–son duo of lawyers living in Dehli, were sentenced in 2005 to life in prison for the 1981 arsenic poisoning of the son's wife after she failed to meet their demands for a dowry. The coaccused mother-in-law died during the course of the lengthy trial.[48]

5.4.9 Marie Besnard

We have seen that the Marsh and Reinsch tests had their problems when first introduced into forensic science. This fate also awaited neutron activation analysis (Section 4.8). In 1949, Marie Besnard was accused of poisoning 11 people, including two of her husbands. All died. She became known as the Black Widow, likening her to the spiders that kill their mates once they have served their purpose. She was also known as the Good Lady of Loudun. The case lasted 12 years and she was in prison for five of these. She had three trials and was a victim of the inquisitorial French legal system which allows the process to be drawn out. Scientific evidence was first used to get an arrest, then finally used to get an acquittal.

Marie Davaillaud was born in Loudun in 1896 to a farming family. After her first husband died in 1927, she married Léon Besnard, a Loudun rope seller in 1929. Léon died of heart disease in 1947. At the time a former German prisoner of war was in their employ, and the locals – who suspected an affair between Besnard and the German – accused her of poisoning her second husband because of his jealousy. She was also accused of poisoning her first husband to get property. On May 11th, 1949, the bodies of her two husbands were exhumed and the analyst reported large quantities of arsenic in the remains. Besnard was arrested on July 21st, 1949.

The judge then ordered the exhumation of every member of the family who had died within the previous 25 years. Ten more samples were sent to Marseille and arsenic was found in all. Besnard was charged with poisoning 11 individuals but her first trial did not take place until February, 1952. She was vilified by the press and public, and treated roughly by the police who were overly keen to get a conviction. The evidence against her was not strong. There may have been arsenic in the bodies but there was no connection between it and Besnard.

At the trial her lawyers showed that more sample containers had been received by the analyst than were originally sent off. They also showed that the court approved analyst Dr. Béroud was incompetent. He was handed six glass tubes and asked to identify which of them contained an arsenic mirror as would be produced from a Marsh test (Section 5.4.2). Béroud singled out three of the tubes only to be told that none contained arsenic. There was an uproar in the court, the trial was adjourned and Béroud disgraced. The charges were reduced from 11 to five for the second trial that began in Bordeaux in March 1954. Besnard had been kept in prison in spite of the doubts raised.

The hearsay evidence was discredited easily, but there was new scientific evidence in the form of results obtained by using an "atomic pile and a Geiger counter." This evidence was given by Mr. Griffon, the director of the Toxicological Laboratory of the Police Prefecture in Paris, who had developed the technique of neutron activation analysis for detecting arsenic in hair (Section 4.8). Besnard's lawyers attacked this new evidence and accused the prosecution of incompetence and falsifying evidence. French and British experts became involved and questioned the methodology. Once again the trial was adjourned on the pretext that the court did not have enough data to pronounce judgment. After being in prison for five years, Besnard was granted bail at last.

The third and final trial took place seven years later, in November 1961. During this time, Besnard was free on bond. Professor Frédéric Joliot-Curie, a Nobel Prize winner, was of the opinion that the results had been falsified, but by the time the trial resumed he had died.

By 1961 the case was a judicial scandal. The prosecution wanted to save some face, but the defence wanted blood and an unconditional acquittal. The defence produced Griffon's laboratory notes that were confused and showed signs of alteration. The defence was able to demonstrate that all the expert evidence against Besnard was worthless. They argued that the arsenic in the bodies could have come from the soil. On December 12th, 1961, Marie Besnard was finally acquitted, but the public had abandoned her, and she lived out the rest of her days alone dying in 1980. She did not receive any compensation for her imprisonment.[4,50]

5.5 Public Perceptions

The trial of Mary Blandy in 1752 was the first court drama to be brought to the eager attention of the British public through the development of the print media. The broadsheets of the day pandered to the basest instincts of their readers and ensured that all details of criminal activities, whether fact or fiction, became immediately available. Even if the literacy rate was not high at the time, there were always eager listeners. By the mid-1800s newspapers, particularly the *Illustrated London News*, had developed the means of collecting and disseminating crime reports. "The policy was to give the public what it wanted – and what it was wanted was murder."[32]

Madame Tussaud's Wax Works, which opened in England in 1802, gratified the same macabre need. The Chamber of Horrors – for which there was an extra charge – featured many poisoners who used arsenic. The early displays were very current at the time and included authentic relics: "An executed murderer's clothing, by long usage the perquisite of the hangman, no longer was disposed of, as had been customary, to the Jews; now the hangman had regular and far better paying customers in the representatives of Madame Tussaud."[32]

Many have claimed, probably correctly, that these revelations inspired and instructed some poisoners. The situation has not improved over the years. Details of the 2007 murder trial of Cynthia Sommer, accused of the arsenic poisoning of her husband (Section 5.9), were widely covered in the media and the trial itself could be seen live on TV and on Internet video clips.

5.5.1 Arsenic and Old Lace

A major reinforcement of the public's association of the A-word with murder resulted from the play "Arsenic and Old Lace" by Joseph Kesselring, first performed on Broadway in New York in 1941. A very popular movie version was made the same year but, because of some legal problems, it had to wait until 1944 for release. The piece is a perennial favourite: in November, 2006, the play was being performed on more than a dozen stages in the United States and the UK.

In the story, Mortimer Brewster, a Broadway critic, discovers that his two maiden aunts are serial killers.

Mortimer Brewster: "But there's a body in the window seat."
Aunt Abby: "Yes, dear. We know."
Aunt Martha: "Well, dear, for a gallon of elderberry wine, I take one teaspoon full of arsenic, then add half a teaspoon full of strychnine, and then just a pinch of cyanide."
"We put it in the wine because it's less noticeable. When it's in tea, it has a distinct odour."

The play was inspired by the case of Amy Archer Gilligan, who in the early 1900s was accused of poisoning residents of her old-age home in Windsor CT, with arsenic, after collecting $1000 (US) from them for life care. According to one account, she did not keep the money for herself, but donated it to a local church for its altar fund. Mrs. Gilligan was originally sentenced to be hanged but this was later changed to life imprisonment. She died in 1962 in the state mental hospital in Middletown.[51]

The phrase "arsenic and old lace" has become a cliché and has spawned books and articles with titles such as *Arsenic and Old Lead* (environmental cleanup), *Arsenic and Old Mustard* (chemical weapons), and *Arsenic and Old Myths* (history of chemistry). We have encountered instances in which the A-word is used to sell fear and we will see a few more examples. The word has such an emotional wallop that it can even be used to sell goods such as dolls and adult clothing and even rock bands.[52,53] Then we have the blog www.arsenic.net.

5.5.2 Crime Fiction

Mystery stories involving poisonings have long been popular with the public, beginning with Poe's *The Murder in the Rue Morgue* in 1841. Charles Felix introduced antimony as a poison in 1865 in *The Notting Hill Mystery*, as did Wilkie Collins in *The Law and the Lady* (1875).

Two writers became prominent in this genre in the 1900s: Dorothy Sayers and Agatha Christie. In 1930, Sayers wrote the classic arsenic-centred *Strong Poison* excerpted in Chapter 1 (Section 1.3), and at least one of her short stories *Suspicion* features arsenic. Sayers, who was a Dante scholar and theologian, was not trained as a chemist but her books reveal a wide knowledge of the science of the time.[55] Arsenic appears as a poison, suspected poison, reference or joke in 25 per cent of all Christie's novels. It was a murder weapon in *The Lernean Hydra* (1927), *The Tuesday Night Club* (1932), *Easy to Kill* (1939), *The Cornish Murder* and *They Came to Baghdad* (both 1951), and *What Mrs. McGillicuddy Saw* (1957).

Christie worked as a hospital pharmacy dispenser during both World Wars, and made literary use of the knowledge she gained. She did not like guns: in more than half of her 61 novels, at least one of the corpses is the victim of an overdose of poison, drugs or other chemicals.[54] The arsenic that Christie chose to employ her novels was usually in the form of arsenic oxide, which is colourless, odourless and tasteless and could be administered on food such as trifle, cake or coffee. In *The Mirror Crack'd* the poisoner claimed the coffee tasted bitter and did not drink it, but Miss Marple knew that arsenic had no taste, so clearly the rejection was for other reasons. (There was a discussion about the taste of arsenic during the trial of Madeleine Smith (Section 5.4.5). The expert witness Dr. Christison claimed it was "very slight indeed – if anything sweetish, and all but imperceptible." This was established by holding 120 mg samples as far back along the tongue as could be done with safety, for around two minutes before spitting them out, and washing the mouth carefully."[31])

In a departure from normal practice in Christie's *They Came to Baghdad*, copper arsenite (Section 3.2) added to food caused severe gastroenteritis, but the victim survived. Christie would have liked the following real life story: In 1968, William Waite – married, with two children and personal chauffeur to Lord Leigh of Stoneleigh Abbey – had access to the estates gardening supply of Paris green (copper acetoarsenite, Section 3.2). He slowly began to add the pesticide to his wife's food and drink because he had fallen in love with another employee, Judith Regan, 20 years his junior. Mrs Waite eventually died. Her doctor was suspicious and an autopsy revealed she had been given large doses of the arsenical immediately prior to her death. The jury found Waite guilty and he was sentenced to life imprisonment.[56]

A recent survey of 300 crime writers in Britain revealed that poisoning was their favourite instrument of death.[57] In a review of the cause of death in 187 works of fiction, cyanide at 13.4 per cent, beat arsenic at 13 per cent, and mushrooms came in third.[9] The A-word made it to the title of *The Arsenic Labyrinth* by crime-writing lawyer Martin Edward, released in March, 2007.[58] The most famous detective of all, Sherlock Holmes, had little to do with poisons, but in a collection of Sherlock Holmes pastiches *Armchair mysteries of Sherlock Holmes*, edited by Alan Downing, the detective reveals a considerable knowledge of contemporary arsenic chemistry.[59]

Opera is another medium that favours death by poisoning but composers and librettists don't usually worry about the chemical composition. In Verdi's *Simon Boccanegra* set at the time of the Italian poisoning excesses (Section 5.3.1), Simon was given a "slow poison" in Act II. His death brought down the curtain at the end of Act III. However, one modern opera *Lives of the Great Poisoners*, with music by Orlando Gough, details the exploits of famous poisoners and their poisons of choice. Madame de Brinvilliers (Section 5.3.2) is in the cast, as is Dr. Crippen who probably used hyoscine to murder his wife. Thomas Midgley is there, billed as one of the greatest poisoners of all because his invention of leaded gasoline and the fluorocarbons that depleted the ozone layer "had more impact on the atmosphere than any other single organism in the earth's history."[60] At the other end of the scale, in the musical *Chicago* with music by John Kander and book by Fred Ebb and Bob Fosse, first performed on Broadway in 1975, one incarcerated murderess describes her crime during the musical number *Cell Block Tango*:

"I met Ezekiel Young from Salt Lake City about two years ago and he told me he was single and we hit it off right away. So, we started living together. He'd go to work, he'd come home, I'd mix him a drink, we'd have dinner. Well, it was like heaven in two and a half rooms. And then I found out: Single he told me? Single, my ass. Not only was he married..oh, no, he had six wives. One of those Mormons, you know. So that night, when he came home, I mixed him his drink as usual. You know, some guys just can't hold their arsenic."

In Cantonese opera, admittedly a very different art form, there are many specific references to the A-word; one work is named *Smiling to Drink Arsenic*.

5.5.3 Portrait of a Poisoner

In real life, poisoners come from all sectors of society. Most poisoners that have been discovered have been male, and not – as commonly held – female. According to some experts[9,12,35] poisoners are for the most part cunning, avaricious, cowardly and physically nonconfrontational. They are childlike in their fantasies, and somewhat artistic in their planning. Poisoners are permanently immature and they rarely confess. Poison is regarded as a coward's weapon because it can be administered by stealth and over a long period of time, in full perception of the victim's prolonged suffering. Poison is often selected as a murder weapon because there is a good chance that the crime will not be discovered. Arsenic ranks as the No. 1 poison in general use, possibly because of its availability. It is followed by cyanide, then strychnine.[9,12,35]

The readers of the newsletter of the Toxicological History Society, *Mirthradata*, were invited to come up with a list of poisoning's biggest villains and the biggest heroes in combating crime and ignorance. The villains: Locusta (*ca.* 55 CE); Catherine de' Medici (*ca.* 1532); Cesare Borgia (*ca.* 1490); La Spara (*ca.* 1700); La Toffana (*ca.* 1650); Marie de Brinvilliers (*ca.* 1670); La Voisin; (*ca.* 1680); Donald Harvey (*ca.* 1970); Anna Maria Schonleben (*ca.* 1907), William Palmer (*ca.* 1853); and Mary Ann Cotton (*ca.* 1871). Apart from Locusta who may have used plant derived poisons – her victims died quickly – all the women used arsenic, as did the males Borgia and Harvey. Professor Taylor found some antimony in the stomach of one of Palmer's victims but came to the very controversial conclusion on minimal evidence that strychnine was the murder weapon. The jury agreed.

The chosen heroes: Dioscorides (*ca.* 15 CE); Galen (131–220); Maimonides (1135–1204); Paracelsus (1493–1541); Orfila (*ca.* 1814); Marsh (*ca.* 1836); and Hugo Reinsch (*ca.* 1842). The last three were involved in the development of tests for arsenic.

5.6 Some Serial Killers

Mary Ann Cotton
She's dead and she's rotten
She lies in her bed
with her eyes wide open.
Sing, sing oh what can I sing?
Mary Ann Cotton is tied up wi' string.
Where, where? Up in the air
Sellin' black puddens a penny a pair.

Children's nursery rhyme *ca.* 1873

5.6.1 Mary Ann Cotton, Britain's First Serial Killer

In 1872 Mary Ann Cotton, living in Auckland, County Durham, sent her stepson Charles to the local chemist to get a small amount of arsenic. Arsenic was a common household item that some people combined with soap to use in scrubbing floors. The chemist would not supply the poison to the child, so a neighbour obliged (Figure 5.1). Charles died soon after of gastric fever, arousing suspicion in the village and Mary Ann was refused payment on his

Arsenic and Crime: The Law of Intended Consequences 193

FATAL FACILITY; OR, POISONS FOR THE ASKING.

Child. " PLEASE, MISTER, WILL YOU BE SO GOOD AS TO FILL THIS BOTTLE AGAIN WITH LODNUM, AND LET MOTHER HAVE ANOTHER POUND AND A HALF OF ARSENIC FOR THE RATS (!) "
Duly Qualified Chemist. " CERTAINLY, MA'AM. IS THERE ANY OTHER ARTICLE ? "

Figure 5.1 Punch 1849, with permission from the Bodleian Library, Oxford.

insurance policy. The doctor, who had kept samples from the stomach of the corpse, found arsenic. Mary Ann was tried in March 1873 for the murder of the child and the prosecution revealed a long list of other gastric fever victims and arsenic purchases. The defence claimed that the arsenic in the child came from dust originating from the wallpaper in the house (Section 3.5). This novel idea was not accepted by the jury and Mary Ann was hanged slowly and painfully on March 24th, 1873. The hangman had misjudged her weight.

It is estimated that she poisoned 21 people: 10 of her own children, three husbands, five step-children, her mother, her sister-in-law, and her lover, thus becoming the first serial killer to be caught in Britain. Her reasons appeared to be greed and the desire to remove obstacles from her life.[12,61]

5.6.2 The Black Widows of Liverpool

Margaret Higgens and her sister Catherine Flanagan were both hanged on March 3rd, 1884, and were also honoured with a place in Madame Tussaud's

Wax Museum. They were the ringleaders in an insurance scam operating in an impoverished area in Liverpool, England. One of their victims was Higgens's husband, Thomas, who was poisoned by arsenic obtained from flypapers. At the time of his death, his life was insured by five different companies for a total of 108 pounds and 4 shillings, then a considerable sum. There were many more victims and murderers. A lodger or relative would be chosen to be insured and neighbours would pay the weekly premiums, even helping out with the payments if one family was short of cash. One of the many insured would eventually be selected for elimination, and the investors would receive their dividend. All the participants in the scam, except the two sisters, got away with murder.

The insurance industry was complacent. The trial judge laid most of the blame on the practice of paying the agents a commission on the premiums collected. He said little about the practice of persons insuring the life of another in whose affairs they have no interest.[62]

5.6.3 Vera Renczi

Another Black Widow, but one who killed for jealousy and not personal gain, Renczi was born in Hungary in 1903. She murdered 35 individuals including her husbands, lovers and a son. Her son was trying to blackmail her, but the rest she poisoned with arsenic because she had become convinced that they were either no longer interested in her or were unfaithful. She was finally discovered after the wife of one of her lovers became suspicious when he did not come home. Vera admitted the crimes and let the police into the basement of her home where the men's remains were on display in expensive zinc coffins.[12]

5.6.4 Madame Popova

Madame Alexe Popova was a serial murderer for profit from 1879 to 1909, who undertook her work as much from sympathy as for the minor fees she charged. Although technically not a Black Widow, she certainly got rid of the husbands of a lot of other women. She was an advocate of women's liberation long before the cause was recognised. A native of Samara, Russia, she was so distressed by the situation of peasant wives held "captive" by their brutish husbands that she volunteered inexpensive, lethal remedies, some involving arsenic. For 30 years before her ultimate arrest in March 1909, she ran a small disposal service for her female neighbours, picking up spare change and carrying out her commissions with dispatch. A client, suddenly remorseful, turned her in to the police, and Madame Popova confessed to "liberating" some 300 wives in her career. In custody, she boasted of the fact that she "did excellent work in freeing unhappy wives from their tyrants." In her own defence, Madame Popova told her captors she had never killed a woman. Czarist soldiers saved her from a mob that sought to burn her at the stake, and she was unrepentant as she stood before the firing squad in 1909.[61]

5.6.5 Johann Hoch

A stylographic pen, one of the first mass-produced fountain pens, was found in the room of Johann Hoch when he was arrested for murder. The reservoir was filled with a white powder, believed to be poison. The powder was later confirmed to be arsenic trioxide.[63] At the time, Chicago police and journalists dubbed him "America's greatest mass murderer" and the "Stockyard's Bluebeard." Hoch bigamously married at least 55 women between 1890 and 1905, bilking all of them of cash and poisoning many, but the final number of his victims is a matter of conjecture. He managed to hide many of the bodies and police were only certain of the identity of 15 who received proper burials. In the end he went to trial (and to the gallows) for a single homicide.

Hoch selected his victims from newspaper "lonely hearts" columns. He avoided detection because of the widespread use of arsenic in embalming fluids and he also gutted the body of his first victim to ensure there would be no vital organs to analyse. He poisoned his final victim and buried her, without any qualms, very soon after their marriage. But times had changed, and although there was now no arsenic in the embalming fluid, there was plenty in the exhumed body. Hoch was hanged in 1904 at the age of 51. He claimed the arsenic was in the pen because he might need it to commit suicide.[12,64]

5.6.6 The Arsenic Gang

Insurance was also the cash generator for a crime syndicate operating in Philadelphia, PA during the 1930s; the group was dubbed the Arsenic Gang because many of its victims were poisoned.

The gang may have caused the deaths of 1000 individuals mainly from the Italian community in Philadelphia and the surrounding area. Gang members posed as faith healers offering their services to Italian immigrants seeking medical and spiritual help. The healers were well established in the community and their services were reasonably priced. Once the customer's confidence had been obtained, the gang purchased life-insurance policies on one or more of his of her relatives. They collected the proceeds after the insured had been murdered. The arsenic was derived from ant poison and was often administered by a compliant relative who also stood to gain from the target's passing. As many as 10 policies were purchased on any one individual. Because the policies often paid double for accidental deaths, some of the murders were orchestrated to appear accidental. The operation ended in several high-profile court trials and 24 defendants were found guilty to some degree of murder. The majority of the cases involved arsenic poisoning.[65–67]

5.6.7 The Grandmothers of Nagyrev

The trial of four grandmotherly women began on December 13th, 1929, in the village of Nagyrev, in the remote region of Tiszazug, Hungary. One of the women, Rosalia Holyba, had dispatched her rude, drunkard husband with arsenic in his coffee. She and the others also used arsenic to get rid of a village shrew ("better off without her") and one male who was lame ("sent to a better place").[68]

The practice was to call a doctor before administering the poison. Usually the victim was suffering from something, so death was not a surprise. The chief instigator and source of the poison seemed to be local midwife Suzanne Fazekasne, who later committed suicide but never showed remorse. The poisoners got their arsenic from flypapers and rat poison. One of the accused said: "It is true that I gave Maria Kotelec the flypaper-liquid. She had been complaining tearfully about her husband who was cruel to her and treated her very badly. I only know that I felt sorry for that poor woman. Besides, I took the water from an old, much used fly-paper – not even the flies died anymore. I'm sure that liquid could not harm anybody."[68] Another of the accused baked flypaper biscuits for her mother to get her inheritance early. Later, when she tired of both her husband and her lover, she eliminated them as well.

The police became suspicious and 50 graves in the village were opened – 46 of the bodies were found to contain lethal doses of arsenic. Initially the women did not know that arsenic could be traced, but once alerted, they tried to interfere with the exhumations and to hinder the identification of the bodies. There were two more trials and more than 50 additional women, aged 53 to 70, were accused of murdering their husbands, sons, lovers, fathers and mothers-in-law, among others. The total number of victims was about 140. The Hungarian government decreed that no corpses that had been interred for more than 20 years could be exhumed. This was because the possibility existed that the elimination of unwanted villagers had been going on for centuries and they did not want to know about it. It had certainly been the case for at least two decades. Some of the accused were acquitted and others hanged, rather arbitrarily. Five women committed suicide before the trials began.

The population of Nagyrev in general supported the accused and seemed to be saying this is the business of our community. We do not want your interference. These are our murders and these are our victims. No word of these mass murders reached the outside world until years later and no satisfactory explanation was ever offered.[12,68] The events have been presented as a play *To Have* and a film *The Angel Makers*.[69,70] The word arsenic does not appear in the play.

5.6.8 Dr. Michael Swango

Swango trained as a medical student at Southern Illinois University. While there, the high death rate of patients in his care led him to be known as "Double-O Swango – Licenced to Kill." One of his sources of amusement was to nonfatally poison acquaintances, including his fiancée. He was convicted of nonfatally poisoning fellow paramedics in Quincy IL, in 1985, after they became suspicious: it seemed that whenever Swango was left alone with their food or drink, someone became ill. They set a trap and found that the unsweetened tea set out as bait suddenly became sweet. Terro Ant Killer, essentially arsenic with sugar, traced to Swango was the cause.

After a short time in prison he was easily able to find medical jobs in the United States and eventually in Zimbabwe and Zambia. Whenever he was suspected of poisoning patients he was dismissed and not charged, leaving him free to take other positions and kill some more. He was a psychiatric resident in New York when he murdered three of the victims mentioned in his guilty plea. He never expressed any remorse.

Swango eventually admitted to killing four people, but it is likely that the number is much greater: probably between 30 and 40, and at one time the Federal Bureau of Investigation suggested it might be 60. He was sentenced in September 2000, to life in prison.[71,72]

5.6.9 Donald Harvey

Harvey, who is on our original list of villains (Section 5.5.3), is a self-professed "Angel of Death" and was, until the unmasking of Dr. Harold Shipman, one of the most prolific serial killers of all time. Harvey claimed to have murdered 87 people and was arrested in April 1987. (Shipman, who had a medical practice in Hyde, England, committed at least 250 murders from 1975–1998.) Harvey worked mainly as a nurse's aide in a number of institutions and killed patients by a variety of methods, including smothering with a wet towel, sprinkling rat poison on food, adding arsenic and cyanide to orange juice, and injecting cyanide directly into their buttocks. Arsenic in food was his major weapon.[61,73]

5.7 Delivery Systems

5.7.1 Food and Drink

Food and drink are the usual vehicles for delivering poison to a victim, as we have seen above. In general the delivery of arsenic is not difficult because it is colourless, odourless and tasteless, although massive doses probably need to be hidden in vehicles such as cocoa (Section 5.4.5). However, the use of arsenic provides an interesting option to a poisoner: Character A starts eating arsenic and builds up a tolerance. Characters A and B then share the same arsenic-loaded meal. Character B succumbs but Character A can plead innocence – the arsenic must have come from somewhere other than the meal prepared by Character A (see *Strong Poison*[74]). The flip side of this modus operandi is the sudden withdrawal of arsenic from an established arsenic eater. The Styrians tell us that this action will bring on all the effects of acute arsenic poisoning. One of the two arsenic-related plot lines in the book *If I'd killed him when I met him* involves this "withdrawal" ploy.[75] After the American Civil War, a young southern woman was accused of poisoning her older carpetbagger husband. Arsenic was found in his body, but all those present at the fateful soirée ate the same pastries that were believed to be the source of the arsenic. So how did she do it? The solution: he was an arsenic eater who each day, to support his habit, consumed arsenic on the pastries baked by his wife. When she substituted sugar for the arsenic oxide so that all present could eat the pastries, he suffered withdrawal poisoning. He died, and she was acquitted.

The fear of mass food poisonings has been exploited over the years to promote a variety of causes and we occasionally encounter headlines like: "Animal Rights Group Poisons Turkeys." This particular act, real or imaginary, forced the North American grocery chain, Safeway, to take the birds off the shelves, for the fifth time in a decade, before Christmas, 2004. The grocery chain was advised an unnamed animal rights group had poisoned the turkeys with arsenic. The industry offered a reward of \$100 000 (Cdn) for information leading to an arrest.[76]

5.7.2 The Poisoned Shirt

The poisoned shirt was a favourite medium of the poisoners of the 17th century. The would-be poisoners soaked the shirttails of their targets in a solution of arsenic. Arsenical concoctions such as the Borgias' La Cantarella were used. The dried garment was then given to the victim to wear in the belief that horseback riding would result in chafed skin and provide a path for the poison to do its work.[15,77] Poison was commonly also dusted into the gloves of intended victims in anticipation of the same results (see also Section 3.7).

Queen Elizabeth I was almost the victim of a similar crime. A number of attempts were made on her life with the backing of the King of Spain. The Spanish Church gave Edward Squyer a mineral poison that would exert its effect through the skin. He found an opportunity to smear the poison on the pommel of the Queen's saddle just as she was about to mount and cried out "God save the Queen!" to escape suspicion. He was later caught, tried, convicted and executed.[19]

This incident was romanticised in the movie *Elizabeth I*. One of her maids-in-waiting together with her lover, were first pictured in ecstasy – but clothed – then she was in agony, and finally dead. The cause of death was implied to be a poisoned dress meant for the Queen but worn by the maid.

5.7.3 Application via a Prophylactic

In the 18th century, a Frenchman was hanged following the death of a number of young and beautiful wives. He evidently ensured a succession of spouses without a succession of children by coating the outside of his goat-skin condoms with white arsenic.[78] A classic book on medical jurisprudence recommends that the condition of the genital organs should be examined in all cases of suspected poisoning.[37]

5.7.4 The Poisoned Maiden

Legends in many cultures involve fatal poisonings delivered through skin contact, usually through kissing, and may be a reference to the transfer of venereal disease. Some versions of the legend have involved the oral transfer of poison in capsules. In one example from the 15th century, a court physician is said to have put poison on the lips of the daughter of the King of Naples. The King kissed the daughter, who also happened to be his mistress, and died.[78]

5.7.5 The Poisoned Ring

The signet ring worn by Cesare Borgia contained a secret receptacle that was supposedly designed to hold a substance, such as arsenic, for ready delivery;[79] however, in the long history of poisoned rings it is the wearer who usually ends up dead.[80] Leonora in Verdi's *Il Trovatore* makes use of such a loaded weapon. She refers to it as an "exquisite jewel" and describes the effect in a long aria as she dies. The theme has a variation in which Cesare's ring is made from two lion's heads with sharp teeth. A warm welcoming handshake led to a bloody hand and a painful, poisoned, departure.[6]

5.7.6 The Poisoned Candle

Toxic green candles were described in Chapter 3 (Section 3.8) although they may not have been used to cause deliberate harm. Pope Clement VII (Giulio de' Medici) is believed to have been deliberately poisoned in 1534 by fumes from a candle he carried in a religious procession. Arsenic had been blended into the wax and was volatilised on heating.[6] The same *modus operandi* may have been used to remove Leopold I of Austria.[5] These stories do seem to be a little far-fetched, although it has been recently found that emissions from candle wicks containing lead present an unacceptable risk to human health; and burning incense during church services can briefly expose worshippers to high levels of unhealthy particulate matter that could cause respiratory, pulmonary and cardiovascular disease.[81,82]

5.8 Public Arsenic Attacks

Malicious mass poisonings may be far more common and deadly than generally appreciated.

5.8.1 Japanese Curry

In July, 1998, four people died and 65 were hospitalised after eating curry laced with arsenic at a festival in Wakayama Prefecture. A rash of copycat incidents followed, involving – among other things – cyanide and azide in tea, and resulting in about 34 deaths during the following year. The population suffered much anxiety and links were made to deeper problems in Japanese society.

Evidently there had been a similar string of poisonings in the 1970s.[83,84] However, in 1998 the reasons were more prosaic.

Volunteers had set up food stalls in Wakayama and the chairman of the local residents' association, Takatoshi Taninaka, was one of the first to eat, but soon he and others started feeling queasy. Many returned home, embarrassed to appear sick in front of their neighbours, which may have been a mistake. Unaware that anything was wrong, the housewives kept serving the curry. By the next morning, Taninaka was dead, along with his deputy and two unrelated children. Sixty-five others had been rushed to the hospital.

The cause was discovered some time later, by accident, when two other individuals were diagnosed with arsenic poisoning. The two regularly dined at the home of a local couple. Another man, who was an employee of the couple, had died in 1985 under mysterious circumstances. More suspicion was heaped upon the couple (who once used arsenic in their termite-extermination business) after press reports named them as beneficiaries of insurance policies taken out on the lives of each of the three men. The wife, Masumi Hayashi, a 41-year-old insurance broker, was eventually charged with the poisonings. She claimed her motive for poisoning the curry was anger toward her neighbours. The objective seems to have been to kill as many as possible, for she had insured many of the afflicted.

There was huge media interest in the affair, with hours of live TV coverage and 5000 people turning up for the 80 seats in the courthouse on the opening day of the trial in 1998. In 2002, Hayashi was sentenced to death by hanging before a crowd of 2200. A powerful piece of evidence was matching the identity of the arsenic sample found in the couple's home with the arsenic in the curry. The antimony and bismuth impurities were the same as determined by using Japan's then new synchrotron source (Section 7.2).

Hayashi, still on Death Row in 2007, continues to appeal her conviction, and there may be some hope. According to *Go Kinjo Zankoku Jikenbo* (Cruel Neighbourhood Crime Files) a woman, who was a schoolgirl at the time of the poisoning and who said she saw Hayashi pour a cup of poison into the curry, has now recanted and admitted that she made up her statement.[85]

5.8.2 Campus Coffee

In 2000, 27 members of Laval University, in Laval QC, were poisoned by drinking arsenic-contaminated coffee from a vending machine. The onset of their symptoms began half an hour after they drank the coffee, which was the only activity the unusually large group of sick individuals had in common. A white substance, later shown to be arsenic, was detected in the coffee grounds. There were no fatalities reported and no threats were made. Because most of the incapacitated had connections with the university's Department of Animal and Food Science, there was the possibility that animal rights activists might be involved, but no one has claimed responsibility for the attack, and authorities have remained silent.[86]

5.8.3 Church Picnic

In 2003, arsenic-laced coffee was served to a Lutheran Church congregation in New Sweden ME. One person died and 15 became ill. Daniel Bondeson, a well-known local marathon runner and a member of the church, committed suicide a few days after the poisoning. He left a note that convinced investigators he was involved in the poisoning and may not have acted alone. More than two dozen churchgoers said they drank the contaminated coffee and said it had a bitter taste. Some three years later, the investigation was closed when police declared that Bondeson had acted alone in retaliation for imagined slights from his fellow parishioners. He had emptied the contents of a can from his farm into the coffee urn, but only intended to give them a fright.[87]

At the time, health officials said the severity of the attack was reduced because of the availability of arsenic antidotes that had been purchased with the aid of federal antibioterrorism grants, made available after the September 11th, 2001, terrorist attacks.[89,90]

Box 5.4   Organoarsenicals

Murder by organoarsenicals is rare, but Dr. Pierre Bougrat attained notoriety during the 1920s when he was convicted of murdering a close friend Jacques Rumèbe by injections of salvarsan. The doctor had already achieved fame in France because he was awarded both the Military Cross and the Legion of Honour during WWI.

After the war, Bougrat set up practice in Marseille, where he showed great empathy for the down and out, the prostitutes, pimps and drug traffickers, often providing his services without charge. He was also a playboy who loved women and spent lavishly. One of his patients was Jacques Rumèbe who was afflicted with syphilis and was being treated with salvarsan (Section 1.6). The preparation of the drug prior to injection required that the doctor take great care. One day, Rumèbe, a few hours after having received his treatment from Bougrat, returned in a panic from a drunken visit to a brothel claiming to have lost his satchel, which was full of money. Feeling sick, he asked his friend to go back to where he had been and try to find the money. Bougrat did as asked but when he returned empty-handed he found his friend dead.

Bougrat declared at the time of the investigation, that he had panicked, thinking that he would be accused of stealing the satchel and killing his friend to cover the crime. So he hid the body in a cupboard in order to give himself time to think. By coincidence, that same day the police arrived to arrest Bougrat for passing rubber checks to cover his gambling debts and they discovered the body of Rumèbe. The doctor was charged, found guilty of murder by a vote of six to five, and condemned to death. The sentence was commuted to 25 years of hard labour because of his war service and because no one decorated with the Legion of Honour could go to the scaffold.

> **Box 5.4 Continued.**
>
> Bougrat arrived at the remote French penal colony in the Bay of Cayenne at the end of 1926, and was immediately appreciated for his medical skills. After six months' incarceration he became one of the few to escape the institution and live to tell about it. He eventually arrived in Iripa, Venezuela, in the middle of an epidemic, where he tended to the population with much skill and devotion. The authorities turned a blind eye to his situation and he continued to practice there until his death in 1936.[79,89]

5.9 Two Ongoing Cases

5.9.1 A Political Poisoning

The use of poison for political purposes did not stop with the Borgias. In 2004 Viktor Yushchenko, then a contender for the Ukrainian presidency, was poisoned with dioxin.[91] When former Russian spy Alexander Litvinenko died in London on November 23rd, 2006, he was initially reported to have been poisoned with thallium, but the lethal agent turned out to be radioactive polonium (Po-210), an alpha-particle emitter. This murder was particularly clumsy and according to one crime writer, even arsenic would have been more "professional".[92–94]

The poisoning death of Munir Thalib continues to attract attention. On September 7th, 2004, the 38-year-old Munir, a human-rights activist who had provided legal council to victims of Indonesia's Suharto regime, was aboard a Garuda flight from Jakarta to Amsterdam. Shortly after takeoff he began to suffer from diarrhoea and acute vomiting, and was dead on arrival at Amsterdam.

The autopsy conducted in Amsterdam found more than 460 mg of undigested arsenic in his stomach. The in-flight orange juice was the suspected source, although another report suggests the arsenic was delivered in some fried noodles.[95] The Indonesian government dragged its feet in the subsequent investigation and there were accusations of a coverup. A fact-finding team concluded the US CIA was involved, resulting in calls for justice from the US Congress and the European Union.

Indonesian pilot, Pollycarpus Priyanto, who had links to Indonesia's intelligence agency, was traveling on the plane, but not on duty, and gave his business-class seat to Munir. Pollycarpus got off the plane when it stopped in Singapore. He was belatedly charged in 2006 with Munir's murder, convicted, and sentenced to 14 years in jail; however, this was quashed on appeal to the Indonesian Supreme Court, prompting more protests and demands from Munir's widow and supporters for a full investigation. Munir's widow is seeking $1.4 million (US) in damages from the airline for negligence.[95–97] The Indonesian government announced in February 2007, that it would undertake a re-examination of Munir's body organs to confirm "the poison type, when the poison was administered, and how the poison came to be in Munir's body."[98] These objectives are well beyond what can reasonably be expected from another analysis; however, the announcement generally pointed to more foot-dragging by the local police.

With the inquiry in its third year, the latest news is that more suspects have been announced, mainly connected to the airline. The lawyer for one of those arrested, Indra Setiawan, reported that his client went missing from his cell without notification. His guards said he had been "borrowed" by investigators.[99] According to the BBC news, October 9th, 2007, a former president of the airline and a deputy are now facing charges that they allowed Pollycarpus the opportunity to poison Munir.

5.9.2 Cynthia Sommer

In this case, the charge was arsenic poisoning and the victim was Cynthia Sommer's husband Todd, a US Marine, who died on February 18th, 2002. The initial diagnosis was a heart attack and the body was cremated, but two sets of samples of his liver, kidney, heart, urine, and brain were saved : supposedly one set of samples was frozen fresh and the other fixed in formaldehyde. On the basis of the analysis of these samples conducted in 2003, that were said to reveal more than 1000 times the acceptable level of arsenic in the liver and 230 times in the kidney, Cynthia Sommer was charged with her husband's murder in November 2005.[100] Because of the uncertainty in the "acceptable levels", the measured liver concentration could be somewhere around 20 ppm which might be believable to 460 ppm which would be unbelievable (Section 5.4.6). Surprisingly, the arsenic concentrations in other samples such as muscle and urine were normal.

Todd Sommer was violently ill on February 8th, 2002; 10 days before his death. He recovered and returned to work, but complained of heart palpitations the night before he died. It is very unlikely that an acute dose of arsenic on February 8th would have such a delayed effect, yet this is what the prosecution maintained. The tissue samples were sent to the Armed Forces Institute of Pathology (AFIP) for analyses where they ended up in the Environmental Division rather than the more appropriate Forensic Division. They also ended up in the care of Dr. Jose Centeno who had no experience with human tissue samples, and who performed the analyses on a newly acquired instrument (ICP-MS, Section 7.2.2). He originally suspected the samples had been contaminated because of the high concentrations[101,102a] but eventually he was willing to put his name to the report: other colleagues were not convinced.[102b] The actual arsenic concentrations found in Todd's organs were not quoted in the news media although some values were posted on the internet.[103] (The numbers have since been confirmed in court documents).[102b] A liver arsenic concentration of 95 ppm (ww) and 15 ppm (ww) in the kidney suggest that Todd Sommer must have ingested a large quantity of arsenic – enough to lead to a very rapid death (See section 5.4.8) Dr. Centeno also reported that the arsenic species in the liver was dimethylarsinic acid, another oddity, because evidence was given that inorganic arsenite was the main species found in a suicide's liver (120 ppm (dw), approx 24 ppm(ww)).[44a] Arsenate is the main arsenic species found in normal human liver and kidney tissues followed by arsenite and smaller amounts of methylarsenic species.[44a,44b]

Prosecutors argued that Cynthia Sommer wanted a more luxurious lifestyle than she could afford on her 23-year-old husband's $1700 monthly salary and saw his life insurance policy as a way to "set herself free".

There was nothing linking Sommer with arsenic or its administration, but this did not stop the prosecution from maintaining that she was the only person with the motive and access to carry out the poisoning.[104] Special Agent Mary Jane Lewis of the Naval Criminal Investigative Service was produced to testify that she paid $25 for seven vials of arsenic she found on eBay, implying that Sommer could have done the same. With the same objective, another Navy investigator testified that he found ant poison, Grants Kills Ants Stakes, for sale in a military store on base.[105] The arsenic from two and a half ant stakes is enough to provide a near fatal dose of 100 mg of arsenic trioxide. Cynthia Sommer would have had to obtain the arsenic from approximately 100 stakes and persuaded her husband to eat this in order for him to accumulate 95 ppm in his liver.

Nevertheless, the jury found Sommer guilty of murder, despite insubstantial evidence against her – an incompetent attorney did not help. It seems that, like Florence Maybrick a century earlier, she was punished for her lifestyle.[106,107] Sommer hired a new defence lawyer, Alan R. Bloom, and appealed the decision on two main grounds: the legal sufficiency of the evidence; and the fact that the judge allowed prosecutors to introduce evidence of the defendant's sexual conduct after her husband's death.

In May 2007, the sentencing hearing was postponed for a second time over the issue of possible contamination of the samples and the lack of proof that the samples came from Todd Sommer. DNA testing is now being done.[108a] It is very possible that the samples not only were sent to the wrong laboratory for analysis, but were mixed up in the process so that the liver and kidney samples actually analyzed were not from the fresh frozen batch but from the "fixed in formaldehyde" batch. Apparently dimethylarsinic acid is commonly added to the formaldehyde,[108b] which would explain both the amount of arsenic in the liver and kidney samples and the speciation.

On Friday November 30th, 2007, Judge Peter Deddeh took the very unusual step of granting a new trial, but no bail, for Cynthia Sommer. It seems that evidence will be presented to assert that her husband Todd Sommer died of something other than arsenic poisoning-possibly from taking the now-banned weight loss pill Ephedra.[102b] The first trial aired live on Court TV and for a small sum could be viewed all over North America. Message boards such as www.Courttv.com provided additional routes for the public to become engaged. The public's fascination with poisonings and poisoners does not appear to have diminished.

5.10 Bezoars, Unicorns and Food Tasters

The Greek natural philosopher Pliny (Section 1.2) believed that some rocks had medicinal properties and that diamonds were an antidote against all poisons. This protective role was eventually taken over in Arabia by bezoar stones, around the year 1100. Bezoar was once translated to mean "the wind of the breeze of poison", nowadays "counterpoison" or "antidote" does the job. The stones are calcareous growths that accumulate around irritant particles in the intestines of animals, and are found in a range of sizes and colours (Section 1.2).[5]

The belief in the curative power of stones such as emeralds and bezoars was widely held, as exemplified in the writings of Maimonides. Moses Maimonides was born in Cordova in 1135 and died in Egypt in 1204 when he was personal physician to the Sultan of Egypt. Maimonides, one of the "heroes" listed in Section 5.5.3, recommended that ground emeralds taken in cold water or wine were a simple antidote against poisons such as arsenic.[109] (Stones such as pearls and diamonds are still part of Chinese and Ayurvedic medicine.[110b])

The most prized bezoar stones came from the bowels of the Persian wild goat and were worth 10 times their weight in gold during the Renaissance. Some cultures swallowed the stone as part of the cure but the patient was not allowed to go far until the stone had been recovered. Other stones were simply waved over food and drink. They were fashioned into jewellery and worn to protect the wealthy from harm. Bezoar stones were included in the fifth London Pharmacopoeia in 1746 but disappeared soon after.[5,110a]

Bezoars still feature prominently in traditional Chinese medicine where, for example, they are ground and mixed with arsenic sulfides (Section 1.2).

The fabled unicorn horn had similar properties.[111] An active trade in unicorn horns existed as early as 1126. At the height of the trading period, in the 15th and 16th centuries, there were probably about 50 complete horns in all of Europe and they were very costly, being worth more than twice their weight in gold. When Queen Elizabeth I ascended to the throne of England in 1558, the Royal Treasury had one horn, then valued at £10 000. In 1577, British explorer Martin Frobisher described a sea unicorn in the form of a great fish with a single spiral horn, confirming at least that unicorns existed in the form of the narwhal. (The narwhal's horn, which can reach a length of nine feet, is in fact a very sensitive sensory organ[112]). But many began to doubt the existence of terrestrial unicorns, and some also went so far as to doubt the effectiveness of all horns against poison; a development that was not encouraged by those who had invested in human gullibility.[5,111]

Terra Sigillata was another universal antidote available in the 5th century. The clay came from the Island of Lemnos and was sold in tablet form, stamped with the name and sealed to ensure authenticity. Later, similar clay was found in different parts of Europe, and in the 16th and 17th century it was used to make mugs that could be drunk from without fear of being poisoned.[5]

The same belief in the protective power of clay held in some parts of India, where the nobility ate from special ceramic plates in the belief that poisoned food would not harm them. In addition, a superintendent always was present in the royal kitchen and food tasters were employed, just to be sure.[5]

Not much has been written about the lives of food tasters apart from cartoons (Figure 5.2). The position of food taster to the Lord of Castle Mandawa in India's Thar Desert recently became vacant. For three generations the family of Mathura Prasad held the job. "Food was kept under lock and key," he said. "The cook would bathe and change into different clothes, and guards would check his pockets and turban. When the food was ready, some from each dish would be fed to a dog. Next I would taste, then the guards. The food would go to the table under armed escort. Several trusted generals would test it. Finally the lord and his guest would exchange bits of each dish, just in case."[113]

Given the job description, human applicants are hard to find in these more ethical times, so it should not be a surprise to hear that a mouse was the official food taster for a recent Asia–Pacific Economic Co-operation (APEC) summit meeting in Thailand: "All food will be tasted in order to thwart any attempt at biochemical assassination...The mouse will be observed to see if it suffers or

"As my new official food-taster, you are continuing the proud tradition of your 16 predecessors."

Figure 5.2 Herman by Jim Unger. Reprinted with permission from LaughingStock Licencing Inc. Ottawa, Canada.

drops dead: if there is a poison such as arsenic the rodents will die within seconds."

Twenty-one leaders, including US President George W. Bush, were to eat five meals that involved use of the mice as testers, including the summit gala dinner.[114]

5.11 Some Historical Connections

Professor Taylor wrote: "In all instances of sudden death, there is generally a strong tendency on the part of the vulgar to suspect poisoning. They never can be brought to consider that persons may die a natural death suddenly, as well as slowly; or, as we shall presently see, that death may readily take place slowly as in the case of disease, and yet be due to poison."[37] We have noted in Chapter 4 that a number of people believe that the death of Napoleon was the result of slow poisoning, although in general poisoners don't have the luxury of an imprisoned victim to observe.

Huo Yuan Jia, a famous Chinese martial arts teacher in the 1900s, died at the age of 42, possibly of pulmonary tuberculosis, but many believed he had been poisoned by the Japanese as revenge for their defeat at a judo competition. Others speculated that he was poisoned by European colonists who felt threatened by the rise of Chinese nationalism. Still others believed that he may have died as a result of changes made to his arsenic-based medicine by his new Japanese doctor.[115,116]

Harry Houdini, the legendary escapologist died in 1926, supposedly of a ruptured appendix. His grand nephew plans to exhume his body to determine if he was poisoned by arsenic, because he publicly debunked the claims of spiritualists that it was possible to contact the dead.[117a] The results of the analysis may not be known for some time. A book published in 2006 entitled, *The Secret Life of Houdini*, claims he received death threats from mediums towards the end of his life.[117b]

5.11.1 Zachary Taylor

The body of Zachary Taylor, 12th President of the United States, was exhumed in 1991 because a biographer named Clara Rising became convinced that he had been poisoned with arsenic. She must have been very persuasive to get Taylor's descendants to agree to this because, although arsenic poisoning was suspected at the time of death, the autopsy done soon after gave no indication that this was the case. His antislavery beliefs were cited as a possible motive. A career officer in the US Army, Taylor became a national hero after the war with Mexico. He took office in 1849 and soon thereafter his normally good heath failed; so much so that he took ill during a political meeting on July 4th, 1850, and had to be taken indoors. There he was given some cherries and buttermilk. He began to vomit and died a few days later, possibly of cholera. Samples of hair and finger nails taken from the body in 1991 were subjected to neutron activation analysis at the Oak Ridge National Laboratory, but no elevated concentrations of arsenic were found. However, dissenters remain.[118,119]

These negative results prompted an essay in *Time Magazine* protesting the practice of exhumation of celebrities' bodies on flimsy evidence.[120]

5.11.2 Wolfgang Amadeus Mozart

Mozart's brief life had few periods of good heath and his final illness lasted 15 days. It began with swelling of the hands and feet, leading to almost complete immobility after a period of prolonged vomiting. He died on the morning of Monday December 5th, 1791; undoubtedly the bloodletting then *en vogue* hastened his demise. The doctors saw nothing unusual in the death and the disease followed a well-recognised pattern. However, the local newspapers reported on the swelling of the body and speculated that this indicated that Mozart had been poisoned. The final diagnosis was "military fever", a non-specific term used to denote a fever with rash – probably rheumatic fever.[121–123]

His wife Constanze seemed to have been convinced that there was a conspiracy against Mozart. A biography of the composer written shortly after his death suggests that he told Constanze he believed he was writing the *Requiem* for himself, and that he had been poisoned. A later version of the same story indicates that some six months before his death he was consumed with the idea that he was being poisoned. "I know I must die," he exclaimed. "Someone has given me aqua Toffana (Section 5.3) and has calculated the exact time of my death, for which they have ordered a Requiem. It is for myself I am writing this."[123]

Soon after the funeral the public declared Italian composer Antonio Salieri, the prime suspect in Mozart's death. Salieri, who was working in Vienna at the time, had no good reason to want Mozart dead, and denied any involvement. Yet the story had legs and led to an opera by Rimsky-Korsakov *Mozart i Salieri*, a setting of a story by Pushkin, that was first performed in 1898. In the opera, Salieri fears Mozart's gift will destroy music tradition and is also jealous. He poisons Mozart's wine and Mozart dies. Salieri is relieved but sad. The same plot is found in Lortzing's *Szenen aus Mozarts Leben*. A modern version of the story, containing many factual errors and much speculation, is presented in the play *Amadeus* by Peter Schaeffer, and its film adaptation.[124]

There are a number of other versions of the cause of Mozart's death, some involving bizarre conspiracy theories.[122,125] However, a recent examination of all the evidence concludes that Mozart was not poisoned and suggests that death was probably a result of a systemic autoimmune disease that caused nephritis and followed a streptococcal infection.[126] But a good story never goes away and an interesting case has been made that antimony in the form of an overzealous use of tartar emetic – in common use at the time to treat depression, fever, and fatigue – may have contributed to his demise. This diagnosis was based on the putrid odour of the body, the swollen hands and feet, fainting, depression and exhaustion. Mercury, again in use as medicine, is also offered up for consideration.[127,128]

5.11.3 Pyotr Ilyich Tchaikovsky

Tchaikovsky was born in the remote Russian town of Votkinsk in 1840. He was sent to boarding school at the age of 12 in order to prepare for life as a civil servant. Graduates from the St. Petersburg Imperial School of Jurisprudence filled the upper levels of the civil service. He spent a few years in government employ before he decided to become a full-time musician.[129]

Tchaikovsky's death occurred at the height of his musical achievements. He had just finished his sixth symphony, the *Pathétique*, which he dedicated to his nephew Vladimir Lvovich (Bob) Davidov. His attraction to Bob, and the possibility of its development into something sexual, was a source of angst for the composer. The conventional story (according to his biographer and brother, Modest Tchaikovsky) goes as follows: After an evening performance of a play in St. Petersburg, Tchaikovsky and friends, including Bob went to a restaurant for supper where the composer drank unboiled water in spite of his knowledge that there was cholera in the city. He was reported to die of the disease a few days later, on November 6th, 1893, suffering greatly from severe diarrhoea and vomiting. His funeral was a national event, with students from his old school playing a major role, and he was buried next to Borodin and Mussorgsky. However, other theories regarding his death have been circulating with one consistent theme: Tchaikovsky committed suicide for one reason or other, and arsenic was involved.[129]

In one version, in 1893 Tchaikovsky had an affair with the 18-year-old nephew of a count who was close to the Tsar. The count wrote a letter to the Tsar and used Nikolay Jacobi, who had been at school with the composer, as the messenger. Jacobi learned of the potential scandal and convened a court-of-honour to protect the name of the alma mater. The court, which consisted of eight school friends living nearby, delivered an ultimatum to the composer: he was required to kill himself by taking arsenic. Arsenic was chosen because its symptoms mirrored the effects of cholera. This account of his death was disseminated by Alexandra Orlova, who spent time working in the Tchaikovsky archives before escaping from the Soviet Union to the United States. She published this theory in 1980 and it has received considerable publicity in the media in Europe and the United States (not everyone agrees, needless to say). It was written up in the *New Grove Dictionary of Music and Musicians* 1980 and also in the *Encyclopedia Britannica* in 1989.

In an earlier version of the suicide story, Tchaikovsky had a homosexual affair with a member of the royal family. When this was discovered, the Tsar gave him a revolver and a poisoned ring filled with arsenic. He chose the arsenic and emptied it into a glass of wine.[129]

Tchaikovsky's opera *The Enchantress* ends with the female lead being tricked into taking poison and the male lead going mad. The work was not well received and is not performed, although there is an ironic connection with the stories of his death.

Anthony Holden[129] wrote a BBC documentary on the subject entitled *Who Killed Tchaikovsky* but the issue remains in dispute. Well-respected modern biographies of the composer (*e.g.* by Poznansky[130]) have concluded that his death was the result of natural causes. Poznansky argues that any coverup on the scale needed by the circumstances would have required the complicity of physicians, relatives, priests, journalists and many others. "No report from serious or tabloid newspapers contains even the remotest hint of any activities behind the scenes."[130]

This story has a modern parallel in the death of the mathematician Alan Turing, the "father of the computer." He played a major part in breaking the Nazi Enigma code during WWII, leading to the Allied victory in the battle of the Atlantic. Turing, a homosexual, was later perceived as a security risk during the Cold War. He was convicted on a charge of gross indecency and was offered at sentencing, the choice between chemical castration by means of hormone injections or prison. Turing chose oestrogen, which proved to have disasterous side effects. He was found dead in his home in 1954 with a cyanide-laced apple by his bedside. The verdict was suicide but some believe it was state-sanctioned murder.[131]

References

1. J. P. Griffin, *Brit. J. Clin. Pharmacol.*, 2004, **58**, 317.
2. D. Shiflett, in *Vancouver Sun*, 2004, Nov. 12, p. D17.
3. M. A. Babcock, *The night Attila died. Solving the murder of Attila the Hun*, Berkley Books/Penguin, Canada, 2005.
4. A. Aggrawal, in *Mithridata*, 2002, **XII**, p. 13.
5. W. Tichy, *Poisons: antidotes and anecdotes*. Stirling Publishing Co Inc., New York, 1977.
6. P. Cooper, *Pharmaceutical J.*, 2002, **269**, 901.
7. S. Klein, *The most evil women in history*, Michael O'Mara Books Limited, London, 2003.
8. R. Lucanie, in *Mithradata*, 1998, **VIII**, p. 4.
9. J. H. Trestrail, *Criminal poisoning*, Humana Press, Totowa NJ, 2000.
10. B. Johnston, in *Chicago Sun Times*, 2002, Oct. 2.
11. N. Cheetham, in *The Oxford dictionary of popes*, Oxford University Press, Oxford, 1993.
12. M. F. Farrell, *Poisons and poisoners*, Bantam Books, London, 1994.
13. D. Williams, in *Globe and Mail*, 2004, Feb. 21, p. A19.
14. (a) F. Mari, A. Potettini, D. Lippi and E. Bertol, *Brit. Med. J.*, 2006, **333**, 1299.
    (b) M. L. Carvalho, F. E. Rodrigues - Ferreira, M. C. M. Neves, C. Casaca, A. S. Cunha, J. P. Marques, P. Amorim, A. F. Marques and M. I. Marques, *X-Ray Spectrom.*, 2002, **31**, 305.
15. W. B. Johnson, *The age of arsenic*, Chapman and Hall Ltd., London, 1931.
16. A.-L. Shapiro, *Breaking the codes: Female criminality in Fin-de-Siecle Paris*, Stanford University Press, 1996.
17. R. Oliver, *History Today*, 2001, **51**, 28.

18. W. Roughead, *The trial of Mary Blandy*, William Hodge and Company, London, 1914.
19. L. O. Pike, *A history of crime in England: illustrating the changes of the laws in the progress of civilization, written from the public records and other contemporary evidence*, Patterson Smith Publishing Corporation, Montclair, New Jersey, 1968.
20. N. Potter, MD thesis, University of Pennsylvania, 1796, Library of Congress, Washington DC.
21. A. G. M. Madge, in *Pharm. Historian*, 1983, Vol. 15, p. 12.
22. J. Marsh, *Edin. New Philos. J.*, 1836, **21**, 229.
23. A. Curry, *Poison detection in human organs*, Charles C. Thomas, Springfield, Illinois, 1969.
24. S. Smith, *Mostly murder*, Charnwood, Leicester, 1959.
25. E. Saunders, *The mystery of Marie Lafarge*, Clerke and Cockeran, London, 1951.
26. P. Edwards, in *Toronto Star*, 2006, July 29.
27. E. H. W. Myerstein, *A life of Thomas Chatterton*, Russell and Russell, New York, 1972.
28. P. Bartrip, *Medical Hist.*, 1992, **36**, 53.
29. Anon, in *Pharm. Historian*, 1994, Vol. 24, p. 2.
30. L. Rosenfeld, *Clin. Chem.*, 1985, **31**, 1235.
31. F. T. Jesse, *Trial of Madeleine Smith*, New Edition William Hodge & Company, London, 1927.
32. R. D. Altick, *Victorian studies in scarlet*, J. M. Dent & Sons Ltd., London, 1972.
33. Anon, in *Chambers' Journal*, 1856, July.
34. *The New Zealand Herald*, 1955, Oct. 8, p. 14.
35. D. Gee, *Poison the cowards' weapon*, Whitcouls Publishing, New Zealand, 1985.
36. L. A. Parry, *Trial of Dr. Smethurst*, Canada Law Book Company Limited, Toronto, 1931.
37. A. S. Taylor and T. Stevenson, *The principles and practice of medical jurisprudence*, J. & A. Churchill, London, 1883.
38. H. B. Irving, *The trial of Mrs. Maybrick*, Wm. Hodge and Co Ltd., London, 1912.
39. F. Jones, *Murderous women*, Key Porter Books Limited, Toronto, 1991.
40. S. Harrison, *The diary of Jack the Ripper, the discovery, the investigation, the debate*, Little Brown, 1994.
41. F. Young, *The trial of Herbert Rowse Armstrong*, William Hodge & Company Limited, London, 1927.
42. J. N. Pirl, G. F. Townsend, A. K. Valaitis, D. Grohlich and J. J. Spikes, *J Anal. Toxicol.*, 1963, **7**, 216.
43. *Richard Alan Abood Lyan vs the State of Texas* No 08-92-00165-CR, Court of appeals of Texas, Eighth District, El Paso, 1994.
44. (a) L. Benramdane, M. Accominotti, L. Fanton, D. Malicier and J.-J. Vallon, *Clin. Chem.*, 1999, **45**, 301.

(b) M. Yukawa, M. Suzuki-Yasumoto, K. Amano and M. Terai, *Arch. Environ. Health*, 1980, **35**, 36.

(c) T. Lech and F. Tvela, *Forensic Sci. Internat.*, 2005, **151**, 273.

45. (a) P. Zang, C. Chen, M. Horvat, R. Jacimovic, I. Falnoga, M. Logar, B. Li, J. Zhao and Z. Chai, *Anal. Bioanal. Chem.*, 2004, **380**, 773.

(b) R. Zeisler, R. R. Greenberg and S. F. Stone, *J. Radioanal. Nucl. Chem.*, 1988, **124**, 47.

46. M. Beales, *Dead not buried*, Robert Hale Publisher, 1995.
47. E. Grice, in *Daily Telegraph*, 1995, May 11.
48. News, in *Delhi Newsline*, 2005, Nov. 24.
49. M. Besnard, *The trial of Marie Besnard*, Heinemann Ltd., 1963.
50. C. Franklin, *World famous acquittals*, Odhams Books, Feltham, Middlesex, 1970.
51. D. Jernigan, in *The Connector. Connecticut State Library News Letter*, 2001, April, vol. 3.
52. Apparel, in *Hamilton Spectator*, 2006, May 10.
53. Anon, in *Globe and Mail*, 2004, Jan. 24.
54. M. C. Gerald, *The poisonous pen of Agatha Christie*, University of Texas Press, 1993.
55. N. Foster, in *Chemistry and crime*, ed. S. M. Gerber, American Chem. Soc., Washington DC, 1983.
56. G. Lloyd, *The passion killers*, Robert Hale Limited, London, 1990.
57. Associated Press, 2000, findarticles.com/p/articles/mi_m1571/is_5_16/ai_59410426.
58. M. Edward, *The arsenic labyrinth*, Allison and Busby, 2007.
59. A. Downing, *Armchair mysteries of Sherlock Holmes: The doctor*, www.murraymoffatt.com/books-sherlock-holmes.html.
60. C. Churchill, O. Gough and I. Spink, *Lives of the great poisoners*, Methuen Drama, 1993.
61. Crime Library, *Mary Ann Cotton*, www.crimelibrary.com/nortorious-murders/women/cotton/l.html.
62. A. Brabin, *History Today*, 2002, **52**, 40.
63. Kamakura, *Hoch arraigned*, www.kamakurapens.com/Archive/ThePoison Pen.html.
64. T. Taylor, *Dead men do tell tales*, www.prairieghosts.com/bluebeard.html.
65. G. Cooper, *Poison widows – A true story of witchcraft arsenic and murder*, St. Martins Press, New York, 1999.
66. M. Greenberg, in *Mithridata*, 2000, Vol. X. p. 7.
67. R. Bentley and T. G. Chasteen, *Chem. Educator*, 2002, **7**, 51.
68. F. Gyorgyey, *Caduceus*, 1987, **3**, 40.
69. J. Valley, in *The Scotsman*, 2005, Nov. 9.
70. J. Hay, *To Have*, 1995.
71. J. B. Stewart, *Blind eye. How the medical establishment let a doctor get away with murder*, Simon and Schuster, New York, 1999.
72. D. Woods, *Brit. Med. J.*, 2000, **321**, 657.
73. Wikipedia, *Donald Harvey*, en.wikipedia.org/wiki/Donald_Harvey.

74. D. Sayers, *Strong poison*, Victor Gollancz Ltd., London, 1930.
75. S. McCrumb, *If I'd killed him when I met him*, Random House, 1995.
76. Canadian Press, in *Vancouver Sun*, 2004, Jan. 17, p. B2.
77. J. W. Mellor, *A comprehensive treatise on inorganic and theoretical chemistry*, Longmans, Green and Co, London, 1929.
78. C. D. Paloucek, F. D. Leskin and J. B. Hoppe, in *Mithridata*, 1999, Vol. IX, p. 3.
79. I. Mellan and E. Mellan, *Dictionary of poisons*, Philosophical Library, New York, 1956.
80. P. Halsall, *Medieval sourcebook: Accounts of the Arab conquest of Egypt, 64*, www.fordham.edu/halsall/source/642Egypt-conq2.html.
81. P. D. Thackler, *Environ. Sci. Technol.*, 2006, Sept. 1, p. 5167.
82. M. v. Alphen, *Sci. Total Environ.*, 1999, **53**, 243.
83. T. McCarthy, in *Time South Pacific*, 1999, Feb. 22.
84. D. Macintyre, in *Time Asia*, 1998, September 14, vol. 152.
85. R. Connell, in *Mainichi Daily News*, 2006, Aug. 1.
86. T. T. Ha, in *Globe and Mail*, 2000, June 2.
87. P. Belluck, in *New York Times*, 2006, April 19.
88. R. Claiborne, *Shattered idyll. Arsenic poisoning, fatal shooting rock Maine church community*, ABC News, New Sweden, 2004, May 2.
89. R. Furneaux, *The medical murderer*, Elk Books, London, 1957.
90. UFE, in l'Union des Franc, ais de l'Etranger, 2002, Jan–March, p. 4.
91. W. Knight, in *New Scientist*, 2004, Dec. 13.
92. T. Gerritsen, in *Washington Post*, 2006, Dec. 17.
93. J. Vasagar, I. Sample and T. Parfit, in *Globe and Mail*, 2006, Nov. 20, p. A2.
94. D. Saunders, in *Globe and Mail*, 2007, May 23, p. A11.
95. Reuters, in *The Age*, 2006, Oct. 10.
96. A. Deutch, in *Washingtonpost.com*, 2006, Oct. 8.
97. National News, in *Indonesian National News*, 2006, Dec. 15.
98. Tempo, in *Tempointeraktif.com*, 2007, Feb. 6.
99. A. D. Soedarjo and R. M. Sijabat, in *Jakarta Post*, 2007, May 29.
100. E. Spagat, in *Washington Post*, 2006, March 11.
101. L. Sweetingham, www.courttv.com/trials/sommer/012307.
102. (a) L. Sweetingham, www.courttv.com/trials/sommer/011007_ctv.html.
 (b) A. R. Bloom, *People of the States of California versus Cynthia Sommers*, Case No. SCN145202, Notice and Motion for New Trial, May 31st, 2007.
103. Courttv, boards.courttv.com/showthread.php? pagenumber=1&threadid= 285667.
104. M. Caruso, in *Daily News West Coast*, 2007, Jan. 14.
105. J. Randerson, in *Guardian*, 2007, March 29.
106. A. Hoffman, in *Associated Press*, 2006, July 11.
107. A. Hoffman, in *Associated Press*, 2007, Jan. 30.
108. (a) K. Wheeler, www.nctimes.com/articles/2007/06/16/news/sandiego.
 (b) W. Bloch, C. Hoffmann, E. Janssen and Y. Korkmaz, *Practical methods in cardiovascular research*, Springer, Berlin, 2005.

109. S. Muntner, *Treatise on poisons and their antidotes*, Vol. 2, J. B. Lippencott Company, Philadelphia, 1966.
110. (a) R. Porter, *The greatest benefit to mankind. A medical history of humanity*, Harper Collins, London, 1997.
    (b) P. Paranjpe, *Ayurvedic medicine. The living tradition.* ed. Chaukhamba Sanskrit Pratishthan, Delhi, 2003.
111. J. C. Giblin, *The truth about unicorns*, Harper Collins, 1991.
112. W. Broad, in *New York Times*, 2005, Dec. 13.
113. Anon, in *National Geographic*, 2005, May.
114. GuardianNews, in *Globe and Mail*, 2003, Oct. 18, p. A11.
115. Wikipedia, *Huo Yuanjia*, www.answers.com/topic/huo-yuanjia#wp-_note-3#wp-_note-3.
116. Anon, *Northern Shaolin*, www.liuhopafa.com/shaolin.htm.
117. (a) L. McShane, in *Vancouver Sun*, 2007, March 24, p. D6.
    (b) W. Kalush and L. Sloman, *The secret life of Houdini*, Simon and Schuster, 2006.
118. D. Zebra, *Zachary Taylor The health and medical history of a president*, www.doctorzebra.com/prez/t12.htm.
119. News, *Anal. Chem.*, 1991, Sept. 15, p. 871A.
120. D. Rosenblatt, in *Time Magazine*, 1996, July 8, p. 52.
121. D. Weiss, *The assassination of Mozart*, Cornet Books, London, 1972.
122. P. J. Davies, *Mozart in person. His character and heath*, Greenwood Press, New York, 1989.
123. H. C. R. Landon, *The Mozart compendium*, Schrivner Books, New York, 1990.
124. J. Warrack and E. West, *The Oxford dictionary of opera*, Oxford University Press, Oxford, 1992.
125. W. Stafford, *Mozart's death. A collected survey of the legends*, McMillan, London, 1991.
126. J. O'Shay, *Was Mozart poisoned. Medical investigations of the lives of the great composers*, St Martin's Press, New York, 1990.
127. E. N. Guillery, *J. Am. Soc. Nephrology*, 1992, **2**, 1671.
128. P. A. Wolf, *Clin. Chem.*, 1994, **40**, 328.
129. A. Holden, *Tchaikovsky. A biography*, Random House, New York, 1995.
130. A. Poznansky, *Tchaikovsky. Last days. A documentary study*, Clarendon Press, Oxford, 1996.
131. A. Hodges, *The Alan Turing home page*, www.turing.org.uk/turing/.

CHAPTER 6
Arsenic at War: Mass Murder

6.1 Introduction

The United Nations' definition of a chemical-warfare agent is: "any chemical which, through its chemical effect on living processes, may cause death, temporary loss of performance, or permanent injury to people and animals." The term is broadened in the minds of many to include agents such as defoliants that indirectly harm people by destroying their source of food. However, we will see that there are political reasons for keeping the official definition as narrow as possible. Many chemical substances such as dynamite, TNT, BTX, TATP (triacetone triperoxide), gunpowder and napalm act as explosives, propellants and incendiaries, and cause injury to individuals by way of bullets, shells, shrapnel, burns and shock, but by long-established convention these materials are not considered chemical-warfare agents.

Although the record is not extensive, chemical warfare was employed long before the introduction of gunpowder drastically changed the nature of armed conflict. The Greek historian Thucydides reports that the Spartans used smoke containing arsenic against besieged cities during the Peloponnesian war, 431–404 BCE.[1,2] A Syrian engineer named Callinicus is said to have invented "Greek fire," which seems to have been a primitive form of napalm, used around 637 CE. Its actual formulation remains unknown although there is speculation that it was made from a volatile hydrocarbon with a mixture of arsenic sulfide and saltpeter (potassium nitrate) as igniter.[3] Saltpeter is formed during the decomposition of organic matter by soil bacteria.[4] This ignition mixture was discovered by the Chinese as they tried to develop an "elixir of immortality." They found that a mixture of saltpeter, honey, sulfur and arsenic sulfide spontaneously erupted in smoke and flame. The discovery was refined during the Sung dynasty (960–1278 CE); the honey was replaced with charcoal, and the mixture of sulfur, charcoal and saltpeter became known as "fire drug." At this time none of the mixtures exploded because none contained the necessary 75 per cent saltpeter to achieve detonation. Arsenic compounds were added to the slow-burning gunpowder

---

Is Arsenic an Aphrodisiac? The Sociochemistry of an Element
By William R. Cullen
© William R. Cullen 2008

mixture to give the product more "oomph," and this arsenical smoke was used in battle.

The first true gunpowder formula was published around 1040. Three combinations of sulfur, charcoal and saltpeter were developed, producing, respectively, an explosive bomb, a burning bomb and a poison smoke ball. Delivery devices were known by such colourful names as the "Poison Fog Magic Smoke Eruptor," and the "Nine-Arrow Heart-Piercing Magic-Poison Thunderous Fire Eruptor," which was an early version of a gun. Gunpowder was introduced to Europe around the mid-1200s, possibly by the Arabs. The Oxford University philosopher Roger Bacon is said to have been keenly interested in gunpowder and its use.[4,5] Early European gunpowder recipes included arsenic-containing minerals.

In 1485, Leonardo da Vinci, in a letter to the Duke of Milan requesting a commission, outlined many different services he could provide, one of these being a poison "gas" that was, in practice, a fine powder made from lime, arsenic sulfide and verdigris. He also advocated the military use of smoke from feathers, sulfur and arsenic sulfide, a tactic that was used by the German armies of the time.[6] By 1575, European armies were using arsenic "smoke balls" delivered via catapults. The ball concept evolved into barrels containing, for example, one part antimony, one part arsenic, one part Armenian earth and two parts asafoetida (a gum that smells like garlic), all mixed with the strongest gunpowder. Sand and limestone were also added and the device was ignited prior to delivery.[6] The arsenic and antimony were probably sulfides.

Some of the mixtures contained plant poisons such as aconite: spiders and toads and other "repulsive material" were occasionally added as well. Large-scale chemical attacks were made by filling the mixtures into barrels that were loaded onto carts, ignited; and these so-called petards were then pushed toward the enemy camp when the wind was favourable.[6] The phrase "hoisted by one's own petard" refers to these.

As in North America, the arrival of European settlers in Australia around 1830 had a profound effect on the native Aborigines. In the Manning Valley their natural food supplies were almost exhausted, and they attacked the settlers. The response was quick and deadly. One tactic, known as the "Harmony Policy," involved giving the starving natives gifts of food laced with arsenic. Their water holes were also poisoned, and those who survived this treatment were simply massacred. The total aboriginal population dropped from 200 000 in 1788 to 22 000 by 1860.[7]

Some years later during World War I, we find the following entry for June 7 1915, in the diary of British 2nd Lieutenant W.V. Hindle, preserved in the Imperial War Museum, London: "At 7 p.m. we returned to the trench in the 'wood of death.' We relieved the 8th Battalion Gordon Highlanders. We soon found that a lovely clear stream ran through our trench. Of course the first impulse is to drink. Hardly had the news of the stream gone along the line when the officer commanding informed us that the water was poisoned with arsenic and that it is kept in a poisonous condition by the Germans. The stream runs through the German lines. Now in spite of that warning, four men of

B Company take the water and drink it. The consequence is fatal to all four and a great lesson to the remainder of the men. The men are buried in the wood with little difficulty."[8]

In Chapter 1 we encountered the vile-smelling, toxic, flammable mixture of arsenicals known as Cadet's fuming arsenical liquor (Section 1.6). This attracted the military eye around 1854, but the British did not elect to develop it into a weapon largely because the technology to deliver the material was not available. At the time they also considered the use of cyanodimethylarsine, cacodyl cyanide, presumable on the premise that the cyanide would enhance the effect of the arsenic.[6] We will meet this notion again (Section 6.3).

In 1887, the eminent German chemist Adolf Bayer proposed that lachrymators – tear gases – could be used for military purposes to produce temporary blindness, and by 1900 Commandant Nicolardot was advising the French military that Germany would use chlorine or bromine in future wars. He recommended that gas masks be developed with sodium thiosulfate as adsorbent.[9]

6.2 The First Chemical Weapons Conventions

The Treaty of Strasbourg, concluded between France and Germany in 1675, outlawed the use of "perfidious and odious" toxic devices (poisoned bullets).[5] The delivery of toxic chlorine gas via shells was contemplated during the American Civil War (1861–65) but was ultimately rejected. In fact, in 1863 the US War Department issued General Order 100, proclaiming "the use of poison in any manner; be it to poison wells, or foods, or arms, is wholly excluded from modern warfare."[4]

The Hague Peace Convention of 1899, convened by Tsar Nicholas II, sought to ban the use of projectiles, "the sole objective of which was the diffusion of asphyxiating or deleterious gases." The United States was the only nation of the 28 participants not to sign.[10] Later, the US had a change of heart, and the rules of war were codified in the Hague Convention of 1907, convened by Theodore Roosevelt. One annex to which the US was a party, Number IV, October 18th, contains the following: "It is especially forbidden to employ poison or poisoned weapons or arms, projectiles or material calculated to cause unnecessary suffering." Although these words had some moral force, no provisions were put into place for verification or monitoring.[6]

6.3 World War I: The Gas War

By the turn of the 19th century the centre of the European chemical industry, which began in Britain in 1856 with William Perkin's discovery of synthetic dyes (Section 3.3), had shifted to Germany.[11–13] Most of the world was hungry for the products of the six major chemical manufacturers that were later organised in 1916 into the I.G. (Interessen Gemeinschaft) for wartime efficiency. Some of these are still active today: *e.g.* Badische/BSE, Bayer and Hoechst. Dye manufacture involved reactants and intermediates that could be used as chemical-warfare agents once the moral barriers to their deployment were overcome, and some production methods were easily adapted to the preparation of new chemical agents such as arsenicals. This same industry spawned a monopolistic pharmaceutical arm that was already producing arsenicals such as salvarsan (Section 1.6).

The language of the Hague Convention did not prohibit the development of chemical weapons, and a number of the European powers began to dabble in chemical-warfare research. So almost immediately after Germany declared war on France on August 3rd, 1914, the French were able to deploy grenades containing a tear gas, ethyl bromoacetate. Their intended German targets hardly noticed the change in tactics, but did report some temporary discomfort such as eye irritation.[14]

The war was not going as planned and Germany soon became seriously short of conventional artillery ammunition.[1] On the suggestion of Professor Walther Nernst they retaliated at Neuve Chappelle, on October 27th, 1914, with 3000 shrapnel shells containing another chemical irritant, but this time it was the French soldiers who were barely distracted.[14] Other irritants were tested on the eastern front, again meeting with little success.[14]

Around this time, Professor Fritz Haber (Box 6.1) became involved in the development of chemical-warfare agents; and the offensive use of chlorine gas, a lung irritant that could easily be compressed to a liquid, moved from concept to realisation.[15] (The use of chlorine and sulfur dioxide had been considered by France, but was rejected.[6]) Successful field trials were made, and General Berthold von Deimling was eventually persuaded to deploy the new weapon by the argument that the use of the gas might shorten the war and thus save many lives.[13,14]

Germany first used chlorine gas on April 22nd, 1915, during the second battle of Ypres. The green toxic cloud, code name "Disinfection," was released from 6000 steel gas cylinders, sited along a 6-km front, near the village of Langemark in Belgium. The cylinders had been in place since February waiting for a favourable wind.[16] About 150 tons of the gas, roughly half Germany's supply, were released late in the day to drift over French territorial divisions who panicked and retreated. Tear gas was also deployed.[17] In Haber's opinion the German attack was badly handled: the military did not use enough gas and lacked imagination in capitalising on the enormous advantage that was achieved by first use of a new weapon on totally unprepared opponents.[1,14] The German army was impressed and chlorine production was stepped up.

There was an immediate international outcry against this development; however, moral indignation was not going to win the war and there were calls for immediate retaliation in kind. The British Cabinet dithered, but on May 18th, 1915, finally decided to use gas as a weapon. The decision was made a little easier by the knowledge they too were short of conventional shells. Major Charles Foulkes, an engineer, was appointed the Army's Gas Advisor. The Cabinet requested that the Kastner Kelner factory in Runcorn, Cheshire, increase production of chlorine from five tons per week to 150. It was agreed that British gas would not be used until a large-scale surprise attack could be launched, with a target date of September 1915.[18] In the meantime the Allies suffered heavy casualties from numerous German chlorine gas-cloud attacks. In May 1915, the Germans launched a major gas attack against Russian troops on the Eastern front deploying 264 tons of chlorine from 12 000 cylinders along a 12-km front.[19]

Initially there was little protection from the chlorine gas, even for the advancing German troops. The Allies resorted to improvised face masks made from cloth soaked in urine in the belief that urea would neutralise the chlorine, although water probably would have been equally effective.[18,20] A very desperate and shocked British military command hurriedly issued gauze pads, treated with sodium thiosulfate, glycerin and sodium bicarbonate that were held over the nose and mouth. Protection improved with the introduction of a gas helmet – actually a chemically treated flannel sac with a mica window in it – that covered the head. An improved version, the PH helmet (so named because it was treated with sodium phenate plus hexamine) was in the field by March 1916, but it severely restricted the wearer's vision. The idea of the box respirator came from Russia and was adopted in May 1916.[18,21] A mask covered the wearer's face and eyes, and the wearer breathed through a tube connected to a box containing filters and charcoal coated with alkaline permanganate to absorb the gas. Air was inhaled through the box and exhaled through an outlet valve, not through the box. The final version of the British design, the small box respirator (SBR) was assembled from 150 parts. It was made at the rate of millions per month and required thousands of tons of charcoal.[17] General issue started in August 1916, but the British army was not fully equipped with them until well into 1917.[9,18] During this time, the PH helmet was the only protection available. Nearly 20 million helmets and 500 000 horse helmets were manufactured, plus protective masks for dogs and bags for carrier pigeons. These undoubtedly saved thousands of lives.

The German respirator was less efficient but lighter and consisted of a small drum attached directly to the face-piece. The drum, which had to be changed frequently, was filled with layers of charcoal and pumice plus hexamine, as well as baked earth soaked in potassium carbonate. There was no mouthpiece.[21]

In response to the deteriorating tactical situation on the Eastern front and in spite of their own weakness, the British launched their first chlorine gas-cloud attack from 5000 cylinders as a major part of their battle plan at Loos, September 25th, 1915. The day before, the British also deployed the lachrymator ethyl iodoacetate, SK – so named because it was developed at Imperial College, South Kensington, London.

Robert Graves,[22] in his autobiography *Goodbye to All That*, gives a depressing account of that battle, which he describes as one "bloody balls up." Among the many things that went wrong, the fickle wind changed direction and British soldiers were attacked by their own gas. The official British military account was not so negative, although the reports do state that "the mechanical arrangements were ill adapted to the requirements," which translates to mean that many of the supplied wrenches did not fit the gas cylinders, so they could not be opened without scrambling for replacements. The battle cost the British 59 000 casualties, 7766 deaths, and any territorial gain was lost within a week.[23] The Germans named the battle "the field of corpses ... Never had the machine gunners such straightforward work to do, nor done it so effectively."[20]

By early 1916, chemical warfare was integrated into all aspects of military planning and training. Ethics had long ago ceased to be a consideration. Chemical substances were being investigated rather haphazardly for their warfare potential in about 35 British laboratories.[15] To consolidate this research effort and provide a secure area for trials, the British government purchased property, known as Porton Down, on the Salisbury Plain, in January 1916. The Germans had extensive testing and filling facilities in the Munster area[24] and the French used a number of sites including ones near Sartory (Versailles), Fontainebleau and Bouchet. Up to this point, more than 3000 compounds had been identified as potential chemical agents.

One of the early candidates was calcium arsenide, which releases toxic arsine on exposure to water. Arsine gas itself was also studied: it is both toxic and dense (Section 3.1). Weapons were constructed in which the ingredients for the rapid preparation of arsine, calcium arsenide and hydrochloric acid, were kept apart until they reached their target. These were forerunners of the modern binary weapons that contain the reagents for producing nerve gas that are mixed during flight. Eventually only 30 chemicals were assessed as suitable for field use. Twelve were adopted and used intensively, but only six really met expectations.[25]

The British continued to think in terms of gas-cloud attacks that depended on the favourable direction of the prevailing wind in France.[9] The French and Germans developed other delivery methods. The first shells containing a toxic gas, phosgene, were fired by the French on February 22nd, 1916, in the warm-up to the battle of Verdun. The phosgene was weaponised as a solution in arsenic trichloride (50 per cent) under the code name CBR. This was arsenic's first real appearance on the battle field. Arsenic trichloride was added to keep the highly toxic phosgene gas from dispersing too quickly, and to provide smoke to mark the point of impact of the shell.[26] Germany countered with similar gas shells soon after. (Old chemical weapons, filled with arsenic trichloride, have been unearthed in Britain, and Japan used the compound as a shell filling in WWII (Section 6.8.5).)[2,23]

The use of shells by France and Germany was clearly a major escalation in chemical warfare and a breach of the Hague Convention since shells are projectiles and they were undoubtedly delivering a poison. At the time, it had been coyly argued by Germany that the release of gas from cylinders was not covered under the Hague Convention and therefore permitted.[14]

Soon after Verdun came the first Battle of the Somme, beginning July 1st, 1916. The British had SK available in shells but were forced to borrow guns and ammunition from the French to deliver a lethal gas. This time they adopted a French filling named Vincennite (VN) that was a mixture containing mainly hydrogen cyanide and arsenic trichloride.[27] These lethal shells did not prevent the loss of 60 000 British soldiers to German machine gun fire on the opening day.[23]

The pace of the gas war picked up in 1917. A shell filling known as JBR was introduced. This consisted of arsenic trichloride mixed with chloroform and hydrogen cyanide, and jellified with cellulose triacetate. The very effective use of these shells and CBR during the battle of Arras (April 4th, 1917) was cited as a major factor in the initial British successes and the associated Canadian capture of Vimy Ridge, on April 9th. But like many of Field Marshal Douglas Haig's previous offensives, the initial success was followed by heavy losses.[9] At Vimy, the British First Army had a large inventory of gas shells that included 12 000 SK, 20 000 CBR and 8000 JBR rounds for their 4.5-inch howitzers, but few were used because of the rapid advance of the Canadian troops.[23] As for Vimy itself, what began as a minor but successful battle for a French hill was "transformed into Canada's defining moment of nationhood."[28,29]

A gas-warfare stalemate resulted once both the entrenched sides were equipped with efficient respirators and practiced good gas discipline. The British required troops to carry their respirators when within 12 miles of the front. Between 12 and 5 miles a man could remove the respirator box in order to sleep, but within 5 miles he must wear it constantly. Within two miles it was to be worn in the alert position (slung and tied in front). When the alarm was given he was to get the respirator on within six seconds.[21]

The stalemate could only be broken by the introduction of chemical agents that would penetrate or disable the mask, or would attack the exposed skin, rendering the mask relatively useless. The British were unable to meet this challenge. Germany did, however, and continued to introduce new agents as they were developed, although some were launched without complete field trials.

6.3.1 Mustard Gas

A successful German attack across the river Yser in Belgium, beginning on July 6th, 1917, saw the introduction of a new shell that caused sneezing and irritation of the nose and chest, but the gas was not identified. A few days later, on the night of July 12th at Ypres – where the British were openly preparing for a major offensive action, and where the first chlorine gas-cloud attack had been launched in 1915 – the Germans used phosgene in a heavy pre-emptive bombardment together with a new gas shell that was marked with a Yellow Cross. The identification of this new agent – bis(2-chloroethyl)sulfide – took about three weeks (Table 6.1).[9] It became known as mustard gas because of its yellowish-brown colour and its odour, reminiscent of mustard plants.

The blistering properties of this persistent oily compound had been known for many years, but it was first suggested for use as a chemical agent by two prominent chemists, Lommel and Steinkopf, who are immortalised in the German mustard gas code name LOST. Mustard gas, in reality, is a liquid that is slow to take effect. It seemed so mild in action that many of the exposed troops did not bother to put on gas masks, with calamitous results. Pain in the eyes, throat, and lungs and partial blindness, were coupled with blistering of the buttocks, genitals and armpits. Many casualties developed bronchitis two days

Table 6.1 Principal arsenical WWI chemical-warfare agents.[52] LCt_{50} is the median concentration in mg min/m^3 that will result in death of half the exposed population. TCt_{50} is the median concentration in mg min/m^3 that will incapacitate half the exposed population.

Formula	Name/Code	Rate of reaction	Class	LCt_{50}	TCt_{50}
$(C_6H_5)_2AsCl$	DA, Clark I	very rapid	vomiting/riot control	15 000	12
$(C_6H_5)_2AsCN$	DC, Clark II	very rapid	vomiting/riot control	15 000	8 to 22
$HN(C_6H_4)_2AsCl$	Adamsite, DM	very rapid	vomiting/riot control	10 000	20 to 30
$C_6H_5AsCl_2$	Pfiffikus, PD	immediate	blister	2600 inhalation	16 vomit 1800 blister
CH_3AsCl_2	MD, methyldick	rapid	blister	3000 inhalation	25 inhalation
$C_2H_5AsCl_2$	ED, Dick	immediate	blister	3000 inhalation	5 to 10 inhalation
$ClCH=CHAsCl_2$	Lewisite, L	rapid	blister	1200 inhalation 100 000 skin	300 eye 2000 skin
$(ClCH_2CH_2)S$	Mustard, H, HS, HD	delayed, hours to days	blister	1500 inhalation 10 000 skin	200 eye 2000 skin
$(ClCH_2CH_2)S/$ $ClCH=CHAsCl_2$	Mustard/Lewisite HL	rapid	blister	1500 inhalation	200 eye
AsH_3	Arsine, Arthur, SA	delayed: 2 h to 11 days	blood	5000	2500

after exposure and some died from inflammation of the lungs. The severe long-term effects of exposure were not revealed until much later (Section 6.10). Mustard gas lingered for a few days even in the warmest weather, while in cold damp conditions it was dangerous for a few weeks. In still colder weather it could be dangerous for months.

Haig ordered his troops to attack through a water-logged battlefield saturated with mustard gas: kilted regiments were particularly vulnerable. The high casualty rate – 14 296 in the first three weeks following the first attack – ensured that field hospitals were overcrowded, and the inability of the Allies to respond in kind provided Germany with a huge psychological advantage. The Allies had nothing comparable. The German troops referred to mustard gas as Hunstoffe, "hot stuff," which was abbreviated to HS by the Allies, and eventually to H.

The Germans initially did not realise that mustard gas had made such an impact, but once they saw how effective it was, they employed it lavishly on both military and civilian targets. Ironically, the increasingly pessimistic Haber had advised against the use of mustard gas unless he could be convinced that the war would be over within a year. This was his estimate of the time it would take for the Allies to be able to respond in kind. Germany's resources were being rapidly used up and he knew that within a year they would lose any advantage they had in the chemical war. His time estimate was close, but ultimately other factors contributed to the outcome.[30] The British initially rejected using mustard gas as a weapon because it was not lethal enough.[31]

6.3.2 Blue Cross

Germany soon reintroduced the sneezing gas first deployed on July 6th, 1917, and unexploded shells marked with a Blue Cross were found on July 28th. Each contained a bottle filled with an off-white solid identified about a month later as the arsenical chlorodiphenylarsine. The official Gas Warfare Monthly Summary of Information for August 1917, referred to new enemy chemical shells containing phosgene, diphosgene and chlorodiphenylarsine, capacity up to 11 litres, marked with a Green Cross; and another marked with a Blue Cross containing the same arsenical dissolved in coal tar. "The identification of chlorodiphenylarsine (Table 6.1), explains the frequent occurrence of sneezing during recent gas shell bombardments."[32]

The heat of the explosion of the shell was expected to transform the chemical filling into a cloud of very fine particles, a smoke or aerosol that during trials remained suspended in the air for a long time, and was able to penetrate all gas masks being used in battle at the time. The following is a report from a human trial: "The irritation begins at the nose as a tickling sensation, followed by sneezing, with a flow of mucus like a bad cold. The irritation spreads to the throat and coughing and choking sets in: finally, air passages and lungs are also affected. Headache, especially in the forehead, increases in intensity until it becomes almost unbearable – there is pain in the chest, the joints and teeth, and shortness of breath. The victim has unsteady gait, a feeling of vertigo, and is trembling all over." A gas-mask wearer became so distressed that he had to remove this protection, whereby becoming exposed to more of the same agent or another more lethal agent that was part of the same package, or delivered simultaneously. A concentration of 0.1 ppm in air can result in vomiting troops driven mad by pain and misery. The arsenical also has a vesicant action on the skin at high concentrations, but not as severe as mustard.[31]

The British were lucky that the full anticipated effect of Blue Cross was not achieved partly because, by luck, their respirators had just been fitted with additional filters, and partly because aerosol formation was inefficient (it is also possible that they were forewarned about the new agent).[1] Nevertheless, the arsenical was a major psychological weapon often fired at the beginning of a bombardment to induce sneezing and inhibit the adjustment of gas masks. British attempts to design an appropriate filter for their respirators were hampered by the lack of test samples, and an effective design was not put into production until September 1918, more than one year after the arsenicals made their first appearance.

The German respirator was modified to allow their troops to follow up arsenical bombardments, but the additions had the effect of making breathing so difficult that the mask became almost unusable.[1,33] Here, Haber faced a dilemma because the use of this weapon caused his own troops considerable difficulty.

Haber became aware of the inadequacy of the smoke production from Blue Cross shells and tested prototype devices to disperse the arsenical into the air without using an explosive charge. These "Gas Büchsen" (gas boxes) were given a field trial but did not work well and the project was abandoned. Nevertheless, in spite of their inadequacy, the Blue Cross shells were kept in the arsenal, and about 10 million of them were fired in 1917–1918. Initially, the arsenical was dissolved in phosgene and then poured into shells but the mixture was too corrosive. Ultimately the arsenical was melted and poured into glass bottles that were inserted into the shell and surrounded with TNT. This enormous effort was not very productive because the payload was only about 1 to 2 per cent of the weight of the shell.[13]

A German chemist described these developments as follows: "The German front would never have succeeded in withstanding the powerful onslaught of the concentrated forces and war material of almost the whole world, if German chemists had not at that moment held the protective shield of the Yellow Cross substance before the German soldiers and at the same time thrust into their hand a new sharp sword in the form of the Blue Cross substance."[34]

6.3.3 Arsenical Agents: The Second Generation

German chemists were asked to provide a nonpersistent, toxic vesicant and came up with dichloroethylarsine as early as 1916. This same problem was addressed by Winford Lee Lewis in the United States and he came up with

Lewisite (Section 6.4). The new German arsenical, known as "Dick", was first deployed in March 1918, again at Ypres, in a shell marked Yellow Cross 1. It was soon reclassified as Green Cross 3 because of its lethal nature. The March Gas Warfare Monthly Summary of Information describes two newly encountered shells. One marked Yellow Cross containing 98 per cent *sym*-dichloromethyl ether and 2 per cent dichloroethylarsine, and the other marked Green Cross containing 18 per cent *sym*-dichloromethyl ether, 37 per cent dichloroethylarsine and 45 per cent dibromoethylarsine. It is said that allied forces had learnt of the use of dichloroethyarsine through their examination of earth samples before the blind shells were obtained for confirmation.[35]

The German military touted Dick as a super-gas, one that would penetrate all gas masks then in use. Haber was cautious, however, and others – including Wachtel, one of Haber's staff – were negative. The vesicant action of Dick was only one-sixth that of mustard and it did not penetrate clothing, so it turned out to be a disappointment. Wachtel wrote, "Toward the end of 1917 the bureaucratic tendencies in the German Research Institute on Chemical Warfare became more powerful and made it possible for a war gas like Dick to be admitted for use in battle. From then on, the efficiency of the German Chemical Warfare Service began to decline. Bureaucracy and mediocre scientists of the type of selfish careerist succeeded – even in this field of warfare – to lead Germany toward defeat."[1]

Cyanodiphenylarsine, another arsenical in the Blue Cross family, was introduced into battle on May 26th, 1918. This was probably the most powerful irritant used in the war. It was expected to have the additional feature of releasing toxic cyanide but failed in this respect. The gas was particularly effective against artillery because only a few shells forced men to put on masks, thus lowering their fighting efficiency.

Dichloromethylarsine, or "methyldick," was also unleashed in May 1918, as was dibromoethylarsine in September. Dichlorophenylarsine, another vesicant, also made its appearance that month. The last arsenical to be introduced onto the battle field was dibromophenylarsine.

6.3.4 Tactics of Chemical Warfare

Mustard gas proved to be the most effective of all the chemical agents used in the war and was deployed ultimately by both sides. It was a major defensive weapon because of its persistence and no area was shelled with mustard within two to four days of an advance. Thus, Yellow Cross was not used on fronts to be attacked, but was employed heavily against artillery batteries, machine-gun nests and villages. Mustard was well suited for use in the tactical procedure, known as "Buntkreuzscheißen" – the simultaneous use of Green, Blue and Yellow Cross shells, respectively, lethal gas, blister (vesicant) and vomiting agents. This tactic was devised by Haber and the greatly respected artillery expert, Lt.-Col. Georg Bruchmüller, during the final days of the Russian campaign, where one counterbattery bombardment involved 90 000 rounds of

Blue Cross out of a total of 120 000 rounds.[9] There were thousands of Russian casualties mainly because their respirators were almost useless.[10]

In general, Yellow Cross was used to deny occupation of an area – "to seal the edges" – which was achieved by barrages of about 21 000 shells per square kilometre. Blue and Green Cross gases were used to fill the gaps.[18] Gas was used against artillery and infantry on a scale that had never been seen before, to kill, injure, or incapacitate the opposition physically and psychologically.[34] Blue Cross shells were often fired at the beginning of a gas bombardment because they could not be distinguished from other high explosive shells until sneezing began. Blue Cross in high concentration was able to penetrate the French gas mask: the British mask was less vulnerable. Green Cross shells were fired next, because the sneezing caused by Blue Cross might prevent proper adjustment of the respirator and increase exposure to the lethal agent. By and large the German troops attacked immediately after a Blue Cross bombardment if this was the only gas being deployed, they held off for one to two hours after a Green Cross bombardment.[9]

The British artillery used similar tactics, especially in counterbarrage operations, although they were usually without mustard and arsenicals, apart from arsenic trichloride. The tear gas SK was the main weapon at their disposal, and although not a lethal vesicant like mustard, it was persistent and forced the enemy to work with masks on at a greatly reduced efficiency.[23] Professor Fritz Haber wrote following the war that, "The Germans never had any trouble in making war gases, the French a good deal, and the British (except for liquid chlorine) all the time."[13]

The Allies began to have major successes in the Battle of the Somme that began on August 8th, 1918. The British army attack with 450 tanks led to "the blackest day of the German army in the history of war".[36] Both sides engaged in all-out chemical warfare. In 1918, 28 per cent of all German shells contained chemical agents.[13] The number of rounds fired was limited only by their availability. The final battles were fought mainly between Germany and the UK and colonial troops, with help in the later stages from the United States. The French forces were not significantly engaged in this action. The battlefield was saturated with chemical agents and many died after drinking water from shell holes. Food, water, tobacco, and equipment were contaminated. The gases were very corrosive and damaged ammunition and guns, so decontamination was required immediately after exposure. The retreating Germans poisoned water supplies with gas. The use of gas increased the misery of war, and morale on both sides plummeted.[18]

Germany ran out of mustard gas near the end of the war and resorted to saturation shelling with Blue Cross.

6.4 The US Enters the Fray

The American people knew little about the chemical war taking place in Europe at the time when the US entered the fray. News had been heavily censored by the US government.[10] The War Department and General Staff essentially ignored the chemical war and did little to prepare the army for the realities of France. President Woodrow Wilson was finally forced to acknowledge the importance of chemical warfare in June 1918, by creating the Chemical Warfare Service as an independent Department of the Army, under Major-General William L. Sibert.

The first 25 000 US troops sailed for France in May 1917, under the command of General John Pershing. They had no artillery and no gas masks. Most US soldiers arrived in France without adequate instruction in gas defence. They initially made do with refurbished British gas masks and attempted to use snails in cages as gas detectors. The snail experiments were soon abandoned.[5] Although General Pershing and his staff were not big gas enthusiasts, the US artillery ended up firing about one million chemical rounds that included shells loaded with chlorodiphenylarsine, phosgene and mixtures of the two.[1,34]

6.4.1 Lewisite

As had other countries, the US initiated a search for new chemical weapons and established laboratories at Yale University and the Catholic University of America in Washington DC. The army attempted to contract out production, but companies such as Dow Chemical Corp. declined the invitation to participate because they saw no market opportunities after the war ended.[37] Consequently, the army decided to build its own gas-manufacturing and shell-filling facility in Aberdeen MD. Construction began in December 1917, on what was to become, at a cost of about $35.5 million (US), the Edgewood Arsenal. By August 1918, Edgewood was producing chlorine, phosgene, chloropicrin and mustard for shipment to France, where the shells were filled.[10] Chlorodiphenylarsine was made in Croyland PA.[34] All US chemical agents were produced by military personnel.

The laboratories at Catholic University became involved in studies of a new arsenical agent, β-chlorovinyldichloroarsine, soon to be named Lewisite, after research leader Captain Winford Lee Lewis (Table 6.1). Like Haber, Lewis was looking for a highly toxic, nonpersistent, fast-acting vesicant. He isolated and characterised the arsenical in 1918. Lewisite may have been first prepared by Father J. A. Nieuwland at the same university in 1904, as described in his PhD thesis, "Some Reactions of Acetylene." Nieuwland was repelled by the noxiousness of the product of the reaction between acetylene and arsenic trichloride, and was actually hospitalised for a while. He elected to "forget" about the experience, leaving the "rediscovery" to Captain Lewis who had access to the thesis. There is good evidence that Nieuwland did not actually make Lewisite, although he was always pleased to take the credit, which seems strange for a man of the church. Lewis claimed full credit for the discovery and never contacted Nieuwland.[37,38]

Trials in Washington that involved human subjects revealed the compound to be a potent vesicant. (The family of one US senator, whose home was near the production facilities, was gassed when 10 pounds of the material were accidentally released.) The US government decided to manufacture Lewisite in a highly secret plant in Willoughby OH. The production details are still under wraps, although we know the plant had to manufacture its own starting materials, arsenic trichloride and acetylene, and that five steps were involved. The plant ran for only four months before the Armistice was signed; then all action ceased and the exhausted soldiers went home.[37]

Because of the extreme secrecy surrounding the operation – German spies were said to be everywhere – there is a lot of confusion about what happened to the plant and any product after November 11th, 1918. In one version the plant had made 150 tons of Lewisite at a rate of 10 tons per day. This was filled into 55-gallon drums, sent by rail to Baltimore, and then barged 50 miles offshore to be dumped in deep water. Another version has the same cargo on its way to France with the ship being scuttled after the news of the Armistice. There is also a lot of confusion about the cleanup of the plant. Some residues were buried nearby and some were probably dumped in Lake Erie.[34,37,39]

We have seen that the German response to the same challenge faced by Lewis was to come up with dichloroethylarsine. Haber and his colleagues had actually prepared and rejected Lewisite in favour of mustard in 1916.[1] They concluded that Lewisite was not a reliable war gas even though it was more potent than any other arsenical agent. According to one expert this was a major tactical error.[34]

Lewisite lost favour with the US military during the course of WWII when the allies refrained from the use of gas although it was deployed by Japan in China during this time (Section 6.8.5). It may have been used on the battlefield in Ethiopia in 1936 (Section 6.8.4) and again in Iraq as recently as 1988 (Section 6.15.6). Some nations still have stockpiles. Mustard was regarded as superior to Lewisite for delivery from the air, either as spray or bombs, and would be more efficient for disabling defence industries and workers. Because of its lower persistence, Lewisite would be preferred if the target were to be attacked again soon after the gas delivery.

6.4.2 Phenarsazine Chloride

This arsenical, known as Adamsite, is closely related to chlorodiphenylarsine in structure (Table 6.1; Figure 6.4), and was first prepared by German chemists in 1913.[40] Major Roger Adams, working in the US at the University of Illinois in 1918, noticed its potential as a chemical agent and lent his name to the compound.

A British team working independently also discovered the compound, and developed a "gas box" of their own, the M-generator, to produce arsenical smokes particularly from Adamsite and chlorodiphenylarsine. The aerosols were expected to be able to penetrate all available gas masks and to give a big advantage to the Allies, but the war ended too quickly for deployment of the new "super-weapon." It was claimed that if properly used and exploited, the M-generator "would have had more important bearing on the course of the war than any other measure that was put to a practical trial on the battlefield or that was even considered."[34] It is possible that Adamsite was used in the field by the Italian army.[41]

Other nightmarish suggestions for 1919 included a highly improbable plan that involved sending Adamsite down the Moselle River in floating pans. The arsenical was to be ignited by use of incendiaries to release a toxic cloud once the pans reached Metz (an important war centre). There was also talk of the "Kettering Bug": an unpiloted biplane capable of flying 50 miles to deliver 200 pounds of Lewisite. The plans were resurrected during WWII but they never got off the ground.[37]

Box 6.1   Professor Fritz Jacob Haber

Out of the industry which gave the world its dyes
I can chemically concoct a shock from the skies.
I'm off to devise a little surprise
Something that's certain to stun
All those who thought that war's only fought
With things like that vulgar gun.
My gas will break the deadlock, make the war much shorter
And therefore save millions from the slaughter.
From *Square Rounds: A Musical Play.*[42]

Fritz Haber was born in Breslau, Silesia, in 1868. He was a Jew who converted to Protestantism at the age of 24 to smooth out the progress of his scientific career. He played a major role in WWI, a war in which science came to the fore for the first time. Haber showed an aptitude for chemistry at a young age and eventually began his serious studies under Robert Bunsen (Section 1.6). Haber's first academic appointment was at the Technische Hochschule Karlsruhe. In 1905 he started work on nitrogen fixation, the process of converting the abundant supply of nitrogen in the atmosphere to ammonia, a much more useful substance, and by 1909 he demonstrated success by constructing a pilot plant. [The fact that we would all be dead without the nitrogen in our atmosphere seems to be ignored when the usefulness of the gas is considered.] Dr. Karl Bosch took the process to the industrial scale and a plant was functioning by 1913. Without this Haber–Bosch process, the world's agriculture would not be capable of supporting its ever-growing population. He was a brilliant chemist, and – as a result of his growing fame – he moved to Berlin in 1911, aged 42, to become director of the newly established and privately endowed Kaiser Wilhelm Institute for Physical Chemistry and Electrochemistry. He was very patriotic and ambitious and at the outset of WWI he enthusiastically threw in his lot and his laboratory with the government. This arrangement was formalised in 1916 resulting in the concentration of war-related research in one place, the Kaiser Wilhelm Institute for Chemical Warfare.[43]

Box 6.1 Continued.

Germany was not prepared for chemical warfare and Walther Nernst was the first chemist of importance to be approached by the military for help. Like Otto Bayer before him, Nernst suggested that chemical irritants be deployed, but these proved useless on the battlefield. In 1914 Haber was handed the task to improve the concept and his staff began looking for suitable chemical agents. He pushed them hard. One experiment, involving the arsenical cacodyl chloride, blew up, killing his assistant, so that line was temporarily abandoned.[13] In another account of the same accident with the same outcome, the arsenical was cacodyl oxide, which was being mixed with phosgene. Under Bunsen, Haber had became aware of the properties of these arsenicals.

A major development was the realisation that a mass discharge of a gas from containers would be effective in dislodging troops from trenches. Chlorine, which the chemical industry produced in abundance, was chosen and Haber was put in charge of the operation. We do not know if Haber made the first suggestion;[30] however, according to his son Ludwig, Haber believed his greatest triumph was the concept of the gas-cloud attack.[13] Professor Haber did not concern himself with international law and regarded himself as an army officer whose duty was to advise and obey those responsible for policy. He actively and intensively promoted the use of chemical weapons and did not regard their use as inhumane.[30] He was intimately involved with the development of all chemicals for delivery by shell and mortar, including phosgene, mustard gas, and the arsenicals. More than 100 arsenic compounds were investigated. Haber slightly gassed himself during a field trial involving chlorine.

He was placed in charge of chemical supply and his first priority was to provide explosives. This he did via his ammonia synthesis.[4] He quickly developed a gas mask once chemical warfare became an everyday experience on the battlefront, and by 1915 the German soldiers were better protected than the Allied forces.

Haber was promoted to "Royal Prussian Captain Professor Fritz Haber, Chief of the Chemical Section of the War Ministry." He did not attain higher rank, although he wished for it.[30] Nevertheless, his only real superiors were the Minister of War and the Commanders of the German Army. Haber was probably the only person in the history of the Prussian army to be promoted in this way. He was responsible for the activities of 150 university personnel and about 2000 assistants through the Kaiser Wilhelm Institute. His advisory board included Walther Nernst, Emil Fischer and Richard Willstätter: Fischer and Willstätter were Nobel Prize winners, and Nernst received his Prize in 1920 for his work on thermodynamics. Haber was ruthless and autocratic, but a brilliant organiser who believed only in victory. However, during the last few months of the war he was overworked, pessimistic and in despair: his judgment was affected and the final outcome of the war was a great shock to him.

Box 6.1 Continued.

At the end of the war, the Allies attempted to try Haber and Nernst as war criminals. Haber fled to Switzerland in disguise and became a Swiss citizen when the Allies requested his extradition. Carl Druisberg, head of the IG cartel and an enthusiastic proponent of the manufacture and use of chemical weapons, also fled. The hue and cry soon died down, partly because of the international realisation that if Haber was guilty then so were his Allied counterparts: Charles Foulkes of Britain and Amos Fries of the United States. Haber returned to Germany and was awarded the 1918 Nobel Prize for Chemistry at a ceremony in 1920, as a reward for his discovery of nitrogen fixation, the Haber process. There was some public outrage at the award because of his wartime activities, and the New York Times editorial of January 22nd, 1920, asked, if Haber got the Chemistry Prize, "Why the Nobel Prize for idealistic and imaginative literature was not given to the man who wrote [German] General Ludendorff's daily communiqués?"

Even in defeat Haber remained a leader of the scientific recovery and rehabilitation of Germany, but he never gave up on his conviction that chemical weapons were legitimate and relatively humane. On receiving the Nobel Prize Haber said, "In no future war will the military be able to ignore poison gas. It is a higher form of killing."[15] He continued to be involved in the secret manufacture of chemical weapons in violation of the Treaty of Versailles. In this he was aided by his wartime collaborator Hugo Stoltzenberg (Section 6.8.2).[44] He gave up these clandestine activities in 1926, but continued to advise the German government on such matters until he was forced to leave the country in 1933.[6]

In public, Haber worked hard with Einstein to end the animosity that had arisen between German and Allied scientists – for a few years following WWI Germans were not admitted into international professional organisations.[30] Haber was an internationalist who developed a close collaboration with Japan. He turned his Institute into a major international centre for leading-edge science.[45] One project involved pest control by chemicals and he developed a means of taming the toxicity of hydrogen cyanide by adsorbing the gas onto fine particles of diatomaceous earth. One of these formulations achieved infamy as the Zyklon B the Nazis later used to murder inmates of their concentration camps.

Haber resigned his position in 1933 when Hitler came to power, because he could not face the anti-Semitic dictates of the National Socialists. In his letter of resignation he wrote: "For more than 40 years I have selected my collaborators on the basis of their intelligence and their character, and not on the basis of their grandmothers, and I am not willing for the rest of my life to change this method which I have found so good."[45] He initially moved to Cambridge, England, where he worked for a while with Sir William Pope who had been involved with the British development of mustard gas, but his presence was not welcome, so in ill health he moved to Switzerland, where he died in 1934.[43] Haber's public image changed for the

Box 6.1 Continued.

better upon his death and hundreds of wives of German scientists attended his funeral. The Third Reich forbade their husbands to do so.[46]

Haber's wife committed suicide on May 1st, 1915, a few days after the first chlorine gas attack, possibly out of shame for his chemical-warfare activities, but possibly because she was aware of the presence of the woman who became the second Mrs. Haber, and the mother of Ludwig F. Haber who was to become the author of an influential book: "The Poisonous Cloud: Chemical Warfare in the First World War."[13] All the evidence shows Fritz Haber to have been unfeeling and unsupportive to his gifted first wife.[47] His son summed up Haber's contribution to the war effort as follows: "In Haber the German Supreme Command found a brilliant mind, a extremely energetic organiser, determined and possibly also unscrupulous."[13]

Fritz Haber is regarded as one of the two fathers of chemical warfare, the other being Walther Nernst.[48]

6.5 Arsenical Chemical-Warfare Agents

Most of the organoarsenicals that have been employed as chemical-warfare agents are shown in Table 6.1. They belong to two classes: the first, now classified as Blister (vesicant) agents, cause tissue damage, skin blistering, respiratory tract tissue damage, and lung ulceration. They are also nasopharyngeal irritants, known as sternutators, and can be lethal. The second class described as Riot Control/Vomiting Agents are also sternutators, lachrymators (tear gases) and respiratory irritants.[49]

Arsenicals have also been used in mixtures with mustard gas. One eutectic mixture contains about 63 per cent Lewisite and 37 per cent mustard. The mixture freezes at $-25\,^\circ\text{C}$, whereas mustard alone freezes at $-14\,^\circ\text{C}$; thus, the mixture can be used in cold weather and at high altitude because it remains a liquid that can be more easily spread or sprayed than a solid. Arsine oil, a readily prepared mixture that is mainly chlorodiphenylarsine and dichlorophenylarsine, is also used in this way.

It is difficult to establish toxicity figures for gases and even more difficult for smokes and fogs. Haber introduced the concept of exposure time into the determination of toxicity as in the formula: $C \times t = Ct$, where C is the concentration, t is the time of exposure and Ct, the "Haber product", is the toxic or lethal dose. For some gases, such as phosgene, the product is constant and a high dose for a short time leads to death just as a low dose for a long time does. Haber found this to be true for mustard gas, so Germany developed it as a weapon. Britain did not consider the time factor, so by their standards mustard gas was not lethal enough.

Because the susceptibility to chemical agents varies from human to human, it is not possible to specify an exact minimum dose or lethal dose for each agent so exposures are generally given in terms of the "Effect Ct_{50}" (written ECt_{50})

which indicates the exposure that has a probability of producing some kind of effect in 50 per cent of those exposed. If the effect is death it is known as Lethal Ct_{50} (LCt_{50}); if incapacitation, ICt_{50}.[52]

Brigadier-General Augustin Prentiss, in his classic text *Chemicals in war: a treatise on chemical warfare* lists German lethality data for 10 minutes' exposure in this very subjective order:[34]

phosgene ($COCl_2$) = diphosgene (CCl_3COCl) > Lewisite = mustard > dichloroethylarsine ($C_2H_5AsCl_2$) = dichloromethylarsine (CH_3AsCl_2) > chlorodiphenylarsine ((C_6H_5)$_2AsCl$) = cyanodiphenylarsine ((C_6H_5)$_2AsCN$) = hydrogen cyanide (HCN) >> chlorine (Cl_2) >>> carbon monoxide (CO).

About 3870 tons of chemical agents were deployed in 1915; 16 535 in 1916; 35 635 in 1917; and 65 160 in 1918, for a total of 121 200 tons[50] (although another source gives a total of 113 000 tons[15]). These amounts are insignificant when compared with the total of 2 million tons of high explosive and 50 000 million rounds of small arms. The number of chemical-agent-filled shells fired during WWI is given as: France – 16 million; Germany – 33 million; Great Britain – 4 million; US – 1 million (mostly supplied by France).[51] Ludwig Haber estimates that Germany manufactured 8027 tons of arsenicals during the war, and the UK around 100 tons.[13]

It should not be forgotten that the production of respirators continually strained the resources of the belligerents as they attempted to grapple with new developments in the offensive use of gas along the deadly path: chlorine, tear gas, phosgene, hydrogen cyanide, chloropicrin, mustard and arsenicals. The Allies eventually produced about 23 million respirators in addition to 66 million pads and helmets. Germany produced 12 million respirators, but only a small number of pads.[13]

6.6 Casualties of the Chemical War

6.6.1 The Combatants

It is very difficult to estimate the number of combatants who died as a direct result of hostilities during WWI. The numbers have to be guesstimates at best: Britain and her Empire, 900 000 deaths; France, 1 400 000; Germany, 1 800 000; Turkey, 300 000; the US 120 000; Romania 750 000; Italy 650 000; Russia, 2 000 000. The total number of casualties was around 37 million.[53] The total number of WWI casualties of chemical warfare is estimated at about 1.3 million.[15] Ludwig Haber gives a lower number of about 0.5 million,[13] but this figure is the sum of UK, 186 000; France, 110 000; Germany, 107 000; and US, 73 000. Data for Russia are not included in Haber's summary because of this extreme unreliability. The number is expected to be high because their soldiers lacked adequate protection. One estimate is 275 000. Gas casualties amounted to about 3.5 per cent of all casualties and about 6.6 per cent of these casualties died, mostly from phosgene exposure.

The death rate in the first gas attack was about 35 per cent of all casualties, and everyone in front of the cloud – some 15 000 men – became a casualty.[54] Once some protection against gas was available and cloud attacks were abandoned in favour of shell delivery, the rate dropped to 6 per cent of all casualties. The introduction of mustard gas and Blue Cross in July 1917, broke the ensuing gas/gas-mask deadlock, and in the first three weeks the gas casualties reached 14 000, but the number of men killed by gas was relatively low, at 500. In general, during the mustard gas period from July 1917 to the end of the war, 80 per cent of gas casualties were the result of exposure to mustard, 10 per cent to Blue Cross and 10 per cent to Green Cross. The mortality was 2.6 per cent of the gas casualties. No deaths were recorded as a result of Blue Cross exposure but this seems unlikely in view of Germany's resort to saturation shelling with Blue Cross at the end of the war.

During WWI gas produced one casualty for each 0.1 ton of chemical agent, whereas the rate for high explosives was one casualty per 0.5 ton. The average amount of mustard gas deployed per casualty was 60 pounds, which is to be compared with 230 pounds of chlorine-like lung agents, and 650 pounds of arsenicals per casualty.[34]

6.6.2 Civilian Casualties

Civilian casualties were not unexpected because civilians lived close to the front for most of the war. The data are limited but one of the highest counts was from a July 1917, German bombardment of Armentières, in which 3000 mustard shells resulted in 6400 casualties including 675 civilians. Of these, 86 civilians died.[9] Otherwise, the factories and filling plants were major sources of chemical accidents. Mustard gas manufacture proved to be particularly hazardous.[9,13,23]

The arsenicals Adamsite and cyanodiphenylarsine were manufactured in the British national factory in Morecomb. The women who were filling the fine powder into canisters were rested frequently. They wore rubber gloves and worked behind glass screens, but inevitably blisters developed on arms, chests and necks. Poor personal hygiene and general bad working conditions made their contribution, not to mention floors saturated with arsenicals.[55] German records on this and other aspects of arsenical production vanished at the end of the war, but we know that they had particular difficulty with some arsenicals in spite of Haber's boast (Section 6.3.4).

A major production facility near Munster was destroyed by fire October 24th, 1919. Huge quantities of agents were released without serious harm being done to humans, although the environment has yet to recover (Section 6.15.1).[56]

6.7 The Aftermath

6.7.1 The Humane War?

After the Armistice in 1918 there was considerable debate about the legitimacy and effectiveness of chemical warfare, and about the "humaneness" of its use.

Charles Foulkes of the UK was a very vocal proponent, as was his American counterpart Amos Fries. They were not exactly disinterested parties, having been intimately involved in the use of gas during the war. They promoted gas as the super-weapon. Foulkes said: "It was evident that the German troops feared gas above all else, that it was a constant subject of discussion amongst them, and that their losses from it had a profound effect, both morale and material, which seriously reduced their fitness for battle."[23] Fries said: "If gas never killed a man the reduction in physical vigour and therefore efficiency of an army forced at all times to wear masks would amount to 25 per cent at least."[57] Capt. Winford Lewis was also a very active and enthusiastic supporter of chemical warfare in general and in particular of the virtues of Lewisite as a war gas. He said in speeches all over the US: "It is the most efficient, most economic, most humane, single weapon known to military science." Lewis believed that the arsenical delivered from an aircraft would have led to a speedy end of the war and coined the phrase, "Dew of Death." He also argued that "it offsets mere brute weight and should be regarded as a weapon of civilised defence."[37] Just like Lewis, Haber lectured relentlessly on his conviction that chemical warfare was both effective and humane.

The humane argument is mainly based on the low ratio of deaths to casualties.[1] Hersh believes the US data are exaggerated in that the figures do not include malingering, so that the percentage of deaths is actually much higher. Malingering was easily overlooked because of the US policy that all troops who were possibly exposed had to be hospitalised.[54]

Callinicus written by John Burdon Sanderson (JBS) Haldane, British geneticist, biometrician, physiologist, and populariser of science,[31] was very influential and persuasive on the pro-gas side (Callinicus was said to have invented Greek fire and saved Christendom from Islamic domination (Section 6.1)). Haldane wrote that it is "no worse, possibly more civilised to kill or wound a man with chemicals rather than with shrapnel or bullets." Some authors were of the opinion that gas exposure had no long-term effects, and that continuing problems were largely psychosomatic.[58]

Haber's son[13] played down the influence of gas and accused the pro-gas lobby of much exaggeration and self-aggrandisement. But others, especially his father, believed in the strong psychological effect of the weapon.[59] Fear and uncertainty are powerful persuaders and it appears that the use of gas has been a major factor in the outcome of at least three conflicts: in Morocco in 1921; Ethiopia in 1935; and Iraq in 1988.

6.7.2 Public Reaction

The pitiful image of blinded men in single file, holding onto the shoulder of the man in front, being led from the battlefield to a field aid station, is immortalised in the painting by John Singer Sargent (Figure 6.1) that hangs in the Imperial War Museum in London. The painting (1918) had a powerful effect on the public who were repelled by the sight of these wounded men, and transferred this repugnance to chemical weapons and their use. Ludwig Haber writes that

Figure 6.1 "Gassed" by John Singer Sargent. With permission from the Imperial War Museum, London.

the painting is contrived and sanitised and regrets that some artists and writers were so profoundly affected by poison gas that they transformed a relatively feeble weapon into something far greater than it ever was in practice.[13]

There may be some problems with the painting, but few would deny the power of Wilfred Owen's poem that is as much antigas as it is antiwar.

Dulce et Decorum Est

Bent double, like old beggars under sacks,
Knock-kneed, coughing like hags, we cursed through sludge,
Till on the haunting flares we turned our backs,
And toward our distant rest began to trudge.
Men marched asleep. Many had lost their boots,
But limped on, blood-shod. All went lame, all blind;
Drunk with fatigue; deaf even to the hoots
Of gas-shells dropping softly behind.

Gas! Gas! Quick, boys!-An ecstasy of fumbling,
Fitting the clumsy helmets just in time;
But someone still was yelling out and stumbling
And flound'ring like a man in fire or lime...
Dim through the misty panes and thick green light,
As under a green sea, I saw him drowning.
In all my dreams, before my helpless sight,
He plunges at me, guttering, choking, drowning.

If in some smothering dreams you too could pace
Behind the wagon that we flung him in,
And watch the white eyes writhing in his face,
His hanging face, like a devil's sick of sin;
If you could hear, at every jolt, the blood
Come gargling from the froth-corrupted lungs,
Obscene as cancer, bitter as the cud
Of vile, incurable sores on innocent tongues,
My friend, you would not tell with such high zest
To children ardent for some desperate glory,
The old Lie: Dulce et decorum est
Pro patria mori.

Wilfred Owen, 1917

After the Armistice it was easy to recognise the wounded who had lost limbs and suffered other physical damage. Compensation, although not lavish, was available. The victims of gas, however, were much more difficult to diagnose. Weakened lungs resulted in a reduced capacity for work, but was it malingering? Conjunctivitis sometimes resulted in blindness many years later. Very few gas cases, less than 100, received awards greater than 50 per cent disability.[13] Figures on the published British casualty list were kept deliberately low and becoming an official gas casualty required roughly the same amount of verification as winning a medal.[15]

The situation was the same in the US, except that there, follow-up studies of veterans revealed instances of chronic bronchitis, emphysema, bronchial asthma and chronic conjunctivitis some 10 years after the exposure to gas.[60] These results, which were in the public domain, established that – contrary to opinion held by the humane-war wing of the pro-gas lobby – exposure to chemical agents had far more serious consequences than sunburn.[39]

There is essentially no mention of the use of arsenicals in WWI in the Imperial War Museum, London, the National War Museum, Ottawa, or the Liberty Memorial Museum in Kansas City. In fact there is little recognition of chemical agents at all, with the notable exception of the painting by Sargent. Gas is usually "swept under the carpet" and is barely mentioned in Regimental Histories, possibly because only a relatively small number of individuals were intimately involved. There is little glamour associated with gas deployment; and the gunners fired the shells as commanded without too much thought about the charge.[18]

Remarque's novel *All Quiet on the Western Front*[61] provides some real sense of gas warfare: "These first minutes with the mask decide between life and death: is it airtight? I remember the awful sights in the hospital: the gas patients who in day-long suffocation cough up their burnt lungs in clots." There are similar passages in *Deafening*[62] but these works are the exceptions. Charles Yals Harrison's novel *Generals Die in Bed,* written in 1930 and said to be the equal of Remarque's work, is completely silent on the subject of chemical warfare.[63]

The victims were not completely silent. A diary of A. R. Camidge, a sapper,[64] contains entries from 1916–1919 including the following: "June 12th, 1916, a lot of wounded and gassed; July 14th, 1917, Gas shells at 9 am warned at 8 am; Nov. 26th Red Lamp a disgrace, men lined up waiting; Dec. 10th Monday, Fritz sent over gas at night, over 100 gassed; Jan. 1918. Mon 18th gas three times during the night – standing in 3 feet of water, heavy shelling; June 6th Parcel from home. Everywhere reeks of mustard gas and sneezing gas; Friday 7th Believe I caught a little mustard. Feeling queer. Thursday 11th Reported sick at dressing station. Very sore in certain parts, get ointment; Friday 14th Sick, to see doctor tomorrow; 15th Sick-gas; . . . Sat 29th Return to duty."

Others mention the excruciating pain in the head resulting from exposure to sneezing gas but soldiers in general did not distinguish between gases. Gas was just part of the business of war. Comrades died of bullets, shrapnel, gas or drowning. It was all the same.

After the Armistice, gas became the focus of the antiwar movement and pro-gas sentiment was not popular. But gradually the images of Sargent and the words of Owen faded and the world became accustomed to living with the new weapon. Attempts were made to convince the public that war gases had peaceful uses and for a while chlorine was recommended for the treatment of colds – even Lewisite was claimed to be beneficial for the treatment of late-stage syphilis.[37] Entomologist Harold Maxwell-LeFroy was experimenting with Lewisite as an insecticide in London, when, in 1925, he died from inhaling the fumes of one of his concoctions. He was the founder of the well-known pest control firm Rentokil.[37]

6.8 Living with Chemical Weapons

6.8.1 The Geneva Convention

In spite of strong military and political backing, the US Chemical Weapons Service was downsized after the Armistice: its funding was reduced and the work was consolidated at the Edgewood Arsenal in Maryland. In contrast, although there was little open support for chemical warfare in Britain, the work continued, largely in secret. The financing was generous under the mandate: "Studies of chemical weapons against which defence is required."[15] Haber's Institute in Berlin was closed down, but the work continued in secret even though the text of the Versailles Peace Treaty of 1919 contained the section: "The use of asphyxiating, poisoning, or other gases and of all analogous liquids, materials, or devices being prohibited, their manufacture and importation are strictly forbidden in Germany."

This statement, which at the time did not gain the approval of the US, was an important precursor to further negotiations that eventually resulted in the Geneva Gas Protocol of June 17th, 1925, which contains essentially the same language. The document was initially signed by 38 nation-states[15] and came into effect in 1928. As usual, ratification by the individual governments was necessary, and in the US a vociferous campaign was launched against this step by the Veterans' Association, the American Chemical Society, the American Legion and many others. The US politicians sensed that the tide was turning and postponed the process. The protocol was not returned to the Senate until 1970 and it was finally ratified with some major reservations in 1975 (Section 6.12). Japan also waited until 1970. Most nations that ratified did so with the provision that they were bound only with respect to relations with other signatories, and were not bound in regard to any enemy states whose armed forces or allies did not observe the provisions.

The Geneva Protocol was essentially a "no-first-use" treaty. It was weak, it did not ban research on chemical weapons or the accumulation of such weapons, and it had no verification provisions. However, given this apparent desire to control chemical weapons, it may come as a surprise to learn that even before peace negotiations were concluded at Versailles, arsenical agents had been deployed by the British against the Bolsheviks at Archangel. The M-generators, which had been readied for use in the field at the end of WWI (Section 6.4.2) were modified so that they could be thrown from bombers, and between August 27th and September 4th, 1919, six targets were attacked.[55] This is claimed to be the first action in which chemical weapons were deployed from an aircraft; however, there are earlier reports in official WWI records and diaries, of bombs that probably contain an arsenical being dropped from German airplanes.[32,65]

6.8.2 The German Reaction

In Germany, where the humiliated military were already planning their revenge, there was a renewed interest in chemical warfare. Haber continued to espouse the view that chemical weapons were "natural" and "superior," and that to use them required "discipline and intelligence." Hugo Stolzenberg, a good friend and disciple of Haber (he had helped to open the gas cylinders for the first chlorine attack), was given the task to rescue, preserve, and if possible expand on what was left of Germany's capacity to engage in chemical warfare. He initiated aerial spraying tests in Germany under the guise of pest control.[6,66] Other scientists were asked to cooperate and many did in secret. Arsenicals such as cyanodiphenylarsine and Adamsite were designated to receive special consideraon.

Stolzenberg was secretly sent to Russia (and later Yugoslavia) to help with their chemical development. He was very active in Spain (Section 6.8.3). Stolzenberg and the German army were asked to supply chemical weapons to General Franco during the Spanish Civil War, but Hitler denied the request so tear gas was used only in a few instances.[6] Stolzenberg died in 1984 at 91, having sold his factory in Hamburg five years earlier.[67]

The search for new agents intensified under the Third Reich. The nerve agent tabun was discovered accidentally in 1936 and soman in 1938 (Box 6.3).[55] But arsenicals such as Adamsite were still required. Construction of a factory to make arsine oil began in 1935 and production started in 1938. The capacity was 270 tons per month and the plant had produced 7500 tons of agent up until March 1944. After this time chlorodiphenylarsine, about 12 000 tons, was the main product. A factory to produce the arsine precursor Trilon 300 (an alloy of arsenic, aluminium and magnesium) was being built but was not completed by the end of WWII: its planned capacity was 100 tons per month.[66]

One estimate of the total agent production from 1933 to 1945 by the Third Reich gives: mustard – 27 500 tons; arsine oil – 12 600 tons; Tabun (Box 6.3) – 12 000 tons; phosgene – 5000 tons; N-mustard [$(ClCH_2CH_2)_3N$] – 2000 tons; irritants – 8150 tons, for a total of 67 000 tons, more than was produced during WWI if chlorine gas is removed from consideration.[66] The arsenical production can be broken down into: Adamsite – 3900 tons; chlorodiphenylarsine – 1500 tons; cyanodiphenylarsine – 100 tons; arsine oil – 7500 tons; and a small amount of Lewisite. Some of this material was weaponised.

6.8.3 Spain in Morocco

Around 1920 Spain was having trouble with the Riff rebels in Morocco and decided to use gas to supplement conventional weapons. In 1921 France helped to build a filling plant in Melilla, Morocco. Spain also asked Haber for help, and he in turn directed them to Stolzenberg who was more than interested. On June 10th, 1922, Stolzenberg signed an agreement for the construction of a plant to produce mustard gas, phosgene and dichloroethylarsine in Madrid: the Fábrica Nacional de Productos Químicos began production in 1925. Stolzenberg also built a plant in Hamburg, with Spanish financial help, to produce precursor chemicals, especially for mustard gas, that could be shipped to Spain and Spanish Morocco for final conversion (the German army was actively involved at this stage). France also was interested because it had problems of its own in French Morocco.[6]

Spain and France used gas in Morocco in an offensive that began in the summer of 1925 and lasted for about two years. Roughly 10 000 mustard bombs were released on the unprotected villages, to devastating effect. We don't know what happened to the dichloroethylarsine, but it probably was not wasted. This is said to be the first military conflict to be decided by the use of chemical weapons.[6] The Riffs protested to the International Red Cross but got nowhere and the League of Nations was not interested. Much of this military action took place during the time of the chemical weapons negotiations in Geneva in 1925.

6.8.4 Italy in Ethiopia (Abyssinia)

Mussolini, who seized power in Italy in 1922, created the Servizio Chimico Farmaceutico Militare in 1923 and gained some practical experience by deploying gas, possibly Lewisite, in Libya in 1923–24 and 1927–28 around the time that Italy ratified the Geneva Convention.[68]

Mussolini invaded Ethiopia in October 1935. The League of Nations responded with some soft sanctions that did nothing to curtail Italy's ability to wage war.[69] After some initial military setbacks Mussolini telegraphed this message to his field commander Marshal Pietro Badoglio, a survivor of the WWI gas attacks of the Battle of the Isonzo and thus personally familiar with mustard gas and Blue Cross: "I authorise Your Excellency to use all means of war – I say all, both from the air and from the ground." The war was over in seven months.

Aerial delivery of chemical agents proved to be very effective and was used on a massive scale against civilians and troops, as well as to contaminate fields and water supplies. Professor Angelo Del Boca, an Italian, gives the following account of one action as told by an Ethiopian commander:[70]

"The bombing from the air reached its height when suddenly a number of my warriors dropped their weapons, screamed with agony, and rubbed their eyes with their knuckles, buckled at their knees and collapsed. An invisible rain of lethal gas was splashing down on my men. One after another, all those that had survived the bombing succumbed. The gas contaminated the fields and killed animals. The attacks continued for days."

"Strange containers were dropped from the planes that burst open almost as soon as they hit the ground or the water, releasing pools of colourless liquid. I hardly had time to ask myself what could be happening before a hundred or so of my men who had been splashed by the mysterious fluid began to scream in agony as blisters broke out on their bare feet, their hands, their faces."[70] (Almost the same language would be used 50 years later to describe events in the Kurdish town of Halabja, after a 1988 attack by Iraqi armed forces (Section 6.15.6)). Italian troops also sprayed chemical agents by hand from backpacks.[37]

The Italian use of mustard gas in Ethiopia is incontrovertible, but it is difficult to establish if arsenicals were used as well. The records usually list mustard gas (mainly), as well as tear gas, sneezing gas and various asphyxiating gases. The sneezing gas was probably chlorodiphenylarsine, which was the main arsenical used to fill all Italian aerial weapons and grenades. The only exception was an air-bursting bomb that was loaded with mustard.[68,71,72] But what about Lewisite, the "Dew of Death"? One source claims that in the late 1930s Italy was capable of manufacturing 25 tons of mustard and 5 tons of Lewisite per day[15] and it was probably used in Ethiopia.[37] However, as we have seen, the arsenical of choice for weaponisation was chlorodiphenylarsine. In fact, Trammell writes "chlorodiphenylarsine and dichlorophenylarsine were used by the Italian army in the invasion of Ethiopia."[72] The extremely fast reaction to the gas suggests that the mustard, if it was used, would have contained some arsenical.[27] Some have claimed that the effect of mustard gas alone would have been magnified by the hot weather,[6] but this argument seems to be negated by the experience of Australian soldiers who were persuaded to walk over a tropical island that had been drenched with mustard gas, for three hours and 20 minutes, yet did not develop redness until five hours later (Section 6.10).[73]

The League of Nations acknowledged that gas was used but took no effective action. Only in 1995 did the Italian Ministry of Defence admit, largely through the persistence of Professor Angelo Del Boca, that gas had been used in Ethiopia.[70] The parallels here with the use of chemical agents by Spain and France in Morocco, Japan in China, and by Iraq in the Iran/Iraq war are striking: unprepared targets were callously drenched; the world powers were indifferent; and international bodies such as the League of Nations and the United Nations were impotent (See also Section 6.11).

6.8.5 Japan in China

Japan invaded Manchuria in 1931. On July 7th, 1937, fighting between the occupying Japanese troops and their Chinese neighbours broke out near Peiping (now Beijing), starting a war between Japan and China that did not end until 1945.

Japan, who had been preparing for chemical warfare since 1918 sent experts to Europe to study agents and production methods. The small island of Okunoshima, in the Inland Sea in Hiroshima prefecture, was selected to be the full-scale production site and the opening ceremony took place on May 19th, 1929. The first products were mustard and a tear gas. The plant was expanded to permit the manufacture of other agents such as Lewisite, mustard/Lewisite, chlorodiphenylarsine, cyanodiphenylarsine, hydrogen cyanide and phosgene/arsenic trichloride. By 1935 the 225 workers were producing 1 ton of Lewisite per week, plus other agents. By 1937, 2645 workers were producing 2 tons of Lewisite per day, plus other agents.[74] The work force numbered over 6000 during the peak production period of 1937 to 1944, when the plant ran 24 hours a day. The island was erased from Japanese maps in 1938 and not restored until

1945. Absolute secrecy was demanded from all workers who were relatively well paid because of the danger.

Women and children were mobilised. The factory was normally so contaminated that the workers were trained to keep an eye on caged parakeets that dropped dead when gas leaks occurred. Only then could they leave the area. Inflammation of the armpits, genitals, and hips was common to all workers, with women being more affected than men: even the toilet paper was contaminated. A large number of workers were maimed for life as a result of exposure to the agents.[75]

Gas troops were sent to China by order of the Emperor as soon as the war started. They were told that although the use of chemical weapons was forbidden by the Geneva Convention, Japan was not a signatory. Soldiers were given a gas mask, gas tubes, and four or five condoms before operations. Gas was generally directed against civilians lacking any protection, but the Chinese soldiers weren't much better off – only a few were equipped with masks.[74]

During one large-scale operation in Wuhan, from August to November, 1938, a total of 9667 arsenical-loaded gas artillery shells and 32 162 arsenical-loaded grenades were used in 378 actions.[74] Mustard/Lewisite mixtures were delivered in shells against the defenders of the city of Hengyang in 1944.[37] *Time Magazine* stated in July 1944, in a heavily propagandised article that Lewisite "is now a favourite of the Japs."

China, backed up by evidence provided by the International Red Cross, protested to the international community in the late 1930s, but the world looked the other way. Much later, in 1945, the US turned its attention to indicting some Japanese commanders for their use of gas in China and Burma but the investigation was suddenly terminated without explanation. It seems that the United States feared that prosecutions for the use of chemical weapons would lead to demands for prosecutions for the use of biological weapons, which in turn would lead to the questioning of General Shiro Ishii and his staff of Unit 731.

The world did not learn about Ishii and his brutal experiments until 30 years later, but the US government gave him immunity from prosecution because it wanted access to his files that contained descriptions of biological and chemical warfare experiments, such as dropping Lewisite into human eyes.[76,77] At the time the Director of the US Research Facility at Fort Detrick, Frederick, MD, commented in public: "Such information cannot be obtained in our laboratories because of the scruples associated with human trials. . . . It is to be hoped that persons who volunteer such information, are spared any trouble."[6]

The gas production facilities on Okunoshima Island were demolished at the end of WWII and the remaining stocks of chemical weapons were dumped into the Pacific Ocean. Now, more than 50 years later, the island is used as a holiday resort. It boasts a tourist hotel, a large permanent population of friendly rabbits, and a small museum, the Okunoshima Poison Gas Museum, dedicated to Japan's manufacture and use of chemical weapons. Much of the Japanese experience is on display, but there is little recognition of the plight of the Chinese soldiers and civilians during the war.[78,79] There is also no mention of the continuing problem of the cleanup of abandoned chemical weapons in China (Section 6.16).

The general Japanese population did not learn that their soldiers had used chemical weapons in China until 1984. A pamphlet from the Gas Museum deals with this in a telling sentence: "The fact that chemical weapons had been used in warfare was kept secret, and the Japanese people, having not concerned themselves with the former army and navy, have scarcely realised the truth." This seems to be a typical government-sanctioned comment that sidesteps the main issue: that the Japanese WWII experience has not become a subject for public examination and comment, and eventual understanding. The process may be beginning now, but it looks like it will be slow. Some Japanese students in Canada declined to translate some captions on photographs of the Gas Museum exhibits claiming there was no need to know about such matters.[80]

6.9 WWII – The Gas War That Never Happened

6.9.1 The Buildup in Europe

Research on chemical warfare in Britain and France had been somewhat low-key after the end of WWI, although, following the success of the Italian campaign in Ethiopia and the rise of the Third Reich, all Europe assumed that chemical warfare would be inevitable in the future, and took steps to increase their chemical arsenals.

The British Chiefs of Staff in 1937 authorised development of offensive chemical-warfare agents and delivery systems. They discussed the use of Lewisite and arsine in bombs: research at Porton Down had focused a lot of work on arsenical gases, particularly arsine gas generated from solids, and it is said that about 1600 animals were killed in experiments with arsine.[81] British Intelligence reported that Germany was buying up all the available arsenic on the world market.[48] In defence, scientists at Porton produced an arsine detector. In response, Germany became convinced that arsine was one of Britain's new agents and rushed arsine into production. A plant was being built at the end of WWII to produce an arsine precursor in the form of the alloy Trilon 300 (Section 6.8). The British civil defence became concerned about the possibility of the release of arsine and phosgene from ships against coastal towns.[48] The arsenicals Lewisite, Adamsite and dichloromethylarsine went into production. In all, about 60 000 tons of chemical agents, mainly mustard, were produced in the UK during WWII.[82] Ordinary citizens were offered the chance to purchase a gas warning system in the form of a budgie: "Safeguard your family or staff. A pet budgie will definitely warn you of gas long before your air-raid warden can. Healthy budgies 6 shillings each."[9]

Winston Churchill, the British Prime Minister, maintained that chemical weapons would be particularly effective on troops massed on beaches and plans were made to spray and bomb beaches with mustard and with "Arthur," the arsine precursor, in the event of a German invasion. There was little worry about any negative international opinion and little concern over first use. Churchill was also prepared to use anthrax against Germany.[54,55]

Preparations in other European countries followed much the same path although on a reduced scale. In Poland research on Lewisite was conducted in the Anti-Gas Institute in Warsaw. The building was notable because of the large number of geraniums in the garden that were planted to mask the smell of the arsenicals.[83]

6.9.2 Russia

With some initial help from Stolzenberg (Section 6.8) Russia built more than 20 facilities for chemical agent production and operated at least nine test sites. Some production was carried out in major cities – one such facility, the Soviet Union Scientific Research Institute for Organic Chemistry and Technology (GosNIIOKhT) is in the heart of Moscow. Arsine was produced at four locations; mustard and Lewisite were the main products at a major facility at Shikhany, Saratov region. Another facility in the Tomsk region was built in 1942 with the assistance of the US. It made mustard and a whole range of arsenicals.[19,84]

Prior to the beginning of WWII, the Soviet stockpile consisted of about 6000 to 8000 tons of mustard. During the war this grew to 77 400 tons of mustard, 20 000 tons of Lewisite and 6100 tons of Adamsite. Hydrogen cyanide and phosgene were also stockpiled. After WWII the Soviets concentrated on nerve-gas production resulting in another 32 500 tons of these agents: the grand total of all agents was about 164 000 tons.[19,85]

6.9.3 The United States

The budget of the Chemical Warfare Service was very much reduced in the 1920s and 1930s and little research was done on new weapons.

Pearl Harbor was attacked December 7th, 1941, and the US entered WWII the next day. A pamphlet issued later that month by the US Office of Civilian Defence, and prepared by the War Department,[86] urged the Citizens' Defence Corps to be prepared for gas attacks, delivered without warning from aircraft: the agents to be expected included mustard, chlorodiethylarsine, Adamsite and chlorodiphenylarsine. Test kits were provided in order that the Corps members could become familiar with the odour of the agents: samples of mustard and Lewisite were included. The pamphlet stated that no requests for gas masks should be made at this stage.[86]

The Redstone arsenal – which was set up in 1941 near Huntsville AL, to manufacture phosgene, Lewisite, white phosphorus and iron carbonyl – deviously supplied Britain with gas prior to the Japanese attack on Pearl Harbor, which then legitimised all such actions.[48] Between 1942 and 1945, the US opened 13 new chemical weapons facilities.[15,87] One of these at Pine Bluff, AR, had 10 000 employees and produced thousands of tons of mustard and

Lewisite. The decision to manufacture Lewisite ignored information, available since 1936, that questioned the battlefield effectiveness of the agent[37] but by 1943 the US military had reconfirmed what they should already have known: "Compared to mustard, it was not a profitable material to have." Production was stopped on November 15th, 1943. The Edgewood Arsenal in Maryland also produced large amounts of Adamsite.[82]

6.9.4 Canada

At the beginning of WWII Canada had no chemical-warfare capability and had to start from scratch. By the end of 1940 the military was contemplating the large-scale acquisition, via purchase or manufacture, of war gases, particularly mustard, cyanodiphenylarsine and Arthur.[88] This activity was not only to aid the Allies in the event that chemical warfare was initiated in Europe, but also to ensure that chemical agents would be available should Japan attack the west coast of North America. In one notable piece of legislation, the use of white arsenic in glass manufacture was restricted to conserve this resource, and a similar order went into effect in the US (Section 2.8).

The problem of space for trials emerged early in WWII. The British carried out some full-scale experiments in a French station in North Africa, but as the war progressed this ceased to be an option. They then requested to establish a full-scale experimental station in Canada operated jointly by the two governments.[89] Canada agreed and in 1941 provided 1000 square miles of Alberta countryside for training and testing: it is situated near Red Deer and is known as the Defence Research Establishment, Suffield, or DRES or Suffield for short. This is where Canada stored US-made agents including 2000 tons of Lewisite and 10 000 tons of mustard.[81] This is also where Canada carried out most of its animal exposure experiments, as well as human trials on Canadian and British soldiers. In addition to individual patch tests (Section 8.10), groups of volunteers were sprayed with mustard from high altitude, and the casualties were severe.[81]

In the South Pacific the Australians were painfully aware that Japan had used chemical agents in China and feared that the Japanese might follow this path in an attempted invasion of Australia (some of these fears were later justified when there were indications that mustard/Lewisite had been used in New Guinea). Australia began to make gas masks in 1940, as did Canada, but unlike Canada they did not manufacture any chemical agents. They looked to the UK and the US for help and were supplied with mustard, phosgene, Adamsite, Lewisite and CN tear gas (Box 6.3).

Box 6.2 British Anti-Lewisite

The population of Britain expected to be drenched with Lewisite when WWII began, so there was much urgency in the quest for countermeasures.

Box 6.2   Continued.

Bleach was the chemical most often used for mitigating the effects of chemical agents during WWI. Thus, there was "Ointment, antigas, No. 1" which consisted of one part of bleach to one of white petroleum jelly by weight. An improved formulation, "Ointment, antigas, No. 2", was composed of chloramine-T in a vanishing cream base. This could be rubbed into the hands as a prophylactic against mustard gas and Lewisite, but it could not be used on more tender parts of the body because it had an irritant action on the skin.[90]

British scientists began their search knowing that in the 1920s it had been established that arsenic compounds reacted with the sulfur-containing constituents of cells (Box 1.3). In 1935 they found that the enzyme pyruvate oxidase was especially sensitive to arsenic compounds.[91] This enzyme, which they isolated from the brains of pigeons, is involved in the Krebs citric acid cycle that is fundamental to the metabolism of animals and plants and many microorganisms. Peters and his colleagues argued that Lewisite might be toxic because it reacts with sulfur-containing groups on the enzyme. They also argued that because Lewisite has two reactive arsenic–chlorine bonds it could react with two closely situated sulfur containing groups, a dithiol, to form a relatively stable arsenical ring. Thus, efficient protection against the arsenical would be afforded by the presence of a competing compound, another dithiol, capable of forming a more stable ring compound with the arsenic compound. At the time these were very novel suggestions.[90–92]

A search for a suitable dithiol rapidly led to the relatively nontoxic 2,3-dimercaptopropanol, chemically related to glycerol, which the Americans – who were involved in similar work – dubbed British Anti-Lewisite (BAL).[92] The name stuck. The well-accepted mode of action of BAL is shown in Figure 6.2. The first step shows the binding of Lewisite to the enzyme via the two thiol groups. The second step shows the displacement of the enzyme-bound Lewisite by BAL to restore the function of the enzyme and form the stable Lewisite-BAL compound. The third step shows how BAL can neutralise the effect of Lewisite by forming the Lewisite-BAL compound before the Lewisite has a chance to react with an enzyme.

Trials on guinea pigs (real and human) and rats revealed that BAL was effective as an ointment against the vesicants Lewisite and dichlorophenylarsine; however, it proved to be unstable in aqueous solution. The problem of producing it in an injectable form was solved in the US: BAL could be injected into muscle if it was dissolved in benzyl benzoate and peanut oil.[93]

Although BAL may not have been used in a war setting, it found immediate application in the treatment of arsenic-induced medical problems, and was used to treat workers accidentally exposed to Lewisite and Adamsite during manufacture. At the time neoarsphenamine was still being used to treat cases of syphilis and some patients reacted badly to the drug, developing an arsenical dermatitis that could sometimes be cured by BAL.[94]

Box 6.2 Continued.

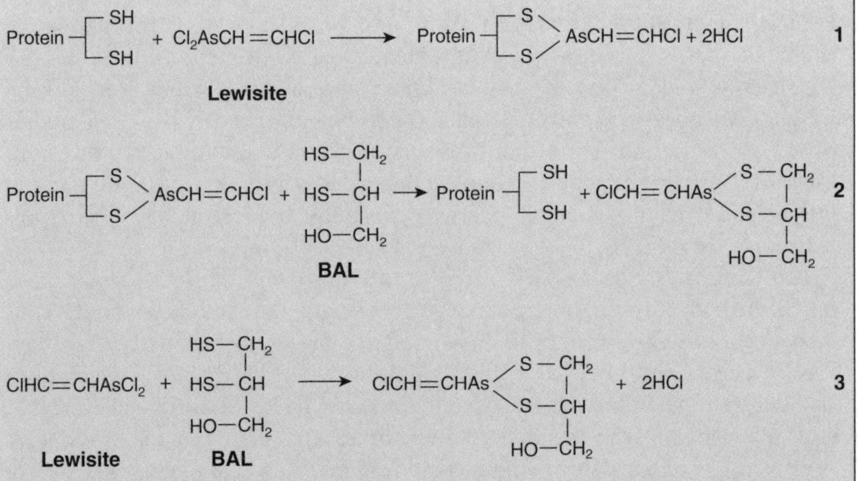

$$\text{Protein} {-}{\Big[}{ -SH \atop -SH} + Cl_2AsCH=CHCl \longrightarrow \text{Protein}{-}{\Big[}{ -S \atop -S}{\Big\rangle}AsCH=CHCl + 2HCl \qquad 1$$

Lewisite

$$\text{Protein}{-}{\Big[}{ -S \atop -S}{\Big\rangle}AsCH=CHCl + {HS-CH_2 \atop HS-CH \atop HO-CH_2} \longrightarrow \text{Protein}{-}{\Big[}{ -SH \atop -SH} + ClCH=CHAs{\Big\langle}{S-CH_2 \atop S-CH \atop HO-CH_2} \qquad 2$$

BAL

$$ClHC=CHAsCl_2 + {HS-CH_2 \atop HS-CH \atop HO-CH_2} \longrightarrow ClCH=CHAs{\Big\langle}{S-CH_2 \atop S-CH \atop HO-CH_2} + 2HCl \qquad 3$$

Lewisite **BAL**

Figure 6.2 Proposed mode of action of British Anti-Lewisite (BAL).[92]

Some say that the discovery of BAL was a major factor in the early 1940s decision of the US and Britain not to use Lewisite as a weapon because they assumed that in spite of the high secrecy of their operations, Germany had developed BAL as a countermeasure. But it seems much more likely that the decision to abandon Lewisite was made on the basis of agent instability. Nevertheless, because other nations might use the arsenical – a problem that is still with us – it was still necessary to have supplies of BAL on hand. In the 1980s, a Baltimore company Hynson Westcott and Dunning manufactured millions of doses of BAL, now also known as dimercaprol or sulfactin, for the US military. These doses were made up in peanut oil and had a one-year shelf life, so business was good.[95]

The discovery of the arsenic-binding abilities of BAL led to the development of a number of other similar compounds, such as DMPS (2,3-dimercapto-1-propanesulfonic acid), and DMSA (*meso*-dimercaptosuccinic acid) that can be used for treating victims of exposure to arsenic, mercury etc., with fewer side effects than BAL (Section 1.11).[72]

Box 6.3 Nerve Gas

At the beginning of WWII, the Allies believed that there was little chance that new chemical agents would be found that would be more toxic than those already weaponised,[1] so the victors were "very surprised" to find

Box 6.3 Continued.

German shells marked GA and GB at the end of the war. This was the first unveiling of the nerve gases (Figure 6.3).

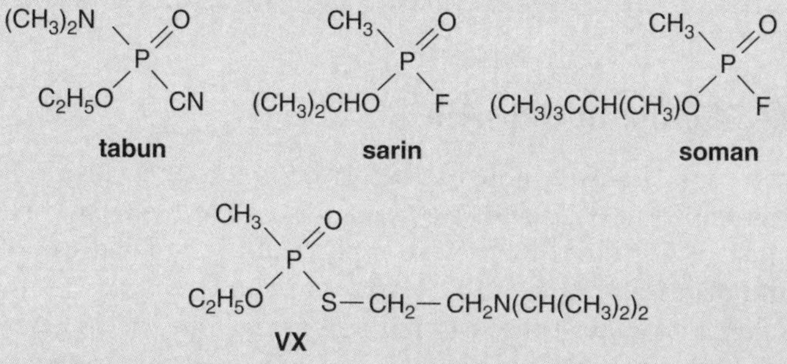

$$(CH_3)_2N\underset{C_2H_5O}{\overset{\diagdown}{\underset{\diagup}{P}}}\overset{\diagup O}{\underset{\diagdown CN}{}}$$
tabun

$$CH_3\underset{(CH_3)_2CHO}{\overset{\diagdown}{\underset{\diagup}{P}}}\overset{\diagup O}{\underset{\diagdown F}{}}$$
sarin

$$CH_3\underset{(CH_3)_3CCH(CH_3)O}{\overset{\diagdown}{\underset{\diagup}{P}}}\overset{\diagup O}{\underset{\diagdown F}{}}$$
soman

$$CH_3\underset{C_2H_5O}{\overset{\diagdown}{\underset{\diagup}{P}}}\overset{\diagup O}{\underset{\diagdown S-CH_2-CH_2N(CH(CH_3)_2)_2}{}}$$
VX

Figure 6.3 Nerve gas.

The first of these – tabun, code name GA – was discovered in 1936 by Dr. Gerhard Schrader who was engaged in routine pesticide research for IG Farben in Leverkusen, Germany. Unlike other chemical agents that have no specific biological target, this substance was found to inhibit an enzyme, cholinesterase, that controls muscular function by breaking down acetylcholine. Acetylcholine buildup results in continuous stimulation of the nervous system, leading to convulsions, loss of consciousness and rapid death. Germany later developed even more powerful nerve gases such as sarin, code name GB. These nerve agents are about 2000 times more toxic than mustard by skin absorption, and 300 times by inhalation.[55,96]

Germany invested a lot of effort and scarce resources into nerve-gas production, in the belief that they had a weapon that could alter the course of WWII. Trials at Munster revealed an "effective factor" for the main agents in the increasing order: mustard, factor 1; phosgene, factor 7; tabun, factor 17; sarin, factor 33. The effective area for a mustard bomb was $600\,m^2$; for sarin it was $10\,000\,m^2$ (lethal) and $20\,000\,m^2$ (incapacitating).[19] Germany had large stocks of the KC 250 bomb. It weighed 250 kilograms and could carry a load of about 80 liters of the chemical agent.

It is remarkable that Germany kept its secret for eight years, bearing in mind that plant construction and operation relied on forced labour. At war's end the Germans tried to destroy the production facilities in Dyhernfurth and its contents, but the Russian troops arrived sooner than expected and captured the facility together with some of the product. The very surprised Soviets thus gained a considerable advantage in the next phase of the chemical weapons standoff. The US troops shipped home any nerve gas that they confiscated, labelled as chlorine. Believing the worst, the US and

> **Box 6.3 Continued.**
>
> Britain began to stockpile nerve agents themselves. The US solved a major manufacturing problem in 1954, and began an enormous program to build up reserves of the more toxic sarin and later VX.

6.9.5 The European Experience

Fritz Haber's son, Ludwig, noted that for the first time in history, a weapon that was developed in one major war was not used in the next.[13] However, at the beginning of hostilities there were reports of the isolated use of chemical weapons in Poland.[6,19,97] In his novel *Mila 18* Leon Uris describes the German use of gas grenades against the Jews in the Warsaw Ghetto.[98] There is a report from a Jewish survivor that individuals were poisoned with arsenic while being transported to German concentration camps.[99] Germany used Zyklon B, hydrogen cyanide adsorbed onto a solid support of silica (the "B" indicates the gas loading) to murder millions of Jews and others in their death camps. Carbon monoxide was the major lethal agent in Treblinka, where the death toll was 1.5 million. But even this massive use of chemical weapons did not result in Allied reaction. Ignorance cannot be used as an excuse because the Joint Chiefs of Staff certainly knew about the genocide at the time.[100a,100b]

> **Box 6.4   Revenge by Jewish Terrorists**
>
> Joseph Harmatz was the head of an organisation known as Din ("judgment"), made up of Jewish survivors who had escaped from the Vilna ghetto in Lithuania. At the end of the war they vowed to take revenge: one German life for every Jew who had been slaughtered during the Holocaust.
>
> Their first plan, to poison the water supply of a German city, was betrayed and the poison ended up at the bottom of the Mediterranean Sea. Later, other members of Din were able to acquire a small quantity of arsenic and one Saturday night in April 1946, they broke into the Stalag 13 camp at Nuremberg, where German prisoners were waiting for the results of the Allied War trials for war crimes. Harmatz painted 3000 loaves of black bread with the white arsenic trioxide which blended with the white flour residue on the bottom of the loaves. (The group knew that the American guards would eat only the white bread that was made on Sundays.) They hoped to kill about 12 000 inmates.
>
> The newspapers later told of how the US hospital had pumped the stomachs of more than 1000 Germans, but Harmatz was disappointed that only 300 to 400 men were killed. "It should have been more," he said in interviews following the revelation of the story in 1999.[101]

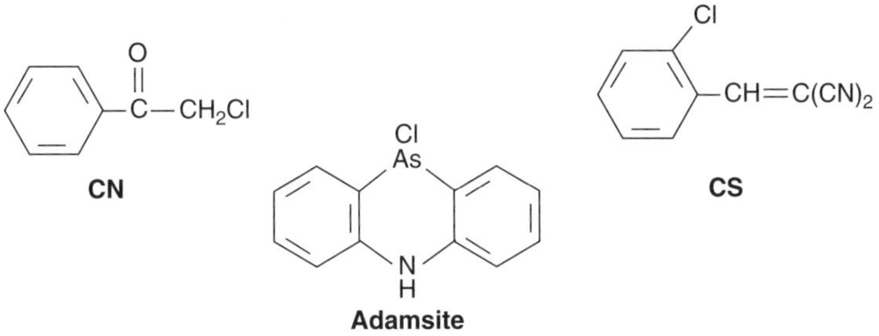

Figure 6.4 Tear gas.

We have seen that Churchill had permission to use gas against an invading force, but he never gave up his conviction that the Allies should strike first with this weapon.[55] In contrast, the British military establishment believed that gas was an unnecessary complication, and its use would bring massive retaliation on cities such as London. They wanted nothing to do with chemical warfare and even requested that General Eisenhower refrain from using phosphorus antipersonnel weapons in Europe because such use would be against the Geneva Convention. Eisenhower replied that the US had not signed the convention and that he was going to do as he wished. And he did.

At the time of the Allies' landing in Europe there was a fear that Germany would use chemical weapons to good effect against the troops massed on the Normandy beaches. But apart from that one instance, the tactical situation was never quite right: the battlefield was too fluid. Moral scruples may have played a small part, fear of reprisal also, because all the belligerents had massive stocks of the tear gas CN (Figure 6.4) that could have been used in minor skirmishes with some saving of life. Hitler came close to unleashing his prized nerve agents (Box 6.3) against the Russian troops when they had advanced to the Vistula River. He thought that, for political reasons, the Western Allies would not be too upset; however, his military commanders vetoed the plan.[19]

6.9.6 The War in the Pacific

Japan had signalled to the Allies in 1944 that it would not use chemical agents in the war, and ceased making them in order to maximise the production of conventional weapons. Japan also believed it had a big enough stockpile of chemical weapons to respond in the event of a first use by the Allied forces.[75] The official US policy at the time, as stated by President Roosevelt, was that chemical and biological agents would never be used by American forces except in retaliation for a chemical/biological attack. General Parker wanted to use gas on Iwo Jima but this was vetoed by Roosevelt.[54,81] That enduring scene of soldiers raising the US flag atop the mountain very nearly didn't happen.

After Japan's defeat, the slaughter on islands such as Iwo Jima – where there were 29 000 US casualties – was pushed as an example of what would not have happened had gas been used.[54] Here, the notion of the humane nature of gas warfare gets a little one-sided: the casualties inflicted on the receivers are not considered. The reduction in casualties of the deliverers and the preservation of real estate become the humane part. Nonetheless, press releases such as the following, extolling the virtues of using gas were not uncommon in 1945: "There is no question in my mind that for the first time in history there is the promise, even the possibility, that war will not necessarily mean death." This fantasy is still being explored (Section 6.16).

6.10 Human Guinea Pigs

6.10.1 The Allies

In the early years of chemical warfare research, huge numbers of animals (cats, dogs, sheep, monkeys, *etc.*) were used as test subjects.[6] However, animals can't speak about how they feel and they often react differently from humans to chemical agents, so scientists indulged in the practice of self-experimentation. This put limits on the work because of the small number of subjects, so volunteers were recruited when necessary. These were obtained with little concern for the ethics of the process. In fact, ethical considerations for human studies by the military and medical profession did not surface until well after the end of WWII. Before then, accountability was not a consideration, especially in a wartime situation where patriotic duty and the urgent need for answers overrode all other considerations. Human experiments started in Porton Down around 1916, and some of the early tests were particularly brutal.[15,48]

Testing became a high priority once WWII started, and in the US about 60 000 human subjects participated in experiments with chemical agents, particularly mustard and Lewisite, according to a special committee that was later struck to investigate alleged abuses of US military personnel.[39] In Britain the number was about 30 000.[37] Most of these volunteers underwent patch tests, although even in these, the levels of exposure to both mustard and Lewisite were high. About 2500 subjects were involved in testing in gas chambers at the US Naval Research Laboratories in Virginia, Illinois, North Carolina, and on San Jose Island in the canal zone. The most common tests were described rather tellingly as "man break." In these, men wearing protective equipment were exposed to gas in the warm humid chamber, to simulate tropical conditions, for periods of up to four hours. They were then required to keep their equipment on for a further period of up to 24 hours and examined for signs of gas penetration such as reddening of the skin. The cycle was repeated until the indications of exposure were very obvious. Some of the volunteers were told they were to test summer clothing and ended up in gas trials. Other "volunteers" were recruited under the threat of court martial and a double dose of gas.[39] In the report "Veterans at Risk" we find: "There is no doubt that some veterans who were involved in the chemical warfare testing program and other circumstances of exposure to mustard agents and Lewisite have been dealing with serious and debilitating diseases for decades. This burden has been further compounded by the secrecy oath taken by the veterans and faithfully kept for nearly 50 years, only to experience the denials of government agencies and their representatives that such tests and activities ever occurred."[39] Finally, in 1994 the US Department of Veterans' Affairs published regulations concerning compensation for disabilities or death resulting from chronic exposure to mustard and Lewisite.[102]

The situation was much the same in Australia, where Major Freddie Gorrill and his team from England discovered that the effect of gas on the Australian volunteers, and by accident on themselves, was greatly magnified by the heat and humidity of the tropics, and that protective measures then in place, such as uniforms impregnated with charcoal were largely useless.[73] The nursing staff were "astonished at the huge blisters that appeared on the mustard and Lewisite victims, who suffered dreadfully." Their endurance astonished the British and US observers who were convinced that their own troops would not have suffered so much, so cheerfully. "But at no time did these same scientists and observers express any remorse over the injuries inflicted on the volunteers or any interest in their subsequent wellbeing."[73]

More than 2000 Australian volunteers signed on. Some ended up in hospital for many weeks: others were injured for life. They were told to remain silent about their ordeal for 30 years, which complicated requests for compensation and disability pensions. To make matters worse, the Australian government also denied that such trials ever took place.[73]

A similar story played out in Canada. About 3500 Canadian volunteers were involved in trials. Tests were conducted with mustard and Lewisite to produce blisters and the subjects were given one dollar per exposure as well as their normal pay. "In addition to the above amount the medical officer i/c Physiological tests may authorise a maximum individual payment of $20.00 for any severe lesion resulting from a physiological test."[103] These and other trials were conducted under a veil of secrecy: the victims were warned that they could be prosecuted under the official Secrets Act.[104]

Canada eventually admitted to the testing in 1989 and a formal acknowledgment of the volunteers was made in May 2000, but there was no apology and no attempt to provide financial compensation. In arguing for compensation, Canada's Military Ombudsman André Marin made the following observations: "For more than forty years these men had to suffer alone, without any official acknowledgement that the tests even occurred, and were under pain of prosecution were they even to speak about it."[104]

In 2004, Canada's Department of Veterans' Affairs agreed to contact gas veterans to inform them they may apply for disability pensions.[105,106]

It should not be forgotten that around this time a number of US citizens were being secretly injected with plutonium without their knowledge or consent but with the consent of the government and its doctors.[107]

6.10.2 Japan and Germany

As mentioned previously, the notorious Unit 731 of Japan's General Ishii carried out chemical testing on some Chinese prisoners as a sideline to its main focus on biological weapons. Some victims were staked out in mock war zones and exposed to live chemical shells. Others were made to drink mustard or Lewisite, known as "crude water"; still others had this "crude water" dropped into their eyes. All experiments were carefully recorded.[74]

During WWI, Germany recruited soldiers as volunteers for experiments with irritant gases and later for gas-chamber experiments to determine limits of tolerability to arsenicals and others agents. Their approach was claimed to be painstaking and methodical.[1] A different picture emerged during the Nuremberg trials after WWII. In addition to the programs of mass slaughter of Jews and others with Zyklon B and carbon monoxide, inmates of concentration camps were also used as test subjects and forcibly exposed to chemical agents, including nerve gas, often with lethal consequences.[6,83,108] A Porton Down report about Dr. Karl Wimmer, who "vaccinated" some inmates of the Sachsenhausen concentration camp with Lewisite to test an experimental ointment, is revealing: "Although we now have a record of German human experiments with mustard gas, phosgene, Lewisite, *etc.*, no records could be recovered of any experiments with the nerve gases. Wimmer is probably one of the few surviving individuals who may be able to provide such information."[55] A search was conducted for Dr. Wimmer but he was never found; or if he was there is no record of him coming to trial. At least one commentator is of the opinion that there is little reason to believe that the Allies, their doctors and commanders, were not capable of similar acts.[107]

6.10.3 The Nuremberg Code of 1947 and its Aftermath

One of the outcomes of the Nuremberg trials was the realisation that it was time to put into place some guidelines for experiments involving human subjects, no matter what the circumstances. The result was the Nuremberg Code, the main features of which are:[39]

- The voluntary consent of the human subject is absolutely essential.
- Full disclosure of all aspects of the tests must be made, including all inconveniences, hazards and possible health effects.
- There should not be any unnecessary physical or mental suffering, especially the possibility of death, and the subject should be free to terminate the experiment if continuation seems out of the question.

This code should have provided a basis for chemical agent testing in the years after WWII, when the emphasis shifted away from mustard and Lewisite to other chemical agents such as nerve gas and hallucinogens. However, this was not the case in either the US or the UK.[15,39,109–111]

In 2005 the US EPA was prepared to allow the use of human subjects, including pregnant women and children, to test the toxicity of pesticides. After a rapid retreat following a lot of negative publicity, new regulations were formulated that required compliance with the "current ethical standards for research conducted by the federal government."[112] A few months later the new regulations still allowed tests involving pregnant women and children. In addition, although an advisory board will be set up to review study protocols, the board will have no power to veto studies or to require changes.[113]

Box 6.5 Tear Gas and Harassing Agents

Some of the soldiers who returned to the US at the end of the "war to end all wars" believed that their experience with chemical warfare could be applied to the war against crime. One advocate from the "humane gas school" claimed: "In these days of civilised customs, the use of firearms cannot be tolerated in dealing with civil disturbances. Such harsh measures are really no longer necessary, for modern technology has provided tear gas munitions. Just as important, the average citizen can obtain devices for his own protection and the collective protection of his family and property."[114] The agent of choice in the beginning was chloroacetophenone (CN) (Figure 6.4), a very powerful lachrymator developed by the US army late in 1918. The limit of tolerability was estimated at about 4 mg/m^3. CN is a solid that can be delivered as a dust in a micropulverised state, and by shells to give a smoke when mixed with an explosive. In solution in ethyl bromoacetate, it becomes "Blind-X." In "CN-KO" it is mixed with Adamsite (Figure 6.4). CN was manufactured in large quantities by all the belligerents during WWII.

The tear gas was loaded into hand grenades in the 1920s but later was packaged in all manner of devices, ranging from fountain pens and mechanical pencils to truncheons. Guns were also developed to fire gas cartridges/canisters: banks and armoured cars were fitted with gas-releasing systems, and tear gas quickly became a part of the culture. In the 1933 movie *King Kong* (giant ape escapes and carries blonde up a skyscraper) the giant ape was first anesthetised and captured with the aid of hand-thrown gas grenades.

For serious situations when gas of a more punishing physiological effect was required, Adamsite was selected mainly because of the ease of manufacture. The product was marketed as a nauseating agent, and classed as nonpersistent if delivered via a smoke candle or grenade. Micropulverisation of Adamsite, as in "CN-KO," was found to result in a product that was more active than that released from generators or explosive devices: in this form it remains active for weeks.

Adamsite was used in law enforcement in the US long after its use had been terminated by European countries in the late 1930s because of its high toxicity. Adamsite was said to be the preferred agent if the tactics required a crowd to be moved and then immobilised for a long time.[114]

Box 6.5 Continued.

Another tear gas, ortho-chlorobenzylidinemalononitrile (CS), was developed by the British in the 1950s when they were enthusiastically looking for a replacement gas for CN: something more potent with fewer side effects. The US military were impressed and adopted it for active service. It causes an extreme burning sensation of the eyes and copious tear flow at low concentrations; at higher concentrations exposure leads to coughing and difficulty in breathing, as well as chest tightening, nausea and vomiting. It is effective at concentrations as low as $0.05\,\text{mg/m}^3$.[96,115]

A wide variety of aerosol-releasing devices containing weak solutions of either CS or CN are now marketed for personal protection under the trade name Mace.

Post-WWII, the British use of tear gas – particularly CS – in Cyprus and later in Northern Ireland initiated an era in which the use of chemical weapons against civilians became an accepted part of policing. As we shall see, the outcry against the use of gas in Vietnam was fleeting, and today the general public seems willing to accept the use of "riot-control gas" by appropriate authorities against their fellow citizens. There has been a move away from CS-based munitions to those containing a less toxic agent commonly known as pepper spray (OC, oleoresin capsicum).[116]

Tear gas and pepper spray have been used in massive amounts to quell protesters and rioters all over the globe: gas is a routine weapon in the Israeli–Palestinian conflict.[117–119] Exposure to these gases can result in severe health problems.[120] The 1925 Geneva Protocol did not specifically preclude the use of CS against a nation's own civilians, or against individuals in a nation with whom it is not technically at war.[48,121]

The British used CS in massive amounts in Northern Island in attempts to quell the outbreaks of sectarian mayhem known as "the troubles." Another lachrymatory agent, known as CR (dibenz[b,f][1,4]oxazepine), was also developed by British scientists. It is 10 times stronger than CS, causing intense skin pain, temporary blindness, coughing and panic. It is toxic, although less so than CS, is a suspected carcinogen and can be lethal. It is likely that CR was used in October 1974 against about 800 republican inmates of the Long Kesh prison who had set fire to the premises. This gas was something never before experienced: the effects were far worse than CS, and many victims have since developed lung problems and cancer. The British government eventually acknowledged that CR gas had been shipped to Long Kesh but initially denied it had been used.[122] Former Republican and Loyalist prisoners are planning legal action against the British government for using the gas.[123,124]

Water cannons were used to deliver CR and CS to quell ANC-led protests in South Africa[125] and there are unsubstantiated reports that CR caused many fatalities, particularly among children.

6.11 The Vietnam War

The United Nations was founded in San Francisco in May 1945. In the first session of the General Assembly it was agreed that all members would eliminate weapons of mass destruction such as atomic bombs, but no serious consideration was given to dealing with chemical or biological weapons. The 1925 Geneva Convention was seen as a necessary pillar of society, and zero tolerance of chemical warfare was the easy and obvious path. However, even before the atomic bomb was dropped on Hiroshima in August 1945, the Allied governments were persuaded that Russia had invested a huge effort in the development of chemical and biological weapons, so secret research was intensified at Porton Down in England and at Fort Detrick in Maryland.[15]

President Roosevelt had made a commitment that the US would not use chemical or biological weapons unless provoked and stated that the US would honour the Geneva Convention of 1925. This was not necessarily the position of his military commanders, as we have seen, and by 1956 Roosevelt's policy had been completely reversed. According to US Army Field Manuals 27-10 and 101-40, "The United States is not a party to any treaty, now in force, that prohibits or restricts the use in warfare of toxic or nontoxic gases, or smoke, or incendiary materials or of bacteriological warfare." It is clear that the US military felt free to wage chemical or biological warfare on a first-strike basis during conventional warfare.[54,96]

But the British were the first to use herbicides as "environmental warfare" agents. In the late 1950s and early 1960s, they sprayed areas of Malaya that were controlled by Chinese guerrilla forces, with a mixture of the herbicides 2,4,5-T and 2,4-D. Their objective was the destruction of jungle-grown crops.[6,126] The Kennedy administration in the United States approved a request from the South Vietnam Government to conduct similar trials against insurgent forces.[54] This led to Operation Trail Dust, the overall herbicide program, and to Operation Ranch Hand, the major component of Trail Dust run by the US Air Force, which was responsible for more than 95 per cent of all herbicides sprayed on Vietnam between 1962 and 1971.

6.11.1 Agent Blue

The defoliation program in Vietnam remained largely experimental in the early 1960s, but by 1967 the operation was in full swing costing about $60 million (US) per year.[54,127] Agent Orange, the 1:1 mixture of 2,4-D and dioxin-contaminated 2,4,5-T was the main defoliant used between 1965 and 1970. The well-recognised name comes from the coloured identification bands painted on the 208 liter (55-gallon) containers. The literature on defoliant use in Vietnam usually fails to make the distinction between Agent Orange (and the chemically related herbicides Agents White, Pink, Purple and Green) and the chemically different Agent Blue, an arsenical that was sprayed on the rice crops.[128] Overall about 70 million litres of herbicide were sprayed on an area of 2.6 million hectares (6.4 million acres, 15 per cent of South Vietnam), and more than 10 per cent of that area was sprayed at least 10 times. Agent Orange accounted for about 46 million litres, and Agent Blue accounted for about one-tenth of this, 4.7 million litres. The arsenical was sprayed mainly in the years 1964–1971.[126]

Agent Blue was sprayed as a solution of dimethylarsinic acid (cacodylic acid) and its sodium salt in water that was sold by the Ansul Chemical Company as Phytar 560-G. The mixture was developed as a plant desiccant and insecticide in the 1920s (Section 2.6) and was first used in Vietnam on crops in November 1962. The choice of name, with its unfortunate connection to the WWI arsenical gas Blue Cross (Section 6.3), is presumably accidental. High application rates were needed to destroy food crops such as rice, and to control tall grasses such as sugar cane; and repeated applications were necessary to ensure continuous plant kill. Agent Blue does not harm most broad-leafed species, but it is effective against vegetation alongside roads. It acts quickly as a contact desiccant, and the foliage dries and shrivels within a few days. In contrast Agent Orange has the effect of speeding up the natural process of leaf loss: "spray a banana tree one day and the next day the leaves fall off and the bananas balloon to an unappetising size."[115,129]

At the time, the US army considered Agent Blue to be relatively nontoxic although they advised against drinking it or leaving it on the skin. The herbicide has recently been deregistered for use in the US because of toxicity considerations (Section 2.6).

It seems that the defoliation program was not very successful and in fact it aided the guerrilla forces by opening lines of fire. Crop killing was easier, and this program became increasingly important as a means of moving the civilian population away from enemy-controlled areas.

Food denial was not a new strategy: one of Haber's lieutenants wrote in 1941 that the destruction of plants and animals leading to diminution of the food supply would be one of the future developments of chemical warfare;[1] however, there is little evidence that crop destruction, which was mainly applied to the densely populated fertile Mekong Delta – "the rice bowl of South East Asia" – had much effect on military operations. It just contributed to the malnutrition of the general population. The military attitude is captured by this quotation from the *New York Times*:[54] "What is the difference between denying the Viet Cong rice by destroying it from the air or by sending in large numbers of ground forces to prevent the enemy from getting it? The end result is the same; only the first method takes far less men."

The use of sodium arsenate and other chemicals to sterilise soil was also part of the American plan for the region, but this does not seem to have been implemented. Agent Blue was used to a minor extent in Cambodia and Laos.

There are modern parallels in the on going US war on drugs, particularly in Colombia where it is the war on coca. The US is assisting the aerial spraying of coca plantations in Colombia with the herbicide glyphosate (N-(phosphonomethyl)glycine) which is banned for use in some states. The areas involved are now in the millions of hectares.[130] There are reports that the herbicide in not good for either the people or the environment[131] but one Canadian expert, Keith Solomon of the University of Guelph and a frequent visitor to Colombia, disputes these claims, saying the herbicide is harmless when used properly.[132]

6.11.2 Adamsite and Other Tear Gases

Following WWII, the US focused its chemical warfare attention on two lethal nerve gases: the nonpersistent sarin (LCt_{50} (inhalation) 100 mg min/m^3) and the persistent VX (LCt_{50} 10 mg min/m^3); and one blister agent, mustard gas. Three "riot-control agents" – CN, CS and Adamsite (Figure 6.4) – were officially adopted for military and civilian use. All three are potentially lethal. According to the US Army Field Manual FM3-10, Adamsite was regarded as the most dangerous of the tear gases. In the field it was usually packaged with CN in grenades and used for "suppressing civil disturbances of a more serious character."

In December 1964, the three "riot-control agents" were used with high secrecy in actions in Vietnam. On March 22nd, 1965, reporter Horst Faas of the London Associated Press revealed that gas warfare was a reality in Vietnam, and that arsenic compounds were being deployed. The news sparked a strong international protest and gas was not used for another six months when a well-publicised exercise was conducted that did not include arsenicals. The troops were instructed to always speak of the weapon as tear gas and never as just "gas." They were told that the use of tear gas was not in violation of any treaty. Observers noted that the temporary absence of Adamsite from the arsenal was tantamount to an admission that the arsenical was not just a tear gas.[54]

By 1966, gas had become an integral part of most operations and was part of the offensive arsenal: there was no pretence that these weapons were used for "riot control" or to save civilian lives. Little attempt was made to discriminate between soldiers and civilians or between soldiers who were surrendering or still fighting. It is likely that Adamsite was reintroduced, but CS became the preferred agent.

6.11.3 Health Effects

Large numbers of Vietnamese civilians appear to have been directly exposed to herbicides, some of which were sprayed at levels at least 10 times greater than those used for similar domestic (US) purposes.[126] By 1970, reports were appearing of an increased incidence of birth defects and respiratory ailments in the local population and Vietnam officials now attribute esposure to defoliants as the cause of cancers and respiratory ailments of 500 000 people, including 300 000 deformed and mentally retarded children born in the 30 years following the withdrawal of US troops.[133,134] These claims are played down by US authorities and companies, although high concentrations of dioxin were measured in blood and nursing mothers' milk in the years immediately following the end of the Vietnam War.[135] The US Government is reluctant to talk about the use of Agent Orange in Vietnam, and denies any connection between reported medical problems and the defoliants, even though it is now compensating many of its own veterans for some medical problems, such as chronic lymphocytic leukemia, soft tissue sarcoma, Hodgkin's disease, and prostate cancer where there is deemed to be sufficient evidence of an association with exposure to Agent Orange.[135] Pockets of high concentrations of the herbicide are still found near former staging areas such as Danang airport.[135,136]

So much attention has been paid to Agent Orange and dioxin that the consequences of spraying the arsenical Agent Blue, now declared to be too toxic for US use, seems to have been ignored, although at the time the presumed toxicity of the herbicide featured prominently in protests against the use of herbicides in Vietnam.[115]

Seymour Hersh reports that the gas attacks were not particularly successful; nevertheless, there was no shortage of civilian gas casualties, particularly among children.[54,115,137]

6.11.4 The Public Reaction

There was an immediate international response to the revelation that chemical weapons were being deployed in Vietnam, and much speculation about the lethal nature of Adamsite. The US government was very defensive and argued that the use of "riot-control agents" was not prohibited by the Geneva Convention.[54,115,138] More recently, Robert McNamara – who was the Defense Secretary under President Nixon – was very vague during his interviews for the Oscar-winning documentary film "Fog of War"[139] about who made the decision to use chemicals for offensive purposes in Vietnam. He seemed reluctant to discuss the consequences of their use, although in general he said he had come to believe that the Vietnam War was a serious mistake.[140]

Following the first phase of the deployment of Adamsite, CN and CS in Vietnam, some organisations such as the American Chemical Society argued for forthright use of chemical and biological weapons on the guerrillas. On the other hand, thousands of scientists were actively protesting the decision to engage in chemical warfare and the widespread devastation of the land.[96,115,141] It should be noted that the general public was not against the use of chemical agents in Vietnam. Any public protest was mainly against the war itself. Students focused on the Dow Chemical Company for supplying napalm to the military (Dow also manufactured Agent Orange), but there was no protest directed against Federal Labs. Inc., the company that provided the CS.[54]

Tear gas and "incapacitating chemicals" were used frequently and forcefully in the US to quiet the many civil protests. In one instance, students protesting the Vietnam War occupied land belonging to Berkeley University. The state governor, Ronald Regan, tried to clear "Berkeley People's Park" on Bloody Thursday, May 15th, 1969. About 3000 students confronted the police who used gas, buckshot and bullets on the crowd. On one occasion an army helicopter was used to spray the gas, and charges were made that "CS, CN, nausea gas, blister gas and Mace" were used.[115] This seems to be the only report of the possible use of Adamsite in the US post-WWII.

6.12 The Chemical Weapons Convention

At the time of the Vietnam War, there was strong consensus that the defoliation of vegetation and the destruction of crops, especially by aerial spraying, were prohibited by international convention.[96] There also seemed to be little doubt that the use of chemicals that could be classified as irritants (lachrymatory, sneeze-producing and blistering) was banned as well. There was apprehension that, once begun, chemical and biological warfare would open up a highly unpredictable dimension of warfare involving soldiers and civilians. Not much has changed since then.

In 1969, after acrimonious debate, a United Nations resolution formally extended the Geneva Protocol to include tear gas and herbicides. The vote was 81 in favour to 3 against, with 36 abstentions. The negative votes were cast by the US, Australia and Portugal. This UN vote was taken a few weeks after President Nixon's unilateral, but conditional, decision on August 19th, 1970, to send the Geneva Protocol to the Senate for ratification. "This will be done on the understanding that the protocol does not prohibit the use in war of riot-control agents and chemical herbicides. Smoke, flame and napalm are also not covered by the protocol." *The New York Times* commented that "President Nixon deserves two cheers for finally sending to the Senate his request for the approval of the Geneva Protocol of 1925, but it is unfortunate that he has been persuaded to cling to a reservation that will downgrade his decision around the world and may even put ratification in jeopardy."

In the early 1970s, the US and USSR began to work toward chemical disarmament, and a verifiable treaty banning chemical weapons, known as the Chemical Weapons Convention, was eventually opened for signature in Paris on January 13th, 1993. It was signed by 160 countries almost immediately, and came into force April 29th, 1997.

Both the American Chemical Society and the Chemical Manufacturers' Association, who worked with the negotiators through the protracted negotiation process, changed their longstanding positions and strongly urged that the US ratify the Chemical Weapons Convention.[142]

During the Senate debate the questions on the minds of many Senators were: "Do we believe that international conventions and conferences keep us safe at night? Or do we believe that a strong national defence is what keeps us safe at night and what has served us so well this century?"[143] This debate continues in the US; nevertheless, the Senate voted to ratify the Chemical Weapons Convention just five days before the treaty came into force in 1997. The Russian Duma also ratified the convention in 1997, after President Yeltsin announced that one-fifth of the costs, estimated to be $7.5 billion, would come from outside Russia, but the Russian vote was taken after the deadline.[143,144] By 2007, 182 nations had signed and ratified. Six had declared chemical weapons arsenals (Albania, India, Libya, Russia, South Korea and the US) and 13 nations – including Egypt, North Korea and Syria – had yet to sign-on, and remain suspect. North Korea is now said to be manufacturing mustard and Lewisite.[38]

The full name of the Chemical Weapons Convention is "The Convention on the Prohibition of the Development, Production, Stockpiling and Use of Chemical Weapons and Their Destruction." In addition to the obvious prohibitions on the development and use of chemical weapons we find that: each state party undertakes to destroy all chemical weapons it owns, possesses or are under its jurisdiction; to destroy all chemical weapons it abandoned on the territory of another state party; to destroy any chemical weapons production facilities it owns, possesses or has under its jurisdiction. Further, each state party undertakes not to use riot-control agents as a method of warfare, although Article II of the treaty explicitly allows toxic chemicals to be deployed for "law enforcement, including domestic riot-control purposes."[145]

The Chemical Weapons Convention requires the declaration of all chemical-weapon storage, production and research facilities and their destruction according to a prescribed schedule. There are adequate verification provisions allowing for inspection and sampling. Snap inspections are permitted to keep everybody honest.

There is a special Annex dealing with old and abandoned chemical weapons. Old chemical weapons are chemicals produced before 1925, or weapons produced between 1925 and 1946 that have deteriorated to such an extent that they can no longer be used. Abandoned weapons are defined as chemical weapons, including old chemical weapons, abandoned by a state after January 1st, 1925, on the territory of another state without the consent of the latter.

Under Article III of the Chemical Weapons Convention, States are not required to declare chemical weapons stocks that were buried before 1977, or dumped at sea before 1985. These dates, the outcome of closed meetings, are believed to have been chosen primarily to accommodate Moscow.[146] For example, at least 3000 tons of Adamsite are known to have been buried at Shikhany in Russia (more accurately dumped into a ravine) but this need not be declared under the Chemical Weapons Convention because burial took place before January 1st, 1977. This arsenical was manufactured in the State Scientific Research Institute for Organic Chemistry and Technology, GosNIIOKhT, in central Moscow. The Institute had earlier disposed of mustard gas locally, by simply pouring it into a hole in the ground.[84]

At the time of the first review conference on the Convention in The Hague, April 2003, about 900 inspections had taken place in chemical weapons production facilities and about 550 in industrial facilities. One delegate said: "it is very clear that destroying chemical weapons is an expensive, complicated, and somewhat unpredictable process." The definition and deployment of nonlethal but incapacitating agents caused the delegates some problems. Was Russia's use of the anaesthetic fentanyl allowable to end the standoff with Chechen separatists in a Moscow theatre? The consensus was yes. There was less enthusiasm for the proposal by the US that riot-control agents be permitted to be used in battle.[147]

6.13 The Cleanup
6.13.1 The Early Years

An estimated 15 to 30 per cent of the munitions used during WWI failed to function, which means that 13 to 20 million chemical rounds were left on the battlefields.[2] In addition, the retreating German army left behind huge munitions dumps, which compounded the burden. The war-weary and often destitute populations attempted to salvage the metal for sale, but identification of chemical weapons was difficult, and the outcome was frequently tragic.

Chemical munitions are generally assembled from two incompatible components: a toxic chemical filling that should be handled in a facility behind at least one airtight barrier; and an explosive charge that requires handling in as open a space as possible. So, when disposing of extant chemical weapons, extreme care must be taken to follow the life-preserving sequence: detection; recovery and assessment of the explosive status; cleaning; transportation; chemical identification; storage; dismantling; destruction; and waste disposal.[51]

In the early days after WWI, identification and dismantling were major obstacles but modern technology makes the task a little easier. Identification that once depended on exterior appearance and markings and "local knowledge" is now aided by using neutron activation analysis, X-ray analysis and photon-induced neutron spectroscopy (PINS), which is able to establish the presence of the elements phosphorus, sulfur, fluorine and arsenic within the munitions.[2,25] The viscous mustard fillings, usually mustard/arsenical, still pose the most problems.

In Belgium, where many chemical attacks took place around Ypres, the first officially approved method of destruction of chemical agents involved crude incineration. By 1922, 5350 tons of Blue Cross had been destroyed in this way, but not without considerable environmental damage.[16]

The French government faced a similar quandary at the end of WWI, but delayed action by fencing off nearly 16 million acres behind what was known as the Cordon Rouge. The French estimate that there are 12 million shells from WWI alone in the soil near Verdun; many more were added during WWII. The cleanup, which is still going on, began with the establishment of the Département de Déminage in 1946. About 900 tons of shells are unearthed per year, roughly 30 tons of which are toxic. Until 2000 they were delivered at night, in sealed trucks, to the Le Crotoy demolition centre on the seaside in northern France. Workers first placed the "toxic ones," then the rest, in a hole dug in the extensive tidal flats, and at high tide the batch was detonated beneath the sea. Only a depression in the sand was visible at the next low tide. The French claimed that because there is not normally much life in the tidal flats no great damage was done to the environment; nevertheless, the practice ceased once the Chemical Weapons Convention came into force.[148]

The Belgian army used similar disposal techniques after WWI but in view of the many difficulties it is not surprising that dumping at sea became one of the preferred options.[51] In 1918 Belgium dumped about 35 000 tons of munitions into the sea near Zeebrugge. The site remains a problem for shipping and coastal development to this day, but the authorities have decided to leave well enough alone, as have others facing similar situations.[16]

6.13.2 Post-WWII

After WWII the world was awash in chemical munitions and chemical agents in spite of their absence from the battlefield. The Allies were faced with the problem of dealing not only with their own stockpiles, but also with the approximately 300 000 tons of chemical weapons (about 30 000–75 000 tons of which were chemical agents) they found on German territory, distributed roughly as follows: US zone – 94 000 tons; British zone – 123 000 tons; French zone – 9000 tons; and Soviet zone – 63 000 tons.[146] The Joint Chiefs of Staff devised a general policy on the disposal of these stocks that essentially said: you dispose of everything you find in your own zone.

A lot of this material was processed on site by whatever method was most convenient. Environmental protection was not a consideration. One chemical munitions production and filling base in Loecknitz, near the German/Polish border, fell into Russian hands in April 1945. The contents of the storage tanks – 3000 tons of arsine oil, chlorodiphenylarsine and mustard – were poured into concrete basins, covered with lime and burnt. The area was then sealed off. Recent studies (1999) show the whole area, 1 km^2, is badly polluted. The arsenic concentration ranges from 25 per cent to the low parts per million and phenylarsenic compounds can be extracted from the soil.[24,149] In general, much of the stockpile of chemical weapons was dumped into the sea, and at the time there were no international laws against this practice. The 1997 Chemical Weapons Convention now prohibits ocean dumping and land burial as a way of disposing of any chemical weapons.

It is instructive to look at the US/European zone disposal record: 32 500 tons were burned, detonated, or landfilled; 27 000 tons were shipped to Italy and the UK; 12 000 tons were shipped to the US; and 31 000–39 000 tons were dumped at sea in an operation known as "Davey Jones' Locker." The nerve gas and mustard were shipped back to the US and UK. The nerve gas was regarded as a prize asset that could be used against the Japanese if necessary.[146]

Operation Davey Jones' Locker ran from June 1946 to August 1948. German prisoners in Nordenham, Germany, were used to load chemical munitions onto nine ships, old, damaged, or unfinished, that were later scuttled. The work was dangerous and workers carried canaries as gas detectors. One ship's stoker on the loaded *Jantje Fritzen* wrote about waiting in harbour. " . . . one day yellow smoke started pouring from one of the holds. We were all brought to the hospital. Some of the crew died from their injuries. The ship was later scuttled in Skagerrak."[146]

Most of the US and British sea dumping took place in the Norwegian Trench, 25 nautical miles south of Arendal, Norway; and in the Skagerrak Strait, 25 miles west of Måseskär on the Swedish coast.[146]

There are no exact data on the French disposal operation, but it seems that two ships loaded with a total of about 1500 tons of arsenical agents were sunk in Skagerrak.[150]

As for the Soviet share of the German munitions, 35 000 to 50 000 tons of weapons were dumped in the Baltic Sea northeast of the Danish Island of Bornholm and southeast of the Swedish island of Gotland. This included 7600 tons of mustard and about 1600 tons of Adamsite, as well as "80 tons of other agents, including Zyklon B used in the German concentration camps."[150] The cargo was often thrown over the side of the ships *en route* to the designated final resting place. No ships were scuttled, so these chemicals have become widely dispersed in the relatively shallow waters – the average depth of the central Baltic Sea is 65 m.[151] The remainder of the Soviet share of the captured German stockpile was not accounted for but was probably part of the 357 000 tons of munitions, mainly of its own manufacture, the Soviet Baltic Fleet continued to dump in the same areas until the late 1980s. Included were 71 469 aerial bombs filled with mustard, 17 543 aerial bombs filed with Adamsite and other agents labelled as arsine.[146]

It is difficult to obtain information on all these activities, and surprises are not uncommon. A 1971 report of high concentrations of arsenic in the Baltic Sea led to the discovery that 7000 tons of unidentified arsenicals had been dumped about 40 years earlier in concrete containers.[146,152,153] In general, the existence of the dump sites was only acknowledged when a chemical weapon or clump of congealed agent was washed up on a beach or caught in a fisherman's net.

A 2005 overview prepared for the OSPAR Commission for the Protection of the Marine Environment of the North East Atlantic[154] lists 148 munitions dump sites. Chemical weapons are to be found in about 30 of these, including the two main sites, the Bay of Biscay and Skagerrak. Belgium, whose inventory of old weapons was getting out of hand largely because of discoveries made during motorway construction – on average each year in Belgium two people die from live WWI ordnance[155] – dumped munitions encased in concrete into the Bay of Biscay as late as 1980. There were some objections particularly from Germany, who worried about the arsenicals.[16]

The chemical and physical state of the dumped weapons is largely unknown. Where possible, inspection of the sites reveals everything from intact munitions to corroded empty casings. Some fillings that end up on the sea floor as large gooey blobs still contain active agents under a thin skin. The German 10-liter spray canisters that were filled with thickened mustard/arsine oil are a major source of these blobs which are particularly hazardous when caught in a net or washed up on a beach, sometimes as far away as Britain.[24]

The general consensus is that little can be done about the recovery and safe disposal of these toxic residues because of the risk involved in moving them from the sea floor. Another aspect of the problem is the cost of recovery, which would be probably about 10 times the cost of manufacture. One possible consolation is the calculation that the maximum amount of inorganic arsenic that the breakdown of the dumped arsenicals could release into the Baltic seawater amounts to about 1 per cent of the total amount naturally found at any time.[151] However, Kiel University scientists reported in 2007 that the arsenic content of fish found in the Baltic near Kiel is around 50 ppm and 10 times greater than normal (Section 7.2).[151a]

Fishing on the dumps is discouraged, but reduced fish stocks have led to aggressive bottom trawling in the areas surrounding the dumps, and occasionally rotting and leaking munitions and clumps of agents from surrounding waters come up in the nets, resulting in injury to the crew. Danish fishermen, who comprise 75 per cent of the Baltic fishing fleet, are now compensated for each chemical weapon they catch provided it does not come from a dumping ground.[156] Other nations in the area, Norway, Sweden, and Germany, are contemplating a similar plan. It seems that without the possibility of compensation the munitions just get kicked back into the sea, spreading the problem.[157]

6.13.3 Japan

At the end of WWII, Japan's stockpile of chemical munitions was eliminated under orders of the US occupational force. Although safe sites were selected in the ocean, much of the cargo was dumped in shallow water close to shore, in areas that were used for commercial fishing. Now fishermen commonly find chemical weapons in their nets, often with unfortunate consequences. Japanese labour was used during dumping and there were many accidents during collection and transport.[75]

6.13.4 Domestic Ocean Dumping

During WWII the US opened 13 new chemical weapons plants, and at the end of the war most of the product was dumped into the ocean. Ten thousand tons of Lewisite were dumped off the South Carolina coast in 1946, and about 3150 tons of mustard bombs and 448 tons of containers of Lewisite were sunk 129 miles off the coast of San Francisco in 1958. Lewisite was also dumped off the Alaskan coast.[37] About $500 million (US) worth of mustard/Lewisite produced at the Rocky Mountain Arsenal in Commerce City CO, was mixed with sodium hypochlorite and dumped in the Gulf of Mexico. Barge-loads of chemical weapons were also dumped off the coast of Panama in 1947.[158]

In 1965, Britain – the first major power to declare it had stopped work on chemical-warfare agents – disposed of most of its stock of agents and weapons inside ships that were scuttled off the west coast of Scotland in Beaufort's Dyke. This deep trench located between Scotland and Northern Ireland, is said to have received one million tons of munitions since the early 1920s; although these were mainly conventional weapons.[15,159]

Russia dumped mustard and Lewisite in the Pacific, the Arctic Ocean and the Black Sea. One such operation in 1950 has been linked to the death of 6 million starfish, 7500 crabs, 30 seals and 10 belugas in the White Sea.[84] The specific cause may have been 50–60 railway wagons full of Lewisite-loaded aerial bombs.[151]

It is estimated that more than 1 million tons of chemical weapons were dumped into global seas, oceans, and lakes between 1945 and 1970. Russia dumped about half of this (its own and captured) and the US about one-tenth.[160] This is one legacy of the arms race that will not go away; another is the need to dispose of weapons found on land.

6.14 Disposal of Stockpiles
6.14.1 Russia

At Chapayevsk, in the Samara region, shells were filled from open vessels until the 1950s: "During the war young men unfit for the front line and girls arrived by the trainload – and just as quickly became ill and died."[84] At war's end the stock of mustard and Lewisite was poured into trenches, covered with lime and bleach, and buried. The factory was converted to produce fertiliser and other chemicals. The arsenic content of the soil at the plant is 8500 times the permitted level; in the town it is up to 10 times greater. Only 8 per cent of the children in the region are said to be born without any kind of defect. In 1985, three years after the Soviet Union voluntarily began destroying its chemical weapons, an incineration plant was ready at Chapayevsk, but it was never put into operation, mainly because of well-coordinated objections from the public who had environmental, economic and safety concerns, and who had not been consulted.[161]

When Russia signed the Chemical Weapons Convention in 1997 it did so on the understanding that there would be considerable financial assistance from other nations such as USA, UK, Denmark and Germany. It was seen to be in the best interests of the Western nations that alternative employment opportunities be created for the approximately 3500 Russian scientists involved in the production and stockpiling of chemical weapons. Fear of a brain drain and the spread of these weapons to rogue states prompted the setup of grant assistance programs such as the US Co-operative Threat Reduction Program, but very little of this assistance went to chemists. Most went to the thousands more involved in the nuclear and biological weapons business.[162]

Russia inherited responsibility for a multitude of production and testing sites following the breakup of the Soviet Union in 1991 and admitted to possessing a stock of 40 000 tons of chemical agents on ratifying the Chemical Weapons Convention in 1997. There are those who say this figure is low and that the real amount is much greater.[84] In 1997, the cost of future chemical-weapon destruction was estimated at $5 billion (US) for Russia, to be compared with about $12.4 billion for the US.[163] The 2003 estimates were $6–10 billion for Russia and $24 billion for the US.[164] These numbers may seem high but to put them in perspective the cost of cleaning up unexploded conventional ordinance left on military bases in the US is estimated to be in the hundreds of billions of dollars.[165]

Lewisite (6400 tons, comprising 15.9 per cent of the total Russian stock of chemical weapon) is stored at Kambarka, Udmurtia Republic, in containers made out of 11-mm-thick steel. Mustard, Lewisite and their mixture, making up 29 per cent of the Russian stock, is stored at Gorny, Saratov district. There is some Lewisite at Kizner, Udmurt region, and some mustard/Lewisite mixture at Maradykovsky, Kirov district, but nerve gases such as sarin are the main problem at these last two sites. In all, there are 6800 tons of Lewisite, 210 tons of mustard/Lewisite and 680 tons of mustard, mostly in bulk storage and mostly stored in a very small area in the Ural Mountains. This material has been around for a long time and there is real concern about the reliability of the storage containers.[161,166] Environmental groups claim that people living near these facilities are sick because of exposure to the chemicals.[167]

In the 1990s, the USSR, like the US, contemplated using underground nuclear explosions to destroy chemical weapons. Various, less drastic thermal processes were also examined, including incineration, but this was rejected mainly because of public reaction. The Russians finally settled on using two-stage processes that usually involve initial chemical neutralisation followed by secondary treatment.[168,169]

The technology currently being used to destroy Lewisite employs chemical neutralisation to remove the organic group from the arsenic, followed by electrolysis of the residue to produce elemental arsenic. The Russians estimated that the arsenic produced in this way from Kambarka has a value of around $230 million (US) if highly pure and $5 million if not.[168] However, these figures are difficult to justify because there is essentially no market for the output. The world has an arsenic surplus (Section 2.1).[170]

A similar two-stage method has been proposed for the destruction of mustard/Lewisite. The arsenic is isolated from the first neutralisation step, with the residue to be further degraded by bacterial action.[161,171,172] By 2003, the only facility actually operating was in the smallest of the storage sites at Gorny, which finished all the first-stage treatment in time for the 10th anniversary of the Chemical Weapons Convention in 2007, but had not started on the second stage. The Lewisite at Kambarka was also being treated by 2007, but only to the first stage.

The overall Russian record is not good. Russia claims to have destroyed 17.5 per cent of its stock but, given the current rate of progress and the need to construct four more treatment facilities, there is little chance that it will meet the 2012 deadline for complete destruction (10 years plus a five-year extension). The US is also behind schedule. The cost of treatment for the Russians has now escalated to more than $7 billion (US), with $2 billion of this to come from the US, Canada and the EU.[173]

Poland was the beneficiary of 9300 tons of Adamsite at the end of WWII. This unpleasant surprise was found in 219 steel drums on a railway siding in Lodz.[83] Arsenic recovery was also incorporated into the disposal process.[174]

6.14.2 United States

The Rocky Mountain Arsenal, located 10 miles from downtown Denver, CO, was built in 1942 on a 27-square-mile site. Immediately after the end of WWII, to reduce operational costs and maintain the facilities for national security, fully functional chemical-agent-production areas of the Arsenal were leased to companies such as Shell Chemical Ltd. for pesticide manufacture. Shell was also contracted to manufacture nerve agent precursors.[175,176] Much of the resulting sarin and VX was later incinerated on site. Lewisite, produced there in 1943, ended up in the Gulf of Mexico as mentioned above. Between October 1956 and March 1966, the US army demilitarised 33 538 500-pound M-78 bombs filled with phosgene and Adamsite. A total of 65 139 pounds of bulk Adamsite and 156 583 Adamsite grenades and canisters, were incinerated on site in the late 1950s; and 36 700 pounds of chemical identification kits met the same fate there in 1982.[176,177]

The area was added to the US EPA National Priorities List in 1987 (see Superfund Box 7.4) and the cleanup of the whole 17 000-acre site is slated for completion in 2011. The plan is to transform it into a wildlife park. By 2006, 10.5 million gallons of liquid waste had been incinerated and 2 million cubic yards of contaminated soil treated. These and other actions cost $1.3 billion (US), but the situation looks hopeful: no chemicals remain on the site. The public is now being wooed with inducements like: "Watch for the opening date for catch-and-release fishing. Offered each weekend."[175]

Most of this post-war activity went unannounced, and therefore unnoticed. The general public became sensitised when it was revealed that the US had sunk a ship loaded with nerve-gas rockets in the Atlantic Ocean in 1970.[152] The negative reaction led to a recommendation by the National Academy of Science in 1984 that incineration be used to dispose of chemical weapons.[177]

In 2006, speaking on the impact of dumped munitions off the coast of New Jersey, Congressman Robert Andrews (D-NJ) said: "It's very important. There's arsenic 10 miles off the coast of my state. That's as important as it gets."[178] Yet another example of the power of the A-word.

In 1986, the US Congress mandated the destruction of all stockpiles of lethal unitary munitions and bulk agents. They could be selective then and keep the binary rockets, but everything became fuel for the fire after the US ratified the Chemical Weapons Convention. The army proposed to burn some of its existing stock of around 31 000 tons of chemical weapons at nine locations. The first incinerator, on Johnson Atoll in the Pacific, had a startup cost of $811 million and was slated to begin operation in 1989. It started up a year later and operated, not without trouble, for 10 years.[179]

These plans met with considerable opposition from national organisations such as the Vietnam Veterans of America Foundation and, as in Russia, from citizens living near the storage sites.[180] (The battle against incineration in particular continues to this day.[181,182]) Faced with these objections, the US Congress requested that the army consider other options such as encasement in cement, bacterial decomposition, chemical conversion to harmless products or chemical conversion to useful products.[148,180]

The incinerator at Toole Army Depot in Utah, the second to begin operations, was and continues to be a constant source of political problems and safety concerns. Whistleblowers, such as Dugway Proving Ground chemist

David W. Hall, who raised the alarm about how Lewisite and related chemicals were handled, were fired.[37,183,184] The Depot held a stockpile of 30 000 tons of nerve and mustard gas, 42 per cent of the total US stock, in addition to some bulk Lewisite.[39] In June 1997, arsenic was found in the incinerator ash, indicating that the stockpile of Lewisite was being burned in violation of the plant's environmental permit. The army denied that this was so.[181] However, there was some good news: by 2002 all the sarin on site (6000 tons or 75 per cent of the US sarin stockpile) had been incinerated.[185] But in 2006 the citizens of Utah protested the Army's request to burn a mustard agent that contained about 1 ppm mercury.[182,186] A plant was built in Toole to dispose of 15 tons of Lewisite based on a method developed at the Defense Research Establishment, Suffield, Canada.[37,187]

By the beginning of 2007, the US had destroyed 40 per cent of its 30 000-ton declared arsenal of chemical weapons, at a cost of \$14 billion, and was on the way to meeting the 45-per cent destruction requirement of the Convention by the end of the year. However, all hopes of meeting the 2012 deadline for complete destruction have faded. At the Newport IN, site where the local population rejected incineration, the army is in the process of neutralising 1269 tons of VX nerve agent. The resulting solution was to be taken off-site for further treatment at an E. I. du Pont de Nemours and Co. facility in New Jersey – another controversial process. However, in January 2007, du Pont pulled out of the deal because of the local opposition. Another company stepped in, but opposition remains, so a secondary treatment facility will probably have to be built on site. Similar problems are expected to be encountered at facilities yet to be built at Pueblo CO, and Blue Grass KY. The optimistic finishing date for complete destruction is now 2023.[173,188]

It seems that, apart from Russia and the US, all nations will meet the 2012 deadline for complete chemical-weapon destruction. Currently, there are no provisions in the Convention for late completion: nobody thought it would be so difficult.

While this commentary has focused on the chemical agents it should not be forgotten that the destruction of the production capabilities of a state, party to the Chemical Weapons Convention, is a major part of the mandate: this process is also expensive and time consuming. "The destruction of chemical weapons production facilities or their conversion for peaceful purposes is emerging as a problem of technical complexity comparable to that of the destruction of chemical weapons."[159] All the US production facilities have now been demolished.[173]

6.15 Some Special Problems

6.15.1 Munster, Germany

Germany continues to find old chemical weapons throughout the country, particularly at former production, filling and storage sites. Between 1978 and 1994, 10 559 chemical munitions and 78 370 conventional weapons were retrieved. The Munster test and production site that comprises about 25 000 acres, includes vast areas – many fenced off – contaminated with inorganic and organic arsenic compounds. Much of this contamination was the result of an accidental explosion on October 24th, 1919, that destroyed 48 buildings. The explosion involved one million shells, 230 000 land mines, 40 railway tank wagons filled with agents and more than 1000 tons of bulk agent. Hurried attempts to dispose of the remaining chemical munitions by open-pit burning further contributed to the problem.[66] The British Royal Engineers blew up new facilities post-WWII, adding more material to the toxic burden.[189] There are layers of thickened mustard, much like peat, below the surface of some areas, and crystals of arsenical agents are to be found in the soil.

Conventional incineration started in Munster in 1980. Arsenic oxide released during the burning is trapped in scrubbers, then converted to a solid that is stored in drums in a former salt mine. A second incinerator that uses a plasma torch is now in operation in Munster but, in a recent departure from incineration, 16 800 hand grenades filled with Adamsite, nitrocellulose and CN were destroyed in the "PYROCAT" hydrothermal reactor. The arsenic is recovered as the element which makes for easier disposal.[190] The cleanup of the whole Munster site will take at least another 15 years. If all costs are counted, the cost of disposal again far outweighs the cost of production.[56]

6.15.2 Spring Valley, US

Spring Valley in northwest Washington DC, is a very exclusive and expensive neighbourhood. It is the home of 13 000 people, including many members of Washington's diplomatic corps, and the home of the American University. In 1916, the campus became the site of the American University Experimental Station, where chemical agents including mustard and Lewisite were investigated. Lewisite had been "discovered" across town at the Catholic University of America (Section 6.4). The chemical agent production facility was very extensive, employing 1200 chemists and engineers. An area was also set aside on the campus for field tests involving agents such as Lewisite and tethered animals. At the time the region was known as "Arsenic Alley," but following cleanup in 1919 the more euphonious name Spring Valley was adopted and its history was essentially forgotten, or at least not disclosed to the new inhabitants.

All was well until January 1993, when construction workers unearthed a cache of weapons and chemical agents resulting in "Operation Safe Removal," which lasted two years. However, there were indications from archival photographs that there were other burial sites, including one at the bottom of the garden of the Korean Ambassador. The army reluctantly started digging there in February 1999 and found more chemical weapons and related material – 623 pieces in all – and soil that was badly contaminated with arsenic. The property next door yielded bottles of mustard and Lewisite and some contractors were exposed to the leaking agents. Needless to say the community became very agitated: people living in the vicinity were reporting unusual health problems, and the army was accused of a coverup. Documents show the army was reminded of the situation by the American University in 1986, and their overall reluctance to take action did not help their case. Extensive soil sampling was undertaken[160,191,192] and the finding of high arsenic levels near the University child-care centre raised the level of concern. Hair samples from residents were collected and analysed, but the arsenic levels were normal for both children and adults. More than 100 homes were found to have soil with an arsenic concentration of more than 20 ppm, and the army agreed that this soil would be removed and replaced. The work is scheduled to be finished in 2008, by which time $125 million (US) will have been spent.[193–195] Ferns are being planted in some areas to see if they will take up the arsenic from the soil (Section 7.2.3).[37] Not everyone was pleased with all the publicity about the cleanup. The very obvious presence of army personnel in protective clothing and gas masks was certainly bad for property values.

During the operation of the American University Experimental Station from 1917 to 1920, research was conducted on many chemical agents. A long list that was released to the public included all the usual organoarsenicals such as Lewisite and dichloromethylarsine, and inorganic compounds, such as calcium arsenide.[196] It is clear that arsine was being seriously evaluated as a chemical agent at the time.[197]

6.15.3 Bowes Moor, UK

In 1985 Britain's Ministry of Defence began the review of six sites in rural England that had been used to store chemical agents and fill chemical weapons during WWII. The land had been fenced off and warning signs posted. It seems the authorities hoped that the problem would disappear, but it didn't.[198] One site, Bowes Moor in the Pennines, was used as a store for mustard, Lewisite and phosgene and to carry out experiments such as flying aircraft through the smoke of a mustard gas fire. The phosgene was sent to India in the late 1940s to become part of that country's declared stockpile: some of the remaining stock was sent to the US, some was dumped in the sea, and the rest was burnt in the open. Animals died accidentally as a result of these activities and the government requisitioned some land in 1946 because it had become contaminated with arsenic. Forty years later the government declared its willingness to look at the problem, resulting in a 1998 release from its Select Committee on Environmental Audit that stated, "The task at Bowes Moor is lengthy and no estimation for completion is available currently."[199]

6.15.4 China

Japan did not officially admit that it had left chemical munitions in China until 1995. A year later the two countries agreed to begin the cleanup.[200] Negotiations foundered on the demand that munitions be taken to Japan for neutralisation, according to the provisions of the Chemical Weapons Convention. China eventually relented and agreed that destruction could take place in China and a memorandum of understanding was signed in 1999.[201,202]

China claims that at least 18 dumping sites, 2 million pieces of chemical munitions, and 120 tons of bulk agent (predominantly mustard and Lewisite) are spread out over at least 10 Chinese provinces, but mostly in northeast China. Japan disputes these numbers, and estimates the number of weapons as closer to 700 000. In reality the numbers and amounts will only be known after the stockpiles are unearthed.[203,204] Experts agree that the task is formidable even with hired help from other nations. The time scale, and the costs, will very much depend on whether the abandoned weapons are declared usable – a decision that is ultimately made by the Organization for the Prohibition of Chemical Weapons, OPCW. If usable, the weapons have to be destroyed according to the normal 10-year schedule. If most weapons are unusable, there could be more flexibility in the timetable.

The deployment of chemical weapons by Japan during the war with China resulted in about 80 000 Chinese casualties and 10 000 fatalities. In the following 50 years another 2000 injuries have been linked to the abandoned chemical weapons.[203] This is a continuing saga. A 2003 report from the *China Daily*[205] states that about 36 people were hospitalised after some rusting barrels of chemical-warfare agents were unearthed on a construction site in Qiqihar, Heilongjiang province, northeast China. One schoolgirl who walked past the site had blistered feet, and two workers who tried to sell some of the metal for scrap became critically ill – one later died from his injuries. In September 2003, a Tokyo court ruled that the Japanese government was responsible for the injuries and fatalities, and should pay compensation as well as remove the munitions, but the decision was reversed in 2007.[206] The Japanese government remains in denial but surprisingly gave $2.56 million (US) to 44 victims. In a move that is said to have considerable political significance, some Japanese Friendship groups set up a fund to provide cash to compensate victims.[207]

Meanwhile, in Japan, citizens of Kimisu-cho, Ibaraki prefecture, have encountered serious health problems as a result of drinking water contaminated with diphenylarsinic acid. This arsenical probably comes from old agents stored or dumped on a site that was once an airport of the Japanese Imperial Air Force. Similar problems are being encountered elsewhere in Japan.[75,208]

6.15.5 Albania

Enver Hoxha – the paranoid Communist dictator who led Albania for 40 years until his government collapsed in 1991 – feared invasion from all sides, and during the 1970s and 1980s he imported large quantities of chemical weapons mainly, it seems, from China. He hid this stockpile in one of the 750 000 bunkers he had built as shelters for the population throughout the country, and then forgot to tell anyone their whereabouts. The Albanian government signed the chemical weapons convention in 1993 and embarrassingly discovered the chemical agents in 2004, close to the capital of Tirana. Albania was faced with the problem of disposal of Adamsite, Lewisite, mustard/Lewisite and mustard: in all about 16 tons of bulk agent. The stock was heavily guarded for fear that it could fall into the hands of terrorists. Incineration of the stock began in 2005 in a unit supplied by the US and on July 12th, 2007, OPCW announced that Albania was the first nation to completely destroy all of its chemical weapons.

6.15.6 Other Recent Deployments of Chemical Weapons

Iraq signed the Geneva Convention in 1931, but when Saddam Hussein came to power he began to develop a chemical arsenal with the help of many European countries.[211] Iraq went to war with Iran in 1983 and the first confirmed, but ineffective, use of chemical agents in the field against Iranian troops took place in August 1983, near Haij Umran.[212] The Iraqi troops subsequently used chemical weapons on at least nine targets. The best known of these was the attack on Halabja in March 1988, which resulted in the deaths of 5000 Kurds.[213] The town had been occupied by Iranian and allied Kurdish troops for 24 hours before the three-day-long chemical bombing attack by the Iraqis. The long-term effects included high cancer rates, miscarriages and birth defects in the surviving population.[214–216]

In general, the nations of the world and their people chose to ignore this blatant abuse of the Geneva Convention and the predicament of the survivors. Some years later, around 2002, the massacre at Halabja became a rallying cry for US President George W. Bush, but at the time Washington and then-president George W. H. Bush averted their eyes. Aid agencies shunned the region and the situation has not improved much.[214] In 2006, the surviving citizens took to the streets in protest on the anniversary of the attack.

It was not easy to identify the chemical agents used in those remote regions.[217] There seems little doubt that blister agents such as mustard and "dusty mustard" (mustard adsorbed on silica particles), nerve gases including VX, and hydrogen cyanide were used.[213] One source suggests that Lewisite was part of the package. When taken to London for examination, one of the injured Iranians presented all the classic symptoms of exposure to the arsenical.[37] German suppliers of DMPS, the improved version of British Anti-Lewisite, claimed at the time that arsenicals were being used in Iraq and that their product was being used for treatment.[95]

Given this background, there was every expectation that Iraq would use chemical weapons during the Gulf War of 1991 following Hussein's "metamorphosis from tolerated thug, to Washington's arch-nemesis."[218] US veterans of the war testifying in 1993 before a Congressional committee recounted experiencing "confirmed gas attacks" and being commanded to go to the highest level of chemical protection. Their detectors, which included some very sophisticated instruments, identified mustard and Lewisite.[219,220] Army officials claimed that printed records of the identification have been lost, although the veterans claim "cover-up."[221,222] The official position is that chemical weapons were not used on the battlefield. The United Nations Special Commission on Iraq, UNSCOM, later found and destroyed bulk quantities of mustard gas and nerve agents stored at production facilities, but no arsenical agents or precursors were reported.

Chickens, one named Geraldine of Arabia, and pigeons were available to US troops as a backup to their M22-enhanced automatic detectors during the invasion of Iraq in 2003.[147,223] The British would have been well advised to follow suit: their stock of 4000 chemical vapour detectors was said to be unserviceable.[224] Following the terrorist attacks in the United States on September 11th, 2001, the justifiable paranoia of the citizens was revealed in a rush to buy canaries to provide an early warning of a chemical attack. The customers did not worry that the female birds did not sing: they just wanted an alarm.[225]

Egypt used a variety of chemical agents in the Yemeni civil war, from 1962 to 1967, in aid of the pro-republican rebels, and was the first to use nerve gas in a war.[212] These incidents, although confirmed by the Red Cross and reported to the United Nations, were largely ignored by the international community.[15,37]

Colonel Gaddafi of Libya may have used mustard in Chad in the early 1980s.[212,226] He was recently rewarded with the renewed friendship of the UK and US when he voluntarily scrapped his chemical weapons program.[227] Russia may have used mustard, Lewisite, and nerve gas during its invasion of Afghanistan in 1979; however, this rumor has not been substantiated.[37] The Sudanese government probably used mustard in 1995 and, more recently, mustard and Lewisite, during the longstanding civil war between the northern Arab minority and the black majority. This conflict, now described as the world's worst humanitarian crisis, is a further example of the impotence of the UN.[228,229]

The use of gas, including Adamsite, against UNITA troops in 1987 was described during the 1997 trial of Dr. Wouter Basson, a Brigadier in the South African Defense Force and dubbed Dr. Death by the press.[230] Basson claimed to have visited gas victims in hospital. Other reports question that chemical weapons were involved at all in this engagement, known as the battle of Cuito Cuanavale.[231] Tear gas, CS and possibly CR, was in routine use in South Africa under apartheid to control the indigenous population (Box 6.5).[125]

6.16 Conclusions

We have seen that during WWI scientists developed chemical agents and defences against them, and some even spoke about the humane nature of the weapon. Once nerve gas became available, scientists became more cautious, but this didn't stop governments from stockpiling chemical agents for "retaliatory use". Chemical agents became unjustifiably labelled as "weapons of mass destruction" and "the poor man's atom bomb". The world worried about the possibility that any nation could start a club of chemical bullies on the cheap, in parallel with the already established nuclear club.

The end of the Cold War led to the Chemical Weapons Convention and the programmed removal of large-scale chemical threats. But the search for military advantage never really stops. In 2002, the US National Academy of Science released "An Assessment of Non-Lethal Weapons, Science and Technology".[232] The document came out strongly in favour of supporting the development of nonlethal weapons such as sticky foams, riot-control agents, malodorants and calmatives (sleep-inducing or mind-altering).[233] Rather ironically the report became public some two weeks after the Russian army used the opiate fentanyl against Chechen rebels and their hostages in a theatre in Moscow. This resulted in the deaths of at least 120 out of the 750 people present.[234,235]

There are real problems with these developments in light of the Chemical Weapons Convention and many argue that the convention puts an outright ban on the use of the riot-control agents and calmatives, with the malodorants being somewhat suspect. Others are prepared to argue that the wording of the convention is subject to interpretation and can be bent to fit the use of these weapons: what is an "international armed conflict," what is "self defence"? Still others, such as retired army Colonel John B. Alexander, argue that "Treaties are meaningless. . . . They did not foresee the technological advances that have taken place or the devolution of the world into a war on terrorism."[235] This was the same position that the US representatives adopted at the Hague Peace Convention of 1899 (Section 6.2).

The events of September 11th, 2001, resulted in an increased concern about the use of chemical weapons by terrorists, and a heightened level of security surrounding access to information about chemical weapons. Documents about the chemical agents of WWI suddenly became unavailable, even to the individuals doing the Spring Valley cleanup in Washington DC. Lewisite use is still feared by the US military, and training sessions involving simulated attacks have been carried out in many cities, including Tacoma WA, and Lexington KY.[37] The whole range of arsenical agents is listed under headings such as "terrorist weapons of choice" on web sites.[236] Advances in manufacturing methodology, such as miniaturisation, now allow the safe preparation of toxic chemicals in small batches that would be ideal for terrorist purposes.[237] Indeed, the anticipated use of chemical weapons in a terrorist action became reality during an attack on the Tokyo subway on the morning of March 20th, 1995, when members of the Aum Shinrikyo cult released sarin nerve gas on five lines of the Tokyo subway. Miraculously, there were only 12 deaths but 5000 commuters were hospitalised.[238] But belligerents don't have to be too sophisticated: chlorine gas was deployed by insurgents against civilians and US troops in Iraq in March 2007, and there were 10 similar attacks over the next two months.[173,239]

Any good feelings regarding the Chemical Weapons Convention should be tempered by the thought that this should also be the case for the Nuclear Non-Proliferation Treaty. However, the nuclear club of five is not playing by the rules and is hanging on to power. It is not destroying its stockpiles as promised in the Treaty, which requires "the cessation of the manufacture of nuclear weapons, the liquidation of all their existing stockpiles, and the elimination from national arsenals of nuclear weapons and the means of their delivery."[240] In 2007 the Bush administration was even seeking funding for the development of a new generation of nuclear warheads.

It is not surprising that other nations may feel compelled to build nuclear weapons without really worrying about receiving the blessing of the founding members of the club. John D. Holum, the director of the US Arms Control and Disarmament Agency, said in 1994, "To make chemical weapons is a waste, to keep them an affliction, and to use them an abomination." This has some ring of truth, but the statement has much more applicability to nuclear weapons. Nuclear weapons are useless against terrorists and their existence poses a risk to global security: "The ultimate horror scenario is nuclear weapons in the hands of terrorists."[241]

The world is in a precarious situation at the moment, full of violent acts and thoughts. "Deterring and countering such violence is a very sophisticated and specialised activity. It involves intelligence, politics, military skills, and science as well as commitment and good organisation. It is also an arena in which an overwhelming military superiority is of little significance."[242]

References

1. C. Wachtel, *Chemical warfare*, Chemical Publishing Company. Inc., Brooklyn, 1941.
2. R. G. Manley, in: *Arsenic and old mustard: chemical problems in the destruction of old arsenical and 'mustard' munitions*, ed. J. F. Bunnett and M. Mikolajczyk, Kluwer Academic Publishers, 1998. p. 1.
3. J. Emsley, *Elements of murder: A history of poison*, Oxford University Press, Oxford, 2005.
4. J. Kelly, *Gunpowder: alchemy, bombards and pyrotechnics: the history of the explosive that changed the world*, Basic Books, New York, 2004.
5. *A Brief History of Chemical and Biological Weapons*, www.cbwinfo.com.
6. D. Martinetz, *Vom Giftpfeil zum Chemiewaffenverbot*, Verlag Harri Deutsch, Frankfurt, 1996.
7. M. Leon, *The History of the Worimi People*, www.tobwabba.com.au/worimi.htm.
8. W. V. Hindle, *Diary of Lieutenant Hindle*, Imperial War Museum, London, 1915.
9. W. Moore, *Gas attack! Chemical warfare 1915–18 and afterwards*, Leo Cooper, London, 1987.
10. C. Heller, *Chemical warfare in World War I: The American experience, 1917–1918*, Combat Studies Institute, 1984.
11. J. P. Murmann, *Knowledge and competitive advantage*, Cambridge University Press, Cambridge, 2003.
12. V. Lefebure, *The riddle of the Rhine. Chemical strategy in peace*, E. P. Dutton and Company, New York, 1923.

13. L. F. Haber, *The poisonous cloud: chemical warfare in the First World War*, Clarendon Press, Oxford, 1986.
14. U. Trumpener, *J. Mod. Hist.*, 1975, 460.
15. R. Harris and J. Paxman, *A higher form of killing: the secret story of gas and germ warfare*, Chatto & Windus, London, 1982.
16. J. P. Zanders, in *The challenge of old chemical munitions and toxic armament wastes*, ed. T. Stock and K. Lohs, Stockholm International Peace Research Institute, Oxford, 1997, p. 197.
17. N. M. Christie, *Gas Attack! The Canadians at Ypres, 1915*, UBC Press, 1998.
18. T. Cook, *No place to run The Canadian corps and gas warfare in the first world war*, University of British Columbia Press, Vancouver, 1999.
19. T. Stock and K. Lohs, in *The challenge of old chemical munitions and toxic armament wastes*, ed. T. Stock and K. Lohs, Stockholm International Peace Research Institute, Oxford, 1997, p. 34.
20. J. Ellis, *Eye-deep in hell: life in the trenches 1914–1918*, William Collins Sons & Co. Ltd., Fontana, 1976.
21. S. J. M. Auld, *J. Ind. Eng. Chem.*, 1918, **297**.
22. R. Graves, *Goodbye to all that*, Anchor Books Doubleday, London, 1957.
23. A. Palazzo, *Seeking victory on the western front: the British Army and chemical warfare in World War I*, University of Nebraska Press, 2000.
24. H. Martens, in *Arsenic and old mustard: chemical problems in the destruction of old arsenical and 'mustard' munitions*, ed. J. F. Bunnett and M. Mikolajczyk, Kluwer Academic Publishers, 1996, p. 33.
25. D. Froment, in *Arsenic and old mustard: chemical problems in the destruction of old arsenical and 'mustard' munitions*, ed. J. F. Bunnett and M. Mikolajczyk, Kluwer Academic Publishers, 1998, p. 17.
26. K. Lohs and T. Stock, in *The challenge of old chemical munitions and toxic armament wastes*, ed. T. Stock and K. Lohs, Stockholm International Peace Research Institute, Oxford, 1997, p. 15.
27. P. Kopecz and J. Thieme, *Review of suspected warfare-related environmental damage in the Federal Republic of Germany Part 3: Chemical Agents Dictionary*, Federal Environment Agency, Berlin, 1996.
28. M. Valpy, in *Globe and Mail*, 2007, April 7, p. F4.
29. P. Burton, *Vimy*, McLelland and Stuart, Toronto, 1986.
30. G. E. Coates, *J. Chem. Soc.*, 1945, 1642.
31. J. B. S. Haldane, *Callinicus; A defence of chemical warfare*, Kegan Paul, London, 1925.
32. General Staff, *Gas warfare monthly summary of information*, August, October, 1917, Imperial War Museum, London.
33. General staff, *Gas warfare monthly summary of information*, January, 1918, Imperial War Museum, London.
34. A. M. Prentiss, *Chemicals in war: a treatise on chemical warfare*, McGraw Hill, New York, 1937.
35. General Staff, *Instructions on the Use of lethal and lachrymatory shell*, National Defense Library, Ottawa, 1918.

36. B. Brodie and F. M. Brodie, *From crossbow to H-bomb*, Indiana University Press, Bloomington, 1973.
37. J. A. Vilensky and P. R. Sinish, *Dew of death. The story of Lewisite, America's World War I weapon of mass destruction*, Indiana University Press, Bloomington and Indianapolis, 2005.
38. J. A. Vilensky and P. R. Sinish, *Weaponry: Lewisite – America's world war I chemical weapon*, /www.historynet.com/wars_conflicts/weaponry/3035881.html?showAll=y&c=y.
39. C. M. Pechurr and D. P. Rall, ed., *Veterans at risk; the health effects of mustard gas and Lewisite*, National Academy Press, 1993.
40. F. Bayer and Co., German Patent Application 281049, July 1913.
41. S. Franke, *Manual of military chemistry*, Institute of Applied Technology, Berlin (East), 1967, Vol. 1, p. 62.
42. T. Harrison, *Square rounds*, Faber and Faber, London, 1992.
43. R. Hoffman, *The same and not the same*, Columbia University Press, New York, 1995.
44. M. F. Perutz, in *New York Times Review of Books*, 1996, June 20, p. 31.
45. K. L. Manchester, *Endeavour*, 2002, **26**, 64.
46. N. B. Jackson, in *Chem. Eng. News*, 2002, Jan. 21, p. 46.
47. R. Hoffman and P. Lazlo, *Angew. Chem. Int. Nat. Ed.*, 2001, **40**, 4599.
48. J. E. S. Parker, *Killing factory: the top secret world of germ and chemical warfare*, Smith Gryphon, London, 1996.
49. K. E. Jackson, *Chem. Rev.*, 1935, **16**, 251.
50. E. M. Spiers, *Chemical warfare*, Macmillan, London, 1986.
51. F. Guir, in *The challenge of old chemical munitions and toxic armament wastes*, ed. K. Lohs and T. Stock, Stockholm International Peace Research Institute, Oxford, 1997, p. 156.
52. J. A. F. Compton, *Military chemical and biological agents*, Telford Press, Caldwell, NJ, 1987.
53. *Casualties: First world war*, www.spartacus.schoolnet.co.uk/FWWdeaths.htm.
54. S. M. Hersh, *Chemical and biological warfare: America's secret arsenal*, Bobbs-Merrill, London, 1968.
55. A. Thomas, *Effects of chemical warfare: a selective review and bibliography of British state papers*, Taylor and Francis, 1985.
56. H. Martens, in *The challenge of old chemical munitions and toxic armament wastes*, ed. K. Lohs and T. Stock, Stockholm International Peace Research Institute, Oxford, 1997, p. 165.
57. A. A. Fries, *Chemical warfare*, McGraw-Hill, New York, 1921.
58. E. B. Vedder and D. C. Walton, *The medical aspects of chemical warfare*, Williams & Wilkins Company, Baltimore, 1925.
59. D. H. Avery, *The science of war: Canadian scientists and allied military technology during the second world war*, University of Toronto Press, Toronto, 1998.
60. H. L. Gilchrist and P. B. Matz, *The residual effects of warfare gases*, US Government Printing Office, Washington, 1933.

61. E. M. Remarque, *All quiet on the western front*, Little, Brown and Company, Boston, 1929.
62. F. Itani, *Deafening*, Harper Collins Publishers Ltd., London, 2003.
63. C. Y. Harrison, *Generals die in bed*, Potlach Publications, Hamilton, Canada, 1974.
64. A. R. Camidge, *Diary of A. R. Camidge*, Imperial War Museum, London, 1916.
65. E. A. Shephard, *Diary of Lieutenant Shephard*, Imperial War Museum, London, 1915.
66. B. Appler, in *The challenge of old chemical munitions and toxic armament wastes*, ed. T. Stock and K. Lohs, Stockholm International Peace Research Institute, Oxford, 1997, p. 77.
67. B. C. Garrett, in *The Monitor*, 1995, Vol. 1, p. 11.
68. J. A. Davis, *Air and Space power J.*, 2003 Spring, findarticles.com/p/articles/mi_mONXL.
69. R. Pankhurst, *The Ethiopians: a history*, Blackwell Publishing, 1998.
70. A. Del Boca, *The Ethiopian war, 1935–1941*, University Chicago Press, Chicago, 1965.
71. *Enemy capabilities for chemical warfare*, Military Intelligence Service War Department, Ottawa, 1943.
72. G. L. Trammell, in *Chemical warfare agents*, ed. S. M. Soman, Academic Press Inc., 1992, p. 255.
73. B. Goodwin, *Keen as mustard. Britain's horrific chemical warfare experiments in Australia*, University of Queensland Press, Brisbane, 1998.
74. Y. Tanaka, *Bull. Atomic Scientists*, 1988, **44**, 10.
75. H. Kurata, *Lessons learned from the destruction of the chemical weapons of the Japanese Imperial forces*, Taylor and Francis, London, 1980.
76. J. Guillemin, *Scientists and the history of biological weapons: a brief historical overview of the development of biological weapons in the twentieth century* EMBO Rep. 2006, European Molecular Biology Organization, 2006.
77. J. Watts, *The Lancet*, 2002, **360**, 628.
78. H. T. Cook and T. F. Cook, *Japan at war: an oral history*, The New Press, New York, 1993.
79. S. Ienaga, *The Pacific war, 1931–1945: A critical perspective on Japan's role in world war II*, Pantheon Books, New York, 1968.
80. W. R. Cullen, 2005, personal experience.
81. J. Bryden, *Deadly allies: Canada's secret war, 1937–1947*, McClelland & Stewart Inc., Toronto, 1989.
82. R. G. Manley, in *The challenge of old chemical munitions and toxic armament wastes*, ed. K. Lohs and T. Stock, Stockholm International Peace Research Institute, Oxford, 1997, p. 231.
83. Z. Witkiewicz and K. Szarski, in *The challenge of old chemical munitions and toxic armament wastes*, ed. T. Stock and K. Lohs, Stockholm International Peace Research Institute, Oxford, 1997, p. 112.

84. J. Perera, in *The challenge of old chemical munitions and toxic armament wastes*, ed. K. Lohs and T. Stock, Stockholm International Peace Research Institute, Oxford, 1997, p. 121.
85. T. Stock, in *Chemical weapon destruction in Russia: political, legal, and technical aspects*, ed. J. Hart and C. D. Miller, Stockholm International Peace Research Institute, Oxford, 1998, p. 75.
86. War Department, *Protection against gas*, United States Office of Civilian Defence, Washington DC, 1941.
87. G. N. Jarman, in *Advances in chemistry series*, American Chemical Society, 1959, Vol. 23, p. 328.
88. E. A. Flood, National Archives, Ottawa, C-5003, C-5013, 1941.
89. *Minutes of a meeting of the treasury board*, Minister of National Defence, G. N. Jarman, Ottawa, Aug. 26, 1941.
90. D. D. Moir, *Analyst*, 1940, **65**, 154.
91. R. A. Peters, L. A. Stocken and R. H. S. Thompson, *Nature*, 1945, **156**, 616.
92. L. A. Stocken, *J. Chem. Soc.*, 1947, 592.
93. BAL, *Dimercaprol*, www.inchem.org/documents/pims/pharm/dimercap.htm.
94. A. B. Carleton, R. A. Peters, L. A. Stocken, R. H. S. Thompson and D. I. Williams, *J. Clin. Invest.*, 1946, **25**, 497.
95. D. Thomas, 2005, personal communication.
96. S. Rose, ed., *Chemical and biological warfare*, George G. Harrap & Co. Ltd., London, 1968.
97. Z. Wertejuk, M. Koch and W. Marciniak, in *Arsenic and old mustard: chemical problems in the destruction of old arsenical and 'mustard' munitions*, ed. J. F. Bunnett and M. Mikolajczyk, Kluwer Academic Publishers, 1996, p. 91.
98. L. Uris, *Mila 18*, Bantam, 1983.
99. G. Jones, in *Daily Telegraph*, 2006, Dec. 30.
100. (a) A. M Lilienthal, *The Zionist connection: what price peace?* Dodd, Mead & Company, 1978.
 (b) A. Heller, in *Globe and Mail*, 2008, Jan. 12, p. A14.
101. D. Davis, in *Jewish World Review*, 1998, April 6.
102. Veterans Affairs, *Final VA rule on claims based on mustard/Lewisite exposure*, Department of Veteran Affairs, Washington DC, 1994, Aug. 18.
103. National Defence, *Regulations governing the use of volunteers for physiological tests*, Department of National Defence, 1945.
104. A. Marin, *Complaints concerning chemical agent testing during World War II*, National Defence and Canadian Forces Ombudsman, 2004, Feb.
105. D. Ljunggren, in *Reuters*, 2004, Feb. 19.
106. G. Galloway, in *Globe and Mail*, 2006, Nov. 23, p. A11.
107. E. Welsome, *The plutonium files: America's secret medical experiments in the cold war*, Dial Press, 1999.
108. K. Szarski and Z. Witkiewicz, in *The challenge of old chemical munitions and toxic armament wastes*, ed. K. Lohs and T. Stock, Stockholm International Peace Research Institute, Oxford, 1997, p. 112.

109. *Cold war at Porton Down*, University of Kent Porton Down Project. www.kent.ac.uk/porton-down-project.
110. A. Heller, in *Globe and Mail*, 2008, Jan. 12, p. A14.
111. R. Evans, in *The Guardian*, 2000, March 10.
112. B. Hileman, in *Chem. Eng. News*, 2005, July 4, p. 9.
113. Concentrates, in *Chem. Eng. News*, 2006, Jan 30, p. 32.
114. T. F. Swearengen, *Tear gas munitions*, Charles C. Thomas, Springfield, 1966.
115. J. B. Neilands, G. H. Orians, E. W. Pfeiffer, A. Vennema and A. H. Westing, *Harvest of death; chemical warfare in Vietnam and Cambodia*, Free Press, New York, 1972.
116. F. Czarnecki, *Occupat. Environ. Med.*, 2001, **3**, 443.
117. I. James, in *Vancouver Sun*, 2004, Sept. 10, p. A12.
118. J. Sallot, in *Globe and Mail*, 2005, Nov. 5, p. A17.
119. S. Rose, in *The Guardian*, 2007, Dec. 12, p. 41.
120. E. J. Olajos and H. Salem, *J. Appl. Toxicol.*, 2001, **21**, 355.
121. H. Hu, J. Fine, P. Epstein, K. Kelsey, P. Reynolds and B. Walker, *J. Am. Med. Ass.*, 1980, **262**, 660.
122. B. McCaffrey, in *Ireland's OWN* 2004, Oct. 16, p. 1.
123. S. Millar, in *Sunday Times-Ireland*, 2005, Feb. 20.
124. *Republican News An Phoblacht*, 2002, Oct. 31.
125. C. Gould, in *Disarmament Diplomacy*, 2000, Nov.
126. J. M. Stellman, S. D. Stellman, R. Christian, T. Weber and C. Tomasallo, *Nature*, 2003, **422**, 681.
128. A. W. Galston, *BioScience*, 1971, **21**, 891.
127. C. D. Harnly, *Agent Orange and Vietnam: an annotated bibliography*, The Scarecrow Press, Inc., Metuchen NJ & London, 1988.
129. A. H. Westing, *BioScience*, 1971, **21**, 893.
130. H. Bronstein, in *Reuters*, 2006, April 17.
131. N. P. F. Castro, in *The UNESCO Courier*, 2001, May 21.
132. K. Solomon, 2006, personal communication.
133. C. Scott-Clark and A. Levy, in *The Guardian*, 2003, March 29.
134. S. Berger, in *Vancouver Sun*, 2005, May 3, p. A9.
135. A. Schecter, Chemical Society of Canada: Annual Meeting, Halifax NS, 2006.
136. L. W. Dwernychuk, *Chemisphere*, 2005, **60**, 998.
137. Testimony, *Weapons panel, Part 1*, Winter Soldier Investigation, Detroit, Jan. 31–Feb. 2, 1999.
138. R. Clark, *The silent weapons*, David McKay Company, New York, 1968.
139. *The fog of war*, E. Morris director, 2003.
140. D. Saunders, in *Globe and Mail*, 2004, Jan. 24, p. F3.
141. P. M. Boffey, *Science*, 1971, **171**, 43.
142. M. Jacobs and L. Ember, in *Chem. Eng. News*, 1997, April 14, p. 6.
143. M. Heylin, in *Chem. Eng. News*, 1997, May 12, p. 27.
144. G. York, in *Globe and Mail*, 1997, Nov. 1, p. A12.

145. *Convention on the prohibition of the development, production, stockpiling and use of chemical weapons and on their destruction*, www.opcw.org/html/db/cwc/eng/cwc_frameset.html.
146. F. Laurin, in *The challenge of old chemical munitions and toxic armament wastes*, ed. K. Lohs and T. Stock, Stockholm International Peace Research Institute, Oxford, 1997, p. 263.
147. L. Ember, in *Chem. Eng. News*, 2003, March 10, p. 12.
148. D. Webster, *Aftermath: the remnants of war*, Pantheon, 1996.
149. F. A. Pitten, G. Muller, P. Konig, D. Schmidt, K. Thurow and A. Kramer, *Sci. Total Environ.*, 1999, **226**, 237.
150. J. Matousek, in *The challenge of old chemical munitions and toxic armament wastes*, ed. K. Lohs and T. Stock, Stockholm International Peace Research Institute, Oxford, 1997, p. 104.
151. (a) G. P. Glasby, *Sci. Total Environ.*, 1997, **206**, 267.
    (b) S. Westall, Reuters Berlin, 2007, Nov. 22
152. O. Schachter and D. Sewer, *Am. J. Int. Law*, 1971, **65**, 84.
153. H.-J. Heintze, in *The challenge of old chemical munitions and toxic armament wastes*, ed. K. Lohs and T. Stock, Stockholm International Peace Research Institute, Oxford, 1997, p. 255.
154. OSPAR Commission, *Overview of past dumping at sea of chemical weapons and munitions in the OSPAR maritime area*, 2005, Pub. No 2005/222.
155. G. Reid, in *Globe and Mail*, 2001, Nov. 10, p. A17.
156. HELCOM, *Final report of the ad hoc working group on dumped chemical munitions (HELCOM CHEMU)*, Helsinki, 1995.
157. M. Simons, in *New York Times*, 2003, June 20, p. A8.
158. J. Lindsay-Poland, in *Revista Envio.*, 1998, Sept.
159. R. G. Manley, *Overview of the status of the chemical demilitarisation worldwide and the way ahead*, Organization for the Prohibition of Chemical Weapons, 2000.
160. J. B. Tucker, *Bull. Atom. Sci.*, 2001, **57**, 51.
161. A. Chimiskyan, in *Chemical weapon destruction in Russia: political, legal, and technical aspects*, ed. J. Hart and C. D. Miller, Stockholm International Peace Research Institute, Oxford University Press, Oxford, 1998, p. 75.
162. L. Ember, in *Chem. Eng. News*, 1999, Dec. 13, p. 9.
163. NATO, in *Science and Society*, Newsletter, 1997, 2nd quarter.
164. L. Ember, in *Chem. Eng. News*, 2003, May 26, p. 25.
165. J. MacDonald and C. Mendez, *Unexploded ordnance cleanup costs*, RAND Corporation, 2005.
166. V. Kolodkin, in *Chemical weapon destruction in Russia: political, legal, and technical aspects*, ed. J. Hart and C. D. Miller, Stockholm International Peace Research Institute, Oxford, 1998, p. 94.
167. Green Cross, globalgreen.org/media/events/Kambarka_Memo_28_February_2006.pdf.

168. N. Kalinina, in *Chemical weapons destruction in Russia: political, legal, and technical aspects*, ed. J. Hart and C. D. Miller, Stockholm International Peace Research Institute, Oxford, 1998, p. 1.
169. V. Sheluchenko and A. Utkin, in *Chemical weapon destruction in Russia: political, legal, and technical aspects*, ed. J. Hart and C. D. Miller, Stockholm International Peace Research Institute, Oxford, 1998, p. 113.
170. W. R. Cullen, P. Andrewes, C. Fyfe, H. Grondey, T. Liao, E. Polishchuk, L. Wang and C. Wang, in *Environmental aspects of converting chemical warfare facilities to peaceful purpose*, ed. R. McGuire, Kluwer Academic Publishers, Dordrecht, The Netherlands, 2002, p. 45.
171. L. Ember, in *Chem. Eng. News*, 2003, Dec. 1, p. 28.
172. S. P. Harvey, T. A. Blades, L. L. Szafraniec, W. T. Beaudry, M. V. Haley, T. Rosso, G. P. Young, J. P. Earley and R. L. Irvine, in *Arsenic and old mustard: chemical problems in the destruction of old arsenical and 'mustard' munitions*, ed. J. F. Bunnett and M. Mikolajczyk, Kluwer Academic Publishers, 1998, p. 115.
173. L. Ember, in *Chem. Eng. News*, 2007, April 30, p. 23.
174. J. Bunnett, 2000, personal communication.
175. Rocky Mountain Arsenal, RMA, *From Weapons to Wildlife*, www.epa.gov/superfund/accomp/success/rma.htm, 2006.
176. Global Security.org, *Rocky mountain arsenal (RMA)*, www.globalsecurity.org/wmd.facility/rocky.htm.
177. R. G. Sutherland, in *The challenge of old chemical munitions and toxic armament wastes*, ed. T. Stock and K. Lohs, Stockholm International Peace Research Institute, Oxford, 1997, p. 141.
178. J. M. R. Bull, in *Dailypress.com*, 2006, May 11.
179. L. Ember, in *Chem. Eng. News*, 2000, Nov. 13, p. 23.
180. L. Ember, in *Chem. Eng. News*, 1996, April 22, p. 32.
181. CWWG, *Chem-weapons disposal chronology*, www.cwwg.org/Chronology.html.
182. L. Ember, in *Chem. Eng. News*, 2006, April 17, p. 27.
183. CWWG, www.cwwg.org/pr_08.13.02hallwin.html.
184. L. Ember, in *Chem. Eng. News*, 2005, May 2, p. 8.
185. Concentrates, in *Chem. Eng. News*, 2002, March 25, p. 29.
186. L. Ember, in *Chem. Eng. News*, 2005, March 7, p. 31.
187. J. McAndless, SR 626, Defense Research Establishment Suffield, 1995, Sept.
188. L. Ember, in *Chem. Eng. News*, 2007, Jan. 15, p. 14.
189. J. Matousek, www.prague2003.fsu.edu/content/pdf/232.pdf.
190. K. F. Koehler, www.arofe.army.mil/Conferences/CWD2001/Koehler.htm.
191. H. Jaffe, in *Washingtonian Online*, 2000, Dec. www.washingtonian.com/health/groundzero.html.
192. B. Plaisted, in *The Corps'pondent*, 2000, December.
193. Fact sheet, in *The Corps'pondent*, 2000, July.
194. S. Levine, in *Washington Post*, 2006, October 12.
195. K. Ginzinger, in *The Common Denominator*, 2006, April 11.

196. ARSDR, *Potential contaminants in soils at American University, Washington, DC*, Agency for Toxic Substances and Disease Registry, 1998.
197. US Army, *Spring Valley area cleanup and safety suggestions*, Public Affairs Office US Army Corps of Engineers, Baltimore, 2002.
198. J. Perera and A. Thomas, in *New Scientist*, 1986, Feb. 13, p. 18.
199. UK Parliament, *Select committee on environmental audit. Minutes of evidence notes and documents promised during oral evidence to the committee 7 April 1998*, House of Commons, London, 1998.
200. I. R. Kenyon, *Quarterly Journal of the Harvard Sussex Program on CBW Armament and Arms Limitation*, 1997, Issue 35, March.
201. R. Whymant, in *London Times*, 1997, Nov. 12, p. 15.
202. SIPRI, *Memorandum of understanding between Japan and China*, Ministry of Foreign Affairs, 1999.
203. NTI, *Abandoned chemical weapons (ACW) in China*, www.nti.org/db/china/acwpos.htm.
204. J.-F. Tremblay, in *Chem. Eng. News*, 1997, June 22, p. 30.
205. Chinadaily.com, in *China Daily*, 2003, Aug. 10.
206. H. Ozawa, in *Agence France Presse*, 2007, July 20.
207. S. Shangwu, in *China Daily*, 2004, Feb. 23.
208. Y. Shibata, K. Tsuzuku, S. Komori, C. Umedzu, H. Imai and M. Morita, *Appl. Organomet. Chem.*, 2005, **19**, 276.
209. Concentrates, in *Chem. Eng. News*, 2004, Nov. 1, p. 19.
210. J. Warrick, in *Vancouver Sun*, 2005, Jan. 22, p. C4.
211. E. Oziewicz, in *Globe and Mail*, 2005, Dec. 24, p. A12.
212. CBWIinfo.com, www.cbwinfo.com/Chemical/Blister/HD.shtml.
213. State Department, *The Lessons of Halabja: An Ominous Warning*, www.usinfo.state.gov/products/pubs/iraq/warning.htm.
214. G. York, in *Globe and Mail*, 2004, March 6, p. A4.
215. G. Roberts, *Poisonous weapons*, www.crimesofwar.org/thebook/poisonous-weapons.html.
216. R. Bucher, *Anniversary of the Halabja massacre*, www.state.gov/r/pa/prs/2001/1322.htm.
217. H. Hu, R. Cook-Deegan and A. Shukri, *J. Am. Med. Assn.*, 1990, **263**, 1065.
218. P. Koring, in *Globe and Mail*, 1998, Dec. 19, p. A14.
219. H. L. Arison, *The cover-up of Gulf War syndrome.* www.warvets/arison/gws.htm.
220. Concentrates, in *Chem. Eng. News*, 1996, Aug. 26, p. 25.
221. Gulflink, www.gulflink.osd.mil/news/na_injured_marine_21mar00.html.
222. Gulfwarvets, www.gulfwarvets.com/testimon.htm.
223. L. Ember, in *Chem. Eng. News*, 2004, Oct. 25, p. 40.
224. J. Ingham and K. Walker, in *Daily Express*, 2003, Dec. 12, p. 6.
225. N. Warshott, in *Vancouver Sun*, 2002, Nov. 2, p. A10.
226. M. R. Gordon, in *New York Times*, 1987, Dec. 24.
227. P. Reynolds, BBC News/Africa, news.bbc.co.uk/2/low/africa/3336493, Accessed Oct. 28, 2007.

228. J. Manthorpe, in *Vancouver Sun*, 2006, March 7, p. A11.
229. L. Polgreen, in *New York Times*, 2006, Oct. 9.
230. J. Warrick, in *Vancouver Sun*, 2003, April 23, p. A13.
231. H. Purkitt and S. Burgess, Annual Meeting of the International Studies Association, 2001, Feb.
232. NRC, *An Assessment of non-lethal weapons science and technology*, National Research Council, National Academies Press, Washington DC, 2003.
233. L. Ember, in *Chem. Eng. News*, 2002, July 15, p. 28.
234. M. Bellaby, in *Vancouver Sun*, 2002, Oct. 30.
235. L. Ember, in *Chem. Eng. News*, 2002, Nov. 4, p. 6.
236. N-B-C-Warfare, *Chemical weapon attacks*, www.n-b-c-warfare.com/chemical.htm.
237. Concentrates, in *Chem. Eng. News*, 2002, Nov. 18, p. 54.
238. F. Guterl, in *Discover Magazine*, 1996, Jan. 1.
239. Concentrates, in *Chem. Eng. News*, 2007, June 11, p. 24.
240. P. W. Roberts, in *Globe and Mail*, 2006, March 18, p. F4.
241. M. Heylin, *Chem. Eng. News*, 2004, Oct. 11, p. 36.
242. M. Heylin, *Chem. Eng. News*, 1998, May 11, p. 26.

CHAPTER 7
Arsenic and the Environment

7.1 Introduction

Although people generally associate arsenic with deleterious effects and sound the alarm whenever it is detected in their environment, the concentration of the element in the Earth's crust is surpassed by that of many other potentially harmful elements, including lead and chromium. With a concentration of about 5 ppm in the earth's crust, arsenic is considered to be a trace element and it is about the 54th element in terms of relative abundance. Elements with a similar abundance to arsenic include uranium, bromine, tin, germanium, tungsten and molybdenum.[1,2] That said, arsenic can become concentrated in some parts of the globe because of natural mineralisation, where it is found in close association with metals such as iron, copper, lead, cadmium, gold, silver, tungsten, and molybdenum – and because of unnatural anthropogenic activities such as mining and pesticide manufacturing. Largely as a result of these latter activities the US Agency for Toxic Substances and Disease Registry (ATSDR) ranks arsenic as No. 1 on its list of priority hazardous substances because of both its prevalence in contaminated environments and its toxicity. This ranking has not changed for many years and the latest, posted in 2005, shows arsenic as No. 1 followed by lead, mercury, vinyl chloride, polychlorinated biphenyls (PCBs), polycyclic aromatic hydrocarbons (PAHs) and cadmium.[3]

But the impact of arsenic on an ecosystem cannot be judged solely on the basis of concentration: the chemical species must be considered as well. We have seen that the properties of compounds of the element depend very much on their chemical constitution. When this is taken into consideration the picture rarely gets worse and often changes a lot for the better from a human point of view, whether it concerns arsenic in a slag heap or arsenic in an oyster.

The concentration of arsenic in the universe is estimated to be 0.008 ppm, and in the sun, 0.004 ppm. Closer to home arsenic is found everywhere: coal can have arsenic concentrations of up to 10 000 ppm; shales and phosphate rocks, up to 10 ppm; and oil up to 1.6 ppm. Volcanoes, both terrestrial and submerged, liberate about 2×10^7 kg of arsenic per year, and processes such as sedimentation return about 9×10^7 kg of it per year to the world's lakes and oceans.[4]

Is Arsenic an Aphrodisiac? The Sociochemistry of an Element
By William R. Cullen
© William R. Cullen 2008

7.1.1 Arsenic in the Atmosphere

Because of the larger land mass of the northern hemisphere and its associated industrial activity, there is an uneven distribution of arsenic in the atmosphere. The concentration ranges from 0.5 ng/m^3 to 2.8 ng/m^3 above land, and occurs mostly as particulates of oxides and sulfides from industrial emissions (mainly from copper smelting, coal combustion, and municipal incineration). The contribution from biological processes is uncertain but may be in the range 0.2 to 26×10^6 kg/year. Special conditions exist in urban and industrial areas where 20 per cent of the arsenic comes from high-temperature combustion and 55 per cent from motor traffic.[4] The mean air arsenic concentration in some US cities is about 1–2 ng/m^3; in Athens the levels are around 6.0 ng/m^3 and in the heavily polluted Shanxi province of China the range is 100–920 ng/m^3. The American Conference of Government Industrial Hygienists' (ACGIH) guideline for eight-hour exposure to inorganic arsenic is 1000 ng/m^3.[5,6]

A lot of the arsenic in the North American atmosphere comes from the deserts and industrial cities of Asia.[7,8] An international team of atmospheric scientists report that much of the arsenic in Nevada air come from Mongolia.[9]

7.1.2 Arsenic in the Pedosphere

Soils that make up the pedosphere, the surface soils that humans are most likely to come into contact with, contain variable amounts of arsenic. The arsenic is usually found bound up as arsenate to iron and manganese oxy(hydroxy) species and is not readily mobile or bioavailable (Section 7.4) so is not a threat to human health. The average soil arsenic concentration in the US is about 7.5 ppm, while the global average is about 5 ppm. Higher concentrations are found in soil that is rich in organic matter. Some natural values as high as 100–115 ppm have been recorded in lower Austria, but anthropogenic influence can result in localised arsenic concentrations of 250 ppm to 10 000 ppm.[4]

7.1.3 Arsenic in the Hydrosphere

The natural background arsenic level for rivers and lakes is about 0.1 ppb to 1.7 ppb; however, as we will see in Chapter 8 some groundwater can have concentrations in the parts-per-million range. In some lakes in unique geological areas, such as Searles Lake CA, we find 225 ppm arsenic.[10] Seawater has a natural concentration of 1 ppb to 4 ppb with an estimated median value of 3.7 ppm.[4] All this arsenic exists essentially as the inorganic, biologically accessible species, arsenate and arsenite. In water that is rich in sulfur, sulfur can displace oxygen from these species so that thioarsenates and thioarsenites become the predominant arsenicals. Some methylarsenic species are encountered, usually in low concentration, where there is biological activity such as algal growth.[11,12]

7.2 Arsenic in the Biosphere

Animals and plants cannot escape exposure to arsenic and have learned to cope in order to survive. The worm *Lumbricus rubellus* lives in the polluted soil of what was once the site of the 19th-century Devon Great Consols mine in Tavistock, UK. The mine used to be the world's biggest source of arsenic (Section 3.13). The concentration of the element in the surrounding soils can reach 9800 ppm and the concentration in the worms 18–37 ppm (dry weight basis). (Because the water content of living things can be appreciable and variable – commonly around 90% of the total weight – it is usual to remove all the water from the sample before determining the arsenic concentration. The result is recorded in ppm (dw; dry weight basis). If the sample still includes any internal water we see, for example, ppm (ww; wet weight basis). In the following text dry weight is assumed unless otherwise indicated.)

The worm has developed a high tolerance to arsenic that is indicated by a colour change in the organ that processes their soil diet. Normally greenish brown, it turns bright yellow in the tolerant species. The tolerance and colour are transmitted from parent to offspring.[13] Another worm from the marine environment, *Tharyx marioni,* is capable of accumulating arsenic up to a concentration of 13 000 ppm. The fern *Pteris vitata* accumulates inorganic arsenic up to a concentration of 23 000 ppm (Section 7.2.4).

One of the outcomes of the 1901 Kelvin Commission (Section 3.12) was the recommendation of limits for the amounts of arsenic allowed in food and drink. These were, for a liquid, 0.143 ppm; and for a solid 1.43 ppm[14] and were put into law in 1928 (Section 3.8).[15] In the early 1900s it was assumed that arsenic could only get into food by adulteration; however, it soon became apparent that the concentration of arsenic in some food could be greater than 1 ppm yet it could be eaten without noticeably harming the consumer. One early paper from Mr. A. Chaston Chapman on the subject stands out and is worth examining in some detail.[16]

7.2.1 Arsenic in Seafood

Chapman initiated this study to ascertain the reason for unusual mortality among oysters in English beds. He found the "best British" oysters contained up to 10 ppm (ww) arsenic and Portuguese oysters up to 70 ppm (ww). In addition scallops, mussels, whelks, lobsters, prawns, shrimps and crabs were all high in arsenic. Fish had appreciably lower arsenic concentrations – up to 7 ppm. He also noted that freshwater fish and crayfish, and terrestrial snails had lower concentrations of arsenic than their marine equivalents, at less than 1 ppm (ww).

Chapman examined the urine of an individual who had eaten a meal of lobster. For the first 12 hours afterwards the arsenic concentration was on average 7.2 ppm but it dropped off rapidly over the next three days. At the time, urine concentrations of this magnitude were thought to be associated only with acute arsenic poisoning. Chapman concluded that the arsenic present in the urine is not arsenious oxide, is "not toxic," and is present as a more or less complex organic substance or mixture of substances that is excreted in urine without change. The arsenic is not detectable by the Marsh test (Section 5.4) without prior treatment with, for example, concentrated acid, but it is extractable from the lobster with alcohol. He also remarked: "It is not impossible that the unpleasant consequences which occasionally follow the consumption of crustaceans and shell fish may not be unconnected with the presence of these compounds."

This is how matters stood for some time because the methodology was not available for isolating and identifying the arsenic species. It is worth noting that Chapman was a very experienced scientist whose results, which are "elevated" compared with modern values, were accepted without question. At the same time others with the same level of competence were probably providing results of similar quality for forensic purposes where elevated numbers could have unfortunate consequences (Section 5.4).

7.2.2 Analysis of Arsenic Species (Speciation) in Living Organisms

Marsh's method for analysing arsenic was effective because the arsenic was removed from the sample as arsine gas. This allowed its detection and quantification in the absence of interference from the rest of the sample. The basic method is still in use today under the name "hydride generation," although the detectors have become better and gas production has been simplified.

The first big breakthrough in distinguishing between arsenic species – arsenic speciation – followed the discovery of a convenient new reducing agent, sodium borohydride. This proved to be more effective than Marsh's reagents (zinc plus acid) for generating arsine, because the new reagent was water soluble, easy to use, and afforded reproducible results. Two researchers from Florida, R. S. Braman and C. C. Foreback, applied this new method to a range of samples collected from the environment and found that the generated arsine gas often was contaminated with other arsenic-containing compounds. These unexpected gases were identified as methylarsines – arsenic compounds in which the hydrogen atoms of arsine, AsH_3, were replaced by one to three methyl groups (Figure 7.1). Braman and Foreback separated these gases and quantified them. They interpreted their results as a measure of the concentration of compounds in the original sample that possessed from zero to three methyl groups attached to arsenic[17] (The figure in Box 3.2 shows the pathway from the acid on the top line, to the corresponding arsine on the bottom line). These methylarsenic species, initially studied by Bunsen and later synthesised for use in human medicine and as agricultural chemicals (Section 2.6), were found everywhere: in water, food extracts and urine, which meant that arsenic was intimately involved in many life process and not just the obscure microbial transformations described in Chapter 3. Thus was born an age in which arsenic in the environment was thought of in terms of inorganic arsenic, monomethylarsonic acid, dimethylarsinic acid, and trimethylarsine oxide, which are the steps along the Challenger pathway (Section 3.9). Hydride generation became the most commonly used method for arsenic speciation.[18]

The hydride generation method follows this sequence:

generation of gases → separation of gases → detection and quantification of gases.

The separation was originally accomplished in 1973 by first freezing the gas sample and letting it warm slowly; in which case the gases separated themselves according to their boiling points. Detection was accomplished by using atomic absorption spectroscopy, an optical technique that in this application "sees" only the compounds in the gas stream that contain arsenic. A few years later chromatography (Box 7.1) was introduced to improve the gas-separation step.

Sometimes, the method did not account for all of the arsenic known to be in a sample. One Japanese study of the urine of 102 subjects found on average: inorganic arsenic 11.4 ppb, monomethylarsonic acid 3.6 ppb, dimethylarsinic acid 35.0 ppb, and total arsenic 121 ppb, leaving 71 ppb unaccounted for.[18] This situation caused some consternation for a while but a look at Chapman's results discussed above would have provided the reason. The arsenic species in urine after a meal of lobster did not give a Marsh test; in other words, it was not easily converted to a gas, so the same species or something similar must be present in the urine of the Japanese volunteers. The hunt was then on to find these species.

Box 7.1 Chromatography

Chromatography involves, first, dissolving a sample in a mobile phase that may be a gas, a liquid or a supercritical fluid. Second, the mobile phase is forced through an immobile, immiscible stationary phase. The phases are chosen such that components of the sample have differing solubilities in each phase or have differing binding strengths with respect to the stationary phase. A component that is quite soluble in the stationary phase (or strongly bound to it) will take longer to travel through it than a component that is not very soluble in the stationary phase (or weakly bound to it) but very soluble in the mobile phase. As a result of these differences in mobility, sample components will become separated from each other as they travel through the stationary phase.

Techniques such as gas chromatography (GC) and high-performance liquid chromatography (HPLC) use "columns" which are narrow tubes packed with a stationary phase, through which the mobile phase, a gas or liquid, is forced to travel. The sample is transported through the column by continuous addition of the mobile phase. This process is called elution. The average rate at which the substance being analysed moves through the column is determined by the time it spends in the mobile phase: this is called the retention time, which is specific to each substance.

In the 1970s, Australia had a good economic reason for trying to account for the undetected arsenic because the Australian export market for lobster tails was in jeopardy. Japanese buyers, who did not concern themselves with the niceties of speciation, were worried about the high arsenic content of the western rock lobster. John Edmonds and Kevin Francesconi, two young scientists working at the Western Australia Marine Research Laboratories in North Beach, solved the problem with the help of Jack Canon at the University of Western Australia. The team started work on 4 kg of tails and was eventually able to isolate the major arsenic species arsenobetaine (($CH_3)_3As^+CH_2COO^-$), an essentially nontoxic compound that had been previously prepared in the 1930s for medical studies (Figure 7.1).[19,20] They analysed human urine collected during a backyard lobster barbecue, and verified the presence of arsenobetaine: it passes through mammals rapidly and without modification. It has been established that arsenobetaine is the most abundant arsenic species found in marine animals, and is the source of most of the arsenic that is not detected when the conventional hydride generation sequence is used.

Next, the Australian group turned their attention to the arsenic species in marine algae; Chapman had reported high arsenic concentrations here as well.[16] It was soon established that arsenobetaine was not present, although again the extracted species were "invisible" to hydride generation. In one experiment, Edmonds started with 32 kg of the macroalga *Ecklonia radiata* collected from a beach after a storm. The soluble arsenic compounds were painstakingly isolated using very careful chromatography. The identification of these compounds was not a trivial exercise, but with the help and advice of Don Cameron of Melbourne University, and Alan White of the University of Western Australia, they established that the compounds have an arsenic atom bound to a ribose (sugar) group. For this reason they are referred to as arsenosugars. The most commonly encountered examples are shown in Figure 7.1.[20–22] These compounds are generally regarded as being non-toxic to humans, being metabolised rapidly to dimethylarsinic acid prior to excretion; however, the reduced species (remove the O from the As=O group) that are readily produced by reaction with metabolically important thiol-containing compounds such as cysteine, exhibit increased toxicity.[23]

Around this time, analytical methodology advanced on two fronts. First, there were developments in separation methods, notably high-performance liquid chromatography (HPLC), that allowed the efficient separation of arsenic species in a solution, and second, the direct detection of these species was made possible by the availability of a new but expensive element-specific detector, the inductively coupled plasma mass spectrometer (ICP-MS), which had the ability to detect very small amounts of the arsenic compounds following their separation on the chromatography column. This analytical method, known by the acronym HPLC-ICP-MS, is now in widespread use. The detector tells us that the components of the solution being analysed contain arsenic, and the signal intensity can be used to calculate how much. Unfortunately, no molecular information is provided, so identification of the species responsible for the signal is possible only if standards are available for comparison: identical retention times probably indicate identical compounds.

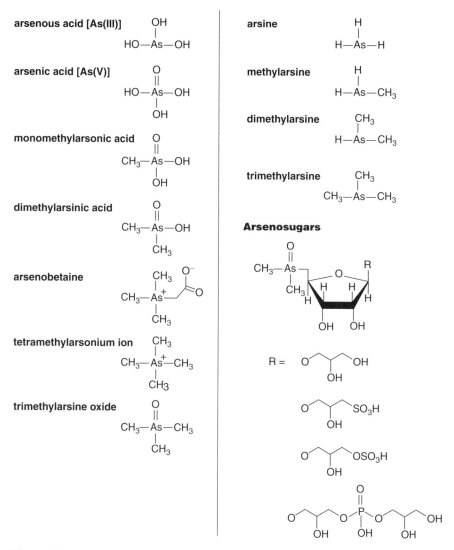

Figure 7.1 Arsenic compounds commonly found in the environment.

This new methodology ushered in a very productive era in which arsenic species could be determined in a wide variety of samples provided they could be easily extracted from the sample.[24] This proviso is necessary because the aim is to extract the species using mild agents, such as water and methanol, in order to maintain the chemical integrity of the species during extraction. Most of the arsenic species in fish can be extracted; for mussels the extraction efficiency is lower, at about 58 per cent, and for many plants speciation of only 10 per cent of the arsenic in the sample is possible because the remaining 90 per cent of the arsenic is tightly bound inside the cells or on the cell walls in a form that is not extractable without changing its chemical makeup. In such a situation, new methods are needed that provide speciation information on the whole sample (Box 7.2).

Increases in the sensitivity of the element-specific detectors have resulted in finding more compounds in samples: one extract of a scallop kidney contained 23 arsenic species, 16 of which were not identified.[25] To cope with this situation, element-specific detectors are being replaced by molecule-specific detectors in the form of mass spectrometers (MS) that provide information about the molecular formula and structure of the extracted species. Species can be identified even if they are not completely separated during the chromatography step. Thus, 15 species were identified in one HPLC-MS study of the water extracts of the kidney of a giant clam, four of these being previously unknown. The kidneys are particularly rich in arsenicals, presumably because of the activities of symbiotic unicellular algae, zooxanthellae, within the tissues (The arsenic concentration in the kidney of the giant clam *Tridacna maxima* is around 1000 ppm).[20,26–29]

7.2.3 Distribution of Arsenic Species in the Living Environment

In the early days of arsenic-speciation studies, there seemed to be a clear line separating the species found in the terrestrial environment and those found in the marine environment. It seemed that terrestrial species comprised inorganic arsenate and arsenite, and some methylated arsenicals including the gaseous arsines, all of which could be nicely accounted for by the Challenger sequence (Section 3.9). There were also a few oddities whose origins are still obscure, such as phenylarsonic acid isolated from shale oil, and some gaseous arsenicals containing ethyl groups found in natural gas.[12] In contrast, the marine environment provided a bonanza of arsenic species: the number is now around 40, but the most commonly encountered organoarsenicals are arsenobetaine and the four arsenosugars in Figure 7.1, all of which are relatively innocuous as far as human consumption is concerned. Arsenobetaine is the major arsenical in fish, lobsters, prawns, crabs, snails and marine mammals. In bony fish where the total arsenic concentration can range from 0.1 ppm to 166 ppm (ww), the proportion of arsenobetaine ranges from 50 per cent to nearly 100 per cent. Arsenobetaine shares the stage with arsenosugars in shellfish: in mollusks, the proportion of arsenobetaine can be as low as 12 per cent but it is essentially 100 per cent of the species in some whelks that accumulate arsenic up to 1360 ppm. The balance shifts almost entirely to arsenosugars in algae.[30–32]

Recent speciation studies reveal that the terrestrial/marine species distribution is not so clear-cut: arsenobetaine, arsenocholine and arsenosugars are being found in terrestrial plants and animals. Some discoveries are clearly the result of the availability of improved instrumentation that allows speciation studies on samples that have lower total arsenic concentrations than encountered in their marine equivalents. Arsenosugars as well as arsenobetaine are found in fresh water fish, but extraction efficiencies are much lower; consequently, we know little about the speciation of the bulk of the arsenic. The total arsenic concentration in freshwater bivalves is much the same as in marine species, but the major extractable species are arsenosugars and, unlike marine bivalves, arsenobetaine is only occasionally found in these animals.[33,34] The same is true for freshwater crayfish, where arsenobetaine is a minor component of the arsenic in the tails: the most predominant species is an arsenosugar.[35]

Arsenobetaine is found in high concentrations in mushrooms, as is the closely related arsenocholine, although the latter is only found at trace levels in the marine environment. The arsenobetaine content of mushrooms depends on their taxonomic position, with the more highly evolved mushrooms containing the most arsenobetaine. The beautiful and marginally poisonous "fly agaric" mushroom (*Amanita muscaria*), unmistakable with its bright red cap covered with white scale, can accumulate 22 ppm arsenic as arsenobetaine plus simple methylarsenic species. The family Agaricaceae, which is the source of many of the edible mushrooms, carries most of its arsenic in the form of arsenobetaine – up to 9 ppm.[36]

The use of mushrooms in ancient and modern Asian medicine is attracting attention in the rest of the world. The arsenic content of the mushroom extracts could be a concern to those who aren't aware of the importance of speciation.[37] Likewise health-conscious people who eat fish oil for its omega-3 fatty acid content should be aware that there is an issue regarding the arsenic content and possibly the PCB content of the fish that needs to be evaluated. Sales of this vile-tasting product have tripled over the past 10 years.

The arsenicals in terrestrial mammals and birds are also more complex than first thought. Birds occupy an important niche in the food chain because they feed on things that may have taken up arsenic (seeds, insects, worms, animals, *etc.*) and are themselves a food source. The American tree sparrow that lives in the arsenic-rich area of Yellowknife (Section 7.6) has 45 ppm of arsenic in its liver, which is on a par with marine levels. It could be argued that the birds have adapted to the local arsenic-rich environment by becoming arsenic-rich themselves, but the species is migratory and eats mainly local seeds that are low in arsenic, so how to explain the arsenic levels?

The arsenic species in the American tree sparrow and a related species, the dark-eyed junco, are arsenite, arsenate and simple methylated species, but two non-migratory birds – gray jays and spruce grouse – have considerable amounts of arsenobetaine in their tissues; indeed, it is the dominant compound in breast tissues and liver. Worms are the only terrestrial creature that could act as a source of arsenobetaine for the birds, but these are not commonly found in cold climes.[36,38] The gray jay does consume fungi, and arsenobetaine containing mushrooms have been identified in Yellowknife, but the grouse only consumes plant material that is unlikely to contain significant quantities of this arsenical. Perhaps the birds actually metabolise simple arsenicals to arsenobetaine for their own purposes.

We have seen that arsenic in human hair can be used as a biomarker of exposure. The feathers of the birds in Yellowknife have relatively high arsenic concentrations, up to 23 ppm in the American tree sparrow, and the main extractable species are inorganic arsenic and dimethylarsinic acid, just as in human hair (Section 1.9.2).[38] (Birds exposed to the pesticide monomethylarsonic acid have elevated arsenic levels in their feathers (Section 2.6).[39])

Both arsenocholine and arsenobetaine are major species in the liver and muscle of Yellowknife hares and squirrels, and both these arsenic species are found in the local red fox. Arsenobetaine is found in the animal's tongue, muscle, lung and diaphragm.[40]

7.2.4 Where do Arsenosugars and Arsenobetaine Come From?

The source of the methyl group in the Challenger pathway outlined in Box 3.1 is S-adenosylmethionine (SAM), and one of these transfers, affording trimethylarsine oxide, is shown in Figure 7.2. It seems that SAM also has the capability to transfer the sugar-containing group (the five-membered D-ribose ring) to arsenic as shown. It is not hard to believe that some chemical processes could lead to the formation of arsenosugars from this product. The pathway to arsenobetaine is not so obvious, because SAM can supply only methyl and sugar groups. There was early speculation that arsenobetaine was derived from arsenosugars, and there is now evidence suggesting that bacteria are able to carry out this transformation, so the pathway arsenosugar to arsenobetaine (Figure 7.2) might be valid in some instances.[41,42]

The fundamental mystery, however, is why these compounds are produced at all. We saw in Chapter 1 that methylation of arsenic was once viewed as a detoxification until the revelation of the toxicity of the methylarsenic(III) species forced a re-evaluation of this hypothesis. Most primitive organisms get rid of "ingested" arsenic, which is usually arsenate, by reduction to arsenite prior to elimination (Section 7.11). In mammals, methylation mainly leads to the formation of dimethylarsinic acid that is then eliminated. To take this a step further by putting a sugar group onto the arsenic seems to be a useless extravagance, especially if the only purpose is to end up with arsenobetaine, which at best may serve a useful purpose as an osmolyte, as has been suggested to account for its presence in the marine environment.[32]

The common button mushroom *Agaricus bisporus* is able to accumulate about 3.7 ppm of arsenobetaine from growth media containing either arsenate or Yellowknife mine waste. The arsenic in the mine waste is considerably less bioaccessible (Section 7.4) than the arsenic in an arsenate solution, yet the concentration of arsenobetaine is independent of the source of the arsenic, suggesting that the mushroom might have a good reason for the accumulation. A clue to the puzzle seems to be offered by the XANES (X-ray absorption near-edge structure) spectra of the fruiting body (Box 7.2) that show that arsenobetaine is the only arsenical in the cap and has a major presence in the stalk edge. This distribution would fit a role as an osmoregulator: something to maintain the structure of the fruiting body.[43a]

But even if the osmoregulator scenario is correct it is unlikely to be the complete story: arsenobetaine has been found in animals that live near deep-sea hydrothermal vents and do not have a plant stage in their food chain, and – as we have just seen – arsenobetaine is found in the terrestrial environment in earthworms, in the livers of birds and foxes, and is a major species in mushrooms.

The end members of the Challenger pathway, trimethylarsine oxide and the tetramethylarsonium ion (Box 3.2) are found in some marine animals and

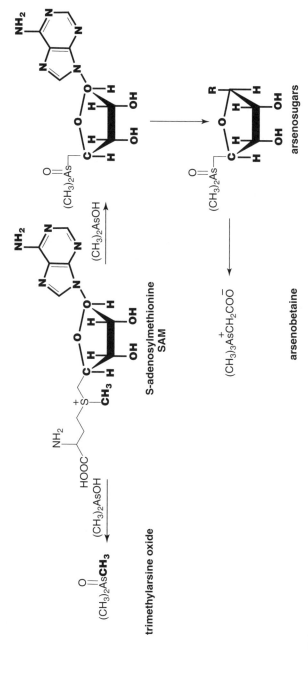

Figure 7.2 SAM as a source of methyl and sugar groups, a possible route to arsenosugars and arsenobetaine.

in mushrooms, so the transfer of the third methyl group is not completely blocked – in fact trimethylarsine oxide is a major metabolite of fungal metabolism. This poses another question: what induces SAM to switch to delivering sugar groups to arsenic after delivering two methyl groups in some organisms but not in others (see Figure 7.2)? The third group on the sulfur atom of SAM is available for transfer[43b] but seemingly not to arsenic.

> **Box 7.2 Speciation Analysis of Solid Samples: X-ray Absorption Spectroscopy (XAS)**
>
> A number of methods are available for determining the total arsenic content of animal, vegetable and mineral samples. These usually involve a total digestion step followed by the application of a method such as the Marsh test or, in modern times, ICP-MS to the resulting solution. Some less-sensitive methods, such as neutron activation analysis and X-ray fluorescence spectroscopy, can be used directly on the samples if the concentration of the element is high enough. If we need to probe more deeply, X-ray diffraction methods are available to determine the structure of crystalline material. Geologists use a battery of tests, such as scanning electron microscopy to establish element distribution in small particles. However, obtaining the same sort of information for living material is not so simple.
>
> A fairly new technique, X-ray absorption spectroscopy (XAS), opens this window a little. In the XANES (X-ray absorption near edge structure) experiment, tunable X-rays of known energy are generated in a synchrotron and then directed at the sample. If the energy of the X-ray beam matches the energy of the core electrons of atoms in the sample, an absorption occurs and the event is recorded. The atoms of each element have characteristic core-electron energies, so as the energy of the X-rays hitting the sample is varied, a response is obtained from the elements in the sample. The core energy for a particular element in a sample is sensitive to factors such as the oxidation state (valence) of the atom, so the method can be used to determine not only the presence of the element but also the presence of particular compounds or types of compounds of that element (species).[44]
>
> By moving the sample in the path of the beam held at constant energy XAS can be used to map the location of say all the arsenic in the sample. This is known as XAS imaging.
>
> X-ray absorption techniques for studying the atomic structure of matter do not require that a crystalline sample be used (www.xafs.org). However, the necessary synchrotron facility is probably the most expensive analytical tool available and there are only about 40 operating in the world. The state of Victoria, Australia, has just unveiled a new synchrotron in Melbourne that cost $207 million (AU): the electron accelerator is 67 metres in diameter.[45]

7.2.5 Arsenic Accumulators and Hyperaccumulators

Generally, arsenic concentrations in freshwater algae are lower than in marine species, but some can accumulate to even higher levels. For example, *Chorella vulgaris* accumulates 20 000 ppm of arsenic when grown in an aqueous medium containing 100 ppm arsenate. The algae grow happily in the presence of 10 000 ppm arsenate. The arsenic speciation information is incomplete because of the low extraction efficiency, but as for other terrestrial algae the main extractable species are arsenate, dimethylarsinic acid and arsenosugars.[46–48]

Edible terrestrial plants take up arsenic from the soil, but it is rare to find the arsenic concentration above 1 ppm. The concentration increases, from about 0.020 ppm in cabbage, in the order: cabbage (*ca.* 0.020 ppm) < carrots < grass < potatoes < lettuce, mosses and lichens < ferns < mushrooms (*ca.* 2.5 ppm.[4] Concentrations tend to be higher in regions with high local soil arsenic concentrations. In Bangladesh, edible arum leaves can reach 4 ppm and the roots, 8.6 ppm.[49a,49b]

Some terrestrial plants hyperaccumulate arsenic. In the 1960s the Douglas fir tree (*Pseudotsuga menziesii*) was discovered to accumulate arsenic, up to 1000 ppm in new growth, and a much smaller herb named sheep's bit (*Jasione Montana* L) growing on mine waste can accumulate up to 6640 ppm arsenic.[50,51] But far more interest has been generated by reports that the Chinese brake fern (*Pteris vittata* L.), which can accumulate up to 64 ppm arsenic in its above-ground biomass during normal growth, is able to accumulate 23 000 ppm (2.3 g/kg) when the plant is grown on arsenic-rich soil.[52] Other ferns from the same family show similar abilities.[53] (There are indications that the plants accumulate arsenic to repel insects.[54])

These concentrations suggest that the ferns could be used for cleaning up arsenic-contaminated ground all over the world. (The general process, known as phytoremediation, is expected to be worth millions of dollars in the near future.) The main requirements for success are: rapid growth, high biomass production, and a high bioconcentration factor (the ratio of the arsenic concentration in the plant to that in the soil). In one report on the brake fern, researchers note the following: "Assuming a constant rate of arsenic accumulation, a best-case unrealistic scenario, it would take four or five harvests (80–100) weeks to phytoremediate a soil contaminated with 100 ppm arsenic."[52] The authors are excited because phytoremediation is normally thought of in terms of a 10-year time frame.

Pilot projects utilising the ferns are in progress on Vashon Island, WA; Spring Hills, Washington DC; (Section 6.15) and at sites in China.[55,56] It is estimated that 20 per cent of China's farmland is polluted with heavy metals, which causes huge crop losses every year. Arsenic concentrations up to 2 000 pm are encountered so that phytoremediation looks very attractive.[57] Vashon Island, located in Puget Sound, was contaminated by fumes from the Asarco smelter in nearby Tacoma (Section 7.7). There, the fronds of the ferns have taken up so much arsenic that the biomass has to be treated as toxic waste and cannot be placed in an ordinary landfill. This is a general and non-trivial problem as the arsenic does not go away; it simply moves from the soil to the plant. Suggested solutions range from collecting and burning the now arsenic-rich harvest, to burial at sea where the resulting increase in arsenic concentration would not be significant.[58]

Genetic engineering is having some success in creating new hyper-accumulators.[59] Subsurface species that can be rotated with rice are also being engineered for phytoremediation. This work would not be possible without understanding the mechanism of the accumulation process. We now know, thanks to XANES (Box 7.2), that the arsenic is taken up into the fronds of the brake fern as arsenate. This arsenate is then reduced to arsenite and stored in the cell vacuoles.[60,61]

The tobacco plant *Nicotina tabacum* is a heavy-metal accumulator.[62] At the time of the Kelvin Commission on Arsenical Poisoning, 1901 (Section 3.12), cigarettes contained 150 ppm[14] arsenic that probably came from the use of arsenical pesticides and from the curing process in which the heat was supplied from arsenic-rich fuels. By 1927 the concentration had dropped to below 29 ppm. Rachel Carson writes that arsenic in cigarettes from North America increased by more than 300 per cent between 1932 and 1952[63] but by the mid-1990s it was less than 2 ppm, and by 2007 the level had fallen to less than 0.5 ppm. Nonetheless the antitobacco lobby still relies on the power of the A-word to shock, and uses it frequently in its advertising: "Cigarettes Contain Arsenic. A RAT POISON."[64] (In 2005 the soil under an abandoned tobacco warehouse in downtown Rocky Mount NC, had to be replaced to a depth of 2 feet because of arsenic contamination.[65])

The arsenic concentration is a concern because exposure to arsenic and other components of cigarette smoke via the inhalation pathway at environmentally relevant levels can act synergistically to cause DNA damage.[66–68]

Another aspect of arsenic in cigarettes deserves mention. The manufacture and distribution of illicit tobacco products is big business. For example, in the UK, 25 per cent of all tobacco sales in 2000–2001 involved illicit products, 20 per cent contraband and 5 per cent counterfeit. In the US, customs seizures of counterfeit American tobacco products amounted to $41.7 million (US) in 2005.[69] There is cause for concern, not only because of the lost revenue and the increase in crime, but also because the counterfeit products contain significantly higher concentrations of metals such as lead and cadmium, and arsenic. Legal cigarettes now contain on average 0.3 ppm arsenic, while the impostors contain 0.9 ppm. Chemical studies indicate that the metals probably come from the use of low-grade phosphate fertiliser available in many developing countries, especially in China where cigarettes are manufactured in slave-labour factories some literally below ground.[70,71] There is a Framework Convention on Tobacco Control in place sponsored by the World Health Organization; however, neither the United States nor the UK has ratified it.

7.3 Arsenic in Our Food and Water

Before we begin this topic, we need to spend some time considering the question – is arsenic an essential element? If it is, we need to be concerned that we are getting enough in our diet. If it is not, we need to be concerned that what we are eating is not harming us.

7.3.1 Essentiality

Of the approximately 90 elements in the Periodic Table that are not man-made or radioactive, about 20 are essential for some life processes. Some major ones are carbon, oxygen, sulfur, iron, nitrogen and phosphorus; the last two being from the same group of the Periodic Table as arsenic. Some less obvious ones are zinc, nickel and selenium. But is arsenic likely to be essential for life when we know that its compounds often have the reverse effect? Such a situation is not unknown – selenium is toxic at high doses, yet it is an essential element at low dose.[72] In this case it is beneficial to recall what Paracelsus said: "The dose makes either a poison or a remedy" (Section 1.2).

For a substance to be an essential nutrient, the criteria are as follows:[18] (1) The substance is present in all organisms for which it is essential. (2) Reduction of exposure to the substance below a certain limit results, consistently and reproducibly, in an impairment of physiologically important functions; and restitution of the substance under otherwise identical conditions prevents the impairment. (3) The severity of signs of deficiency increases in proportion to the reduction of exposure to the substance.

An additional criterion was proposed around 1967: the abnormalities produced by a substance's deficiency should always be accompanied by specific biochemical changes – that is, the biochemical mechanisms of action should be known.

Animals are used for experiments involving essentiality, and the simplest way of gathering evidence is to completely eliminate the substance from their diets and see what happens. But this cannot be done very easily when the substance, such as arsenic, is everywhere. At least three research groups have been seriously involved in trying to get around these difficulties and have studied goats, miniature pigs, rats and chicks to find, for example, that the addition of sodium arsenite to a synthetic amino acid diet stimulated growth of rats. In 1986, a report from a US EPA workshop on arsenic suggested that information from these and other studies "demonstrate[s] the plausibility that arsenic, at least in inorganic form, is an essential nutrient. A mechanism of action has not been identified and, as with other elements, is required to establish fully arsenic essentiality."[73] More recent studies showed that arsenic has a growth-stimulating effect on animals that are subject to additional dietary stress such as an amino acid (methionine) deficiency. Nevertheless, in 1999 another EPA convened committee wrote: "Studies to date do not provide evidence that arsenic is an essential element in humans or that it is required for any essential biochemical process. Arsenic supplementation seems to have a growth-stimulating effect at very high doses in mini-pigs, chicks, goats and rats."[18]

The daily intake of inorganic arsenic that animals "require" is suggested to be 25 to 50 ng/g of diet. If this is extrapolated to humans eating 2000 calories per day, their "arsenic requirement" is suggested to be 12.5 to 25 μg/day, which would not be met by many individuals on a normal diet as described below, and could put hemodialysis patients especially at risk.[74]

7.3.2 Arsenic Market Baskets

The US Food and Drug Administration has conducted "market basket" surveys of food yearly since 1961. Identical baskets of food are purchased by FDA inspectors three to four times a year from retail stores in four geographic regions. The shopping list is chosen to track food availability and consumption patterns. The food is prepared as for consumption, and then analysed for a range of components, one of these being arsenic. However, only total arsenic is measured at present. On this basis, the major arsenic contribution to the diet of all age groups comes from the meat, poultry and fish category (only fish really matter), with rice products coming in second. For 18 baskets in the years 1991 to 1997, seafood products amounted to 85 per cent of the total arsenic intake from food for 16- to 18-year-old girls, and rice products 8 per cent of the total. The situation is much the same for other age groups, although there is a decrease in the relative contribution from rice products with increase in age. The one big exception is seen in the six- to 11-month-old infant group, where rice products account for 31 per cent of the total arsenic ingested, and seafood accounts for 42 per cent. This means that infants get a bigger proportion of their arsenic as inorganic species (Section 7.2.2).[75,76]

The general picture changes if only inorganic arsenic is considered. A recent survey of the arsenic content of 156 representative foods comprising all the food groups (as well as beer and tap water) and prepared for eating if that was usual for the product, found the total inorganic arsenic concentrations that are summarised in Figure 7.3. The magnitude of the intake from all the common sources is shown in Figure 7.4. The major contributions come from food and water. Dietary contributions from air and soil, which are variable and depend on the local environment, are minor.[77] The food contribution in the United States was swamped by that from water at the recently abandoned Maximum Contaminant Level (MCL) for arsenic of 50 ppb, but the two are about the same at the current 10 ppb MCL. It would certainly be hard to argue for further reductions.

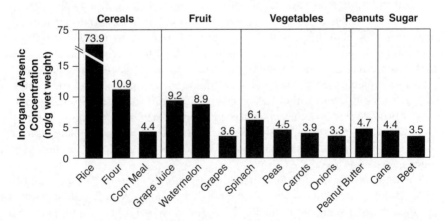

Figure 7.3 Inorganic arsenic content of the food we eat (from Schoof et al. 1999[77] with permission from Elsevier Limited, Oxford).

Arsenic and the Environment 303

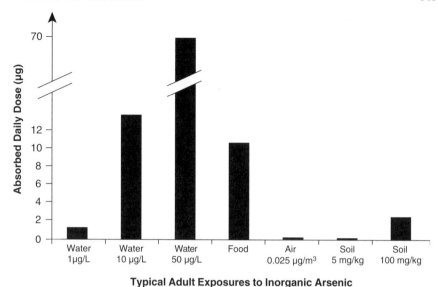

Figure 7.4 Sources of the inorganic arsenic we ingest (adapted from Schoof *et al.* 1999[77] with permission from Elsevier Limited, Oxford). The dose from water assumes an intake of 1.4 liters at the indicated arsenic concentration.

Rosalind Schoof and her coworkers estimated that the average inorganic arsenic intake from water in the US in 1999 was about 5 μg per day; controlled by the intake of around 1.4 liter of water per day and the local water supply, whose concentration ranges from 1 ppb to about 10 ppb. The contribution from food depends on the diet, with rice being by far the main source. So, in spite of what has been said about the high arsenic concentration in seafood, the dietary contribution from that source in the form of inorganic arsenic is too low to make a significant contribution to the average diet. For example, canned tuna has less than 25 ppb (ww) inorganic arsenic although the total arsenic is as high as 1 ppm (ww). These numbers, combined with the appropriate dietary information result in a predicted inorganic arsenic intake of 1–20 μg per day.[77,78]

As indicated above (Section 7.3.1), if inorganic arsenic is shown to be an essential nutrient, the arsenic requirement for humans is estimated to be in the range of 12 to 25 μg per day. If this range is correct, the low intake predicted for much of the US population could indicate the potential for unforeseen problems.[78]

Other studies come to a somewhat different conclusion. The mean arsenic concentration in rice grown in the south-central US is 0.30 ppm, whereas the mean for California rice is 0.17 ppm. The difference is attributed to the historical application of arsenical pesticides and desiccants to the land of the south-central US at the time that it was being used to grow cotton (the rice varieties were chosen for their resistance to arsenic). Estimation of the daily intake of inorganic arsenic from rice and water suggests that, with respect to the US current 10 ppb MCL for arsenic in water: "for Asian Americans, dietary arsenic exposure from rice could be higher than from water." A similar situation is claimed to exist for infants and those forced to eat a gluten-free, rice-rich diet.[79] This work (cf. Figure 7.4) garnered a lot of national and international media coverage. Professor Andrew Meharg from the University of Aberdeen and the lead author of the study, is quoted as saying that he does not let his children eat the suspect rice and suggests that other consumers might take the same decision (30 to 40 per cent of the long-grain rice imported into Britain comes from the US: 10 000 tons a month). The USA Rice Federation and the UK Rice Association were not very impressed: "Enough nonsense about arsenic already!" "Consumers would need to eat an awful lot of rice before nearing a harmful level."[80–83] The arsenic content of 310 foods studied by the UK Food Standard Agency did not show any arsenic concentrations above legal limits.[84]

7.3.3 The Effect of Cooking

The food selected for the market basket studies is cooked if that is the usual preparation method, but the subsequent analysis gives no information on the effect of cooking on arsenic speciation. This is another important consideration because of toxicity and bioavailability concerns, but unfortunately only a few studies are available. The speciation of rice is unchanged by cooking;[85] however, arsenobetaine in seafood is converted into the tetramethylarsonium ion on cooking. The amount of conversion depends on the cooking method and the type of oven. In one sample of sole, the conversion accounted for 10 per cent of the total arsenic in the product.[86a] This results in a significant increase in toxicity because the LD_{50} (Section 1.7) for tetramethylarsonium chloride is 580 mg/kg (mice) whereas that of arsenobetaine is $>10\,000$ mg/kg.

7.3.4 More on Rice

Probably the best way of monitoring food intake, known as a duplicate-plate study, involves the analysis of plates of food that duplicate what is being eaten. A Japanese study of this sort involving four subjects found that the mean daily intake of arsenic was 202 μg (ranging from 31 to 682 μg), of which on average only 6.8 per cent was inorganic arsenic, amounting to an intake of 14 μg per day.

Of the ingested arsenic, a daily mean of 81 per cent of it was eliminated in the urine, and 25 per cent was found in feces. The study ran into unforeseen problems: "Some of the fecal values for the woman of couple B could not be obtained because of constipation."[86b] The fecal elimination route is generally ignored or negated by other scientists. Fastidiousness trumping science? The high amount of arsenic in the Japanese diet is directly related to the amount of seafood, including algae, eaten. In spite of this, the daily intake of inorganic arsenic, mainly from rice, is not excessive.

Professor Meharg and his coworkers are devoting considerable resources to the "arsenic in rice" issue (much of it complimenting the considerable effort already under way in Bangladesh[87,88]). They report that US long-grain rice "has the highest mean arsenic level in the grain, at 0.26 ppm, of which 42% is inorganic. The mean arsenic concentration in rice from Bangladesh is 0.13 ppm (within a big range of 0.10 to 0.95 ppm) that is 81% inorganic. Most of the arsenic in the rice plant is actually in the roots and stems, and the concentration in the rice straw and rice roots can be around 100 ppm. These are used as cattle feed, with unknown direct human consequences;[49a,89,90a] however, Professor Dipankar Chakraborti and colleagues (Section 8.3) report that cow dung cakes burned as fuel in domestic ovens release arsenic into the air, resulting in respiratory problems for the inhabitants (Section 7.10.2).

Rice is the major caloric contributor (80 per cent) to the Bangladeshi diet, and to get this energy the average adult eats about 450 g of the food per day. Similar data from other countries allow estimation of the amount of inorganic arsenic ingested by the population from rice cooked in water containing 10 ppb arsenic. In this model, rice adds another 13 per cent per day to the amount of inorganic arsenic consumed from water in North America: for Japan the increase is 86 per cent; for India 116 per cent; for China 131 per cent; for Thailand 157 per cent; and for Bangladesh 215 per cent. If the drinking water contains more arsenic the figures improve although the relative risk from the arsenic in the drinking water is increased. In Bangladesh (Chapter 8): "With a drinking water intake of 100 ppb, arsenic intake from rice will account for up to 30 per cent arsenic consumption if rice contained typical grain concentrations of 0.2 ppm arsenic."[87,91]

It is clear that some rice-eating nations will be at risk from arsenic in their diet even if they adopt and achieve a drinking water standard of 10 ppb. However, the water contribution is likely to far outweigh that from food for many years to come. Avoiding exposure to arsenic in these countries, if and when arsenic-safe water is available, will require considerable effort. In Bangladesh the arsenic levels in paddy soils, where the main rice crop *Oryza sativa* is grown, ranges from 3.1–42.5 ppm depending on the age of the well supplying the water. Rough estimates indicate that irrigation water at 100 ppb, adds 1 ppm of arsenic to the soil per year,[91,92] but it will be difficult to remediate paddy soils to prevent uptake by the plants. Some relief might be obtained by using surface water from rivers for irrigation. The use of different rice varietals could decrease arsenic uptake and adoption of the developing technology of aerobic cultivation would also help because arsenic is released from soil under the normal anaerobic flooded conditions. An extreme solution would be to shift agricultural practice to growing nonirrigated crops, such as wheat and maize.[93] But even then troubles could keep coming, because it seems that climate change is likely to reverse many of the successes in wheat production that resulted from the green revolution in the 1970s, and a global rice shortage is possible according to the International Rice Research Institute.[94]

7.3.5 Hijiki and Other Algal Products

Many algal products are encountered in food. Some familiar names are nori, also known as purple laver (*Porphyra spp*); kombu or haidai (*Laminaria Japonica*); wakame and quandai-cai (*Undaria pinnatifida*); hijiki (*Hizikia fusiformis*); dulse (*Palmaria palmata*); and Irish moss or carrageen (*Chondrus crispus*). Seaweed is a major source of the alginate that is widely found in food and cosmetic products. Carrageen soaked in whisky was an Irish folk medicine used for treating colds and respiratory ailments. Distributors of algal products, who prefer to refer to them as being derived from sea vegetables rather than seaweed, boast about the high mineral and vitamin contents of their merchandise, but many of these products are marketed as food supplements without any indication of their arsenic content.

Early in 2007 a report surfaced of a woman who believed she was made unwell by the arsenic in her kelp supplements. This prompted the following remarks from Marc Schenker the principal investigator in the study: "It's unfortunate that a therapy that's advertised as contributing to vital living and well-being would contain potentially unsafe levels of arsenic."[95] This very surprising public statement was based on a determination of the total arsenic content of the product, *Laminaria digitata* – which was in the usual range – and ignorance of the speciation of the arsenic, which is expected to be essentially all in the form of arsenosugars. It is true that the patient recovered when she stopped eating the kelp, but the case for arsenic poisoning was never convincing. The American Herbal Products Association challenged the conclusions, noting that the subject had ignored specific label cautions; and suggesting that iodine overdose was the cause of the problems, not arsenic.[96]

Around the same time, Hong Kong newspapers were warning the local population about the arsenic content of popular dried squid snacks. The samples exceeded the permitted arsenic level of 10 ppm, but no consideration was given to the speciation that was probably all arsenobetaine.[97]

It is true that one seaweed species – *Hizikia fusiformis*, used in the food product hijiki – has an unusually high content of inorganic arsenic, commonly around 50 per cent of the total arsenic content, which averages about 22 ppm (ww).[98] The seaweed can be collected from the wild, but most of it is commercially cultivated in Korea. In order to manufacture the product, the harvested seaweed is first dried, then boiled in water together with another macroalgae to give it its prized black colour (the water is not changed very often). The hijiki is then steamed for four or five hours before being laid out to dry in the sun. Prior to use, the black, brittle commercial product is soaked for 10–15 minutes in warm water, drained and then typically stir-fried with tofu and vegetables.[99]

Hijiki has been part of the Japanese diet since ancient times, and is purported to reward the consumer with thick, black, lustrous hair. The product has been available in the West for many years. Canada began worrying about the arsenic content of hijiki in the late 1990s and in 2001 published a consumer advisory: "Consumption of only a small amount of hijiki seaweed could result in an intake of inorganic arsenic that exceeds the tolerable daily intake for this substance. Therefore, the consumption of this type of seaweed is to be avoided."[100] The United Kingdom, New Zealand and Hong Kong followed suit.

Needless to say, the Japanese Health Minister thinks that such action is unjustified,[101] claimings that soaking the hijiki prior to use removes most of the inorganic arsenic. When it is fed to mice, essentially all the arsenic in the washed product is eliminated in the feces.[102] If we had confidence in these results we would be inclined to believe that the arsenic in the product was not

Arsenic and the Environment 307

bioavailable (Section 7.4) and it could be eaten without concern. However, it seems that humans are different from mice, because one *in vitro* study of the bioaccessibility (Section 7.4) of the arsenic in hijiki afforded the following results:[103] (1) Of the total inorganic arsenic (55 ppm) in the raw product, 75 per cent is bioaccessible. (2) In the cooked product, 88 per cent of the total inorganic arsenic (31 ppm) is bioaccessible. Thus, the arsenic in the product does remain available for absorption into human intestinal mucosa. This conclusion is reinforced by other Japanese scientists: "After eating one serving of hijiki, arsenic intake and urinary excretion were at levels similar to those in individuals affected by arsenic poisoning."[104]

The practice of using seaweed for fertiliser is an ancient one and its benefits well known; however, it causes the arsenic content of the soil to increase over time because the arsenosugars are quickly broken down to inorganic species after application. Liquid fertilisers containing 10 to 50 ppm arsenic are also prepared from seaweed, but this should not be a concern because the recommended application rates are two orders of magnitude lower than the maximum permitted sewage sludge load for arsenic in the UK, which is 0.7 kg per hectare per year.[105]

Box 7.3 The Arsenic-Eating Sheep of North Ronaldsay

North Ronaldsay, the most northerly island of the Orkney Archipelago north of the coast of Scotland, is home to a breed of sheep that feed mainly on the seaweed that washes up on the shore. Most of this is kelp (*Laminaria digitata*) with an arsenic concentration of about 70 ppm, mainly as arsenosugars. This feeding practice has a long tradition that can be traced back at least 200 years, but some records suggest that the Vikings may have introduced the sheep hundreds of years earlier.

The little island (8 km by 5 km) has no natural harbour, so during the windswept winters the 50 inhabitants (as of 2007), known as crofters, are confined to the island. The land is fertile but most of it is used for growing barley and maintaining a few cattle: there is little room for any grazing animal that would guarantee a protein source for the crofters during the long winter months. The islanders solved the problem some time ago by adopting unusual animal husbandry techniques. They built a wall around the island to keep a herd of wild sheep on the beach. The sheep evolved to eat the piles of storm-cast kelp readily available during the winter, and were fat and ready for slaughter when other food was scarce. The crofters did make some grass available to the sheep during the lambing season.

Research workers from the University of Aberdeen wanted to study these animals to see if their seaweed diet had any short- or long-term effects, but getting permission was difficult. First, they had to win the confidence of the locals, and the approval of the long-established "sheep court" chaired by the Laird (lord) of the island. Professor Jörg Feldmann and his students

Box 7.3 Continued.

eventually won them over and permission was granted. But the hard part was catching the sheep. The community usually round up the sheep twice a year for culling and shearing. The 3000 or so animals are wild and take any opportunity to escape by running along the long beach or swimming in the ocean. They learned to swim to reach seaweed that is available in tidal pools during the summer months when the beach kelp supply is short.

The researchers eventually managed to heard the sheep into pens on the beach for controlled feeding experiments. The sheep eat about 3–4 kg of seaweed per day, resulting in an intake of about 50 mg of arsenic, so it was necessary to collect urine to study how the sheep metabolised the arsenic. At first the sheep were very shy but the students discovered that they urinated when their noses were pulled. Coordinated action resulted in successful sampling. However, as the study progressed the sheep became friendly with the students, and nose-pulling ceased to be effective, so a new strategy was developed that created a little consternation amongst the islanders: the sheep were fitted with nappies.

The collected urine had arsenic concentrations up to 5000 times of that of control sheep: levels of 50 ppm arsenic were not uncommon. The main metabolite is dimethylarsinic acid, just as it is for humans (Section 1.9).[106]

Arsenic species particularly dimethylarsinic acid and inorganic arsenic, are found in the sheep's wool and horns. The horn of the sheep starts to grow *in utero* and its arsenic burden results from efficient transplacental transfer (see Section 1.10).[107] The horn arsenic concentration can be used to monitor the exposure of other sheep, as well as cattle and deer, that are commonly seen nibbling on seaweed on beaches in Ireland and the west coast of Scotland. Anecdotally, it has been observed that those herds or flocks that have access to seaweed are generally healthier and fatter, just like the horses and the people of Styria of old (Section 1.3).

The meat that was once the main food source for the islanders has become a delicacy because of its "seaweedy" taste, and is now sold for premium prices to up-market restaurants in the UK. The meat has an arsenic content of less than the 1 ppm maximum laid down by the Food Standards Agency.[108]

7.3.6 Bottled Water

In 2005, the US population drank about 8 billion gallons of bottled water, or about 26 gallons per person. The product is distributed mainly by Nestlé, Pepsi and Coca-Cola. In the US, bottled water is considered a food and should be regulated by the Food and Drug Administration even though 25 per cent of it is repackaged tap water.[109] About 65 per cent of all bottled water in the US is packaged in the same state in which it is sold, which makes it exempt from FDA regulation. The FDA monitors the labelling of bottled water, but the bottlers are responsible for testing and one in five states does not regulate bottled water.[109] In 2006 the US House of Representatives voted to remove most of the warnings from food labels, potentially affecting alerts about arsenic content. The food industry is backing the move.[110]

A national advertising campaign for Fiji brand bottled water was based on the slogan: "The label says Fiji because it is not bottled in Cleveland." The name was supposed to make the consumer think of pristine environments and artesian water untouched by anything, including air. However, the Cleveland Water Department was neither impressed nor amused and ran some tests on its city water, Fiji water and other brands. Fiji placed first, but in the highest arsenic content category, at 6.3 ppb. None of the other samples was even close. The ads were pulled and the campaign cancelled.[111]

In 2007, bottled water imported to North America from Armenia was found to contain arsenic in the range of 500–600 ppb: the US permitted level for drinking water was 10 ppb (Section 8.11). Jermuk bottled water ("Jermuk Original Sparkling Natural Mineral Water Fortified with Natural Gas from the Spring") was to the fore. Warnings were issued in Canada and the United States regarding its arsenic content and the product was removed from US shelves. The Armenians were not particularly worried about the lost business; only 4 per cent of its export market went to North America. Most of the sales, 62 per cent, went to Russia. The Armenians claim that the water meets the requirements of *their* national law, passed in 2000, that bottled water should not exceed 700 ppb arsenic. Prior to this, water containing up to 1500 ppb arsenic was available. According to Benyamin Harutyunyan, director of the Resort and Physical Medicine Research Institute of Armenia, the 700 ppb limit allows the mineral water to be marketed as "medicinal table water."[112,113] The water is regarded as a national tonic, second only to local cognac, and the US action was viewed as an insult to Armenians and possibly part of a plot concocted by competing water-exporting nations.

Sales in North America were brisk after the warning because people were worried the supply would stop.[114] One shocked distributor in North Hollywood claimed to drink 10 to 16 bottles a day. His grandfather lived to 98 and his grandmother to 101 and they both drank the water regularly.

ZamZam water, sacred to Muslims, comes from a well in Mecca, Saudi Arabia, and is claimed to have the power to satisfy hunger and thirst and cure illness. It is consumed locally, especially by pilgrims performing the Hajj. The water is not available for export, but pilgrims are allowed to take samples home for personal use and for gifts. In spite of these restrictions, water labeled "ZamZam" is readily available outside Saudi Arabia. Recently, the British Food Standard Agency discovered samples containing about three times the allowed amount of arsenic (30 ppb). The arsenic-rich water available in Britain was sold in "what looks suspiciously like plastic petrol [gasoline] cans." The police became interested because it looked like the water was a local product and, if so, the arsenic was either already in the supply – an alarming prospect – or the arsenic was added to the supply, an even more alarming prospect. One Islamic bookshop was selling around 20 000 liters of the water a week in 2007.[115]

7.3.7 Metabolites

We have seen that humans metabolise ingested inorganic arsenic species mainly to dimethylarsinic acid, which is then eliminated in the urine. Typical human urine contains about 10 to 15 per cent inorganic arsenic, 10 to 15 per cent monomethylarsonic acid, and 60 to 80 per cent dimethylarsinic acid, with the total arsenic concentration below 20 ppb (Section 1.9). We also know that most people eliminate arsenobetaine rapidly in the urine; but what happens to other ingested arsenic species? We know a little about the fate of the arsenosugars, because of volunteers who were willing to eat seaweed products and collect their urine periodically for later analysis. The first observation is that the sugars are rapidly converted to dimethylarsinic acid, which is eliminated in the urine. Thus, high concentrations of this species in the urine do not necessarily indicate high exposure to inorganic arsenic. The second observation is that the rate of elimination of the arsenic species in the urine and their amounts varies from person to person: indeed the urine of some individuals can be essentially arsenic-free even after a meal of seaweed. Some individuals clearly have a different metabolism that may be related to ethnic origin.[116,117] The Japanese study mentioned above found substantial fecal elimination of ingested arsenic (Section 7.3.4), and the high concentration of dimethylarsinic acid in Japanese urine (Section 7.2.1) is certainly the result of arsenosugar ingestion.

Most mammals excrete arsenic much like humans, with the principal metabolite being dimethylarsinic acid. However, some species – such as chimpanzees and marmoset monkeys – seem to lack the ability to methylate arsenic, at least to the extent that methylarsenic species are not present in their urine. Feeding studies with seaweed have yet to be attempted.

7.4 Bioavailability and Bioaccessibility

The toxicity of a substance is usually dependent on the concentration achieved at its site of action. However, toxicity values are based on applied doses; that is, on the dose of the substance ingested, inhaled or placed on the skin. Toxicity values are thus unique to that substance. When a known toxicity value is used in connection with a different material it is implied that the bioavailability of the two are the same. (Bioavailability is the extent to which a contaminant is absorbed into the blood stream from the gastrointestinal tract and redistributed around the body, recognizing that absorption can also occur (but usually to a much lower extent) *via* lungs and skin.) However, the risk associated with ingesting arsenic from arsenic trioxide cannot be the same as ingesting insoluble arsenic-rich rock, or arsenic-rich mushrooms. Nevertheless, regulators commonly apply a single toxicity value in these situations – that of arsenic trioxide – claiming that arsenic is arsenic regardless of its species. Then, when challenged, they say they are being conservative because they need to protect the public. But sometimes the public needs to be protected from this type of risk assessment because it can panic populations into needless remediation activities that can cost them a lot of money. Adjustments must be made to the toxicity values used in risk assessment to take bioavailability into account.

The related term bioaccessibility is defined as the fraction that is soluble in the gastrointestinal environment and is available to be absorbed.[118,119]

There is good reason to extend the concept of bioavailability to food, although this application is in its infancy. In order to evaluate the impact of eating arsenic-rich food, ideally we need to know the arsenic species originally present, the species present after cooking, the species present in the bioaccessible fraction, the species in the bioavailable fraction and the species of any metabolites.

First, how do we measure bioavailability? One commonly used method relies on a mass-balance approach in which urine and feces are monitored for a substance over time, and the percentage of the applied dose excreted through each route is calculated. Animals are usually used for these studies, with immature swine, monkeys, rats and rabbits favoured as test subjects.[120] These experiments take time and are expensive; also, there are questions about the validity of the models and ethical concerns regarding the use of animals. Some generalised information about the bioavailability of different arsenic-containing samples obtained in this manner is shown in Figure 7.5.

The figure shows that the more soluble species such as arsenic trioxide are more bioavailable than "insoluble species" such as the mineral arsenopyrite (FeAsS). Smaller particles that provide an overall larger surface area are more bioavailable; encapsulation of the arsenic species inside an insoluble skin decreases its bioavailability. Thus, in principle we could take a sample of the arsenic-containing substance, determine its chemical and physical makeup by using some of the wide range of techniques now available and, with the help of some common sense, make a very good guess at the bioavailability of the arsenic in the sample. Unfortunately, this approach is not very practical because the instrumentation required is not usually available to regulators; it is sophisticated, expensive, and found only in well-equipped research laboratories. In addition, the approach may be effective for soils and minerals where we are dealing with inorganic species, but not for food, because even if it were possible to establish the speciation we don't know enough at present to estimate bioavailability and the associated risks.

There are alternative and more economical approaches to estimating bioavailability and bioaccessibility that do not require the use of animals. At the simplest level these attempt to establish if it is easy to extract the arsenic from the material and get it into solution; the argument being that unless that can happen, the arsenic, no matter what the species, will not be very bioaccessible. Tests such as the TCLP (toxicity characteristic leaching procedure), used as a measure of the suitability of a material for deposition in a landfill (Section 8.5), are simple examples. (The TCLP uses dilute acetic acid to simulate the conditions that the material might encounter in a landfill.) This thinking also underlies more sophisticated approaches that attempt to mimic the biochemical conditions in the human or animal gastrointestinal tract.

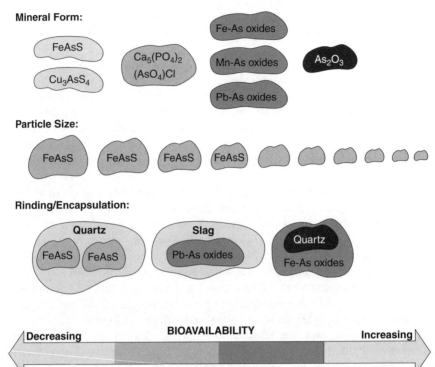

Figure 7.5 Factors influencing bioavailability of arsenic species (Ruby et al., 1999[121] with permission from the American Chemical Society).

7.4.1 Sequential Selective Extraction (SSE)

SSE involves subjecting a soil sample to successive chemical extractions with chemical reagents of increasing "strength", and analysing the extracts for the elements of interest. SSE provides information about the bioaccessibility of the element and its potential for its remobilisation.[122,123] The bioaccessibility of arsenic in soils from Yellowknife NT, as determined by this method, correlates well with results obtained from the gastrointestinal models.[124,125]

7.4.2 Gastrointestinal Models

One commonly applied gastrointestinal model is the Physiologically Based Extraction Test (PBET) which has two stages: an acidic gastric extraction stage, followed by a small-intestine stage at a higher pH. Total element concentrations are measured in solution following the extractions, to give a measure of bioaccessibility.[118] In the case of arsenic in food, speciation of this extract is necessary in order to make use of the information, but to date this has received scant attention.

In one application soil from the mining town of Anaconda MT, was fed in controlled doses to New Zealand white rabbits. The relative oral bioavailability of arsenic in the soil was found to be approximately 24 per cent, indicating that the soil arsenic was in a much less absorbable form than sodium arsenate. These animal models were used to validate PBET results[120,126] which in turn were utilised to set a cleanup level for the site of 250 ppm arsenic, although this decision was not universally accepted. "Rarely are bioavailability studies undertaken simply to improve the accuracy of a risk assessment. Rather, they are performed to justify site cleanup goals that are more financially or technically feasible [and] that involve leaving appreciable amounts of contaminant mass in place while still being protective of public health and the environment."[119]

From 1985–1989, 69 arsenic-contaminated sites covered by the US EPA Superfund project (Box 7.4), which was created to eliminate the health and environmental threats posed by hazardous waste sites, were targeted for cleanup. The goals were spread over a very broad range that reflected local and state practice and prejudices. For residential sites the range for arsenic was 2 ppm to 305 ppm, with a geometric mean of 23.4 ppm, and a calculated risk of 1 in 10^{-6}. For industrial sites the range was 8 ppm to 500 ppm, with a mean of 62.5 ppm. The decisions made at 84 per cent of the sites were risk-driven, and some of them, such as the Anaconda site mentioned above, were based on bioavailability and bioaccessibility considerations. (According to the US EPA, a risk assessment is an analysis that uses information about toxic substances at a site to estimate a theoretical level of risk for people who might be exposed to these substances. The information comes from scientific studies and environmental data from the site. A risk assessment provides a comprehensive scientific estimate of risk to persons who could be exposed to hazardous materials present at the site.)

The remaining 16 per cent of decisions were based on attainment of background concentrations with a narrower range of 8–21 ppm arsenic; it is likely that the use of this default risk variable resulted in over-remediation. "It is incumbent on decision-makers to use common sense when establishing cleanup levels. While the risk assessment process is critical in defining cleanup goals, incorporation of site-specific variables, *e.g.* bioavailability, is necessary to ensure an accurate assessment of potential site risks."[127a] The bioaccessibility and speciation of arsenic in clams was recently evaluated and the results support the decision to close Seal Harbour NS for clam harvesting.[127b]

7.5 Arsenic in the Anthrosphere

Elemental arsenic, although not commonly found, does occur in veins in crystalline rocks and schist, often associated with ores of antimony, silver, and zinc. About 500 arsenic-containing minerals have been characterised, the chief of these being arsenopyrite (FeAsS), also known as mispickel. The two sulfides, realgar (As_4S_4) and orpiment (As_2S_6) are also common arsenic-containing minerals (Section 1.2). There are numerous arsenides such as niccolite (NiAs), and arsenates such as scorodite ($FeAsO_4.2H_2O$). Sulfur "analogues" such as enargite (Cu_3AsS_4) are actually sulphides (Figure 7.5). The oxide occurs in two mineral forms: arsenolite and claudetite. Although not an arsenic mineral, iron pyrite (FeS_2), also known as "fool's gold," can be rich in arsenic: up to 4.4 per cent–it is widely distributed in mineral veins and in coal. This wide distribution of arsenic in the Earth's rocks ensures its presence in all water bodies above and below the Earth's surface (Chapter 8).

Nowadays, most people want as little to do with arsenic as possible, so there is little interest in mining minerals for their arsenic content, although this was not always so (Section 2.1). Arsenic usually makes the business news in stories such as: "At almost lightning speed, the Canadian miner, Goldcorp (the world's third-largest gold mining company), has shot into the ranks of the world's top gold producers. But, last week, an environmental study released in Honduras revealed dangerous blood levels of lead and arsenic among villagers local to the San Martin mine, which has been a target for protests over the past six years. . . . As this news broke, the Honduran president re-confirmed the moratorium on all new mining projects which his government had introduced in January, 2006."[128–131]

Governments normally view mining as an activity that provides economic benefit for their people and legislate accordingly. The World Bank finances a lot of these often problematic operations: "We invest to help reduce poverty and improve people's lives."[132] However, gold mining is arguably the dirtiest industry operating in regions such as Peru, Indonesia, Ghana and North America. In producing about 2600 tons of gold per year, gold mining consumes about 10 per cent of the world's energy while employing 0.09 per cent of the world's work force. About 79 tons of waste, often arsenic-rich, is produced for each ounce of gold and the legacy of open pits, acid mine drainage, and waste deposits pose serious human health concerns. In addition, mines are often developed in otherwise officially protected areas: 50 per cent of all newly minted gold comes from land belonging to indigenous peoples. Critics of the industry are becoming more vocal[132–134] but are not necessarily being heard. Five million cubic metres of mud containing arsenic, copper, and mercury spilled from a tailings pond at the Aznalcollar mine in Spain in 1998 into the environment of the Doñana National Park, the largest bird sanctuary in Europe. An engineer was fired after he had warned the Canadian owners Boliden Ltd. of Toronto ON, about the problem two years earlier.[135]

7.5.1 Gold Prospecting

The strong association between gold and arsenic has been used to good advantage by prospectors: arsenic on the surface could indicate that there is gold underneath. In mining news releases it is common to see statements such as "The property produced some highly anomalous gold results – up to 3.4 ppm – as well as very high arsenic concentrations - up to 16.5%."[136] "Gold values ranged from less than 10 ppb to 90 ppb, and averaged 15 ppb and arsenic ranged from less than 2 ppm to 1800 ppm and averaged 59 ppm."[137]

Because the root systems of plants and trees can be extensive and penetrate well below the surface, they "sample" the mineral composition of the growth area. And although there are no indicator plants that are specific for gold, some may be specific indicators of soil or waters rich in pathfinder elements, such as arsenic. In this way the Douglas fir that grows on the west coast of North America proved to be a useful indicator of gold and other metals because arsenic accumulates in the new growth (Section 7.2.3).[51,138]

7.6 Arsenic Trioxide and the Giant Mine, Yellowknife NT, Canada

The name Yellowknife does not derive from gold, but from the name "copper Indians," given by early fur traders to the local natives because they used yellow copper-bladed knives. The first mineral claim was staked by Cominco Ltd. in the vicinity of the town on the shore of Great Slave Lake, in 1935. But, because the gold was strongly bound up within arsenopyrite and pyrite, it was not until a roaster (smelter) came into service in 1948 that gold production began.

The ore was brought to the surface, crushed, ground to a fine powder and mixed with water. Then the gold-rich material, the sulfide concentrate, was separated from the remainder by differential flotation. The concentrate was roasted to oxidise the minerals and release the gold. This process converted the arsenic to volatile arsenic trioxide and the sulfur to sulfur dioxide gas, both of which are highly toxic. The solid product from the roaster which contained the gold was reacted with cyanide to get the metal into aqueous solution. The gold was then recovered and refined to the product bullion. All solid and water residues were discarded into tailings ponds.[138,139]

The stack gas, mainly arsenic trioxide and sulfur dioxide, was initially released directly to the atmosphere. A few years later, in 1951, an electrostatic precipitator was installed to recover most of the arsenic trioxide. The system was further upgraded over the next two decades so the quantity of arsenic trioxide initially released in 1949 – an estimated 7300 kg per day – had dropped to about 75 kg per day by 1960. By 1991 the daily emissions were 26–59 kg per day, and remained in this range until the mine ceased operating in 1999. The precipitated dust was collected in the bag house on cloth "bags" suspended in the gas stream. These bags were mechanically shaken from time to time to dislodge the dry powder that was automatically piped to the underground storage area. In the late 1990s, eight to 10 tons of mine dust were recovered each day. Employees working in the bag house received "arsenic pay" because of the increased risk of arsenic exposure.

The mining company began storing captured mine dust underground in 1951 in the belief that the storage areas – initially abandoned mined-out sections – would revert to permafrost once human activity ceased, thereby entombing the arsenic in perpetuity. (At the time, climate change was a possibility still to be considered let alone articulated.) Unfortunately, permafrost did not return to the area; however, the practice of underground storage was continued, and eventually special chambers were constructed to hold the mine dust. There are now 15 storage chambers under the mine site with an average size of 20 m × 50 m × 60 m. The largest chamber is the size of a 20-storey office building. The total amount of mine dust now in storage amounts to about 240 000 tons, of which about 85 per cent is arsenic trioxide.

The water that accumulates in the tailings ponds, along with the solid residue, is treated to remove arsenic and other chemical species before being released back to the environment during the summer, after the ice has thawed. At the Giant mine, removal of the arsenic from the water is accomplished by coprecipitation with ferric hydroxide. Consequently, the Giant mine tailings ponds (which cover a huge area) contain unprocessed crushed rock, residues from the flotation process, residues from the cyanide extraction and "ferric arsenate" precipitate (Sections 1.6.1 and 8.4). The treated water ends up in Back Bay of Great Slave Lake, where the arsenic concentration is well over the Canadian drinking water maximum acceptable concentration (MAC) of 10 ppb.

In the early days of the Giant mine's operation, two children from the Dene nation were said to have died after eating snow. Some sled dogs and cattle met the same fate. The family of the children received $10 000 (Cdn) compensation. Some miners developed perforated nasal septa, prompting the decision to clean up the stack gas and residents were warned not to eat vegetables grown in local soil. Suggestions that the cancer death rate was about twice the national average, prompted a federal government study in 1976. Arsenic was found in the snow and in the house dust, presumably from the stack emissions, and there were high concentrations – to a maximum of 372 ppm – in the hair of children, adult residents and miners.[139,140] In response, the water source for the town was changed from the lake to a relatively pristine river. The full report, which failed to establish a relationship between arsenic and cancer incidence, was released some 20 years later.

Newspaper headlines in 1989 read "Giant Yellowknife hopes arsenic will be good for it." Shareholders were told that the stored arsenic trioxide was to become a source of gold and a new business activity. The mining company planned to bring in a sublimation method called the WAROX process (white arsenic oxide) to separate the gold from the dust stockpile. The arsenic trioxide would then be sold into the CCA wood preservative market at a rate of 8000 tons per year (Section 2.5). The project was expected to recover 110 000 to 120 000 troy ounces of gold that was trapped in the arsenic trioxide.[141] The town breathed a sigh of relief and hoped that the arsenic problem would disappear. It didn't.

The last few years of the Giant mine's life were not happy. There was a riot-torn strike lasting 17 months that saw replacement workers trucked in and nine men killed in an underground bombing, on September 18th, 1992. Disgruntled miner Roger Warren was eventually convicted of the crime, the bloodiest episode in Canadian labour history.[142] An uneasy truce was called between the miners and Royal Oak Mines Inc. president Margaret Witte (now Peggy Kent). The mine continued to limp along under a ridiculously low security bond of $400 000 (Cdn) in the 1990s. This bond was later raised to $7 million, but by 1999 the company was in financial trouble. It declared bankruptcy and walked away from the property, leaving the arsenic problem to the Canadian people. The local member of parliament, Ethel Blondin-Andrew, a junior minister in the federal cabinet, "thought that the mine was cursed."[143]

It was later revealed that Royal Oak had paid only $400 000 (Cdn) of the required security bond, and left unpaid $1.6 million in property taxes. Ms. Witte and her associates rewarded themselves with about $4 million in compensation that could not be recovered by the town and shareholders: "Peggy had become Piggy," according to the local press. A few weeks later she was back in business as CEO of Eden Roc Mineral Corp., another mining company.[144-146]

By the end of its life in 1999, the Giant mine had processed 17 435 000 tons of ore to yield 6 954 250 troy ounces of gold.

7.6.1 Giant Mine: An Underground Cleanup?

In the 1990s the mood of the Yellowknife community and regulators was strongly against any plan that would result in arsenic dust being stored underground any longer than absolutely necessary. Getting the dust out of the mine was seen to be a difficult and probably dangerous task, there being no access route to at least three of the storage areas. Nevertheless, the community was adamant: "Time bomb. Make it go away. Ship it north. Ship it south. I am just mad about it" said Dave Nutter from Environment Canada in 2003.

Workshops were held, consultants were hired, and ballots counted for preferred options. The choices were ultimately narrowed to four, one of which involved leaving the dust in place – a choice that had been rejected in the preliminary rounds. In this option, thermosyphons would be used to create conditions of permafrost around the chambers, sealing the arsenic behind a wall of frozen rock and ice. This is possible to achieve because, during the very long and cold Yellowknife winters, the outside temperature is much lower than the temperature around the storage chambers, so that heat will flow to the atmosphere if there is a connection. This connection can be made via long metal tubes containing liquid CO_2 that transfer the heat through the cycle: liquid → gas → liquid. The engineers believe that localized permafrost will be established if enough of these tubes are in place.

The other three choices required taking the dust to the surface with all the attendant difficulties. One option was the yet-to-be-validated WAROX process, but it was immediately apparent that this was not viable because the CCA market was disappearing (Section 2.5). The other two options involved either stabilising the dust via a chemical transformation, or stabilising it by incorporation into a "permanent" matrix, a practice known as solidification/stabilisation.

There are very few arsenic compounds that can be regarded as completely inert under all conditions and because all compounds would have to be prepared from the dust *via* reactions in solution, a water cleanup requirement would be introduced. However, some gold would be recovered. Calcium arsenate produced by "lime neutralisation" is insoluble; hence, the arsenic is much less bioaccessible, but the material slowly releases its arsenic with the formation of calcium carbonate. The mineral scorodite ($FeAsO_4.2H_2O$) is seen as one of the best targets for chemical stabilisation, but its production is difficult and would require the construction of a new plant to carry out the conversion.[147]

Solidification/stabilisation of the dust could be accomplished by encasing it in cement, but a very large amount of cement would be needed since the arsenic leaches out easily at high loadings. The dust dissolves in bitumen, however, and much higher loadings could be used before the arsenic leaches out.[148]

The rough cost estimates for these options varied from $90–120 million (Cdn) for the thermosyphons, $400–500 million for removal and scorodite production, and $230–280 million for removal and cement stabilisation. It should not come as a surprise that the final decision was made in favour of thermosyphons and induced permafrost.[149]

7.6.2 Giant Mine: Surface Cleanup

The Canadian cleanup guideline for soil is 12 ppm arsenic, which means that this concentration should be achieved in a cleanup unless there are good reasons for doing otherwise. It was obvious that the background arsenic concentration in the Yellowknife area was greater than 12 ppm, but how would someone determine a reasonable number? The problem was solved by applying a statistical technique (principal component analysis) to the analytical results for a suite of elements obtained from a large number of samples. Two distinct compartments were readily distinguished. The arsenic concentrations in the mine-impacted samples ranged from 29 to 12 600 ppm, whereas the range for background samples was 2.5 to 300 ppm with the average being around 100 ppm.[150a]

The next requirement was to establish the bioaccessibility of material in the areas slated for cleanup. Sequential extraction studies showed good agreement with results from gastric fluid models; thus tailings and rock samples that had higher arsenic concentration, showed the same arsenic bioaccessibility as organic soils with lower total arsenic concentrations. A human health risk assessment based on these numbers indicated that there was little risk from arsenic exposure to receptors in the Yellowknife area, accounting for the health of the ecosystem and the community.[124,125] This study supports the proposal to clean up the surface area of the mine to an arsenic concentration of 350 ppm and cap the tailings ponds.

It will be many years, possibly decades, before any plans get approval and then many more before implementation. But before we move south of the Canadian boarder, it should be noted that the smelter at Flin Flon Manitoba is currently releasing more arsenic into the air than any other plant in North America. Lead, mercury, cadmium and copper emission also rank number one.[150b]

7.7 American Smelting and Refining Company. Asarco

Asarco, a subsidiary of Grupo Mexico SA of Mexico City, filed for Chapter 11 bankruptcy protection and reorganisation in 2005. The company has been found liable for contaminating 94 sites in 21 US states. The sites, some entire towns, include some of the largest Superfund areas (Box 7.4). The cost of the cleanup is estimated to be well over $1 billion (US).[151]

The company's long involvement with arsenic began in 1899 with the amalgamation of several nonferrous, copper, lead and zinc smelting and refining operations. One major byproduct was arsenic trioxide, which was sold into a welcoming market at the time, and Asarco became the world's largest arsenic producer.

7.7.1 The Everett and Tacoma Smelters

The company operated a lead and arsenic smelter in Everett WA, from 1894 to 1912. The plant was bankrolled in part by John D. Rockefeller and the feed came from his local mines. The town was slated to become "the Pittsburgh of the west" and three smelters were built. (At the time Pittsburgh PA, was producing between a third and a half of the nation's steel, enabled by an influx of thousands of immigrants from Europe.) Production of arsenic trioxide was moved to Denver CO, and Murray UT, and then to Tacoma WA, in 1919.

A passage from Mellor's 1929 textbook describes the production of arsenic trioxide. The technology has not changed for centuries: "The arsenic driven off by the heat of the roasting furnaces was caught and retained as far as possible in long flues, culverts and settling chambers, but although these were often of great extent, reaching in one place about half a mile [0.8 km] and a capacity of 60 000 cubic feet [1700 m^3], yet arsenic was always liable to pass away. In one place small flakes of it were seen falling continually in a mild snow shower."[152]

The work was dangerous and the protection provided was minimal: there were barrier creams against skin rashes and crude respirators made from gauzy material (cf. Box 7.5). The company did introduce urine testing for arsenic, although at the time the action level was 250 ppb, later reduced to 35 ppb. The arsenic concentration in the air at the worksite was around 500 μg/m^3 resulting in a threefold increase in the risk of respiratory cancer death among the workers at the Tacoma site.[153] These tragic figures were much the same elsewhere,[154] and reinforced the decision to declare inorganic arsenic a carcinogen. Naturally, the health of communities living near smelters was also questioned. Asarco funded studies in the 1980s and 1990s that generally came to conclusions such as: "[There is] little evidence that the lung cancer mortality risk among residents of six rural Arizona smelter communities is positively associated with general environmental (residential) exposures to smelter emissions."[155] Nonetheless, the regulators were pressing and the occupational standard was lowered to 10 μg/m^3. This and other requirements made arsenic trioxide production unattractive. In addition, the advent of CERCLA (Comprehensive Environmental Response, Compensation and Liability Act) and RCRA (Resource Conservation and

Recovery Act) in the US in the 1980s meant that the company was required to become involved in site remediation. Asarco ceased producing arsenic trioxide in 1985. The Tacoma smelter became part of the Commencement Bay Superfund site in 1983.[155,156]

When the Everett WA, plant closed, some of the land was made available for housing. Then, in the 1990s, arsenic was discovered. Asarco bought back the land and demolished 22 houses. The total contaminated area amounted to 700 acres. In the worst area, which was fenced off, the ground was chalky white and did not look like regular dirt. The arsenic concentration in those 7 acres reached 70 per cent – the concentration of arsenic in pure arsenic trioxide is 76.5 per cent. Despite demands from many government agencies for a cleanup, little was done, so the city of Everett bought the 7-acre site for $3.4 million (US), and cleaned it up itself at a cost of $7.1 million. Some 126 165 tons of arsenic-laden soil were removed, part of which was destined to be encased in concrete. Half the site was capped with new soil. The city of Everett sold the land to developers and now 90 new townhouses are being built. Meanwhile the state is seeking $13 million from Asarco, now in bankruptcy court, for the cleanup.

The cleanup in Everett was coordinated with similar work being done in Tacoma, WA, where that smelter had also been placed on the Superfund priorities list in 1983. The Tacoma cleanup was still in progress at the end of 2006, with about two thirds of the $180-million project completed. The bankruptcy court allowed the company to sell some land to developers in order to finish the work, with a deadline of 2013.[151]

Health authorities say it is unclear what health effects, if any, the smelter has had on the local population.[157] Nevertheless, contaminated soils will be replaced at schools within a 816-km^2 area where some of the highest windblown concentration of arsenic and lead have been found. To date, 27 schools and one day care have been identified for cleanup.[158]

Box 7.4 Superfund

The Comprehensive Environmental Response, Compensation, and Liability Act (CERCLA), commonly known as Superfund, was enacted by the US Congress on December 11th, 1980. This law created a tax on the chemical and petroleum industries and provided broad federal authority to respond directly to releases or threatened releases of hazardous substances that may endanger public health or the environment. Over five years a total of $1.6 billion (US) was collected, and the tax went to a trust fund for cleaning up abandoned or uncontrolled hazardous waste sites. The law authorises two kinds of response actions: short-term removals, in which action may be taken to address releases or threatened releases requiring prompt response; and long-term remedial response actions, which permanently and significantly reduce the dangers associated with releases or threats of releases of hazardous substances that are serious, but not immediately life-threatening.

Box 7.4 Continued.

Tax collection ceased at the end of 1995 and the money ran out in 2003. Since then cleanup has been paid for entirely from general US government revenues. It should be said that previously general revenues had accounted for roughly 20 per cent of all appropriations from the program. Attempts to reintroduce the tax on corporate profits (0.12 per cent) in order to replenish the kitty are meeting with opposition from the Bush administration; however, there may be some hope of support if the funds can be used to finance cleanup from hurricanes Katrina and Rita.[159–161a] The A-word is being used as an environmental hammer (Section 7.10). In the meantime Superfund cleanups are "plugging along at a slowing pace."[161b]

7.7.2 The Globe Smelter: Some Unexpected Relief

The Asarco smelter in Globeville CO, on the edge of downtown Denver, produced arsenic trioxide from about 1912 until 1927. The State of Colorado sued the company under the provisions of CERCLA (Box 7.4), claiming the smelter had spread arsenic and lead through parts of a 12-km^2 neighborhood. The area was declared a Superfund site in 1993 and the state set a cleanup level for arsenic of 70 ppm. The act states that Asarco is not responsible for cleaning up anyone else's mess, so soil samples were collected for arsenic analysis to determine the extent of the company's liability. But to everyone's surprise, the arsenic concentration did not decrease with distance from the plant: values above 500 ppm were not uncommon and were randomly distributed in some neighbourhoods. Further investigation revealed that the arsenic had been applied to specific residential properties in the 1950s well after the smelter had ceased arsenic trioxide production. The lawn-care product, PAX, turned out to be the source: it contained arsenic trioxide and lead arsenate and was used for crabgrass control not only in Denver but elsewhere in the US. The soil concentrations found in Denver, 100 ppm to 1000 ppm, are consistent with one to 10 applications of the product over a 10-year period.[162]

Asarco was let off the hook, but the smelter site itself remains a problem. It was closed by the company in 2006, much to the relief of local residents.

7.8 A Transboundary Dispute: Teck Cominco vs. US EPA

Teck Cominco Ltd. and predecessor companies have operated a lead/zinc smelter for many years in Trail BC, 20 km north of the US border. Until 1995, the company dumped the slag from the smelter, some 20 million tons, into the Columbia River – the equivalent of one full dump truck every hour for 60 years. Most of the slag, considered to be inert because its 3-per cent metal content is not bioavailable, now coats the bottom of Lake Roosevelt, a 150-km-long reservoir created when the grand Coolie Dam was built in the 1940s. Teck produces 290 000 tons per year of zinc and 120 000 tons per year of lead.

Lake Roosevelt is the western border of the Confederated Tribes of the Colville Reservation. The tribes who own the beaches and campsites and fish the lake, claim that Teck is responsible for decades of pollution. The US EPA issued an edict in December, 2003, to force the mining company to submit to CERCLA. This is the first time US citizens have attempted to use the Superfund law against a Canadian company.

There is no denying that arsenic and other metals have been released into the Columbia River by the smelter. For example, Stoney Creek, on the Teck property, was still releasing 5.8 kg of arsenic per day into the Columbia in April, 1999; however, a survey by the US EPA found that, of a range of elements, only some arsenic concentrations were above the soil health standards applicable in the state of Washington.[163,164]

In the course of negotiation with the EPA, Teck pledged $13 million (US) toward an evaluation of the risk to human health and the environment, and to make reparations if necessary, but that was not good enough. The EPA wanted to do the job itself and follow US regulatory processes. The company did not agree, largely because Teck's operating permits were not issued in the US and consequently they could not be used by Teck in its own defence. Furthermore, if the study were expanded to include organic pollutants such as dioxins, as would be required for a CERCLA investigation, the cleanup costs could be monumental.

The dispute made its way through US courts to the upper levels of the federal government.[165] Finally, an agreement was reached on June 5th, 2006: Teck would fund the required studies with $20 million (US) under the oversight of the US EPA, and with the participation of local government, the US Department of the Interior and the Government of Canada.[166,167] Preliminary studies have found that many beaches are safe.

There is one precedent for successful litigation of a transboundary pollution case. In the early 1920s, farmers in the US – concerned about their crops – complained about fumes from the Trail smelter. The US government took up the torch and the dispute was referred to the International Joint Commission, a rarely used arbitration panel, and then to a separate tribunal convened by the two countries. The US was eventually awarded $428 000 (US) in damages. The tribunal also found that no nation should allow activity in its territory that causes serious damage to another country [!].[168]

There are international aspects to cleaning up the lead, arsenic, and cadmium that Asarco distributed over the US southern boundary region from its smelter in El Paso TX on the US/Mexico border.[169]

7.9 More Woes

7.9.1 Some Other Surfaces Affected by Mining

It is estimated that more than 4000 km^2 of agricultural land of England and Wales is contaminated with one or more metals as a result of historical mining

Arsenic and the Environment

and smelting activities. Operations in Cornwall date back to the Bronze Age, with copper and tin being the main elements extracted until work ceased by the end of the 19th century. There was a longstanding practice of redistributing mine waste as landfill and over private paths and roadways. Apparently cattle can survive on land that is arsenic-rich, to levels of about 400 ppm arsenic in the topsoil and up to 3.6 ppm in the herbage, but humans in the same area can have problems.[170,171] Not because they go around deliberately eating the grass and the soil, but because they use the abandoned mine workings for motor sports. Some dirt bikers end up incapacitated for months. When asked about the contaminated earth, a local barmaid whose business was threatened said: "that's all right – they don't eat it do they."[172]

In Canada, Nova Scotians used similar tailings areas for recreational purposes and it was not unusual to find children eating hotdogs at the same time that they were playing in the "sand." Mobile food vendors did a good business at these popular events. The problem was only recently recognised, and a press release from 2005 advises that people should not let children play on tailings or use tailings as fill for driveways, gardens, or sandboxes. However, "there is no current evidence of Nova Scotians becoming ill from mercury or arsenic from tailings."[173] Car rallies were cancelled. (Mercury was used in Nova Scotia mines for gold extraction, rather than the cyanide process employed at the Giant mine.) Similar problems have recently emerged in southern California.[174]

Another way of spreading the problem of arsenic exposure was the once-common use of mine tailings for building roads and railways. Mines, such as the Bonanza mercury mine in Oregon, crushed cinnabar (mercury sulfide) and produced mountains of red-brown tailings and waste rock that were free for the taking. The material was laced with arsenic, mercury and other heavy metals. "Nothing would grow in it."[175] This waste was used as a foundation for the 27 km-long Red Rock railway from the mine into the closest town. Now hundreds of people live along this track. One study released in 2001 found an elevated cancer risk in people who had been in contact with or swallowed the dirt. In the meantime, it is easy for children to access the track, and the residents wait for a cleanup plan to be announced.[175]

7.9.2 Nickel Arsenide

In 1903, the Mond Nickel Refinery at Clydach in Wales was commissioned to refine nickel by using the Mond carbonyl process. The hydrogen and the carbon monoxide required came from water gas (CO_2 plus H_2) and producer gas (CO), hence the need for Welsh coal to supply the carbon. The process was very efficient and the only byproducts were silicon dioxide (SiO_2) and iron-containing siliceous slag from the furnace. During the first 30 years of operation, the standard mortality ratio for the work force from nasal cancers was above 30 000; by the 1940s it had dropped to zero. The incidence of lung cancer was also high, but this did not diminish as dramatically. Most workers with cancer died shortly after ceasing work, which explains the company practice of placing workers on a pension immediately after the cancer was diagnosed.

The workplace was very unpleasant, described as "awful to appalling,"[176] so recruitment and retention of workers was difficult. Analysis of the employment record suggests that nasal cancer could be brought on after only five years' exposure. The incidence of disease indicated that cancer was a consequence of the processes inside the plant. Only those intimately involved were at risk: labourers, such as those transporting the material around the refinery, were particularly vulnerable.

The feedstock for the refinery in the early years came from Canada in the form of Bessemer matte (copper nickel sulfide). The first roasting of the matte in Wales produced mainly copper oxide and nickel oxide, which were extracted with dilute sulfuric acid. The residue, after extraction, became enriched in cobalt, silver, platinum and arsenic. The arsenic, which amounted to between 1 and 2.5 per cent of the residue, came from the initial matte, which contained about 0.2 per cent arsenic, and from the sulfuric acid that was manufactured from pyrites and contaminated with arsenic. For some unknown reason, the arsenic travelled with the solids, and subsequent processing of this material created new residues that contained up to 10 per cent arsenic of unknown speciation.[176]

The arsenic was seen as a possible source of the cancer problem and during the 1920s, improvements were made to the process. Notably, a switch was made to arsenic-free sulfuric acid prepared by the more expensive contact process. The health benefits soon became apparent.

The residues from the plant have recently been shown to contain small particles of a nickel arsenide mineral – $(NiFeCu)_{4.2}(AsS)_2$ – known as orcelite. Professor M. H. Draper from the Edinburgh Centre for Toxicology, UK, suggests that these particles were carried to the olfactory region to initiate the cancer accounting for the fact that the nasal cancer epidemic lasted only the three decades when the arsenic-rich sulfuric acid was being used.[176]

This interpretation is questioned by some who argue that particle size was probably the major driver. Recent epidemiological investigations conclude that arsenic does not contribute much to the respiratory cancer risk at least in the Xstrata/Falconbridge nickel refinery in Norway.[177,178]

7.10 Arsenic in Energy Sources

7.10.1 Coal

Coal is a fossil fuel formed from plants that grew in a swamp ecosystem, by geological processes. It is a readily combustible black/brown rock composed primarily of carbon and is the largest single source of fuel used for the generation of electricity worldwide. Pyrite is a common impurity and with this usually comes arsenic, at up to 4.4 per cent of the mineral. This is not the only source of arsenic in coal: XANES (Box 7.2) helped to identify arsenate probably as in scorodite, and sulfides as in arsenopyrite. There are also "organic-bound" species, and recently methylarsenic compounds were isolated from coal samples from the Czech Republic.[179] The average arsenic content in US coals is about 20 ppm, although the amount can be as high as 2000 ppm: the worldwide average is around 5 ppm.[180]

Companies that distribute natural gas make the most of this arsenic content in coal when they masquerade as environmentalists to further their own interests. They fund advertising campaigns that feature rhetorical questions such as: "Would you bathe you child in coal? Sprinkle arsenic, mercury and lead on your husband's cereal?" They overlook the fact that their own product bears a considerable load of arsenical gases, up to 1 µg/L.[181,182] Richard Rowe, who worked in the oil industry in the 1950s described how arsenic trichloride in concentrated hydrochloric acid was pumped into the wells under high pressure. His claim was that it functioned as a corrosion inhibitor; but it probably contaminated water wells miles away.[183]

In some abandoned mines, or even in undisturbed pyrite-rich strata, the oxidation of pyrite – which may be accelerated by the presence of bacteria and arsenic – can lead to dispersal of the element into the environment and to enrichment of local sediments up to 180 ppm. The other product of this oxidation is sulfuric acid, which is the root cause of the miners' scourge, acid mine drainage,[184] although the industry prefers to use the term acid rock drainage.

Arsenic is listed as a potential hazardous air pollutant (HAP) in the 1990 US Clean Air Act amendments, so its concentration in coal is something to worry about. Certainly the historical record shows an aerial impact on West Virginia that peaked around 1970.[184] Coal is now often washed to remove sulfur before burning: this process removes some arsenic as well, leaving mercury to be the main element of concern emitted from burning coal. The US EPA estimates that 40 per cent of the mercury that lands on the US comes from overseas; particularly from power plants in China, Korea and other parts of Asia.[185]

During the process of energy production by coal combustion, two types of solid residue are produced: fly ash, fine dust removed during flue gas cleanup, and, bottom ash, the nonvolatile residue in the burning chambers. Roughly speaking, 1 per cent or less of the arsenic ends up in the slag, 85 per cent in the fly ash and 2 per cent escapes in the vapour phase.[186] The ash is usually designated as nonhazardous waste by virtue of the very forgiving TCLP (Toxicity Characteristic Leaching Procedure) test (Section 7.4) and is used in cement, concrete, aggregates and structural fill.[180] Sometimes the coal contains naturally high concentrations of toxic trace elements and the ash may have to be classified as hazardous solid waste. This forces utilities to dispose of the ash in secure landfills, in heap storage facilities, or to submerge it in storage ponds, increasing the possibility of mobilisation and drinking-water contamination.[187]

Not all coal-burning and associated arsenic-spreading is controlled. There are hundreds of coal fires burning around the world: many are underground and cannot be extinguished. One fire in China that was recently quenched had been burning since 1874: it emitted 100 000 tons of gases such as carbon monoxide (CO) and 40 000 tons of ash every year. Coal fires in China consume 200 million tons of coal per year, and their emissions have a major effect on climate change.[188] The overall average arsenic content of Chinese coals is about 4.5 ppm.[189]

7.10.2 Arsenical Peppers

In Guizhou Province in southwest China, there are at least 3000 people who have been diagnosed since 1976 as suffering from severe arsenic poisoning; while an additional 70 000 to 200 000 are at risk. The indicators of arsenicosis are very evident in the population. Hyperkeratosis is prevalent and 17 per cent have dermal lesions. The average hair arsenic concentration is about 9 ppm.[190] The people of Guizhou are also afflicted with fluorosis. This disease is prevalent in the surrounding provinces, where more than 10 million people are afflicted.

The arsenicosis results from eating chili peppers and corn that have been dried indoors over fires fuelled with coal rich in arsenic. Traditionally, these fires are lit inside dwellings that are poorly ventilated and usually without chimneys. The arsenic content in the local coal can be as high as 35 000 ppm, although the normal range is 30 to 534 ppm with an average of 56 ppm.[190,191] The practice of drying food this way developed when trees were available for fuel but when the hills became denuded in the early 1900s, the people turned to local coal with its high arsenic content. The concentration of arsenic on the smoked peppers ranges from about 70 ppm to 500 ppm and the indoor dust can reach 3000 ppm.[191,192] The arsenic air levels reach up to 400 μg/m^3.[192]

The drinking-water concentrations in the region are mostly normal for arsenic, so the intake is apportioned: 50–80 per cent from food, 10–20 per cent from air, 1–5 per cent from water and less than 1 per cent from direct skin contact. That said, exposure to general indoor air pollution from the combustion of solid fuels has been implicated as a causal agent of diseases such as lung cancer, tuberculosis and asthma, and this risk alone is a major cause of global mortality (Section 7.3.4).[5]

The people of Guizhou are extremely poor, but resist change. The local coal mines were closed down to remove the arsenic source but the people now dig their own supply. The government provides chimneys to residents for free, but the population does not install them because historically they have not felt a need for them. The government has also provided free doses of the chelating agents DMSA and DMPS to the population (Section 1.11), but the drugs do little to reverse conditions such as keratosis: as in Bangladesh, these signs persist even after the arsenic source is removed. One not particularly helpful suggestion was to provide the people with free analytical test kits.[192,193]

All the good will in the world will only go so far. At least one experienced research group from the United States will not be going back to Guizhou in spite of the ongoing requests for help from the Chinese government. The group had to bribe the village leader with cash to get permission to do its study: a gift of medical supplies to help the whole village was not acceptable. They encountered similar problems in transporting samples home for analysis.

To end on a positive note: traditional Chinese medicine in the form of Han-Dan-Gan-Le pills seems to improve some arsenic-induced health problems and they have been recommended to the local population.[191,194]

7.10.3 The Oil Sands of Alberta

Oil sands are deposits of bitumen that make up about 60 per cent of the world's total petroleum reserve. The bitumen is a form of oil that does not flow at normal temperatures or pressures. It is difficult and expensive to extract, and specialised facilities are needed to convert it into synthetic crude oil and other petroleum products. The Alberta oil sands deposits contain about 85 per cent of the world's total bitumen reserves covering more than 140 000 km^2: an area larger than the state of Florida.

Bitumen deposits located near the surface can be recovered by open-pit mining techniques, but *in situ* recovery is necessary for most deposits because these are more than 75 m below the ground surface. Steam injection is used to extract the bitumen in Canada's largest recovery project at Cold Lake AB.

Large volumes of water are used in all aspects of the fuel recovery. Two to four barrels of water are required for every barrel of synthetic oil produced in Alberta and 80 kg of greenhouse gases are released into the atmosphere. This water use is causing concern, but new projects are being approved before studies are completed on the cumulative impacts of existing operations and if these can be managed sustainably. Arsenic is the last word that anyone wants to hear amid all this frenetic economic activity. Nevertheless, the mobilisation of naturally occurring arsenic has been observed near some well bores, probably the result of the high temperatures associated with the heated steam recovery processes and casing failures.[195,196]

Dr. John O'Connor, a physician at Fort Chipewyan, had suspected for a number of years that "copious amounts of arsenic dumped into the water by the nearby oil sands development might explain why so many of his aboriginal patients were presenting with cancer." So, in 2006, he was not surprised to learn through the news media that moose meat from the area contained 453 times the acceptable level of arsenic. "They don't know if they should eat their moose meat, and that's really the only thing they have to eat," said Donna Cyprien of the local health board.[197,198]

Health Canada disputed Dr. O'Connor's action when he issued public warnings and they laid a formal complaint that he was causing "undue alarm." New results released in 2007 by Alberta Health showed arsenic concentrations up to 33 times the acceptable level, which – although less alarming – is still inexplicably high, and it seems that the same levels are present in the region surrounding the oil-sands development.[199,200] Although these problem were revealed during an environmental assessment of the region that was required before operations could expand, the Alberta Energy and Utilities Board did not shy away from approving new projects.[201]

Apparently the laboratory that performed the initial analysis of the moose meat samples did not have the capability to detect low levels of arsenic, so it reported that the value was below its detection limit. This was not good enough for the risk assessors who wanted a number for their calculations, so they elected to use a value of 1 ppm, which was half the detection limit.

This make-believe value then became transformed to the actual measured value that was released to the world. Yet another example of the power of arsenic to blunt the human brain.

The locals are not convinced that the whole story is out and still do not know if it is safe to eat their traditional food. One report from the regional Aquatics Monitoring Program found that 7 per cent of the fish in the local river had growth abnormalities. And, yes, a very unhappy Dr. O'Connor left town.[198]

7.10.4 The Sydney Tar Ponds: Arsenic as an Environmental Hammer

In 1899 construction of a new steel plant began in Sydney NS, in the heart of a particularly beautiful region of Eastern Canada. There was ample local coal to feed the 400 coke ovens required to provide the coke for the steel furnaces. In addition to the coke, the ovens produced 330 tons of tar and volatiles, including sulfur dioxide and probably arsenic trioxide, per day. The residue was dumped into a local stream that ran into an estuary, and the whole area rapidly filled up with 700 000 ton of oily residues, consisting of polychlorinated biphenyls, polycyclic aromatic hydrocarbons, arsenic, lead, and raw sewage, to name but a few constituents. This dump is now known as the Sydney Tar Ponds, Canada's worst toxic waste site.[202]

Residents in the area became alarmed when rumours of increased cancer risks surfaced in spite of government attempts to keep a lid on the information. In 1986, funding was announced to clean up the ponds, with the job scheduled to be completed in about 10 years. Incineration and encapsulation were considered, but the tar remained untouched even though the price tag for the study reached $55 million (Cdn). Then, in 1999, came the discovery that arsenic levels were high in the soil and in coal ash, and that a yellow-orange ooze, containing 50 ppm arsenic, was seeping into some house basements in the area. The politicians said not to worry, but the residents and the media seized on this news. Arsenic was the trigger to release another $62 million in funding for tar-sands studies, but no money was allocated for actual cleanup or to relocate individuals. The government did offer to purchase some homes in the area, but the owners said the price was unacceptable.[203]

By 2001, when the plant was closed in the face of huge financial losses, the headlines were: "Arsenic Coming Through the Middle of the Floor," "185 Homes Plus a Day Care Impacted," "Concentration of Arsenic 67 Times Limit." The community was in revolt. More tests were done and five children out of the total of 237 pregnant women and children who were examined, had high arsenic levels. Patricia Fraser was told her 19-month-old son's body contained almost twice the allowable level of arsenic and she decided to leave home with her family. Her house is about 2 km from the toxic site. Again the government officials were not convinced that the arsenic came from the ponds and suggested the need to look for other sources, but the community was intransigent. They had their issue, and the only outcome they would consider was relocation.[204,205] More plans were made and rejected. Another study showed that the house dust in some homes had high concentrations of arsenic and lead.[206]

Finally, in 2007 the Canadian government, in uncharacteristically "green" pre-election mode, made the announcement that solidification/stabilisation technology would be used to deal with the tar-ponds problem. The toxic sludge will be mixed with a "concrete-like substance" and allowed to solidify. The area will then be covered with a plastic sheet, layered with soil and planted with grass. Some local residents have expressed misgivings, saying they expected a cleanup, not a coverup. The project will take seven years and will cost $400 million.[207] Cynics might say that this is a repeat of the process that began in 1986 and will probably have the same outcome. Nevertheless, there is little doubt that the A-word was a catalyst for action – it was an environmental hammer.

It seems that much the same use was made of the A-word during debates on the extent and fate of the environmental contamination in New Orleans in the wake of Hurricane Katrina.[160] And critics of fluoridation were happy to make use of the news that trace amounts of arsenic were getting into drinking water by means of the chemicals used in fluoridation.[208] Paul Beeber, President, New York State Coalition Opposed to Fluoridation, bewilderingly writes: "The NSF sets the allowable level of arsenic in fluoridation chemicals at 2.5 ppb. The maximum contamination level (MCL) of arsenic in treated water is 10 ppb."[209]

7.10.5 Cleaning Up

Arsenic is a contaminant of concern in groundwater at 380 Superfund sites and in soil at 372 of those sites. It is the second most common contaminant of concern at the 1209 sites on the US National Priority List.[210] Cleaning up the environment after arsenic has been let out of the bottle is not an easy task: the element does not go away; and many of the solutions simply transfer the problem to a different environment where new difficulties can be encountered.

We have seen that mining has made major contributions to our environmental problems with arsenic but, apart from smelting operations, much of this was and remains in remote areas. The designated US sites are closely monitored by regulatory agencies and are targeted by statutes. In contrast, arsenical pesticides were made in locations close to population centres and their products were liberally applied to farms and orchards that were once rural but now are being urbanised and used for housing and schools. In Washington State pesticide application resulted in up to 2553 ppm arsenic concentrations in orchard soils; now the land is needed for housing. In North Carolina, a former orchard site now used for housing had soil arsenic levels above 40 ppm in some samples, which resulted in soil being removed from 28 residences.[211] One developer in Pennsylvania is prepared to dig up more than 20 000 dump-truck loads of arsenic tainted soil before building 491 homes on the site of what was once an orchard where lead arsenate was liberally used until the 1950s. The plan is to scrape the top 6 inches of soil off the 187 acres, pile it in two areas capped with 2 feet of clean soil, and mix the remaining soil to a depth of about a foot with clean soil.[212] However, caution is necessary in such operations as there is evidence from the Superfund Basic Research Program at Dartmouth College, NE, that lead arsenate that is immobilised in soil below the surface, is best left that way.[213]

One very relevant study concluded: "There is often the incorrect assumption that a given level of arsenic in surface soils at a hazardous waste site somehow poses more risk than the same level at an equally contaminated residential/public space site."[211] Contrast a designated fenced-off, patrolled, monitored site with the situation revealed by the headline: "Arsenic Levels Force Closure of Baltimore Park," referring to Swann Park in South Baltimore that was found to contain more than 2000 ppm arsenic in the soil.[214] The park, which opened in 1914, is next to a former Allied Chemical Corp. factory that, until 1976, manufactured arsenical pesticides and other agrichemicals. "Arsenic trioxide, a toxic white powder, rumbled in on rail cars and was unloaded next to the baseball field. Workers mixed the arsenic with nitric acid in metal hoppers to make the pesticides." The city brought the land in 1977 and at the time of sale Allied said nothing about the possibility of arsenic contamination.[214] The park is heavily used for recreation and sports, so many people have inhaled and ingested the soil.

The new soil analyses were prompted by the release of confidential 1976 company data showing 6600 ppm of arsenic behind home base of the baseball diamond, and up to 10 000 ppm elsewhere. According to this new information: "The filters that were used in the plant "were constantly developing leaks and arsenic was being discharged into the atmosphere ... In 1962 or 63 a collector bag broke and coated Swann Park – [it] looked like snow."[214]

An EPA-funded study published in 1981 concluded that arsenic was likely responsible for the unusually high lung cancer rate in the neighbourhood. The city was informed but took no action at the time, possibly preferring to maintain good relations with the newly established local industry. The area was known as Cancer Alley, but one resident says that might have been because of the number of residents that smoked (which may have made matters worse (Section 7.2.4)). Now the locals are worried about their health and are outraged that they were not informed of the results of the study. In October 2007, the Department of Health declared that "recent and historic exposure to Swann Park soil is considered a public health hazard."[214] There are high arsenic levels in neighbouring backyards and the lawyers are circling. Peter Angelos, a prominent Baltimore trial attorney and owner of the Baltimore Orioles, the local baseball team, is soliciting residents who live near or visited the park and who believe they may have suffered damage from arsenic contamination. The law firm is offering free consultations for interested parties. Mr. Angelo is described as a class action specialist who has previously built fortunes out of asbestos and tobacco.[215,216] Honeywell, the current owners of the business, proposes to perform limited excavations and to put down a two-foot layer of clean dirt on top of the entire park.[214]

The "safe limits" for cleanup vary considerably from state to state. In Florida, where the residential limit for arsenic is 2.1 ppm, lawyers for more than

30 neighbours of a site that was used to store chemicals for the Hernando County's public works department are going into battle over a report that shows *hot spots* (emphasis added) with 2.15 ppm and 2.31 ppm arsenic on nearby property. The county abandoned the site in 2003.[217] At the moment, both sides of the dispute think they have a case.

This quixotic variability of limits even offers the chance to move goal posts during the game. Here is an example. The Middeltown Town Council favoured raising the acceptable level of arsenic in the soil from 7 ppm to 15 ppm in light of environmental issues that "delayed development of a public park on the 45-acre Kempenaar Valley property". Town Administrator Gerald Kempen said "the town has received estimates ranging from $1 million to $5 million to remediate the property [to 7ppm]." A 15 ppm target would be much less expensive to meet. The issue was resolved by keeping the old limit of 7 ppm in the hope that a cheaper means of remediation will be found.[218]

It is getting harder to find places to put contaminated soil, as is happening in a cleanup near Mount Helena MT, where the waste would have to be trucked through a town full of concerned citizens to a facility that is not ready to receive it. A commercial landfill is available, but the costs are prohibitive.[219] In situations such as these, monitored natural attenuation begins to look attractive.

7.10.6 Monitored Natural Attenuation

The name says it all: do nothing and watch the problem go away. It is well accepted that some problems associated with organic contaminants simply disappear over time as a result of biological processes. This will not happen with inorganic arsenic, so the chances of success are not claimed to be high, according to the US National Research Council.[220] The best that could happen is that processes such as adsorption and precipitation will take place in a reasonable time, so as to mitigate risk by decreasing the arsenic concentration and reducing contaminant mobility. This idea is being actively investigated because the price looks right. The monitoring costs can be high, however; hence the note of caution. One such site in Florida was a cattle-dipping vat during the 1930s and 1940s; part of the Cattle Tick Fever Eradication Program. The surface soil around the vat was removed in 1998 and now the groundwater at the site is being monitored according to a strict schedule.[210]

Nevertheless, the picture is not particularly rosy, even in situations in which we think we are doing a good job. Here are the main conclusions reached at a USGS workshop held in Denver 2003, which focused on the problems of managing arsenic risks to the environment, and the characterisation of waste, chemistry and disposal:[221]

> (1) We know little about the dominant biotic and abiotic processes that control arsenic transformation and speciation in wastes and waste-disposal environments.

(2) Waste is currently categorised as hazardous on the basis of the TCLP test, but the test has little validity when arsenic species are the target, and it certainly has no predictive power.
(3) Few follow-up studies have been carried out in any waste-disposal environments, to monitor the chemical and biological changes that could result in mobilisation, in spite of evidence from some landfills that this occurs.
(4) Microbial process can mobilise arsenic from so-called "inert" materials, such as ferric arsenate.

Recent studies in a landfill in Saco ME, show that arsenic can be released even if none is added. Evidently, the decomposition of garbage results in the release of arsenic from the surrounding soil into ground water by essentially the same process that takes place in Bangladesh (Section 8.2). The authors recommend that landfills be lined, and that arsenic wastes be separated and maintained in a well-aerated oxidising environment away from organic matter.[222]

The town of Saginaw MI, was the home of the Saginaw Plate Glass Co. which closed in the early 1920s. The plant produced about 280 000 m^2 of glass a year and left behind enough arsenic to rank it among the most contaminated properties in the state: (the soil contains 100 ppm arsenic, but the problem was not noticed until 1993). Now the town is practising its own form of natural attenuation and is isolating the site with PVC sheets "Basically, we are building a big bathtub. Whatever is inside will stay inside."[223]

On a mildly positive note, Belluck and coworkers in a literature review, "Widespread Arsenic Contamination of Soils in Residential Areas and Public Spaces: An Emerging Regulatory or Medical Crisis," published in 2003, comment that they did not find any reports showing that elevated arsenic levels in soils cause morbidity or mortality in exposed individuals." However, they do caution that this may simply be a reporting artifact.[211]

7.11 Microbes and Arsenic

Weathering and erosion of minerals is a major route by which arsenic becomes mobilised in the environment. Weathering occurs *in situ*, and involves the breakdown of rocks and soils through direct contact with the atmosphere. Erosion involves the movement and disintegration of rocks and minerals by more physical processes, such as exposure to water, wind, ice, hail and gravity. The liberated arsenic normally ends up as "free" inorganic species, arsenate or arsenite, in aqueous solution and eventually becomes bound up again in metal oxy(hydroxyl) species, usually of iron, manganese or aluminium, in sediments and soils. Arsenite can be oxidised by iron and manganese species.[12,224] These processes are normally thought of as having little connection with biology, but in the case of arsenic, chemical and physical processes encountered in the environment sometimes receive a big assist from the world of microbiology. This mobilisation and interconversion of species is part of a global arsenic cycle.

Arsenic and the Environment 333

Unless they are in an unusual environment, such as a cattle dip or down a gold mine, micro-organisms take up arsenate inadvertently through their phosphate transport systems. The phosphate is needed, the arsenate is not. Arsenate interferes with vital processes such as energy storage by ATP production: arsenato-ADP, if formed, spontaneously hydrolyses, setting up a futile cycle that drains the cell's energy stores (Section 1.7). One common cellular response to the presence of arsenate is to reduce it to arsenite; however, because arsenite is about 50 times more toxic to cells than arsenate, it has to be removed by means of a specific arsenite efflux pump. This efflux mechanism costs the cell energy, but has stood the test of time. Most bacteria obtain this energy from the oxidation of organic carbon compounds using oxygen as the terminal electron acceptor and are described as autotrophic.[225]

The arsC gene has been identified in the DNA of many arsenate reducers. This gene codes for the production of the enzyme arsenate reductase. It is now possible to get a good measure of the total amount of this gene in samples – an indication of the "arsenate reducing potential" – without isolating all the individual bacteria. The entire DNA is extracted from the sample, the arsC gene is isolated, amplified, and quantified.[226]

This is what we know now, but the first indication that some micro-organisms had an unexpected way of coping with arsenic exposure came from Gosio's work on volatile species in the 1890s, and the identification of trimethylarsine as a metabolic product by Challenger in 1933. We have seen that the reduction of arsenate and methylarsenic analogues is an important step in the Challenger pathway (Section 3.9). At the time, Challenger did not know that the cell kicked out arsenite once it was formed. He thought everything from uptake of arsenate to elimination of trimethylarsine took place inside the cell.

But if a cell chose to eliminate arsenite, why would it go to the trouble of taking it up again and attaching methyl groups? The answer is probably that the cell has no choice. Arsenite is taken up accidentally because it looks a bit like glycerol,[227a] and once it is back in the cell methylation processes take over. But again we need to note that most of the arsenic is eliminated from the cell before trimethylarsine is formed. This biomethylation to species with increased membrane permeability[227b] accounts for the presence of dimethylarsinic acid in most environments.

The loss of arsenic to the atmosphere by volatilisation from soils is a minor pathway.[228] However, when the conditions are right – high arsenic concentration and lots of micro-organisms – the production of gaseous methylarsines can be substantial. These conditions are found in hot springs, sewage digesters and landfills.[12,229,230]

7.11.1 More – but very Small – Arsenic Eaters

The discovery in 1918 that arsenite used in cattle dips in South Africa was being oxidised by bacteria was treated as a curiosity, but given more credence when it was reported that Australian cattle dips exhibited similar tendencies. Since then, arsenic oxidisers have been isolated from a range of arsenic-rich environments such as microbial mats in geothermal springs, and gold mines.[231]

Most of these oxidisers seem to use oxidation as a means of detoxification. The why is not known; however, some species of these unusual bacteria are chemolithotropic, meaning that they can derive all their energy needs from the oxidation of arsenite.[232]

Related bacteria are able to dine on arsenic minerals such as the sulfides orpiment and arsenopyrite. Thus *Thiobacillus ferrooxidans* oxidises the latter to arsenate and sulfate in a complicated sequence of reactions, so it should not come as a surprise to learn that this process has been developed as an alternative to smelting to release gold from arsenopyrite prior to cyanidation. Plants are operating in Australia, South Africa, the US, and probably elsewhere. The bacteria in use include *T. ferrooxidans* and *Leptospirillum ferrooxidans*. In these operations the bacteria are provided with nutrients. *T. ferrooxidans* is the organism responsible for "acid mine drainage": it can oxidise a variety of metal sulfides to sulfuric acid, and the acid produced in turn leaches out any acid-soluble metals present in the rock, leading to huge environmental problems.[231,233]

Since 1994, we have known that there are bacteria capable of using arsenate as a terminal electron acceptor to achieve growth; they respire arsenate just as we respire oxygen. The arsenate can even be in the form of the mineral scorodite, ferric arsenate; and as a result these bacteria contribute to the release of arsenic from aquatic sediments.[234] The bacteria, which are phylogenetically diverse and ecologically widespread, are usually isolated from arsenic-rich environments such as sediments, hypersaline lakes and mine wastes.[235] None are obligate, meaning they can use other electron acceptors such as nitrate if necessary. Given the choice, *Desulfotomaculum auripigmentum* utilises arsenate first, followed by sulfate leading to the precipitation of arsenic sulfide. It has been suggested that these biological processes could be used for arsenic mitigation.[236,237] Here is one version of such a process: (1) Oxidise the arsenite to arsenate with bacteria that get energy from this process. (2) Treat the solution with special prepared magnetic beads coated with ferric hydroxide to bind the arsenate. The beads are removed, together with the arsenic. (3) Regenerate the beads by microbial reduction of the arsenate to arsenite in the presence of sulfide (produced microbially as well) and precipitate as arsenic sulfide. Recycle the regenerated beads.

Less-specific biological processes for cleaning up water are well established in the mining industry. One company, Applied Biosciences of Salt Lake City UT, claims their ABMet® process will take water at 500 ppm to below 5 ppm.[238] Following a disastrous water release in 1992 from the abandoned Wheal Jane mine in Cornwall, UK, the arsenic-containing leakage is being treated at enormous cost by a combination of processes, one of which has a biological component.[239] These remediation processes usually end up producing sulfides of arsenic and other metals, which can still be a problem.[240]

The methylarsenic compounds found everywhere in the environment are the product of biological processes, but these same species are not immune to further biological action. Bacteria will eventually remove all the methyl groups from arsenobetaine to produce arsenate and thus start the process all over again. Demethylation is the eventual fate of the methylarsenic pesticides that were applied to fields. The one detailed study of this demethylation process shows that the methyl group of monomethylarsonic acid is removed by the reverse of the oxidative addition process (known as reductive elimination) that put it on (Section 3.9).[241] Arylarsenic compounds are remarkably stable in some environments; witness the recovery of such species from soil 50 years after being dumped in the form of chemical weapons (Section 6.13). But the aryl ring is eventually lost from the chicken food additive roxarsone on composting, and a recent study shows that a species of the bacteria *Clostridium* will convert the compound to arsenate in culture, but the mechanism is yet to be established (Section 2.1).[242] It seems unlikely that the intact aryl ring would be removed by simple carbon–arsenic bond cleavage.

There can be no doubt that the microbial enzymes that reduce arsenate, oxidise arsenite, methylate and demethylate arsenic species play an important role in the mobility of the element in the biosphere and in the formation and decomposition of various minerals.[243] To emphasize this connection, a team of Korean Scientists has just published a paper "Biogenic formation of photoactive arsenic sulfide nanotubes by *Schewanella* sp strain HN-41."[244]

Box 7.5

A Claim Against the Cobalt Silver Smelting Co. for Damage from Handling Arsenic (with Permission from the National Archives of Canada Ottawa, McCormack, 1912[245]):

Attorney-General   Quebec
Government Buildings,   October 30th, 1912
Ottawa, Ont.:

Sir:
Subject of the following is a statement in connection with a grievance, which I have got to make against the Cobalt Silver Smelting Company of Orillia, Ont. The grievance being caused by damages received while working in the Arsenic department of that concern, being employed by the said company.

Last April, I was in the Town of Orillia boarding at a Hotel, called the Robinson House, kept by a man the name of O'Donnell. He also conducted an Intelligence Office for the purpose of providing work for men out of employment. One morning of the same month (April) this man O'Donnell got an order from the company's office, through the phone, to get two men for them to go to work right away in the smelter. I happened to be in the office at the time the order came and in the meantime he was wanting some of the men to go, but they all refused, so then he pressed upon me to go and finally I consented to do so.

He then got his son to write a letter and then gave it to me to hand it in at the office of the Co. After I had done so, I was put to work right away.

Box 7.5 Continued.

I worked there about three weeks loading small cars with Arsenic and charging the coal furnaces with it. I was then obliged to give it up after, on account of a breaking out of sores over my eyes and all about my face especially my legs and feet, caused from the effects of the Arsenic and the smoke coming from it.

The company afforded me but very little protection against the danger of such a deadly poisonous stuff to work in. They only gave me a pair of overalls with a hood attached to it for to put over my head. After working there about three weeks, I got into a very bad condition with sores and sickness, so much so that I was obliged to quit working, and when it came the pay day I was not able to go to the office after my money. I had to send after it.

I was then laid up sick in the Boarding house, so then after a few days I left for the city of Toronto for the purpose of getting medical treatment, and remaining there about three weeks or more, until my money was nearly all out, paying expenses for board and medical treatment, then I had no other alternative, but to go back again and work for the same company as I felt a little better and rested up, so I finally went back again and went to work.

After I had worked two more weeks I was obliged to quit working there again, by reason of a small blood vessel to burst from an abscess or sore, which ever it might be called, causing a terrible profusion of blood to come from it, so much so, that I fell prostrated to the ground with weakness and was compelled to remain in that condition until a horse and rig could be procured to carry me to my boarding house, and in the meantime the Company's doctor was sent for, shortly after he arrived and then he dressed my leg, and finally he gave me permission to go to the Orillia General private Hospital, he being the medical Supt. for it.

After being in the Hospital over three weeks I was getting somewhat better, so with that I came out, then I went to the Office of the Company to report for work and in the meantime I asked if there was any money in the office for me, then I was asked by one of the clerks how much was my Hospital board and medical expenses, then I produced my Hospital Bill for $11.90 and also $1.50 cents for Ambulance, and two weeks board had been paid by them for me. When everything was added up I had only Twenty Cents coming to me; so with that I inquired of one of the company if I could go to work, he told me that I was welcome to come into work the next day.

When I reported at the office for work the next morning I met the foreman there and was told by him that there was no more work for me; that they had to shut down the Arsenic furnaces and some other department of the smelter causing quite a number of men to be laid off. Immediately after I was informed of the affair, I inquired of one of the authorities who was in the office at the time what they were about to do regarding my misfortune on their account. The reply he made concerning my inquiry was, "We have got nothing to do with that, you did not get that there. I know where you got it,

Box 7.5 Continued.

our Doctor says you have got Blood poison, and we go by what he says." With that I told him that I would see about it. After that I was informed by him that I could do my best. Now, when you come to look at this from a logical point of view, what a predication of a few insignificant availing eye uses these are, for any member of such a rich company to set up as a defence against a man in their employment, having the sad misfortune of being the recipient of such a calamity, so much so as to incapacitate him, probably, for the balance of his life from making a living. I have lost all my summer's work on account of it, only the exception of a few days work here and there, being obliged to quit all the time, suffering from an agonising Arsenic Blood poisoning pain in my legs and feet.

A few weeks ago I had to leave a camp, where I was working only for a very short time, on account of the dreadful condition that I am in with my left leg. When I had it examined by the Company's Doctor he ordered that I should leave camp at once. Of late I am in a worse condition that ever, so much so, that I will have to go to the Hospital again. Now, the question is, why did I not bring an action against the Company for damages, the simple reason is this: not having any financial means by which I could put it into the hands of a Lawyer to introduce it into Court.

Now that I am so incapacitated at the present time and a great deal worse than ever, so much so, that I am not able to stand on my legs, I have to keep either sitting down or walking, and then I suffer more or less; instead of getting better I am only getting worse; I shall have to go to Hospital again. But before I do go, I thought that I would be in the right of predimising a statement of the true facts and circumstances of the case and forward it to you for your discrimination, so as that you might put it before the proper Authorities, those whom it may concern, for investigation and finally for a decision as to whether I am entitled to get compensation or not from the Company that has acted so unreasonable and so unsympathising as to utterly refuse to make good the damages, even not to pay my Hospital Board, and finally to refuse me of work, after the day before to tell me that I was welcome to come in and go to work.

To ignore my case in every way possible so as to make it appear, as it were, that they were not under any compliment or obligation whatever to me. This defence was not only got up to break down my confidence in bringing an action against them, but also to try to through sand, as it were, under the eyes of the Law to blindfold. I presume this must be the reason and also to make it an example for others in their employ who might have the misfortune of being the recipient of my condition. However, I think the same Co. would give better protection to other people than to me, by reason that I had been told that I was imposed on by being kept working so long in the arsenic and arsenic furnaces, being only a few days for each man in turn, it being so dangerous to work in it above two or three days at a time.

Box 7.5 Continued.

Now, that I have stated the true facts and circumstances of the case as are here in given, I take liberty of forwarding it to you, you being the Attorney-General, a Government Officer and of high standing in the Law Courts, to put it before the proper Court of procedure, if possible for you to do so.

Company's Office Address
Cobalt Silver Smelting Co.
Orillia Ont
James McCormack
c/o Metropole Hotel
Du Palais St.
Quebec.

November 7th, 1912
1519/12

Dear Sir,

On behalf of the Minister of Justice I beg to acknowledge your letter of the 30th October, together with the statement of facts concerning your illness and resulted incapacity attributed to the handling of arsenic while in the employ of the Cobalt Silver Consulting Co. of Orillia. I thoroughly appreciate the trying circumstances in which you have been placed. I would not however, attempt to express an opinion upon the validity of your claim against the Company. I wish however, to point out that you appear to be under a false impression concerning the prerogatives and duties of the Attorney-General. I think if you could see some trustworthy lawyer, conversant with the law of Ontario, you would upon acquainting him with the details of your case, be advised as to the probability of success if you instituted an action.

Yours very sincerely

James McCormack, Esq.
c/o Metropole Hotel,
Du Palais St.
Quebec, P.Q.

*DESTROYED IN ACCORDANCE
WITH
T.B. 654568, APR. 28, 1966*

SEC. II Sub. Sec. OF
SUBMISSION
Sig._____

References

1. CRC, *Handbook of chemistry and physics*, 87th edn., Chemical Rubber Company Press, Cleveland Ohio, 2007.
2. K. B. Krauskopf, *Introduction to geochemistry*, 2nd edn., McGraw-Hill, New York, 1979.
3. ATSDR, *CERCLA Priority list of hazardous substances*, www.atsdr.cdc.gov/cercla/05list.html.
4. J. Matschullat, *Sci. Total Environ.*, 2000, **249**, 297.
5. G. He, B. Ying, J. Liu, S. Gao, S. Shen, K. Balakrishnan, Y. Jin, F. Liu, N. Tang, K. Shi, E. Baris and M. Ezzati, *Environ. Sci. Technol.*, 2005, **39**, 991.
6. N. Manalis, G. Grivas, V. Protonotarios, A. Moutsatsou, C. Samara and A. Chaloulakou, *Chemosphere*, 2005, **60**, 557.
7. D. W. Griffin, C. A. Kellogg, V. H. Garrison, C. Holmes and E. A. Shinn, *Epidemio-Ecology News*, 2002, **1**.
8. H. Bradisher and D. Barboza, in *New York Times*, 2006, June 11.
9. R. C. Cowen, in *Christian Science Monitor*, 2001, Dec. 13.
10. R. Oremland, J. F. Stolz and J. T. Hollibaugh, *FEMS Microbiol. Ecol.*, 2004, **48**, 15.
11. D. Wallschlager and C. J. Stadey, *Anal. Chem.*, 2007, **79**, 2873.
12. W. R. Cullen and K. J. Reimer, *Chem. Rev.*, 1989, **89**, 713.
13. T. G. Pierce, C. A. Langdon, A. A. Meharg and K. T. Semple, *Soil Biol. Biochem.*, 2002, **34**, 1833.
14. Lord Kelvin, W. H. Dyke, W. S. Church, T. E. Thorpe, H. C. Bonsor and B. A. Whitelegge, *Report of the Royal Commission appointed to inquire into the arsenical poisoning from the consumption of beer and other articles of food and drink*, Vol: I, Houses of Parliament by Command of His Majesty, London, 1901.
15. F. A. Robinson and F. A. Amies, *Chemists and the law*, E. and F. N. Spon, Ltd., London, 1967.
16. A. C. Chapman, *The Analyst*, 1926, **51**, 549.
17. R. S. Braman and C. C. Foreback, *Science*, 1973, **182**, 1247.
18. NRC, *Arsenic in drinking water*, National Research Council, Washington DC, 1999.
19. J. S. Edmonds, K. A. Francesconi, J. R. Cannon, C. L. Raston, B. W. Skelton and A. H. White, *Tetrahedron Lett.*, 1977, **18**, 1543.
20. K. A. Francesconi and J. S. Edmonds, *Adv. Inorg. Chem.*, 1997, **44**, 147.
21. D. P. McAdam, A. M. A. Perera and R. V. Stick, *Aust. J. Chem.*, 1987, **40**, 1901.
22. J. S. Edmonds, PhD thesis, University of Western Australia, 1988.
23. P. Andrewes, D. M. Demarini, K. Funasaka, K. Wallace, V. W.-M. Lai, H. Sun, W. R. Cullen and K. T. Kitchen, *Environ. Sci. Technol.*, 2004, **38**, 4140.
24. X. C. Le, X. Lu and X.-F. Li, *Anal. Chem.*, 2004, Jan. 1, 27A.
25. J. J. Sloths, E. H. Larsen and K. Julshamn, *J. Anal. At. Spectrom.*, 2003, **18**, 452.

26. S. McSheehy, J. Szpunar, R. Lobinski, V. Haldys, J. Tortajada and J. S. Edmonds, *Anal. Chem.*, 2002, **74**, 2370.
27. V. Nischwitz and S. A. Pergantis, *J. Anal. At. Spectrom.*, 2006, **21**, 1277.
28. V. Nischwitz and S. A. Pergantis, *Anal. Chem.*, 2005, **77**, 5551.
29. A. Benson and R. E. Summons, *Science*, 1981, **211**, 482.
30. V. W.-M. Lai, A. S. Beach, W. R. Cullen, S. Ray and K. J. Reimer, *Appl. Organomet. Chem*, 2002, **16**, 458.
31. K. Ebisuda, T. Kunito, R. Kubota and S. Tanabe, *Appl. Organomet. Chem*, 2002, **16**, 451.
32. J. S. Edmonds and K. A. Francesconi, in *Organometallic compounds in the environment*, ed. P. J. Craig, John Wiley and Sons Ltd., Chichester, England, 2003.
33. C. Soeroes, W. Goessler, K. A. Francesconi, E. Schmeisser, R. Raml, N. Kienzl, M. Kahn, P. Fodor and D. Kuehnelt, *J. Environ. Monit.*, 2005, **7**, 688.
34. I. Koch, K. J. Reimer, A. Beach, W. R. Cullen, A. Gosden and V. W.-M. Lai, in *Arsenic exposure and health effects IV*. ed. W. R. Chappell, C. O. Abernathy and R. L. Calderon, Elsevier Science Ltd., Amsterdam, 2001, p. 115.
35. V. Devesa, M. A. Suner, V. W.-M. Lai, S. C. R. Granchinho, D. Velez, W. R. Cullen, J. M. Martinez and R. Montaro, *Appl. Organomet Chem.*, 2002, **16**, 692.
36. D. Kuehnelt and W. Goessler, in *Organometallic compounds in the environment*, ed. P. J. Craig, John Wiley and Sons Ltd., Chichester, England, 2003.
37. Associated Press, in *Vancouver Sun*, 2002, Aug. 21, p. A9.
38. I. Koch, J. V. Mace and K. J. Reimer, *Environ. Sci. Technol.*, 2005, **24**, 1468.
39. C. A. Morrissey, C. A. Albert, P. L. Dods, W. R. Cullen, V. W.-M. Lai and J. Elliott, *Environ. Sci. Technol.*, 2007, **41**, 1494.
40. K. J. Reimer, 2007, personal communication.
41. A. W. Ritchie, J. S. Edmonds, W. Goessler and R. O. Jenkins, *FEMS Microbiol. Lett.*, 2004, **235**, 95.
42. J. S. Edmonds, K. A. Francesconi and R. V. Stick, *Natural Products Reports*, 1992, 423.
43. (a) P. G. Smith, PhD thesis, Queen's University, 2007.
    (b) S. D. Kinzie, B. Thern and D. Iwata-Reutyl, *Org. Lett.*, 2000, **9**, 1307.
44. P. G. Smith, I. Koch, R. A. Gordon, D. F. Mandoli, B. D. Chapman and K. J. Reimer, *Environ. Sci. Technol.*, 2005, **39**, 284.
45. D. Harrison, in *The Age*, 2007, July 31.
46. S. Maeda, in *Arsenic and old mustard: chemical problems in the destruction of old arsenical and 'mustard' munitions*, ed. J. F. Bunnett and M. Mikolajczyk, Kluwer Academic Publishers, 1996, p. 135.
47. V. W.-M. Lai, W. R. Cullen, C. F. Harrington and K. J. Reimer, *Appl. Organomet. Chem.*, 1997, **11**, 797.

48. L. A. Murray, A. Rabb, I. L. Marr and J. Feldmann, *Appl. Organomet. Chem.*, 2003, **17**, 669.
49. (a) H. K. Das, A. K. Mitra, P. K. Sengupta, A. Hossain, F. Islam and G. H. Rabbani, *Environ. Int.*, 2004, **30**, 383.
    (b) M. M. Rahman, G. Owens, M. Mallavarapu and R. Naidu, *Environ. Internat.*, 2007, in press.
50. E. K. Porter and P. J. Peterson, *Sci. Total Environ.*, 1975, **4**, 365.
51. C. H. Haug, K. J. Reimer and W. R. Cullen, *Appl. Organomet. Chem.*, 2004, **18**, 626.
52. C. Tu, L. Q. Ma and B. Bondada, *J. Environ. Qual.*, 2002, **31**, 1671.
53. M. Srivastava, L. Q. Ma and J. A. G. Santos, *Sci. Total Environ.*, 2006, **364**, 24.
54. B. Rathinasabapathi, M. Rangasamy, J. Froeba, R. H. Cherry, H. J. McAuslane, C. J. L., M. Srivastava and L. Ma, *New Phytologist*, 2007, **175**, 363.
55. S. Gordon, in *The News Tribune*, 2005, Dec. 22.
56. Chinaview, in *China View*, 2005, Nov. 6.
57. G. Liu, F. Cheng, S. Gao and M. Mengqiu, *Zhongguo Nongye Kexue*, 1985, **4**, 9.
58. K. A. Francesconi, P. Visoottiviseth, W. Sridokchan and W. Goessler, *Sci. Total Environ.*, 2002, **284**, 27.
59. O. P. Dhankher, B. P. Rosen, E. C. McKinney and R. B. Meagher, *Proc. Nat. Acad. Sci.*, 2006, **103**, 5413.
60. D. R. Ellis, L. Gumaelius, E. Indriolo, I. J. Pickering, J. A. Banks and D. E. Salt, *Plant Physiol.*, 2006, **141**, 1544.
61. N. Lubick, *Environ. Sci. Technol.*, 2006, Aug. 15, p. 4816.
62. W. E. Stephens, A. Calder and J. Newton, *Environ. Sci. Technol.*, 2005, **39**, 479.
63. R. Carson, *Silent spring*, Houghton Miffen Company, Boston, 1962.
64. J. Aubry, in *Vancouver Sun*, 2006, May 23.
65. M. Barrett, in *Rocky Mountain Telegram*, 2005, Oct. 27.
66. A. M. Hays, D. Srinivasan, M. L. Witten, D. E. Carter and R. C. Lantz, *Toxicol. Pathol.*, 2006, **34**, 396.
67. Y. Maruyama, K. Komiya and T. Manri, *Radioisotopes*, 1970, **19**, 250.
68. C. L. Chen, L. I. Hsu, H. Y. Chiou, Y. M. Hsueh, S. Y. Chen, M. M. Yu and C. J. Chen, *J. Am. Med. Ass.*, 2004, **292**, 2984.
69. B. Boothin *Environ. Sci. Technol.*, 2005, **39**, 34A.
70. Phillip Morris, *The illicit trade in cigarettes*, Phillip Morris International, 2007, April 15.
71. C. McGuigan, in *Sunday Telegraph*, 2006, June 18.
72. W. Mertz, *Science*, 1981, **213**, 1332.
73. EPA, *Special report on ingested inorganic arsenic: skin cancer; nutritional essentiality*, EPA, Washington, 1988.
74. E. O. Uthus and C. D. Seaborn, *J. Nutr.*, 1996, **126**, 2452S.
75. S. S.-H. Tao and P. M. Bolger, *Food Addit. Contam.*, 1999, **16**, 465.
76. N. P. Vela and D. T. Heitkemper, *J. AOAC Int.*, 2004, **87**, 246.

77. R. A. Schoof, L. J. Yost, J. Eickhoff, E. A. Crecelius, D. W. Cragin, D. M. Meacher and D. B. Denzel, *Food Chem. Technol.*, 1999, **37**, 839.
78. R. A. Schoof, J. Eickhoff, L. J. Yost, E. A. Crecelius, D. W. Cragin, D. M. Meacher and D. B. Menzel, in *Arsenic exposure and health effects III*, ed. W. R. Chappell, C. O. Abernathy and R. L. Calderon, Elsevier Science Ltd., Amsterdam, 1998, p. 81.
79. P. N. Williams, A. Raab, J. Feldmann and A. A. Meharg, *Environ. Sci. Technol*, 2007, **41**, 2178.
80. G. Smith, in *The Herald*, 2007, March 23.
81. Evening Standard, in *Evening Standard*, 2007, March 24.
82. S. Westcott, in *Sunday Express*, 2007, March 24.
83. B. Booth*Environ. Sci. Technol.*, 2006, April 1, 2077.
84. FSA, *FSA Publishes "Metals in Food" findings*, Food Standards Agency, 2007, www.foodqualitynews.com.
85. K. Sundl, K. A. Francesconi and W. Goessler, ICEBAMO, Crete, 2006, Oct. 10–11.
86. (a) V. Devesa, A. Martinez, M. A. Súner, D. Valéz, C. Amela and R. Montaro, *J. Agric. Food Chem.*, 2001, **49**, 2272.
    (b) T. Mohri, A. Hisanaga and N. Ishinishi, *Food Chem. Toxicol.*, 1990, **28**, 521.
87. T. Roychowdhury, T. Uchino, H. Tokunaga and M. Ando, *Food Chem. Toxicol.*, 2002, **40**, 1611.
88. M. F. Ahmed, M. A. Ali and Z. Adeel, ed., *Fate of arsenic in the environment*, Bangladesh University of Engineering and Technology and the United Nations University, Dhaka, 2003.
89. M. J. Abedin, M. S. Cresser, A. A. Meharg, J. Feldmann and J. Cotter-Howlls, *Environ. Sci. Technol.*, 2002, **36**, 962.
90. (a) S. M. I. Huq, in *Behaviour of arsenic in aquifers, soils and plants: implications for management*, Symposium Dhaka, Jan. 16–18, 2005.
    (b) N. Athey-Pollard, in *ChemScience*, Oct. 5, 2007.
91. A. A. Meharg and M. M. Rahman, *Environ. Sci. Technol*, 2003, **37**, 229.
92. P. N. Williams, M. R. Islam, S. A. Hussain and A. Meharg, in *Behaviour of arsenic in aquifers, soils and plants: Implications for management*, Symposium Dhaka, Jan. 16–18, 2005.
93. J. G. Lauren and J. M. Duxbury, in *Behaviour of arsenic in aquifers, soils and plants: Implications for management*, Symposium Dhaka, Jan. 16–18, 2005.
94. (a) M. Mittelstaedt, in *Globe and Mail*, 2007, Feb. 24, p. A8.
    (b) C. Crimmins, in *Vancouver Sun*, 2007, Dec. 14, p. A12.
95. E. Amster, A. Tiwary and M. E. Schenker, *Environ. Health Perspect.*, 2007, **115**, 606.
96. *AHPA challenges kelp product case*, American Herbal Products Association, Silver Springs MD, 2007, www.ahp.org.
97. M. Lau, in *The Standard, Hong Kong*, 2007, June 17.
98. J. S. Edmonds and K. A. Francesconi, *Marine Pollut. Bull.*, 1993, **26**, 665.

99. K. Hanaoka, K. Yosida, M. Tamano and T. Kuroiwa, T. Kaise and S. Maeda., *Appl. Organomet. Chem*, 2001, **15**, 561.
100. CFIA, *Inorganic arsenic and hijiki seaweed consumption*, Consumer Advisory, Canadian Food Inspection Agency, Ottawa, 2001.
101. *Japan Today*, 2004, July 31.
102. Y. Sugimori, M. Usui, K. Hanaoka, T. Kaise, M. Nagaoka and T. Maitani, ICEBAMO Crete, Oct. 10–12, 2006.
103. J. M. Laparra, D. Velez, R. Montoro, R. Barbera and R. Farre, *J. Agric. Food Chem.*, 2003, **51**, 6080.
104. Y. Nakajima, Y. Endo, Y. Inoue, K. Yamanaka, K. Kato, H. Wanibuchi and G. Endo, *Appl. Organomet. Chem.*, 2006, **20**, 557.
105. H. Castlehouse, C. Smith, A. Raab, C. Deacon, A. A. Meharg and J. Feldmann, *Environ. Sci. Technol.*, 2003, **37**, 951.
106. H. R. Hansen, A. Raab, K. A. Francesconi and J. Feldmann, *Environ. Sci. Technol.*, 2003, **37**, 845.
107. G. Caumette, S. Ouypornkochagorn, C. M. Scrimgeour, A. Rabb and J. Feldmann, *Environ. Sci. Technol.*, 2007, **41**, 2673.
108. J. Feldmann, 2007, personal communication.
109. C. Sass, *Bottled or tap water – which is best?* American College of Sports Medicine Annual Meeting, Dallas, March 28, 2007.
110. Associated Press, in *St. Petersburg Times*, 2006, March 9.
111. P. Leo, in *Pittsburgh Post Gazette*, 2006, July 20.
112. FDA, in *foodconsumer.org*, 2007, March 7.
113. S. Khojoyan, in *ArmeniaNow.com*, 2007, March 16.
114. R.-G. Lin, in *Los Angeles Times*, 2007, April 2.
115. FSA, in *foodstandards.gov.uk*, 2005, Oct. 20; R. Harrison, *Arab News*, 2007, Oct. 16.
116. X. C. Le, W. R. Cullen and K. J. Reimer, *Clin. Chem.*, 1994, **40**, 617.
117. V. W.-M. Lai, Y. Sun, E. Ting, W. R. Cullen and K. J. Reimer, *Toxicol. Appl. Pharm.*, 2004, **198**, 297.
118. J. Wragg and M. R. Cave, *In vitro methods for the measurement of the oral bioaccessibility of selected metals and metalloids in soils: a critical review* P5-062/TR/01, Environment Agency, British Geological Survey, Bristol, 2002.
119. L. J. EhlersR. G. Luthy in *Environ. Sci. Technol.*, 2003, Aug. 1, 295A.
120. M. V. Ruby, A. Davis, R. Schoof, S. Eberle and C. M. Sellstone, *Environ. Sci. Technol.*, 1996, **30**, 422.
121. M. V. Ruby, R. Schoof, W. Brattin, M. Goldade, G. Post, M. Harnois, D. E. Mosby, S. W. Casteel, W. Berti, M. Carpenter, D. Edwards, D. Cragin and W. Chappell, *Environ. Sci. Technol.*, 1999, **33**, 3697.
122. A. Tessier and P. G. C. Campbell, *Hydrobiologia*, 1987, **149**, 844.
123. G. E. M. Hall, G. Gauthier, J.-C. Pelchat, P. Pelchat and J. E. Vaive, *J. Anal. At. Spectrom.*, 1996, **11**, 787.
124. C. A. Ollson, PhD thesis, Royal Military College of Canada, 2003.
125. W. R. Cullen, K. J. Reimer, I. Koch and C. A. Ollson, in *Arsenic metallurgy: fundamentals and applications*, ed. R. G. Reddy and V. R. Ramachandran, TMS Publications, 2005.

126. G. B. Freeman, J. D. Johnson, J. M. Killinger, S. C. Liao, A. O. Davis, M. V. Ruby, R. L. Chaney, S. C. Lovre and P. D. Bergstrom, *Fund. Appl. Toxicol.*, 1993, **21**, 83.
127. (a) A. Davis, D. Sherwin, R. Ditmars and K. A. Hoenke, *Environ. Sci. Technol.*, 2001, **35**, 2401.
    (b) I. Koch, K. McPherson, P. Smith, L. Easton, K. G. Doe and K. J. Reimer, *Marine Pol. Bull.*, 2007, **54**, 586.
128. S. Leahy, in *Mines and communities web site*, 2007, Feb. 24.
129. Editorial, in *New York Times*, 2006, Jan. 9.
130. J. Perlez, in *New York Times*, 2006, Feb. 4.
131. Associated Press, in *Taipei Times*, 2006, Aug. 27.
132. J. Perlez and K. Johnson, in *New York Times*, 2005, Oct. 24.
133. S. H. Ali, *Mining the environment and indigenous development conflicts*, The University of Arizona Press, Tucson, 2003.
134. E. Bellett, in *Vancouver Sun*, 2004, Feb. 12.
135. A. Freeman, in *Globe and Mail*, 1998, May 1, p. A1.
136. R. W. BoyleA. S. DassD. ChurchG. MihailovC. DurhamJ. LynchW. Dyck*Geol. Surv. Can.*, Paper 67–35, 1966.
137. M. Hudson, *Mawson discovers new drill targets at Middagsberget North*, Sweden, Mawson Resources Limited, Vancouver, 2006, www.geonord.org/law/exp/21.html.
138. R. W. Boyle, in *Geological Survey of Canada, Bulletin 280*, 1979.
139. M. A. Rychlo, *The arsenic papers*, Highway Book Shop, Cobalt, Ontario, 1977.
140. R. A. Smith, *Environ. Res.*, 1976, **12**, 171.
141. A. Robinson, in *Globe and Mail*, 1989, May 27, p. B5.
142. J. Gatehouse, in *Maclean's Magazine*, 2002, Aug. 19, p. 20.
143. E. Blondin-Andrew, 1999, personal communication.
144. J. Aldous, *A giant legacy*, CBC Radio North, 1999, Yellowknife NWT, Canada, April 19.
145. P. Kennedy, in *Globe and Mail*, 1999, May 22.
146. A. Robinson, in *Globe and Mail*, 1999, Feb. 16, p. B1.
147. P. A. Riveros, J. E. Dutrizac and P. Spencer, *Can. Metall. Quar.*, 2001, **40**, 395.
148. W. R. Cullen, V. W.-M. Lai, L. Wang, Y. Sun, E. Polischuk and K. J. Reimer, in *Arsenic exposure and health effects V*, ed. W. R. Chappell, C. O. Abernathy, R. J. Calderon and D. J. Thomas, Elsevier, Amsterdam, 2003.
149. DIAND, *Giant Mine remediation project*, Department of Indian Affairs and Northern Development, http://nwt-tno.inac-ainc.gc.ca/giant/index_e.html 2006/06/15.
150. (a) ESG, *Arsenic levels in the Yellowknife area: distinguishing between natural and anthropogenic inputs* RMC-CCE-ES-01-01, Environmental Sciences Group, Kingston, Ontario, 2001.
    (b) J. Skerritt, in *Vancouver Sun*, 2007, Oct. 11, p. A6.
151. Sierra Club, www.commondreams.org/news2006/0412-02.htm.

152. J. W. Mellor, *A comprehensive treatise on inorganic and theoretical chemistry*, Vol. IX, Longmans, Green and Co, London, 1929.
153. P. E. Enterline, R. Day and G. M. Marsh, *Occupational Environ. Med.*, 1995, **52**, 28.
154. I. Hertz-Picciotto and A. H. Smith, *Scand. J. Work Environ. Health*, 1993, **19**, 217.
155. D. A. Robbins, in *Arsenic metallurgy: fundamentals and applications*, ed. R. G. Reddy and V. R. Ramachandran, The Minerals, Metals, and Materials Society (TMS), 2005.
156. D. A. Robbins, 2005, personal communication.
157. D. Chircop, in *The Daily Herald*, 2006, Dec. 28.
158. S. Gordon, in *The News Tribune*, 2007, July 25.
159. C. Hogue, in *Chem. Eng. News*, 2002, Sept. 9, p. 31.
160. F. Barringer, in *New York Times*, 2005, Dec. 2.
161. (a) L. R. Ember, D. J. Hanson, G. Hess, B. Hileman, C. Hogue and S. R. Morrissey, in *Chem. Eng. News*, 2006, Jan. 23, p. 13.
(b) C. Hogue, in *Chem. Eng. News*, 2007, Nov. 12, p. 41.
162. D. J. Folkes, S. O. Helgen and R. A. Little, in *Arsenic Exposure and Health Effects IV*, ed. W. R. Chappell, C. O. Abernathy and R. L. Calderon, Elsevier Scientific, Amsterdam, 2001.
163. Golder Associates, *Aquatic problem formulation report March 2003 to Teck Cominco*, 002-2371, 2003.
164. D. Horswill, *Conference call-US EPA and Lake Roosevelt December 17 2003 Transcript*, 2003.
165. D. Whitley, in *Vancouver Sun*, 2004, p. D5.
166. B. Yaffe, in *Vancouver Sun*, 2005, Jan. 4, p. A7.
167. International Mining, *Teck and US EPA agree on Columbia River assessment*, www.immining.com/Articles/TeckandUSEPAagreeonColumbia Riverassessment.asp.
168. M. Preusch, in *New York Times*, 2004, March 20.
169. D. W. Valdez, in *El Paso Times*, 2005, Nov. 5.
170. X. Li and I. Thornton, *Environ. Geochem. Health*, 1993, **15**, 135.
171. E. Hamilton, in *Chem. Brit.*, 1997, April, p. 49.
172. Betty Barmaid, 1995, personal communication.
173. Nova Scotia, *Nova Scotians advised to avoid gold tailings*, Department of Environment and Labor, Province of Nova Scotia, Halifax, 2005.
174. Associated Press, in *Signonsandiego.com*, 2006, April 26.
175. D. Dietz, in *Eugene Register-Guard*, 2006, June 29.
176. M. H. Draper, in *Carcinogenicity of inorganic substances. Risks from occupational exposure*, ed. J. H. Duffus, Woodhead Publishing Limited, Cambridge, 1997.
177. T. K. Grimsrud, *Epidemiology*, 2005, **16**, 146.
178. E. Nieboer, 2007, personal communication.
179. Z. Slejkovec and T. Kanduc, *Environ. Sci. Technol.*, 2005, **39**, 3450.
180. A. Kolker, C. A. Palmer, L. J. Bragg and J. E. Bunnell, *Arsenic in coal*, Fact sheet 2005-3152, US Geological Survey, 2006.

181. A. Price, in *Austin American-Statesman*, 2007, Feb. 6.
182. K. J. Irgolic, D. Spall, B. K. Puri, D. Llger and R. A. Zingaro, *Appl. Organomet. Chem.*, 1991, **5**, 117.
183. R. Rowe, in *Chem. Eng. News*, 2000, March 13, p. 6.
184. M. B. Goldhaber, E. R. Irwin, J. R. Hatch, J. C. Pashin, A. Grosz, E. C. Callender and J. Grossman, *USGS Workshop on arsenic in the environment*, Denver CO, 2001, Feb. 21–22.
185. Staff, *Briefing report: pollution drift*, Californian State Senate Republican Caucus, Los Angeles, 2005, p. 1.
186. X. Guo, C.-G. Zheng and M.-H. Xu, *Energy Fuels*, 2004, **18**, 1822.
187. V. N. Bashkin and K. Wongyai, *Environ. Geology*, 2002, **41**, 883.
188. Newsscripts, in *Chem. Eng. News*, 2004, Nov. 22, p. 128.
189. K. Luo, *Toxicol. Environ. Chem.*, 2005, **87**, 427.
190. A. Shraim, X. Cui, S. Li, J. C. Ng, J. Wang, Y. Jin, Y. Liu, L. Guo, D. Li, S. Wang, R. Zhang and S. Hirano, *Toxicol. Lett.*, 2003, **137**, 35.
191. J. Liu, B. Zheng, H. V. Aposhian, Y. M.-L. Chen, A. Zhang and M. P. Waalkes, *Environ. Health Perspect.*, 2002, **10**, 119.
192. R. B. Finkelman, H. E. Belkin and B. Zheng, *Proc. Nat. Acad. Sci. USA*, 1999, **96**, 2427.
193. S. S. Simpson, in *Sci. Am.*, 2002, Feb. 23.
194. Y. Lu, T. Lu and M. Cheng, *Chin. J. Hepatol*, 2000, **8**, 108.
195. Government of Alberta, *Oil sands*, www.energy.gov.ab.ca/89.asp.
196. M. Griffiths, A. Taylor and D. Woynillowicz, *Troubled waters, troubling trends*, www.oilsandswatch.org/pub/612.
197. R. Gandia, in *Fort McMurray Today*, 2006, Nov. 15.
198. A. Nikiforuk, in *Globe and Mail*, 2007, May 19, p. F3.
199. R. Gandia, in *Fort McMurray Today*, 2007, April 3.
200. P. Woodford, *Nat. Rev. Med.*, 2007, **4**. March 30.
201. D. Henton, in *Edmonton Sun*, 2006, Nov. 18.
202. M. Barlow and E. May, *Frederick street*, Harper Collins Publishers Limited, Toronto, 2000.
203. Canadian Press, in *Globe and Mail*, 1999, May 29.
204. A. Auld, in *Globe and Mail*, 2001, July 13.
205. K. Cox, in *Globe and Mail*, 2001, July 14.
206. T. W. Lambert and S. Lane, *Environ. Health Perspect.*, 2004, **112**, 35.
207. Canadian Press, in *Vancouver Sun*, 2007, Jan. 29.
208. D. Gutierrez, www.newstarget.com/021796b.html.
209. P. S. Beeber, in *Agric. Environ.*, 2007, March 13.
210. J. Reisinger, D. R. Burris and J. G. Hering, *Environ. Sci. Technol.*, 2005 Nov. 15, 458A.
211. D. A. Belluck, S. L. Benjamin, P. Baveye, J. Sampson and B. Johnson, *Int. J. Toxicol.*, 2003, **22**, 109.
212. J. Hook, in *publicopiniononline.com*, 2006, March 7.
213. S. Knapp, www.dartmouth.edu/~news/releases/2006/02/17.html.
214. T. Pelton, in *Baltimore Sun*, 2007, May 27; Oct. 20.

215. S. Schultz, in *Baltimore Business Journal*, 2007, June 22.
216. T. Pelton, in *Baltimore Sun*, 2007, June 14; Oct. 6.
217. A. Loder, in *St. Petersburg Times*, 2006, July 27.
218. G. Macris, in *The Providence Journal*, 2007, June 15.
219. E. Byron, in *Helena Independent Record*, 2007, May 12.
220. J. A. Macdonald, *Environ. Sci. Technol.*, 2000, 346A.
221. *US EPA Workshop on managing arsenic risks to the environment*, Denver, CO, Oct. 2003.
222. J. L. Delemos, B. C. Bostick, C. E. Renshaw, S. Sturup and X. Feng, *Environ Sci. Technol.*, 2006, **40**, 67.
223. J. Stettler, in *The Saginaw News*, 2005, Oct. 24.
224. V. Q. Chiu and J. G. Hering, *Environ. Sci. Technol.*, 2000, **34**, 2029.
225. S. Silver, L. T. Phung and B. R. Rosen, in *Environmental chemistry of arsenic*, ed. W. T. Frankenberger, Marcel Dekker Inc., New York, 2002, p. 247.
226. Y. Sun, E. A. Polishchuk, U. Radoja and W. R. Cullen, *J. Microbiol. Methods*, 2004, **58**, 335.
227. (a) J. Quin, B. P. Rosen, Y. Zhang, G. Wang, S. Franke and C. Rensing, *Proc. Nat. Acad. Sci.*, 2006, **103**, 2075.
 (b) R. G. Males, J. C. Nelson, P. S. Phillips, W. R. Cullen and F. G. Herring, *Biophys. Chem.*, 1998, **70**, 75.
228. S. Gao and R. Burau, *J. Environ. Qual.*, 1997, **26**, 753.
229. B. Planer-Friedrich and B. J. Merkel, *Environ. Sci. Technol.*, 2006, **40**, 3181.
230. J. Feldmann and A. V. Hirner, *Int. J. Environ. Anal. Chem.*, 1995, **60**, 339.
231. H. L. Ehrlich, in *Environmental chemistry of arsenic*, ed. W. T. Frankenberger, Marcel Dekker Inc., New York, 2002, p. 313.
232. G. L. Anderson, P. L. Ellis, P. Kuhn and R. Hille, in *Environmental chemistry of arsenic*, ed. W. T. Frankenberger, Marcel Dekker Inc., New York, 2002, p. 341.
233. K. S. Fraser, R. H. Walton and J. A. Wells, *Miner. Eng.*, 1991, **4**, 1029.
234. D. Ahmann, L. R. Krumholz, H. F. Hemond, D. R. Lovley and F. M. M. Morel, *Environ. Sci. Technol.*, 1997, **31**, 2923.
235. W. P. Inskeep, R. E. Macur, N. Hamamura, T. P. Warelow, S. A. Ward and J. M. Santini, *Environ. Microbiol.*, 2007, **9**, 934.
236. R. S. Oremland, D. K. Newman, B. W. Kail and J. F. Stolz, in *Environmental chemistry of arsenic*, ed. W. T. Frankenberger, Marcel Dekker Inc., New York, 2002, p. 273.
237. J. M. Santini, R. N. Vanden Hoven and J. M. Macy, in *Environmental chemistry of arsenic*, ed. W. T. Frankenberger, Marcel Dekker Inc., New York, 2002, p. 329.
238. J. Adams, *Technical workshop on arsenic in mining*, Winnipeg, Manitoba, 2002, Nov. 7–8.
239. Environment Agency, *Wheal Jane a clear improvement*, Environment Agency South West Region, Exeter, 2000.

240. J. M. McArthur, 2004, personal communication.
241. C. R. Lehr, E. Polishchuk, U. Radoja and W. R. Cullen, *Appl. Organomet. Chem.*, 2003, **17**, 831.
242. J. F. Stolz, E. Perera, B. Kilonzo, B. Kail, B. Crable, E. Fisher, M. Ranganathan, L. Wormer and P. Basu, *Environ. Sci. Technol.*, 2007, **41**, 818.
243. R. E. Macur, C. R. Jackson, L. M. Botero, T. R. McDermott and W. P. Inskeep, *Envion. Sci. Technol.*, 2004, **38**, 104.
244. J. H. Lee, M. G. Kim, B. Yoo, N. V. Mynug, J. Maeng, T. Lee, A. C. Dohnalkova, J. K. Fredricksonm, M. J. Sadowsky and H. G. Hur, *Proc. Natl. Acad. Sci. USA.*, 2007, **104**, 20410.
245. J. McCormack, 1912, National Archives, Ottawa, Justice Series A-2, Vol. 175, file 1912–1519.

CHAPTER 8
Accidental Exposure to Arsenic: The Law of Unintended Consequences

8.1 Introduction

Screaming, they left their villages in haste –
"The Devil's water is rising, flee!"
They had no wish to touch, nor any desire to taste.
Aid agencies and their "yes men," without a qualm explained,
"The Green Revolution will arrive, so will healthy bodies thrive –
Now come receive in the cup of your hands – this boon the gods have sent."

SOES Jadavpur University[1]

Water is the medium of life, yet as the third millennium begins 1.5 to 2.5 billion people in the world do not have access to adequate supplies of it. In addition, "millions are exposed to arrays of water-borne diseases that countries and financial institutions lack the will and money to completely eradicate by simple sanitation: diseases such as cholera, schistosomiasis, and trachoma. Looming by the year 2025 is an additional 2.5 billion people who will live in regions already lacking sufficient water, much less clean water."[2] When these words were written in 1999 the world was slowly and reluctantly becoming aware of another major water-based problem: arsenic.

The World Health Organization (WHO) first produced its International Standards for Drinking Water in 1958 and categorised arsenic as a "toxic substance that, if present in drinking water supplies, could present an actual danger to health", and established 0.20 ppm, 200 ppb, as a maximum allowable concentration. In 1963 the standard was updated to a lower concentration of 50 ppb. In 1993 it was lowered again to 10 ppb.[3] Arsenic species are naturally

---

Is Arsenic an Aphrodisiac? The Sociochemistry of an Element
By William R. Cullen
© William R. Cullen 2008

present in all waters, normally at concentrations in the low parts per billion (Chapter 7), but at the present time millions of people in particular regions only have access to water that contains arsenic at concentrations that are 10 to 100 times this limit.[4,5] Some of this exposure is the unintended consequence of programs that were undertaken to improve the lot of the local populations.

8.2 West Bengal and Bangladesh: The Devil's Water

8.2.1 The Green Revolution

"Tapas Sarkar, from a village near Calcutta, has supped the Devil's water since birth. As a result, his feet and those of the rest of his family are covered with the giant sores that signify chronic arsenic poisoning (Figure 1.4). Four members of his family have so far died from this cause. The poisoning started 30 years ago when engineers began to sink wells into the flood plain of the River Ganges to irrigate new high-yield strains of rice during the dry season. Until irrigation, farmers grew one rain-watered rice crop per year. Now they can grow [at least one more] through the six-month dry season. These economic gains have had a huge human cost."[6]

Sewage contamination of surface waters, which people used for drinking water and often shared with domestic animals, was thought to be the main cause of diarrheal diseases, a major if not the principal cause of mortality in many developing countries.[7] So, it was easy to believe that the groundwater that was being used on their crops would also be good to drink. The United Nations Children's Fund (UNICEF) was the main proponent of a tube well program to tap into the groundwater. The non-governmental organisation (NGO) provided the design and material (1.5 inch PVC pipe), and the governments of Bangladesh and West Bengal – together with international aid agencies – sank more than four million of these tube wells, which were easy to drill, convenient to use, and hand-pumped. The population got accustomed to the new source of water, liked it, and drilled millions more wells, probably close to a total of 11 million. As a result, by the year 2000 about 97 per cent of the rural Bangladeshi population was drawing its drinking water from tube wells sunk to a depth of 10 to 80 metres. Today four out of every five tube wells in Bangladesh are privately owned. In some areas, this figure is as high as nine out of every 10.

The tube wells provided individual families or groups of families with their own source of "safe" water. They became status symbols, and traditional community practices such as digging shallow-water wells (dug wells) and reserving ponds for drinking water were abandoned. However, water-related diseases remained the major cause of mortality and morbidity and, because nobody checked, the tube wells introduced another health hazard: arsenic.[8]

The first diagnosis of arsenical dermatosis resulting from drinking arsenic-rich tube well water was made by Dr. K. C. Saha in July 1983. This patient, who was admitted to the School of Tropical Medicine in Calcutta, West Bengal, was the first of 1214 that would be examined over the next four years. Saha almost lost his job for reporting his findings.[9–11] He described the clinical features of melanosis and keratosis (Section 1.8) and noted that a small percentage of the patients had developed skin cancer. The arsenic concentration in the water that his patients were drinking ranged from 60 ppb to 1350 ppb, with a mean of 320 ppb. The initial reports of the disaster in West Bengal used the phrase "the biggest arsenic calamity in the world" and detailed urine arsenic concentrations of up to 2116 ppb; hair concentrations of 1.8 ppm to 31 ppm; nail concentrations of 1.5 ppm to 52 ppm – all indicators of high exposure.[12,13] The government of West Bengal, although made aware of the problem, did little in response.[14]

8.2.2 Bangladesh

In 1992, a woman from Bangladesh, but living in West Bengal, was diagnosed with arsenicosis. It was clear that her illness was the result of prior exposure in Bangladesh and further inquiry revealed that she was not unique: many from her district were similarly afflicted. Arsenic was first detected in Bangladesh tube wells the next year and within a few years it was established that the problem was widespread in Bangladesh. The calamity in the region was then redefined as the "largest mass poisoning of a population in history."[15,16] "Arsenic in drinking water poses the highest cancer risk ever found," said Dr. Alan H. Smith, Professor of Epidemiology in the School of Public Health of the University of California, Berkeley in 1998: "Every day that people continue to drink the contaminated water could result in more arsenic-related deaths down the road in five, 10 or 25 years. This is really a major emergency." In some regions such as Chile (Section 8.8) researchers concluded that one in every 10 deaths could be due to arsenic poisoning.[7,17] At the time of its discovery, the scale of the arsenic disaster was compared with the Chernobyl radiation release of 1986, which affected 17 million people to some degree, and the Union Carbide chemical plant explosion in Bhopal, India, which killed 22 000 in total.

In 1997, the British Geological Survey (BGS) was contracted by the government of Bangladesh to conduct a groundwater survey of 41 of the 52 districts in Bangladesh. The area surveyed covered what were believed to be the worst affected parts of Bangladesh. BGS subcontracted Phase I of the work to Mott MacDonald International, Dhaka, because the company had the expertise on the ground. In total, approximately 2000 samples were collected, or one sample per 50 square kilometres. The results of the study showed that the median arsenic concentration was 108 ppb. Of the total samples collected, 25 per cent of the samples had arsenic concentration above 50 ppb, the permitted Bangladesh maximum. The highest concentration found was 1670 ppb and the primary contamination zone lay at a depth of 10 to 70 m. Mott MacDonald estimated the number of people exposed to arsenic concentrations greater than 50 ppb to be in excess of 21 million. By 2000, after Phase II of the work was completed by the BGS – an additional 1500 wells tested – the number was revised upward by another 10 million.[18] The situation in 2004 as seen by Professor Chakraborti and colleagues (Section 8.3) is set out in Table 8.1.[19] They suggest that about 60 million people have been exposed to drinking water above the WHO 10 ppb standard for arsenic, and 40 million exposed above the Bangladesh standard. While others dispute the absolute numbers, nobody disputes the magnitude of the disaster. To add to the problem, there is no clear link between the number of contaminated wells and the prevalence of people showing signs of arsenicosis. Families are routinely encountered in which only one adult shows symptoms, but all drink from the same well.[20]

The general public in the western world has little comprehension of the scale of the calamity in the Bengal Basin. This is also true for the general population of Bangladesh. Bangladeshis in the diaspora essentially ignore the issue, although there was some effort in 2002 by the Bangladeshi–American Foundation Inc. (www.bafi.org) to raise awareness in the United States, and there was a call for action from Bangladeshi scientists living in North America.[21] We will see later (Section 8.4) that there was a burst of national pride when a Bangladeshi was awarded the 2007 Grainger Prize for producing a technology for removing arsenic from well water, but the problem itself took a back seat.

A number of commentators have remarked on the massive spontaneous response to the victims of the Indian Ocean tsunami on December 26th, 2004, when individuals and governments worldwide gave enormous sums of cash for relief, much eventually unused, to be contrasted with the feeble response to other international disasters such as the arsenic crisis and AIDS.[22] As if to emphasize the point, the US House of Representatives was quick to vote 388-0 in support of a resolution expressing sympathy and pledging support for the victims of cyclone Sidr which hit southern Bangladesh November 15th, 2007. According to experts, the "identifiable victim effect" has us sympathising with victims we know rather than the "unknown others," even if they are in more dire straits. And it seems that education is not the answer: the more we comprehend a situation rather than react to it, the less likely we are to help. The fact that many of us are repelled by the sight of the ravages of arsenic exposure does not help (Section 1.8).[23]

The West Bengal Government, which embraces a moderate form of communism, is highly centralised and has a weak commitment to environmental management, so when the extent of the arsenic crisis was first revealed its response was very muted.[14,24] They did ask for $200 million (US) from the central Indian government to pipe water from the Ganges, but this was to be phased in over about 20 years and no relief would be provided in the interim.[25] NGOs in West Bengal in general avoided the arsenic problem. In Bangladesh, a weak, government denied there was a problem but was incapable of doing much in any event.[24] However, unlike West Bengal, a vast network of NGOs in Bangladesh was attempting to fill gaps created by government inertia, and these were the first to confront the arsenic crisis in the country.

The Bangladesh Rural Advancement Committee (BRAC), an NGO founded in 1972 to improve the health of the rural poor in Bangladesh, undertook a major initiative in 1997. They trained women to test water by using field kits. More than 50 000 wells were tested and painted red (don't drink) or green (drink), depending on whether the concentration of arsenic was above or below

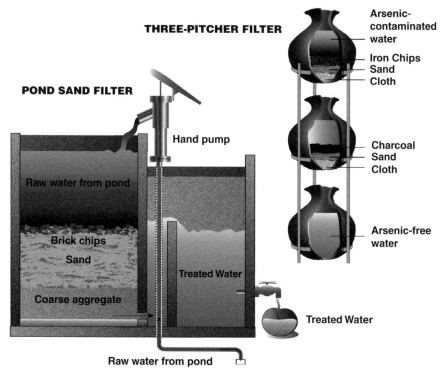

Figure 8.1 Indigenous arsenic-mitigation technologies.

50 ppb. They also trained women to identify arsenicosis patients, and advised villagers about water options. BRAC promoted the use of surface water and pond sand filters, as well as well water filters such as the three-pitcher filter (sometimes three-kolshi) (Figure 8.1). But here they encountered other problems. For example, by this time most of the ponds were being used as fish farms and were polluted with pesticides and other materials used for maximising productivity.[26,27]

With the support of the World Bank, BRAC conducted a major survey of the rural population in 2001 asking what sort of water supply they would prefer, and how much they would be willing to pay for it. Most of the respondents were not aware of the serious health implications of consuming water high in arsenic, although they had some idea that there could be a problem. A strong preference was expressed for community-based systems, but not ones that "went back in time," such as dug wells. The Bangladeshis wanted piped water and were willing to pay for it. The choice had little to do with arsenic mitigation and a lot to do with convenience. Some respondents had been supplied with water from arsenic-safe deep wells dug by the government, but complained about the distance they had to travel to collect water. A piped water system would centre on a treatment facility that would remove arsenic and pathogens and could allow the communities to go back to using surface water.[28] BRAC started to implement this on a village scale, with the communities contributing 20 per cent of the capital costs and 100 per cent of the operations and maintenance costs.[29] Columbia University researchers estimated that it would cost $290 million (US) to provide a deep well in each of the country's 86 000 villages.[30] Such a scheme would need a lot of help from donors and would take time. Also, as we will see, it would probably not be a permanent solution.[27]

BRAC recently received the Gates Award for Global Health – a cash award of $1 million (US). The local newspapers made no mention of BRAC's work on arsenic mitigation when they reported on the ceremony in Dhaka.

UNICEF has tried to distance itself from its role in drilling the wells because of the bad publicity;[31] nonetheless, it did fund a testing program in 1996 that by 1999 had results for 51 000 wells[32] that were similar to those that Mott MacDonald/BGS reported in their Phase I study. The World Bank was also somewhat coy about its role in the provision of arsenic-contaminated water; however, in late August 1998, the bank announced it would provide $32.4 million (US) in interest-free credit for an arsenic-mitigation project.[7,33] This grew into the Bangladesh Arsenic Mitigation and Water Supply Project (BAMWSP) and was jointly financed by the Government of Bangladesh, the Swiss Development Co-operation and the World Bank. This $44 million, 15-year project rapidly ran into problems and required restructuring. There were "fundamental uncertainties in how to proceed," and by June 2001, only $2 million had been spent.[34] Many commented on the lack of urgency of the program.[31] In response, the bank pointed out that resources are limited and that bacteria in water continues to kill about 110 000 children annually in Bangladesh.[17]

8.2.3 The Affected

"Pinjara Begum was married at age 15 to a millworker, she had made a pretty bride. Soon, however, her skin began to turn blotchy, then ultimately gangrenous and repulsive. Her husband remarried. In 2000 she died of cancer, at 26 years of age, leaving three children."[27]

Arsenicosis patients can be categorised in three stages (Section 1.8). One study of 3606 villagers who were drinking from wells with a mean arsenic concentration of 240 ppb found that 10 per cent of the population was suffering from arsenicosis. Most of these patients were between 10 and 39 years old and were in the first and second stages of arsenicosis. Keratosis was seen in 68 per cent of cases. Three patients had skin cancer. Adverse pregnancy outcomes, in terms of spontaneous abortion, stillbirth and preterm birth rates, were significantly higher among the arsenic-exposed group.[35,36]

The crisis affects those in the lowest socio-economic strata, particularly the rural farmers who form the majority of the population of Bangladesh. It has been suggested that women are more often affected because they are more likely to be malnourished. But the societal problems associated with arsenic not only affect the individual, but ultimately the entire family unit and community.

Sometimes the skin disorder associated with arsenic poisoning is mistaken for leprosy or some other contagious disease, resulting in social isolation. The affected are refused water from neighbours' tube wells, affected children are barred from attending school, and adults are discouraged from attending work, going shopping and/or visiting medical professionals in the hospital. Affected young women are condemned to stay unmarried. Family units are often broken up when husbands abandon their affected wives, leaving the women and children destitute. Children are more likely to be taken out of school to assist in supporting ailing family members, thereby reducing the likelihood of breaking the poverty cycle through education.[24]

8.2.4 Where does the Bangladesh Arsenic Come From?

The following account is given by Professor John McArthur of the London Arsenic Group, University College, London.[37]

The delta plain of the Ganges–Brahmaputra–Meghna Rivers is one of the largest in the world. It is built from sand, silt and mud that originated from the Himalayan Mountains to the north. Erosion of the mountains results in small particles that wash into the rivers and are either swept out to sea or deposited in the delta to build up the land. In the weathering process some of the iron minerals in the rocks are turned into iron oxides, chemically related to rust, and this material is found on the surface of some particles. Arsenic, which is also released in the weathering process, has an affinity for iron oxides, so as the iron oxide particles travel from the mountain to the delta they collect arsenic from the river water, soaking it up like a sponge. The rock particles are swept downriver to be deposited in the sea or in the soils of the delta.

The particles are buried as, year by year, new material is carried down the rivers and laid on top of the old, and the citizens of modern Bangladesh and West Bengal build their homes and grow their crops on these arsenic-containing sediments. The overall concentration of the arsenic in the soil is not a threat to human health: it remains firmly fixed to the iron as long as there is some oxygen around.

Unfortunately, lurking in the subsurface of many deltas in Asia are ancient buried wetlands containing deposits of peat, a material composed mostly of dead vegetation and rich in organic matter. This material is slowly rotting and is a source of food for bacteria. However, like humans, these bacteria also need oxygen to process the food. Oxygen is available from the atmosphere if the peat is close to the surface, but the supply becomes depleted with increasing depth below the surface. When this happens the bacteria have to find another source of oxygen and the one they choose happens to be the oxygen bound to the iron – the same iron oxide that binds the arsenic. The bacteria use the oxygen but have no use for the iron and the arsenic, which they leave behind to accumulate in the surrounding groundwater. This is why the Bangladesh groundwater is rich in both iron and arsenic and products of peat decay such as phosphorus, ammonia and sometimes methane (some well water can be "burnt" as it surfaces). Thus the concentration of the arsenic in the groundwater will vary vertically and laterally depending on the location of the buried decaying vegetation.

This explanation is probably close enough to the truth for our purposes (the arsenic and the organic carbon may have been codeposited[38]) and it also has the advantage of being politically neutral: no one is to blame for the arsenic release. Two other possible mechanisms do not meet this requirement, because in each the arsenic in the water could have resulted from human activities. In one of these, phosphate that is used as fertiliser percolates through the sediment and displaces arsenic species bound to the iron oxides; in the other, oxygen gets into an aquifer through drawdown of the water and reacts with the sediment to release arsenic species.[39]

The McArthur model suggests that similar geographic areas could have similar problems, and points to the Red River Basin in Vietnam, the Great Bend basin of the Yellow River in China, and the upper Mekong River in Cambodia (Section 8.7).

Box 8.1   Binod Sutradhar *vs.* Natural Environment Research Council (2003)

"Bangladeshis To Sue Over Arsenic Poisoning," was the eye-catching headline in a 2001 issue of *Nature*.[40] Also, "UK Funds Bangladesh Lawsuit: Fifteen Bangladeshis were granted legal aid from the UK to sue the British Geological Survey (BGS) for negligence.[41]

The BGS, now a research centre within the UK Natural Environment Research Council, was working in Bangladesh on a deep tube well irrigation project in the late 1980s and had some surplus funds. The team used this money to sample 150 tube well sites in central and northeastern Bangladesh and to prepare a report in 1992. The case is set out by Lord Justice Kennedy of the England and Wales Court of Appeal: "The report had limited circulation, but, it is said, it was foreseeable, and indeed it was intended, that it would be made available to the authorities in Bangladesh responsible for ensuring as far as they could that water from the sites sampled by BGS was safe to drink. In fact, the water from those sites on which many poor people depended was contaminated by arsenic. Those who prepared the report appear not to have tested for arsenic, although they tested for 31 other elements, and it is the claimants' case that, even though this study was only a short-term pilot project, having regard to the way in which BGS knew the report would be circulated to and relied upon by authorities who had no possibility of testing for themselves, BGS should in the circumstances either have tested for arsenic or at least have made it clear that they had not done so. . . That resulted in he [the claimant] and many others suffering serious health problems attributable to ingestion of arsenic."[42]

Box 8.1 Continued.

The main legal argument for the BGS (NERC) defence was based on the issue of proximity: the question of whether the defendant owed to the claimant a legally recognised duty of care. Proceedings began on August 8th, 2002, with a test case, and on May 8th, 2003, Mr. Justice Simon of the Royal Courts of Justice concluded that "the issue of proximity is difficult to isolate from the issues of foreseeability... There are various factual matters which are likely to be relevant to the issue of proximity which will have to be determined at trial."[43a] In making his decision for the plaintive, Mr. Justice Simon also agreed that NERC had withheld material documentation. This decision was appealed and on February 20th, 2004, the headlines in the Scottish national newspaper, The Scotsman, read: "Appeal Court Throws Out Arsenic Compensation Case."[42,44] "A legal action which could have cost the British taxpayers millions of pounds in compensation to Bangladeshi arsenic poisoning victims was thrown out by the Court of Appeal." However, a January 25th, 2005, press release from the law firm Leigh Day & Co. of London (cf. Section 4.3) announced that, after some setbacks in the courts, the claims of 400 Bangladeshi citizens against the National Environmental Research Council would proceed to trial before the House of Lords. A senior partner said, "The case raises crucial issues in the context of the developing world, such as what is the extent of the duty owed by an organisation to an impoverished government and its citizens when that organisation has been paid out of aid funds."[45]

The BGS maintains that it did not test for arsenic in water because arsenic was not then a recognised contaminant of groundwater in Bangladesh or regions with similar flood plain geology. In addition it claims the arsenic problem in West Bengal, India had not been reported in the international scientific journals, and only came to international attention during 1995.[46]

There is a subtext to this legal activity. NGOs do not like the threat of legal action hanging over their heads because they say it slows down the flow of donor money, curtailing further development, and they themselves have to be more conscious about what they are doing, "checking with lawyers every day."[47] Some experts also felt that western hydrogeologists may not want to work in the developing world because of the threat of legal action. However, there may be a silver lining: governments and private companies are now more likely to put more constraints on actions likely to impinge on the health of the people and think more about the environmental and social consequences of proposed actions.[43b]

In July 2006, the BBC announced that a panel of five judges at the House of Lords had upheld the earlier ruling: the allegation, which could have cost British taxpayers millions of pounds, was declared "hopeless." Alan Thorpe, chief executive officer of the Natural Environment Research Council, said that the ruling confirms that "Scientists cannot be held responsible for the research they decide not to do."[48,49]

8.3 Professor Dipankar Chakraborti

Professor Chakraborti is the director of the School of Environmental Studies (SOES), Jadavpur University. After graduation, he spent more than a decade in western universities but returned to West Bengal once he became aware of the seriousness of the arsenic problem. He has now been working to mitigate this calamity in West Bengal and Bangladesh for over two decades. In recent years he and his coworkers have discovered that the arsenic problem extends to the neighbouring Indian states of Bihar, Uttar Pradesh, Jharkhand, Assam and Manipur. He fears that more than 500 million people might be at risk from arsenic groundwater contamination.

In the early 1990s, Chakraborti attempted to alert various NGOs and government agencies in West Bengal to the arsenic problem: "Why immediate attention was not paid to assess the gravity of such contamination and why sufficient efforts were not taken to combat the situation by aid agencies really puzzled me."[50] In February, 1995, he organised an international conference entitled "Arsenic in Groundwater: Cause, Effect and Remedy," at Jadavpur University. Representatives of the Government of West Bengal and many scientists took part, including some from the British Geological Survey. Also present were twenty patients from affected villages:18 of these are now dead, 15 of them from various types of cancer.[50] A report of the situation in West Bengal, mentioned that one attendee at this conference was "stunned" by the numbers of victims of arsenicosis that had been identified, but was still able to comment, "if there were a little more scientific data, people would take it more seriously." The report placed considerable emphasis on how useful further studies would be for helping other countries, particularly the United States, sort out their own arsenic problems.[25] One outcome of the conference was the recommendation that "much more geological, geochemical and hydrogeological investigation of the aquifer needs to be made."[51] In spite of this, the BGS did not test for arsenic in 1995 while assessing the water quality in the alluvial aquifer of the city of Hanoi, Vietnam. The presence of arsenic was later revealed in 2001, prompting the statement, "several million people [who were] consuming untreated groundwater might be at a considerable risk of chronic arsenic poisoning."[52]

Chakraborti was incensed when the BGS claimed in 2002[46] during the course of the legal case described in Box 8.1 that there should have been a greater effort to alert the rest of the world to the Bangladesh situation. He pointed out that in addition to Saha's first paper in 1984[9] there had been a number of others, including one with the unambiguous title "Chronic Arsenic Toxicity From Drinking Tube Well Water In Rural West Bengal", published in the *Bulletin of the World Health Organization* in 1988.[53] John McArthur of University College London has also been critical of the BGS, maintaining that a major issue is BGS's failure to test for arsenic in Bangladesh, given that it is listed in the World Health Organization's Guidelines for Drinking Water Quality as a potential threat to health. The BGS *did* analyse for arsenic in a UK survey in 1989.[54] (WHO designated BGS as its international collaboration centre on ground water in 1992.)

Accidental Exposure to Arsenic: The Law of Unintended Consequences 359

A team from the Dhaka Community Hospital visited Chakraborti in July 1996 to learn more about the problem in the Bengal region following their discovery of a number of patients in Bangladesh showing symptoms of arsenic poisoning (Section 8.2.2). Subsequently, the two groups carried out a survey in Bangladesh to determine the severity of the arsenic contamination. They found that the ground water was highly contaminated and recorded skin lesions in the population of 14 districts of Bangladesh. They presented the results to government representatives and to NGOs, but the response was underwhelming. The rest of the world learnt about the disaster through subsequent joint publications.[55] Since then, the Dhaka Community Hospital, under the leadership of Dr. Quazi Quamruzzaman, has taken an active role in publicising the issue and in demanding action from the government. The hospital also treats the victims.

In addition to organising international conferences to draw attention to the issue, Chakraborti takes every available opportunity to speak passionately about the vastness of the arsenic problem, the need for proper care of the victims, and the need for clean water. He travels tirelessly to solicit support.[39] At one meeting on Arsenic Exposure and Human Health Effects, held in San Diego in 2000 and organised by the Society of Environmental Geochemistry and Health, Chakraborti was not very impressed by the ongoing debate regarding proposals for the maximum allowable concentration of arsenic in US drinking water – should it be set at 3 ppb, 5 ppb, or 10 ppb and was the evidence obtained from individuals exposed to high doses of arsenic or low doses (Section 8.11.3) – when his countrymen were drinking water with 100 times those concentrations. A cartoon that appeared on the meeting notice board reflects his thinking (Figure 8.2). Sometimes his outbursts at meetings are not well received – too much

Figure 8.2 Found on a notice board at a conference in San Diego, 2000.

passion. He was not invited to the workshop on arsenic mitigation organised by the Government of Bangladesh in Dhaka January, 2002 (Section 8.4), even though he was in the area.

Chakraborti was particularly angered by the suggestion that, on the basis of the cancer risk, developing countries should adopt a drinking water guideline of 50 ppb rather than opt for the lower WHO level of 10 ppb.[56] (The current guideline for India and Bangladesh is 50 ppb). He asked rhetorically "Are some animals more equal than others?"[19] He suggested that malnutrition enhances the negative effects of arsenic, that keratosis has been observed in individuals whose drinking water contained less than 50 ppb, and that drinking water intake varies with geography so that drinking even a little water at 50 ppb in West Bengal would deliver a much larger dose of arsenic than a lot more water at 10 ppb in the United States. Then there is the arsenic-in-food contribution, which he argues is greater in countries where the irrigation water is arsenic-rich (Section 7.3), so, in his opinion, the arsenic-in-water guideline should be appropriately adjusted downwards, not upwards. It is difficult enough, using the mitigation technologies currently available, to remove arsenic to levels below 50 ppb, let alone 10 ppb, so this is a reality that must be faced – and to be fair, Alan Smith and Meera Hira-Smith[56] wrote that reducing the water concentration guideline to 10 ppb could be self-defeating "if an implication of this policy was that high exposures continue until the long-term solution of achieving less than 10 ppb is eventually in place, perhaps decades from now."

Chakraborti believes that groundwater withdrawal is partly responsible for oxidative release of arsenic from the sediments into the surrounding water, which accounts for safe wells becoming unsafe over time. Thus, he believes that the Green Revolution and the excessive use of groundwater for agriculture are directly responsible for the current crisis. Needless to say, Chakraborti is a strong advocate for the use of surface water. He also believes that the arsenic in the ground water is addictive and millions may actually find it more palatable than clean rainwater.[57]

His research group presented an overview of the arsenic crisis in West Bengal and Bangladesh in 2004 (Table 8.1), along with similar disturbing data from the Indian provinces of Murshidabad and Lakshmipur.[19] We can be sure all the authors are in agreement with Rachel Carson who wrote in *Silent Spring*: "For those in whom cancer is already a hidden or a visible presence, efforts to find cures must of course continue. But for those not yet touched by the disease and certainly for generations as yet unborn, prevention is the imperative need."[58] The immediate provision of safe drinking water should override all other considerations.

8.3.1 Field Testing Kits

The first reaction of the NGOs in the 1990s to the ground water problem, was to begin to test all hand-pumped tube wells for arsenic. The laboratory capacity of the country was not sufficient to deal with collecting, analysing and reporting the

Table 8.1 Physical parameters and arsenic affected areas of West Bengal and Bangladesh.[19]

Physical parameters	Bangladesh	West Bengal
Area in sq. km	147 620	89 193
Population in millions	122	80
Total number of districts	64	18
Number of arsenic affected districts (groundwater arsenic above 10 ppb)	60	14
Number of arsenic affected districts (groundwater arsenic above 50 ppb)	50	9
Area of arsenic affected districts in square km	118 849	38 865
Population of arsenic affected in millions	104.9	50
Total number of hand tube well water samples analysed	50 515	129 552
Per cent of samples having arsenic above 10 ppb	43.0	49.6
Per cent of samples having arsenic above 50 ppb	27.5	24.7
Total number of hand tube well water samples analysed from affected districts	44 696	125 506
Per cent of samples having arsenic above 10 ppb in affected districts	48.5	51.0
Per cent of samples having arsenic above 50 ppb in affected districts	31.0	25.5
Number of arsenic affected blocks/police stations with arsenic above 50 ppb	189	85
Number of arsenic affected villages (approx.) with groundwater arsenic above 50 ppb	2000	3200
People drinking arsenic contaminated water with 10 ppb (in millions)	52	8.7
People drinking arsenic contaminated water with 50 ppb (in millions)	32	6.5
Districts surveyed for arsenic patients	33	7
Number of districts where we have identified people with arsenical skin lesions	31	7
People screened as arsenic patients from affected villages (preliminary survey)	18 991	92 000
Number of registered patients with clinical manifestations, including children	3762 (19.8%)	800 (9.7%)
Per cent of children having arsenical skin lesions based on total number of patients	6.1	1.7

data from 11 million or so water samples. The only way out of this dilemma was to use field testing kits; so the work began and the wells were painted red or green.

The BGS/Mott MacDonald survey in 2001 found that many wells were painted green that should be red.[59] Chakraborti's group also questioned the reliability of these early tests, and concluded that millions of dollars were being wasted in this mammoth project.[60] One test kit was recently singled out for commendation.[61,62] The users of this kit correctly determined the status of wells 88 per cent of the time, although Chakraborti in rebuttal claims that the error is still too much for comfort.[63] In West Bengal, because test kits have been abandoned, only community wells are being tested.

In a recent development, bacteria that have been engineered to glow in the presence of arsenic offer the promise of inexpensive and more reliable test results.[64]

8.4 Arsenic Mitigation in Bangladesh

8.4.1 Dhaka, Bangladesh, January 2002

The Awami League, which formed the Government of Bangladesh in the late 1990s, did not acknowledge the arsenic problem as being of much significance, and had the reputation of turning a blind eye to the many ways in which money from NGOs could be diverted to more deserving causes, and pockets, than mitigating the effects of arsenic ingestion. One aid worker with a lot of experience claims it was the worst country he had ever worked in: corrupt and violent, with a self-serving bureaucracy that had little sense of country. The 1996 United Nation's Human Development Index, ranked Bangladesh 143rd out of 174 countries surveyed. Other indicators, such as the Corruption Perception Index, suggest it is also among the most corrupt: 91st out of 91 in 2001. This perception was unchanged in 2005.

In 2001, Begum Khaleda Zia, leader of the coalition Bangladesh Nationalist Party (BNP), won a very bitter and violent election campaign. The newly elected Prime Minister had made arsenic mitigation a part of her election platform and promised to start the process within 100 days of taking office. True to her word, an International Workshop on Arsenic Mitigation was hastily organised in Dhaka in January 2002. The delegates were briefed on the situation as seen by the government, given lunch at which the national drink, Coca Cola, was served, and set to work to come up with recommendations.[65] The main topic at lunch was the warning that some of the local bottled water was obtained directly from contaminated city outlets. (One of the readily available and genuine local brands of bottled water, MUM, makes the claim on the label that "It comes naturally".)[47]

A reasonably coherent set of recommendations was produced from the workshop, despite discussions that were rarely focused or calm, and often self-serving. This output was edited and polished by the government and the NGOs. The latter were heavily involved because of the real threat that their support could be withdrawn unless the leakage of donor funds was curtailed. Finally, in 2004, the recommendations became a major part of the National Policy for Arsenic Mitigation and its accompanying Implementation Plan. Copies of these documents are now available though the National Arsenic Mitigation Information Center[66] set up recently as part of the mitigation policy.[29]

The major components of mitigation include raising awareness; screening tube wells for arsenic; examining villagers for arsenicosis; and implementing short- and long-term mitigation technologies. The policy states that, other thing being equal, preference will be given to surface water over groundwater as a supply source. In addition, piped-water systems are favoured whenever feasible. Nowhere is there a sense of urgency.

Box 8.2 Delwar Hosen

A few of the 2002 workshop delegates met Delwar Hosen by chance on a field trip into the Bangladesh countryside. He lives in the village of Malipara, in the union of Jampur, in the thana of Sonargaon, and in the district of Narayangong. He asked that this be noted down because he lives in an arsenic-affected area, although not an arsenic-affected village, and he hoped that a report on his plight might result in an improvement in the local situation. Mr. Hosen's village is just northeast of Dhaka and is relatively prosperous because the inhabitants are able to make some money from their weaving and their vegetable crop. The steady beat of the looms provided a constant audio background.[47]

He first showed delegates a tube well that was painted both red and green (Figure 8.3). This well was sunk to a depth of 21–24 m and was first painted green as the initial test showed the water to be arsenic-free. He did not know what this meant in terms of the actual concentration, and he did not know how the test was done. Some months later the same well was tested and high arsenic levels were found, so this time the well was painted red. Mr. Hosen was perplexed.

Figure 8.3 Rainwater storage tank, unused, and tube well, unsafe, in the village of Malipara, Bangladesh.

There is a cement tank adjacent to the well that was used to store rainwater collected from a roof. Mr. Hosen said that this water was smelly and did not taste very good (Figure 8.3). The tank had been built with the help of funds from an Australian aid agency but nobody used the water. Finally, Mr. Hosen pointed out the one tube well painted green that produced good water. He said this well, which was sunk to a depth of 200 m, served a

> **Box 8.2 Continued.**
>
> population of about 10 000 people living within a 5 km radius. There should have been a steady stream of people using this source of drinking water, but this was not the case. It seems that the local population would rather risk drinking from the more accessible but contaminated wells. There was no arsenicosis in Malipara – one case had cleared up when good water became available. Perhaps this village was lucky in having well water that was low in arsenic as indicted by the first test.[47]
>
> As for the less fortunate villagers, one arsenic victim is quoted as saying, "It was hard to fetch water from so far away. One of my daughters-in-law would go and carry back a full pitcher for drinking. But sometimes the path was too muddy. It was knee deep. She couldn't wade through it and she stopped going."[67]

8.4.2 Arsenic-Mitigation Technologies

"Ancient [2000 BCE] medical lore of India advises that impure water should be purified by heating or filtering through gravel and coarse sand."[68]

If an arsenic-mitigation technology is to be used in rural areas, it should be inexpensive, employ local materials, and be user-friendly. This also means in Bangladesh that the women, who do the water collection and carrying, must be comfortable with the process. Electrical power is not readily available, so technologies such as reverse osmosis, which are commonly used in developed countries, are not viable options.

Tube well water in Bangladesh and neighbouring countries is usually rich in arsenic, iron, manganese and phosphate. The very simplest purification method involves exposing the water in a container to air. The iron slowly precipitates as "rust" removing the arsenic at the same time just as Bunsen described many years ago (Section 1.6). Iron can be added to the water if necessary, and light can be used as a catalyst to speed up the process.[69] One advertisement in a trade magazine was pushing capsules "containing 0.6 g of a synthetic mixture of complexing, coprecipitating and adsorbing agent. One capsule can remove 97 per cent of arsenic from 10 litres of water containing 300 ppb arsenic. Now you need not be scared of arsenic poisoning through water."[70] Sounds too good to be true.

Many more elaborate arsenic-mitigation technologies have been suggested, and some have even made it into the field. One of the simplest is the three-pitcher system, which consists of three earthenware (clay) water pitchers arranged one above the other (Figure 8.1). The top two are for filtering and have holes in the bottom; the third pitcher is for storage. Arsenic removal is effected by binding to iron oxides. One system lasts a family for about six months. There are many variations on this theme, the simplest being a bucket of sand and nails.[71–73] Other less obvious filter materials have been suggested, one being powdered roots of water hyacinth.[74]

8.4.3 Verification of Mitigation Technologies

Shortly after the 2002 Dhaka Workshop, the Bangladesh government announced that "all chemical-based arsenic filters or removal plants must obtain certification from the government for being on the open market." This policy has a strong Canadian connection.

In the late 1990s, Roy Boerschke suggested that the Ontario Centre for Environmental Technology Advancement (OCETA) should extend its operations to Bangladesh. (OCETA is licenced by the government of Canada to perform environmental technology verification (ETV) and to certify viable technologies.) OCETA in turn made some inquiries and approached the Canadian International Development Agency (CIDA) for support in launching an ETV program in Bangladesh. CIDA provided $4 million (Can), and the ETV-AM (arsenic mitigation) program was up and running in association with the government of Bangladesh, with Mr. Boerschke as director of the project. In addition to evaluating technologies, the program was intended to establish capacity in Bangladesh to implement the ETV process. The technologies were to be evaluated on technical, social and fiscal parameters.[75–78] A compelling argument for implementing the ETV-AM program was that the Canadian government and especially the NGOs, were reluctant to invest money and effort into installing arsenic-mitigation technologies without a guarantee of their reliability.

The implementation package was finally approved with some changes on February 19th, 2003. In the midst of these preliminaries, the UK Department for International Development (DFID), carried out an independent short-term evaluation of some arsenic-removal technologies. Some of the units gave encouraging results, but bacterial contamination, including faecal coliforms, was found in many instances.[79] Nonetheless, this was not seen as a barrier to the Canadian project.

The sampling program and the analytical work were largely carried out by Bangladeshi personnel and laboratories, although the capacity to do this work was stretched to the limit (Box 8.3). Unexpectedly, the proffered technologies were backed up by very incomplete performance data so major revisions had to be made to the original testing protocols. The costs of the truncated testing program were borne by the ETV-AM program; that is, CIDA. In February 2004, the Government of Bangladesh was able to put a positive spin on developments, announcing that it had approved for commercial sale four technologies to remove arsenic from contaminated water. The press release states that the Bangladesh Council of Scientific and Industrial Research (BCSIR) acting for the government received 17 technologies for evaluation in 2002 and that it was aided by OCETA, a Canadian NGO.[80,81] One of the technologies, the Sidco arsenic-removal plant from Germany, removed arsenic by adsorption onto a bed of granular ferric hydroxide and operated at the community level; the others, such as the Sono 45-25 filter that uses a bed of iron to remove arsenic, were household systems.

Box 8.3 Acid in the Community

To perform the large number of chemical analyses that were an essential part of the ETV-AM program, and as part of the capacity building requirements, Bangladeshi scientists received training in modern techniques and in some cases were supplied with remodelled laboratories and state-of-the-art instrumentation. Associated with this activity was a greatly increased requirement for the laboratory chemicals used in performing the analyses. This requirement caused a hiccup in the program for a while because of government restrictions on importing acids.

On January 16th, 2005, The Independent newspaper in Dhaka reported that Sahir Kaltu was sentenced to life imprisonment for throwing acid on his wife in the middle of the night, causing her severe injuries. This was not an isolated incident. Every year, many people in Bangladesh – mostly women – are attacked with acid. The acid, often concentrated sulfuric, is thrown in the face of the victim. The perpetrators are most commonly spurned lovers or jealous husbands, but in recent years the weapon has been used in other types of domestic disputes. The victims, sometimes children, experience agonisingly painful burns and in most cases are disfigured for life. They are consequently very much devalued in the eyes of Bangladeshi society.[82] The government put import restrictions on acids in an attempt to control these monstrous acts. Canada is one of the countries supplying financial aid to the victims. Its partners are the Dutch government, UNICEF and the Acid Survivor Foundation of Bangladesh.[83] The donors deal directly with the NGOs that are working within the community, an approach that works very well in this instance and bypasses the Bangladesh government.

The Government of Bangladesh also insisted that the technologies would have to comply with the National Waste Management Protocol regarding the safe disposal of the waste generated.[84] These regulations are particularly onerous and are likely to be ignored in the absence of government assistance and inspections: who is going to take the spent media from the filters, encapsulate it with concrete in a PVC pipe, and bury the package away from a populated area?

The disposal of these wastes is also a problem in other jurisdictions. In North America, the classification of this material as hazardous is often determined on the basis of leaching tests, such as the toxic characteristic leaching procedure (TCLP) (Section 7.4) that attempts to simulate the conditions in a landfill and determine if, for example, arsenic would be leached from the material. If the material passes, which means that essentially no arsenic leaches out during the test, it can be disposed of in an ordinary landfill. If arsenic leaches easily, the material is declared hazardous and must be disposed of in a special hazardous-waste facility, which is a much more expensive option. There is reasonable agreement that the TCLP test is not much use for arsenic-containing wastes and all such tests have nothing to say about long-term behaviour (Section 7.10.6).[85-87]

It would be a challenge to implement a collection, testing, disposal program for all the wastes produced by millions of household arsenic-mitigation technologies in any country, let alone Bangladesh.

The next stage of the ETV program was planned to begin early in 2005, the Canadian government having guaranteed further funding. CIDA contributed $14.8 million, which included a grant to UNICEF for the purchase and installation of the approved technologies from Phase I.[88] The agreement to proceed was ready to be signed in Dhaka, and the celebratory dinner arranged, when the chief Bangladeshi representative produced new demands, many self-serving.[47] The celebration turned into a wake as the program was put on hold. A military coup in 2006 was the next roadblock, but the program was up and running again in 2007.[89]

NGOs and government officials all say that the verification program is a valuable contribution and is needed. However, at least one major report is not very enthusiastic about arsenic-removal technologies at the household (tube well) level, arguing that the mitigation process could result in the substitution of other hazards, particularly microbial ones.[90] (In 2005, about 42 per cent of the people of Bangladesh did not use any form of latrine, and 50 per cent lacked adequate sanitation.[91]) Thus, the implementation of any mitigation method would have to be accompanied by an education campaign to promote proper hand-washing practices and thorough cleaning of water receptacles;[92] admittedly something that should be in place anyway.

Other problems with remediation technologies came to light in a recent two-year study conducted in West Bengal. Eighteen technologies were evaluated for arsenic and iron removal. None could achieve the 10 ppb WHO standard, and only two the 50 ppb standard of India. Only three units were consistently in operation over the two-year test period, and four units were officially abandoned during testing. The average price of the units was $1 500 (US) and the cost of the chemical ingredients ranged from $27 to $900 per charge. Only one supplier had made provisions for waste disposal. The villagers' arsenic status did not improve during the test period, as judged by urine analysis.[93] Professor Chakraborti commented that, "Even the God of Wealth would not waste his money on arsenic-removal plants as we are doing."

The United Nations Educational Scientific and Cultural Organization (UNESCO) has recently developed a low-cost arsenic filter based on iron-oxide-coated sand. It will distribute 1 000 units to affected Bangladeshis at no charge. One filter is claimed to meet the needs of 20 people.[94] This filter did not go through the ETV program.

8.4.4 The Grainger Challenge Prize

In spite of political and practical obstacles such as the existence of the Bangladesh ETV program, the US National Academy of Engineering, with the generous support of the Grainger Foundation, offered Grainger Challenge prizes of $1 million, $200 000 and $100 000 for first, second and third place respectively, for the design and creation of a workable, sustainable, economical, point-of-use water-treatment system for arsenic-contaminated groundwater in Bangladesh, India, Nepal and other developing countries.[95]

Fifteen technologies were selected for evaluation from more than 100 entries. The standard water used for testing contained arsenic (300 ppb) and iron, but no manganese or phosphorus, which is commonly found in ppm concentrations in Bangladesh samples. Nevertheless, the winner, announced early in 2007, was the Sono filter. Its coinventors – Professor Abul Hussam from George Mason University, in Fairfax VA, and Dr. A. K. M. Munir, a physician – are both from Bangladesh. The point-of-use filter is based on iron, sand and charcoal and is very similar to the three-pitcher system shown in Figure 8.1. The winners donated their $1 million to the purchase of filters, which cost about $40 (US) each. The second prize went to an international team led by Professor Arup SenGupta of Lehigh University, Bethlehem PA. Each of their units can serve about 800 households, and cost $1200 to $1500 (US). Arsenic is removed by a column containing "activated alumina or hybrid anion exchanger" (HAIX). The third prize went to the Children's Safe Drinking Water program at Proctor and Gamble Ltd., Cincinnati OH. In their process chemicals for disinfection of the water and precipitation of the arsenic are combined in a sachet whose contents are stirred into a bucket containing 10 L of water. Some 20 minutes later, the arsenic – which settles out of solution – can be removed by filtering the water through a clean cloth. The Children's Safe Water Fund has worked with partners to provide 57 million sachets in more than 30 countries in the past three years. Each sachet costs about the same as an egg and has the advantage of providing disinfection as well as arsenic removal.[96] Similar capsules were described above (Section 8.4.2).

The award of first prize to someone born in Bangladesh got a lot of attention from Bangladeshis all over the world, particularly in the United States; although the prize seemed to be more important than the reason for its inception. Dr. Badrul Haque, chairman of the board of trustees of the Bangladesh American Foundation Inc. (BAFI) wrote on the group's web site (www.bafi.com): "Under the overall leadership of BAFI, Prof. Hussam also played a critical role in 2000 to ensure that the arsenic issue in Bangladesh was at the forefront of US decision makers, and then successfully lobbied for a US–Bangladesh Science and Technology agreement, which will foster cooperation between the two countries."[97] One enthusiast was inspired to write in *The American Muslim*: "Will Bangladesh finally be able to filter out the arsenic of corruption, greed, nepotism and misrule, once and for all from its government?"[98a]

Fortunately, the winning filter had also been tested and verified for sale by the ETV-AM program, so the possibility of a diplomatic tiff was avoided. However, the disposals of the spent filter material remains a matter of debate: "The Sono filter is certainly causing more harm than good to the entire nation of Bangladesh . . . " www.sos-arsenicnet/english/mitigation/sono.html. And headlines such as "Bangladeshi scientist lifts curse of arsenic poisoning" are reasons for concern. Whilst this may be true for those lucky enough to be conceived and born into an arsenic free environment, those who have already supped the Devil's water continue to be at risk for many years.[98b]

Staying on the subject of prize winners, the NGO Dalit was one of the runners up in BBC World's *The World Challenge 2006*. The competition seeks to highlight and reward examples of community enterprise and innovation. The NGO is described as fighting the scourge of arsenic contamination by locating and tapping rare pure-water sources, installing filtration systems, and researching herbal remedies for arsenic-related diseases. The project also includes an educational element aimed at acquainting people with the dangers of arsenic poisoning and how to avoid them. The NGO is named after the Hindu Dalit people, who are discriminated against in Bangladeshi society and have suffered disproportionately from arsenic poisoning.[99]

8.4.5 Nanoparticles

We have seen that iron oxides bind arsenic, so research on the magnetic behaviour of ferric oxide nanoparticles – particles with at least one dimension smaller than 100 nm, that act as a bridge between bulk materials and atomic or molecular structures – has been trumpeted as a solution to Bangladesh's problems because the particles not only bind arsenic from solution but also are attracted to a magnet for separation. Particles around 12 nm in diameter removed nearly all the arsenic; however, there are practical problems: "Although the nanoparticles used in the publication are expensive, we are working on new approaches to their production that use rust and olive oil, and require no more facilities than a kitchen with a gas cook top."[100,101] This story, which got a lot of mileage in the world media, some in connection with the Grainger Prize, ignores most of the problems already mentioned such as hygiene and ease of application, while neglecting to mention that nanoparticles can be far more reactive and toxic than a similar mass made up of larger particles.[102] "What do we know about nanotechnology, about its effects on human health and environment? Not much. What are we doing to get these answers? Not enough. Can the existing regulatory system protect the public from potential problems with nanotechnology? Not adequately."[103]

8.4.6 Other Arsenic-Mitigation Methods

Deep wells. Deep tube wells tap into an aquifer, typically more than 150 m beneath the surface, and underneath the generally used source. The water is withdrawn using a piston hand pump that operates above ground, and so far the water is largely free of arsenic. These deep wells are the source of the arsenic-safe water used in many village-scale supply systems in Bangladesh and West Bengal.[27,29] The Bangladesh Arsenic Mitigation Water Supply Project was planning to install 1649 deep tube wells in coastal areas in 2004, with each well expected to benefit about 80 to 100 families.[104] Many have expressed caution, including a panel of experts convened in Dhaka: they say that deep aquifer water should be viewed as part of the solution to providing safe drinking water, not the sole or preferred alternative.[105] Already some deep wells show unacceptable levels of microbial contamination, primarily as a result of poor operation.[106] There are some even deeper aquifers that are arsenic-free, but the water is very old and it is claimed that the water would not be renewed if it were pumped out.[107]

Dug wells. Dug wells are approximately 5 to 10 m deep, depending on the lowest seasonal depth of the aquifer. Dug wells were the water source displaced by tube wells and many individuals and NGOs are urging a return to this source. The main limitations are seen to be the possibility of contamination from surface activities, particularly from fertilisers and animal faeces.

Surface water. The Bengal Basin has an abundance of surface water and precipitation. About 18 per cent of Bangladesh is covered in standing water. This abundance has tremendous potential for fulfilling both the everyday and commercial water needs of rural villagers. But villagers are often unwilling to switch back to using surface water as a drinking water source after being told previously that it was polluted. A typical reserve pond with a slow sand filtration setup is shown in Figure 8.1. The cleaning action includes a biological process in addition to the physical and chemical ones. A sticky mat of suspended biological matter forms on the sand surface, where particles are trapped and organic matter is biologically degraded.[68] The reserve pond ensures that the source water is low in turbidity. It has a historical precedent of use, is easy to operate, inexpensive, and is actively promoted by aid agencies. River water is also being used as a source for the pond, prior to being distributed at the village level.

Rainwater harvesting. At the household level, rainwater is collected from corrugated iron roofs and stored in cement tanks that are enclosed to prevent contamination and to limit exposure to sunlight, inhibiting biological growth. This method is being promoted by NGOs, although as we have seen in the example of Mr. Hosen (Box 8.2), the water quality may not be acceptable.

At the community level rainwater is collected into maintained, cordoned-off, specially constructed reserve ponds to ensure the water quality. The method is being promoted particularly in areas affected by saline intrusion into the aquifer.

Membrane filtration. To date, emphasis has been placed on mitigation technologies that require low maintenance and minimal electrical power. General Electric has recently developed a 200 W low-maintenance, portable unit that runs on solar power, and can provide more than 9 000 L of clean water every day. The units use a membrane to filter out bacteria, parasites and viruses from well or stream water.[108]

8.5 Where Are We Now?

8.5.1 Treatment Options for the Afflicted

Unfortunately there is little that can be done for the severely afflicted, apart from giving them salicylic acid ointment for the relief of keratosis. The best early treatment available is to remove patients and potential patients to a source of arsenic-safe water. Early symptoms can be reversed but this is not possible in the later stages. Dimercaprol (BAL) and DMPS (Section 1.11) may help according to Dr. Saha[10] and his colleagues,[109a] but the related DMSA, meso-2,3-dimercaptosuccinic acid, was not effective in producing clinical or biochemical benefit or any histopathological improvement of skin lesions. Spirulina, a blue-green alga grown in Bangladesh, may have a positive effect on skin problems when taken internally. It is recommended by some dermatologists as a cure for chronic arsenic poisoning.[109b,110] Homeopathic doses of Arsenicum Album (arsenic trioxide) are being investigated by a group of Indian scientists from the University of Kalyani, "to ease the suffering" of arsenic poisoning – but so far only in mice. "Encouraging" human trials are described in Chapter 1 (Section 1.4),[111] but caution is advised.

Vitamins, beta-carotene and zinc may help mitigate affects: selenium and folic acid supplements are also being studied.[112a] Apparently, folic acid reduced the total arsenic blood concentrations in a Bangladesh study population by 14 per cent.[112b] This aspect of the Bangladesh situation has been used to put a public interest spin on stories that might otherwise garner little attention. A recent example: according to the *Saskatoon Star Phoenix* "Fighting arsenic poisoning in Bangladesh is among the projects being funded by a $39.6 million (Can) award to the University of Saskatchewan [to support a synchrotron facility (Box 7.2)]." The article goes on to say that some of the money will be used to see if lentils grown in the province's selenium-rich soil could be the solution to arsenic poisoning in Bangladesh. Then comes the remarkable statement: "It isn't that Bangladesh people are arsenic-rich but that they're lacking in the essential selenium."[113] Whilst it is true that trials with selenium dietary supplements are underway in Bangladesh, we can be sure that the investigators there believe they are dealing with an arsenic problem. In one such study, Wendy Verret and coworkers[114] from the Columbia University School of Public Health, came to the conclusion that supplementation with vitamin E and selenium, either alone or in combination, slightly improved skin lesion status, although the improvement was not statistically significant.

8.5.2 Arsenic Mitigation

The World Bank wrote in 2003, "The potential for widespread arsenicosis is just one of the many health challenges faced by Bangladesh. The main causes of disease and death in the country continue to be malnutrition (which may contribute to susceptibility to arsenic poisoning), acute respiratory infection, tuberculosis and gastrointestinal (diarrheal) infections. Contributing to these challenges are marked gender disparities in health status, low levels of education, poor sanitation, and an inadequate health care system. The delivery of services is impaired by low expenditure, inefficient use of existing resources, and weak institutional and management capacity." More followed in the same vein.[115]

Similar good/bad news was announced by Bangladesh's Director-General of Health at a conference in Dhaka in 2004: the government will soon begin training doctors in rural areas to detect arsenic patients; but the government admits there is no protocol for their treatment.[116] One outcome of this conference was the so-called Fifth Dhaka Declaration that noted that considering the severity and magnitude of this human tragedy, the progress is very slow and inadequate.[117] It seems that few of the recommendations of the 2002 workshop, which became Government policy, have been implemented (Section 8.4). Amazingly, in 2004 the World Bank described the arsenic problem along with HIV/AIDS as "relatively new."[118]

A recent survey by the Dhaka Community Hospital finds that the situation in monitored villages has deteriorated in the years 1997 to 2005.[119] Local government minister Ziaul Haq Zia told the Bangladesh Parliament in 2006 that 29 500 people, from 190 sub-districts, had been struck down by arsenic poisoning in Bangladesh since 1998.[120] Many people say they feel bitter that they seem to be valued only as subjects. In 2005, the Dhaka Independent reported that the NGO Gono Unnayan Prochesta (GUP), working in the three upazillas, examined 15 400 tube wells in 21 unions. They found arsenic in 13 084 of these, and 405 people with arsenicosis. Six people had died of arsenic-related disease since 2001. "The NGO distributed 40 arsenic-free water kit boxes among 40 poor families."[121]

Wells are still being drilled at an ever-increasing rate, which "seems to indicate that people either do not worry about arsenic, or believe that new wells are free of danger." The drillers in some districts have learned to install tube wells to a deeper depth than earlier, 60 m rather than 30 m, into reddish sediments that have lower concentrations of iron and arsenic.[122a,122b] In general, it is true that newer wells are more likely to have lower arsenic concentrations,[123] a fact noted earlier by Peter Ravenscroft of Mott MacDonald (Section 8.2.2); however, all wells should be regularly monitored because there is a serious risk that their arsenic concentration will increase over time.[124]

At the closing of 2006 we could still read in the West Bengal press about the "criminality of a government not bothered about letting people, young and old, be poisoned.[125] "Scientists Urge Bangladesh Arsenic Strategy Change" was the headline on a report criticising the 2004 National Policy for Arsenic Mitigation, which is said to have had little impact and needed major revision. The use of deep wells was still being promoted, but for drinking water, not irrigation.[126,127] But before this quick fix is accepted, there is the always real problem of possible contamination from other aquifers (see Renner, 2004[128], for US studies on aquifers in rock), and whatever is done has to be considered in the light of other national and international problems such as the degradation of agricultural lands and the threat to food production (Section 7.3.4).[129]

It should come as no surprise that according to a UNICEF report: "Bangladesh is likely to miss one of the Millennium Development Goals of bringing 86 per cent of the population under safe drinking water coverage by 2010."[130] Now, in 2007, we learn for the first time that there is good evidence that after exposure to arsenic in water, even at about the Bangladesh standard of 50 ppb, mothers run an increased risk of having a miscarriage or losing their child in the first year.[131] This bad news further emphasises the urgent need for arsenic-free water. The country is yet again under military control following a political crisis, with a paralysed Parliament unable to prepare for general elections and cope with growing militant Islamism. One international crisis group suggests that the country is at risk of floundering because of, for example, "entrenched corruption, weak judicial and law enforcement agencies, ethnic conflict, poverty, and poor development indicators for women."[132]

8.6 Taiwan

8.6.1 Southwestern Taiwan

About 100 000 inhabitants of southwestern Taiwan began using artesian wells in the early 1920s because the water from their shallow wells near the sea coast was often salty. The new wells were deep, with most being between 120 m to 180 m. They were dug into sediment containing black shale, which unfortunately, contained arsenic in the range of 0.51 ppm to 13.7 ppm. Shale is formed from hardened mud and some deposits of volcanic origin can contain over 200 ppm arsenic.[133]

These Taiwanese people were poor and their livelihood came from farming, fishing, and salt production. The first reports of health problems in the population came out in 1968 when W.P. Tseng and his coworkers described a high prevalence of skin cancer, hyperpigmentation, and keratosis.[134-137] The skin cancers were atypical in that about 75 per cent were located on parts of the body not usually exposed to sunlight. Not one case of the characteristic skin diseases was seen in a control population. The water in the wells of the exposed population was found to be high in arsenic, with many samples above 400 ppb, although there were considerable temporal variations. As in Bangladesh, some wells also produced ignitable gas.

The youngest cancer patient was aged 24, the youngest with hyperpigmentation was three years old and the youngest with keratosis was four years old. The prevalence of a particularly nasty illness called blackfoot disease was 8.9 per 1000.

Blackfoot disease results in gangrene of the extremities. Erythematous swelling (an abnormal redness of the skin caused by capillary congestion) is followed, one or two months later, by scattered ulcerations occurring on the lesions and then gangrene develops, eventually affecting the entire limb. The condition is extremely painful and results in amputation of the limb, either surgically or spontaneously. The first case was reported in 1954. It is prevalent in the arsenic-endemic area and is said to be related to arsenic ingestion. However, there are those who believe that humic acid in the water is a contributor if not the cause.[138] A centre for blackfoot victims was set up in Peimen township, in the centre of the endemic area. There were about 1600 registered patients in 1982 and there have been about 50 new patients each year.[139] Safe water was supplied to the region by 1979.

8.6.2 Northern Taiwan

The Lanyang basin of northeast Taiwan is another arsenic-affected area. The residents drank contaminated water from shallow wells from the late 1940s through the mid-1990s. One study of 8102 villagers found a number of cases of urinary tract cancer.[140]

Medical data collected from these unfortunate Taiwanese have been invaluable in developing guidelines for arsenic concentration in the less exposed parts of the world (Section 8.11.2).

8.7 Vietnam and Elsewhere in the East

8.7.1 Vietnam

The first report of an arsenic problem in Vietnam appeared in 2001 when arsenic was discovered in groundwater in the Red River delta, a region that includes the city of Hanoi.[52] As in Bangladesh, UNICEF had promoted the use of tube wells to avoid surface contamination. The levels of arsenic found are similar to those in Bangladesh and its presence can be similarly explained (Section 8.3). Unlike Bangladesh, the well concentrations vary dramatically with the season: the average of 69 wells in September, 1999, was 194 ppb arsenic, the next May it was 52 ppb.

About 10 million people in Vietnam are at risk, but so far there have only been a few incidents of arsenicosis: seven out of a sample of 400 people living in a high-arsenic region. This may be because the exposure times have been relatively short – less than 10 years.[141-143] After an initial period of denial, the government is now encouraging the use of household sand filters for wash and bath water. These remove about 80 per cent of the arsenic and for some reason work better in Vietnam than in Bangladesh.[59] Filters made from charcoal prepared from coconut husks are also claimed to be effective.

The situation in Cambodia is similar.[143]

8.7.2 Nepal

A group from the Massachusetts Institute of Technology is making a major effort to supply inexpensive technologies to the people of Nepal who also have arsenic in their ground water. The simplest of these consists of sand from the river, gravel, iron nails and shards of bricks in layers in a plastic bucket topped with a diffuser. The fledgling NGO is entitled Filters for Families, and its slogan is "For $2 you can save a person for a lifetime."[144]

8.7.3 China and Japan

Nearly 312 million Chinese have no access to safe drinking water. Natural contaminants include arsenic, fluoride and salt, but there is also an abundance of industrial pollution resulting from human activities. There are high arsenic concentrations in the well water of the provinces of Shanxi and Inner Mongolia. But according to the World Watch Institute, an environmental research organisation in the United States, "This natural contamination pales in comparison with that caused by humans. In September, 2006, two chemical plants in Yueyang County of Hunan Province were cited for dumping tens of thousands of tons of wastewater containing high concentrations of arsenic compounds – at more than 1000 times the permitted level – into a major river that serves as the primary water source of some 100 000 residents."[145a] The central government has set a target of providing all of China's rural residents with clean drinking water by 2015.

Industrial activities have also resulted in tragic human arsenic exposure in Japan, notably in the mines of Toroku and Matsuo and in an arsenic sulfide factory in Naniki.[145b] One of the more unusual exposures through tainted milk is described in Box 8.4.

8.8 South America

8.8.1 Argentina

A number of provinces in Argentina have high arsenic levels in their well water. The best characterised is Córdoba, located in the centre of the country, where the water arsenic concentration can reach over 2000 ppb and where the local physicians have documented symptoms of exposure since the early 1900s. The first reference in the literature entitled *Arsenicismo reginal indemico* was written by A. Ayeza in 1917 (*Bol. Acad. Med.* 1917, p. 1). The population at risk is estimated to be about two million, and exposure has resulted in an increased risk of internal cancers.[4,5,146]

8.8.2 Chile

About 400 000 people have been exposed to excess arsenic in Chile. The sources are natural (thermal springs, volcanogenic sediments) and anthropogenic (mining). In one of the northernmost regions of Chile (Region II), one of the driest places on Earth, much of the drinking water comes from rivers that originate in the Andes Mountains. These rivers are rich in arsenic, possibly the result of ancient (Inca and Spanish) and more recent mining activities. The other regions of Chile are not affected by arsenic.[4,5]

The citizens of the city of Antofagasta, population 200 000 in Region II were unlucky to have a water supply that was contaminated by arsenic, at 800 ppb, from the years 1958 to 1970. The water was taken from the Toconce River. The first sign of arsenicosis was seen in children in the early 1960s.[147] A new supply system was introduced that took water from two rivers with arsenic concentrations of 800 ppb and 1300 ppb, respectively, resulting in an *increase* in the arsenic concentration in the city's drinking water, to 870 ppb. A treatment plant was introduced in 1971 and the arsenic concentration in the supply dropped to 260 ppb. Over time the level has improved. It reached 40 ppb around 1990 and is now about 10 ppb. Other population centres around Antofagasta have had a similar historical exposure to arsenic, apart from remote towns such as San Pedro de Atacama where the level remains around 600 ppb.

The citizens of San Pedro were said to be unaffected by the arsenic in the water and, because their ancestors had lived in the area for 11 000 years, there was speculation that they had acquired immunity.[148,149] However, the prevalence of skin lesions among men and children from San Pedro is similar to that reported for individuals with similar exposure in Taiwan and West Bengal.[150]

"Because of the high arsenic water levels in past years, a stable population with widespread exposure to arsenic in drinking water, and good sources of mortality data, Region II was an ideal population in which to search for further evidence that ingestion of arsenic might cause increased mortality from these internal cancers."[151] This statement from Dr. Alan Smith – we met him in Section 8.2 – and his coworkers proved to be prescient. They have used the information available on death certificates to reach a number of startling conclusions about the effects of arsenic exposure, including exposure *in utero*. It seems that arsenic might account for 7 per cent of all deaths among those in Region II aged 30 and over. This impact is greater than that reported anywhere else from environmental exposure to a carcinogen. These risks, which are extraordinarily high, last a very long time after initial exposure and remain a long time after exposure ends. Their findings also "... suggest that exposure to arsenic during early childhood or *in utero* has pronounced pulmonary effects, greatly increasing subsequent mortality in young adults from both malignant and nonmalignant lung disease" (Section 1.10).[151–153b]

8.9 Africa

Ghana produces about one third of the world's gold, so the presence of arsenic in the form of arsenopyrite is not surprising. The water concentration of arsenic is high in the mining region and the villagers have white spots on their skin, but know little about the source of the affliction. They are advised to use iron oxide minerals found in local streams to purify their drinking water. The rocks are crushed and packed into buckets that have a finger-sized hole in the bottom. The water is poured through the bucket and a mosquito net is used as a sieve. One bucket is good for 200 passes and the score is kept on the bucket with chalk marks. Women and children take responsibility for these precautions; the men are said to be useless.[5,154]

In the neighbouring country to the north, Burkina Faso (previously the Republic of Upper Volta), the government has recently closed access to 11 deep wells in an area where water is particularly scarce, and there is the possibility that it might have to do the same for hundreds more. This action was taken in response to the increasing number of skin diseases found to be attributable to arsenic in the water. UNICEF drilled 360 wells in the country's northern provinces and tests revealed arsenic in seven of these. New wells are scheduled to replace all those that are contaminated.[155]

Box 8.4 Morinaga Milk Poisoning

On July 23rd, 1955, 8-month-old Kiyoshi Hisayama was hospitalised at the Department of Pediatrics, Okayama University, with complaints of fever, abdominal swelling and anemia. The child soon developed fever and diarrhea and did not respond to antibiotics. Four other children were soon admitted with similar symptoms, some showing dermal pigmentation. On August 5th, someone (it is not known who) mentioned the fact that all the patients had drunk Morinaga dry milk. By August 20th, there were seven patients – one of whom died, giving the opportunity to perform an autopsy. This was not very informative, leaving mass intoxication as the only conceivable cause for the multiple occurrences of the disease. The presence of arsenic was verified by clinical tests and on August 23rd, Dr. Kanda of the Forensic Medicine Laboratory proved there was arsenic in the powdered milk. (The amount was later found to be 1.5 ppm to 2.4 ppm). At that time there were 197 patients.

The local health department announced these results to the public, after which the number of patients increased to 384. By October 11th, the number had reached 971 and 62 of these had died. BAL therapy (Section 1.11) proved to be effective in alleviating many of the survivors' symptoms. One conspicuous finding was a sudden weight gain concurrent with the beginning of treatment: "Demonstrating that the metabolism which had been destroyed, improved rapidly when the toxic action of arsenic accumulated within the body was eliminated by BAL. Such an obvious nutritional recovery has not been recorded in other diseases."[156]

The victims, who were mainly infants, had been ingesting 1.3 mg to 3.6 mg of arsenic per day. Most developed the disease after drinking at least 10 bottles of milk. Dr. Hamamoto writes: "Surprisingly enough, this is believed to indicate stronger resistance to arsenic intake than in the case of ordinary ingestion of arsenic preparations."[156]

Dr. Hamamoto published these findings in a Japanese journal in 1955 and the outside world would learn very little more for many years.[157]

The source of the arsenic was the disodium phosphate added to the cow's milk as a stabiliser prior to processing. The additive was an industrial-grade byproduct of the aluminium industry, and contained about 6 per cent arsenic as sodium arsenate. At the time, the milk company was manufacturing 200 000 cans of the dried product per month: competition was stiff, and food-safety concerns were minimal. The company showed little remorse or interest in follow-up studies of the victims, so valuable data regarding exposure were not collected to allow for dose/response studies. Dakeishi and coworkers[157] liken the company reaction to that of Chisso Corp. in 1962, following the revelation that its wastes were the cause of Minamata disease (Section 3.7). The Tokushima District Court found the Morinaga Milk Company not guilty and did not award any support to the 12 131 victims, or to the families of the 130 individuals who died as a result

> **Box 8.4 Continued.**
>
> of poisoning. These numbers had been recorded by the Ministry of Health by June 9th, 1956; however, the number of victims is undoubtedly higher because of the difficulty and cost of obtaining the analytical evidence from hair or nail samples discouraged some parents from registering their children. The disappearance of acute symptoms was judged by the authorities to indicate a complete cure.
>
> In 1969, there were reports that the victims had a higher rate of physical and mental complaints than controls, and the occurrence of central nervous system disorders, such as epilepsy and mental retardation, was elevated. A support association was established in 1974, involving the government, the company, and the victims' group. It reported in 2003 that approximately 6000 victims had established contact. Of the 798 victims who received welfare allowances from the association in 2001, 337 suffered from developmental retardation, 129 suffered from various disabilities, 103 from mental disorders, and 33 from epilepsy.
>
> Little effort was made to broadcast these results: the government was embarrassed. The authors of later studies from other parts of the world, who reported similar though less dramatic findings, were largely unaware of the Japanese experience. The dose to generate serious neurotoxic effects in the Japanese infants was estimated to be around 60 mg of arsenic from milk containing up to 7 ppm arsenic. Because similar loadings could be achieved in children drinking contaminated water in countries such as Bangladesh, the Japanese scientists wrote: "The effects of arsenic on central nervous system development therefore need to be further explored and included in future risk assessments."[157]
>
> The fatality rate of the victims has been decreasing to a normal level since the 1990s, when most of the survivors reached their 30s. However, 11 men and four women died in traffic accidents – about twice the rate of the general population, probably a consequence of vision and hearing impairment.[158]

8.10 Europe

The main arsenic hotspots in Europe are Hungary and Romania. The source is sediment deposited by water that originates in the mountains, and the groundwater arsenic levels range from less than 2 ppb to 176 ppb.[4] The problem was first detected in the groundwater in 1981. Currently, 1.4 million people are exposed to arsenic concentrations of between 10 and 30 ppb. The European Union has given Hungary until 2009 to reduce the concentration to 10 ppb. The costs are being paid for partly by the EU Cohesion fund, but it is unlikely that the deadline will be met.[159]

8.11 North America

The concentration of arsenic in water throughout North America ranges from less than 1 ppb to 100 000 ppb, the result of natural (rocks, thermal springs, closed-basin lakes) and anthropogenic (mining, pesticides, arsenic-rich stockpiles) inputs. The first report of groundwater contamination by arsenic in the United States came in 1962 from Lane County where the concentration ranged from not detectable to 1700 ppb. In general, the concentrations in natural groundwater are highest in the western US, but some parts of the Midwest and Northeast have arsenic concentrations above the 10 ppb WHO provisional guideline. The most severely arsenic-tainted groundwater is found in Arizona, California, Idaho, Minnesota, Montana, North Dakota, New Hampshire, Nevada, Oregon, South Dakota and Texas. Less than 10 per cent of the small public water-supply systems in the US were estimated to exceed arsenic concentrations of 10 ppb.[160] However, private wells that are not regulated by federal and state authorities can be a serious source of arsenic. Thus, the US Geological Survey reports that about 41 000 people in three southeast New Hampshire counties are using private wells that contain more than 10 ppb arsenic. This arsenic is naturally occurring and is a consequence of the geology of the area.[161] Mining was found to have a major impact on the stream water in Alaska, where levels of 1300 ppb of arsenic were found in 1978 and domestic wells in Fairbanks had 10 000 ppb.[162]

Levels exceed 50 ppb in many parts of Eastern Canada where 10 per cent of the wells recently sampled in Newfoundland had concentrations above 500 ppb. On Bowen Island, BC, the highest measured concentration is about 600 ppb.[163]

8.11.1 Mexico

Chronic arsenic exposure via drinking water has been reported in seven areas in Mexico: Sonora, Chihuahua, Michoacán, Morelos, Puebla, Hidalgo and Region Lagunera. It is difficult to estimate the number of people affected, but a figure of 400 000 has been suggested for Lagunera where the arsenic concentrations range from 8 ppb to 624 ppb.[4,5] The town supply of Santa Ana, Coahuila, contained 400 ppb arsenic until the early 1990s, when it was reduced to 100 ppb with a change of water supply. One study of individuals in this area with chronic arsenicosis found urinary arsenic levels ranging from 500 ppb to 700 ppb, with a mean of 560 ppb.[164]

8.11.2 The US Standard for Drinking Water

In 1942 the US Public Health Service set an interim drinking water standard for arsenic of 50 ppb which was adopted by the US EPA in 1976. The number was not chosen for any particular reason but it semed to indicate that below 50 ppb "the danger, if any, is so small that it cannot be discovered by available means of observation." There were even suggestions in 1974 that the this maximum contaminant level (MCL) be raised to 100 ppb because no adverse effects were being seen at 50 ppb. Following a comprehensive assessment, the EPA estimated an excess lifetime skin-cancer risk of 3–7 cases per 100 000 population for each microgram of inorganic arsenic per litre of drinking water – about 1–3 in a group of 1000 people drinking water containing arsenic at 50 ppb – and invested considerable effort into trying to improve the knowledge base required for future assessment.[135,165,166] The WHO used the same data, mostly from Taiwan, as the basis of its decision to lower its drinking water standard to 10 ppb, and around the same time (1993) the US Congress directed the EPA to propose a new arsenic MCL by January 2000. But this Safe Drinking Water Act, designed to control pollution in drinking water, is a little unusual in that the MCL must be established on the basis of the scientific evidence and a cost/benefit analysis. This provision was put into law to allow compromise in situations where benefits would be low and costs high.[165–167]

Because of considerable controversy surrounding the risk assessment of arsenic, the EPA asked the National Research Council (NRC) for help, which eventually came in the form of a report prepared by a committee of scientific experts with "diverse perspectives and technical expertise" convened for the purpose.[136] The selection process to join the committee was rigorous, and those that made the cut had to be committed to the task. They were paid only expenses and nothing extra: a spouse paid his or her own way if present for a rare social event. The committee was asked to review all the available relevant data but was not asked to provide a formal risk assessment for arsenic in drinking water. This committee first met on March 26th, 1997, and a final version of a report was approved late in 1998. The committee concluded there was sufficient evidence from human epidemiological studies in Taiwan, Chile and Argentina, that chronic ingestion of inorganic arsenic caused cancer of the skin, bladder and lungs. In the absence of data on the effects of consuming water at the US MCL of 50 ppb arsenic, the committee concluded that ecological studies from Taiwan provided the best available empirical human data for assessing the risks of arsenic-induced cancer.

The report was released March 22nd, 1999, and there was an immediate reaction. Politicians, especially those in states such as New Mexico (Republican Senator Pete V. Domenici) with high arsenic concentrations in their water, were up in arms, as was the American Water Works Association.[168] There were personal attacks on the committee members, suggesting that they were only second-rate scientists who had nothing better to do and needed to boost their egos.

One of the more contentious issues for the committee was the shape of the dose–response curve. Usually, toxic effects of a chemical substance are studied by dosing animals with the chemical in question and establishing whether there is an adverse reaction. The dose is reduced or increased and again the response is noted. In this way a graphical plot of dose *versus* response can be obtained – the dose–response curve. Some chemicals give zero response at a particular concentration, known as the No Observed Adverse Effect Level (NOAEL), and

Accidental Exposure to Arsenic: The Law of Unintended Consequences 381

this dose – determined from animal studies – is converted to a much lower reference dose for the protection of human health. In the case of a chemical that is a known carcinogen, it is assumed there is no NOAEL and the particular carcinogen is considered to be hazardous at all concentrations (chloroform is one exception to date).[169a] Inorganic arsenic compounds are known to be carcinogenic in humans, but no laboratory animals are affected in the same way as humans, so the only way to get a dose–response curve for arsenic/human interaction is to study humans. The NRC committee used the male bladder cancer results from Taiwan and developed a mathematical model to fit the data, thus creating a dose–response curve in which the 1 per cent response matched a 400 ppb concentration of arsenic in water. To estimate the response at lower concentrations, for which there were not enough good data, a straight line was drawn from the 1 per cent point at 400 ppb to zero. The response at 50 ppb can then be ascertained from the line. The procedure is conservative and well established, although not universally accepted.[136,165,166]

The following eye-catching statement is found in the executive summary of the report: "For male bladder cancer, a straight line extrapolation from the 1 per cent point of departure yielded a risk at the MCL (50 ppb) of 1 to 1.5 per 1000. Because some studies have shown that excess lung cancer deaths attributed to arsenic are two- to fivefold greater than the excess bladder cancer risk, a similar approach for all cancers could easily result in a combined cancer risk on the order of 1 in 100."[136] The consensus of the committee is that the MCL of 50 ppb "does not achieve EPA's goal of public health protection and, therefore, requires downward revision as promptly as possible." The risks were found to be comparable with those associated with environmental tobacco smoke. The committee was not asked to make any suggestions about an appropriate MCL. Two members of the committee later expressed the view in public that there was no need to lower the 50 ppb limit.[169b]

8.11.3 Setting the US MCL

The Safe Drinking Water Act directs the EPA to proceed in three steps. First, the EPA is asked to set "Maximum Contaminant Level Goals" (MCLGs) for water pollutants. The goals must be set "at the level at which no known or anticipated adverse effects on the health of persons occur and which allows an adequate margin of safety." In practice, this statutory standard will frequently call for an MCLG of zero, because many contaminants cannot be shown to have safe thresholds, and because the "adequate margin of safety" language will, in these specific circumstances, seem to support an MCLG of zero. In 2000, soon after they had digested the NRC report, the EPA recommended a zero MCLG for arsenic.

Next, the EPA is told in the Act to specify "an MCL which is as close to the MCLG as is feasible." The initial proposal for arsenic, based on the NRC report, was an MCL of 5 ppb, but the EPA asked for comments on 3 ppb, 10 ppb and 20 ppb. The 3 ppb level was considered to be technically feasible, giving hope to the environmentalists, and some 95 000 comments in support of this standard were submitted by the public.[170] On the other hand the American Water Works Association supported an MCL of no lower than 10 ppb because of the high costs of treating the water to such levels.[171] The final requirement of the SDW act is that the EPA undertake a risk assessment for pollutants, discussing the level of the danger and the costs of achieving the requisite reduction. The appropriate cost/benefit analysis was also presented in 2000.[165,166]

This cycle of reanalysis, and the consideration of new data on lung cancer, was abruptly cancelled when the Clinton administration in its final days instituted an MCL of 10 ppb.[165,166,169a] Soon after, in March 2001, President George W. Bush delayed this decision and asked for more study.[172] This action proved to be very unpopular. A survey conducted in April 2001, found that 56 per cent of Americans rejected the Bush decision, whereas only 34 per cent approved of it. *The Wall Street Journal* said "You may have voted for him, but you didn't vote for this in your water."[173] Cartoons appeared such as the ones shown in Figures 8.4 and 8.5. However, one distinguished commentator Cass R. Sunstein, Professor of Political Science and Law at the University of Chicago, argued that the backlash was largely driven by "intuitive toxicologists" who believe that any substance that causes cancer is unsafe and should be banned. He also lamented that during debate on the subject in Congress and elsewhere, one of the most successful arguments "was that other countries regulated arsenic at the level of stringency proposed in the Clinton administration." This was used as a mental shortcut, avoiding the questions about the scientific basis for these practices. The A-word substituted for a close analysis of the situation.[167]

Another NRC arsenic committee was rapidly convened, with some overlap of membership, and its prompt report concluded that the risks were greater than previously publicised. In October 2001, the EPA announced the new standard of 10 ppb would become effective in February 2002, with compliance in 2006.[167,174–176]

Information about the high cellular toxicity of methylarsenic(III) derivatives slowly surfaced while discussions about the MCL were underway, but the result

Figure 8.4 DOONESBURY (c) 2001 G. B. Trudeau. Reprinted with permission of Universal Press Syndicate. All rights reserved.

Accidental Exposure to Arsenic: The Law of Unintended Consequences

Figure 8.5 Dwane Powell editorial cartoon. Reproduced with permission of the Cartoonists Group Seattle, Washington.

that caught the eye of regulators was the discovery that these compounds showed genotoxic (*i.e.* DNA-damaging) effects.[177,178] Some believe that this last result was pivotal in the final acceptance of the MCL of 10 ppb. In recent years there have been a number of reports on finding these methylarsenic(III) species in the urine of arsenic-exposed populations, and methylation is currently regarded as a possible cause of many of arsenic's health effects (Section 1.7).

8.11.4 Cost/Benefit Analysis

The costs of this decision to move to a MCL of 10 ppb, about $210 million (US), were obtained by estimating the number of water systems, the arsenic concentration to be treated, the population served, and the cost of treatment by using the best available technology. In fact, 90 per cent of the 10 683 surface water systems in the United States have arsenic concentrations below 2 ppb, and only 0.1 per cent has more than 50 ppb. The best of the technologies recommended are ion exchange, activated alumina, reverse osmosis and coagulation/filtration. The total annual cost per household estimated to achieve 10 ppb is $357 (US) if the number of users is from 25 to 100: the cost is $18 (US) if the number of users is 100 001 to 1 million.[165,166]

The EPA calculated that, at 10 ppb, the annual total number of bladder and lung cancer cases avoided would be 37.4 to 55.7. The total quantified health benefits at 10 ppb would amount to a savings in the health care system of $139.6 million to $197.7 million, while the cost of implementing the technology would be $180.4 million to $205.6 million; so there is not an overwhelming case to be made in favour of a switch from 50 ppb to 10 ppb if cost/benefit ratios alone are considered. The EPA does point out that there are nonquantifiable health benefits, such as avoidance of skin cancer, liver cancer, cardiovascular effects, neurological effects and others.[174]

The calculation of benefits depends on the number of cases of bladder and lung cancer that have been prevented. Once established, this number is multiplied by "the value of a statistical life," which the EPA estimates to be about $5.8 million. As the Federal Register puts it, if 100 000 people would each experience a reduction of 1/100 000 in their risk of premature death as a result of a regulation, the regulation can be said to "save one statistical life, *i.e.* 100 000 × 1/100 000. If each member of the population of 100 000 were willing to pay $20 for the stated risk reduction, the corresponding value of a statistical life would be $2 million, *i.e.* $20 × 100 000."[165,166] The $5.8 million value used by the EPA is an average of a number of studies, the range being $700 000 to $16.3 million.[174] For nonfatal cancers, the multiplier was $607 000, which is the "shoppers' response to hypothetical questions about how much they would be willing to pay to reduce a statistical risk of chronic bronchitis."[167]

A few critics of the 10 ppb decision were vocal about the cost estimates (*e.g.* Gurian *et al.*, 2001);[179] however, many more were critical of the benefit estimates.[167] The linear extrapolation of the dose–response curve again came under attack (Section 8.11.2). For example, if the curve is sublinear at low doses, *i.e.* dips down rather than goes straight, the response to a given low dose will be less than that predicted by the linear model. In the extreme, this sublinear response has a threshold – meaning that there is a dose below which there is no response. According to one study, linear extrapolation predicts that reducing the MCL to 10 ppb saves about 28 lives per year in the US population but if there is a threshold, this number is decreased to five or six, leading to the conclusion that the costs would exceed the benefits by about $190 million each year – a noteworthy achievement, according to the authors with tongue in cheek.[180] Burnett and Hahn[180] argue that the regulation is likely to produce a net loss of life, rather than a gain, because expensive regulation has been found to have mortality effects by, for example, making less money available for healthcare expenditures.

Another analysis suggests that the number of lives saved can range from zero to 112 and the monetised benefits from $0 to $560 million depending on the amount chosen as the value of a statistical life. "In these circumstances there is no obvious, correct decision for government agencies to make."[167] A few years later, in 2004, the EPA was valuing the lost life of a 35-year-old at $3.7 million and that of a 70-year-old at $2.3 million. The senior discount was later dropped in response to public pressure.[181] One somewhat cynical analysis of the decision from Albuquerque, NM – the state has high arsenic concentrations in its water – suggests that the provision of bottled water by the city to meet the new MCL could lead to about 20 annual human deaths from residential falls resulting from handling and distributing the containers. This would be offset by the avoidance of less than one annual cancer death.[182]

The correct decision was obvious to some, though: "There is not much of a safety margin between the well established 1-in-10 risk of death from cancer from 500 ppb arsenic in water to the probable 1-in-100 from 50 ppb. In the light of these risks excessive debate seems unnecessary."[176]

Once the MCL was established individual states had to comply or set a lower standard. New Jersey was the first to lower its level to 5 ppb. This was to come into force at the same time as the national standard, on January 23rd, 2006. The media commented: "Build a better New Jersey," is the slogan of State Governor James McGreevey as he sets "the strongest limit on the mercury and arsenic that taints our water and land." California was preparing for a similar move when it set a public health goal of 4 parts per trillion for arsenic in drinking water.[183] Fears that small community water-treatment systems would be forced out of business because they could not ever reach the 10 ppb level seem to have been overcome by the introduction of efficient and less expensive removal technologies; however, it is less expensive to remove arsenic from New Jersey's water than from New Mexico's.[75,184] The EPA estimates that 5.4 per cent of ground water community water systems and 0.7 per cent of surface water community water systems had arsenic levels above 10 ppb.[165]

8.11.5 The MCL Revisited

When the MCL of 10 ppb came into force at the beginning of 2006, there were about 10 million Americans drinking water with arsenic above that level. Some states granted exemptions that allowed as much as an extra nine years for systems serving fewer than 3300 people to comply. However, in response to lobbying that began a few years earlier, President Bush is now proposing to permanently allow arsenic levels up to 30 ppb in some cases. Under the Safe Water Drinking Act of 1996, Congress is obliged to consider that it costs small rural towns proportionately more to meet federal drinking water standards. Under the new proposal, compliance could not cost more than $335 per household. So, if compliance is expensive and if the arsenic content is below 30 ppb, the operators of the system need do nothing.[185]

The National Rural Water Association supports the proposal (it lobbied hard for it) but others believe it is an attack on public health. Erik Olson of the Natural Resources Defense Council asserted the revisions "would weaken health protection for millions of Americans and create a two-tiered water system in the United States." The National Drinking Water Advisory Council had earlier written: "The potential acceptance of lower water quality for disadvantaged communities is ethically troublesome."[185,186]

8.11.6 Fallon, Churchill County, Nevada

One magazine article about Fallon NV, begins with the statement "The most beautiful thing about Fallon, population 8300, is how ordinary it is – how old-time and rock-ribbed."[187] But it is remarkable in at least one way: in 2001 Fallon had a greater concentration of arsenic in its drinking water, 100 ppb, than any other town its size or larger in the United States.

Since 1997, 17 children with ties to Fallon have been diagnosed with childhood acute lymphocytic leukemia, and three have died. The US Centers for Disease Control were asked to help find the cause and of the 110 chemicals they measured, only arsenic and tungsten were found to be appreciably elevated in the residents, healthy and sick. (There is a tungsten refinery in the middle of the town.) Urinary arsenic concentrations were above the reference level of 50 ppb in 29 per cent of those tested, and 5.5 per cent were above 200 ppb. But the results of a later study of about 1000 residents, announced by Dr. Rebecca Calderon of the EPA in 2005, indicated that neither arsenic nor tungsten was the cause of the leukemia cluster, which still remains a mystery.[188,189]

Shortly after the rediscovery of the high arsenic levels, the US government ordered the City of Fallon to treat its water or face a fine of $27 500 per day – $10 million per year. Residents say that the city had known about the situation for many years but the mayor and council always insisted that the water was safe to drink.[190] Documents readily available at the local library report arsenic concentrations of more than 4 ppm in the groundwater of the nearby Naval Air station. In fact, the first person to alert the outside world to the arsenic problem in Fallon, Dr. Edward Crippen, was fired years earlier, on February 26th, 1969. City officials knew but were keeping it quiet, but the firing backfired and the report became prominent news all over the nation.[191]

The local population was not worried about their health. They saw no sign of illness and had been drinking the water all their lives. The mayor said: "Arsenic isn't a health problem here. I mean, where are all the dead bodies?"[187] But he did start worrying about federal fines, so the city constructed a treatment plant, first to get the arsenic concentration below 50 ppb and then down to 10 ppb by January 2006. On January 13th, 2006, Fallon's water contained 6 ppb arsenic, for which the city's water users were paying a monthly fee of $20.44 in addition to charges for water consumption.[192] The cost of the plant was $17.5 million, funded largely by a $16 million federal grant.

The county outside Fallon has about 25 000 residents, 13 500 of whom rely on private domestic wells for their water supply. The maximum arsenic concentration in these wells is 2100 ppb and the median 26 ppb, but they are not subject to the 10 ppb arsenic rule. A survey carried out in 2002 showed that 72 per cent of the respondents consumed water from their wells and that among this group a minority applied treatment. Evidently the old-time residents, like their counterparts in Fallon, did not see the arsenic as anything to worry about. Ironically, the survey found that because treatment encouraged consumption, some individuals ended up being highly exposed because their filters did not function properly. Some of the remediation technologies could not cope with the high arsenic and iron concentrations; some were malfunctioning, and some were inappropriate for arsenic removal. The study concluded that, as in

Bangladesh, more had to be done in rural areas to educate private well owners about the risks associated with contaminants in their water.[193]

8.12 Other Small Systems

Small systems, such as the one in Fallon, that serve fewer than 10 000 people have had the most difficulty in dealing with the new MCL. One big problem has been convincing the citizens that their water needs any attention. In their minds, tens of millions of dollars were needed to build plants that offered limited health benefits. And then there were those who suddenly became aware of their health problems. A lawsuit was filed by about 300 residents of Brownsville TX, claiming their water company endangered their health by not alerting them soon enough to the arsenic in their water, which turned out to have a high of 12 ppb. "Many children have suffered rashes and stomach ailments that are unexplained except by the poor quality of the water."[194] Presumably they were in good health when the MCL was 50 ppb.

Sun Groves, a town in Arizona, saved itself $4 million by tapping into an arsenic-free hot spring. Most of the citizens were happy with this "well from hell" but it is hard to have a cold shower when the water comes out the faucet at about 40 °C. One resident complained that she had to put ice cubes in the washing machine to protect delicate fabrics.[195]

8.13 The Canadian Maximum Acceptable Concentration (MAC)

In November, 2004, a document was circulated in Canada asking for comments on a proposal to lower the non-legally binding MAC for arsenic in drinking water from 25 ppb to 5 ppb. The comment period ended May 17th, 2005. A year later Health Canada was able to announce that the MAC had been lowered to 10 ppb.[196] Costs were taken into consideration with the MAC being set as close to the health-based guideline of 0.3 ppb as possible. Benefits seem to have been left out of consideration in this relatively smooth process.

8.14 The Opportunists Knock

The Earth Foundation, a Dhaka based NGO, is currently attempting to market a "new" arsenic filter in Bangladesh. However, the local *Daily Star* reported October 28th, 2007, that the necessary verification certificates offered in support of the filter were fake and were never signed by the named organizations. Furthermore, the chairman and executive director of the foundation had a substantial criminal record. Back in the USA, society's persistent fear of arsenic is being exploited in the aggressive marking of a tabletop-style pitcher water filter that not only removes all the usual water contaminants (lead, chlorine, benzene etc) but also arsenic. "Stop Drinking Poison-How Do You Take Your

Water? With or Without Arsenic? Making Safe Water Also Makes for Career Opportunity" is the heading of one marketing media release. Another warns: "The water you are drinking now could contain 10 parts per billion of arsenic." Nonetheless, the results that are available on the Internet suggest that the filter material can remove arsenic from water that initially contained the element in the range 130 ppm to 1191 ppm to below the "detection limit."[197]

Residents of a southern suburb of Oklahoma City complained to the police after they encountered individuals on the phone or door-to-door who were claiming to be government officials asking to perform water tests or trying to sell them water filters. At the time their water was easily meeting the 50 ppb MCL for arsenic.[198]

8.15 Epilogue

Professor Chakraborti suggests that there are about 500 million people being poisoned by arsenic in their drinking water, in northern India and Bangladesh. A more conservative though still alarming estimate made in 2007 by Peter Ravenscroft of Cambridge University and Mott MacDonald International, puts the number at 140 million who live in 70 countries, with South and East Asia accounting for more than half this number. Even if the numerical truth lies somewhere in between, this is a tragedy that can only get worse as time goes on. The BBC news report was headlined "World facing arsenic timebomb".

The *Journal of Environmental Science and Health*, has recently published a special issue devoted to "Groundwater arsenic contamination and health effects in South East Asia.[199] Chakraborti was involved as guest editor for the project; however, he felt obliged to withdraw his support because the editorial board rejected four submitted papers, even though they had been accepted by the peer reviewers. Apparently titles such as 'Groundwater arsenic calamity in Bangladesh and Dhaka Community Hospital's experience" by M. Rahman and Q. Quamruzzaman from the hospital, may have been seen as a little too political for a scientific journal. It is a pity that some compromise could not have been reached because, in general, scientists are as ignorant of the human dimensions of the arsenic crisis as their fellow citizens. Everyone should know that the child drinking that water today is very likely to develop arsenicosis some time in the future – yet there is no real sense of urgency to deal with the problem. Defusing the "arsenic timebomb" is beyond the efforts of well-meaning individuals and NGOs. Massive response from both the developed world and the local governments is required immediately.

I have presented a lot of material in this book in support of the thesis that the sociochemistry of arsenic is a topic that is interesting and deserves attention in its own right. Some of this information is finding its way into the textbooks currently in use; for example, in a first year university text *Chemistry the Central Science*[200] and in the course *Science in Context*, offered by the UK's Open University, which has a very good unit on Bangladesh.[201] Perhaps the absence of the A-word in other texts is because authors are taking to heart the request from some mothers of school children in Colorado that, because arsenic is so dangerous, it should be removed from the periodic table. Whether this story from the Internet is true or not, it shines light on the main source of the problem that society has with arsenic and its compounds – ignorance—and society will be ill-served if it has to continue to learn about arsenic only from mystery writers and media reporters.

I leave it to the reader to answer the question posed in the title of this book bearing in mind the demonstrated ability of the A-word to befuddle normally rational minds.

References

1. School of Environmental Studies, Jadavpur University, SOESJU, Calcutta, 2007.
2. W. Lepkowski, in *Chem. Eng. News*, 1999, Dec. 6, p. 127.
3. WHO, *Towards an assessment of the socioeconomic impact of arsenic poisoning in Bangladesh*. WHO/SDE/WSH/00.4, World Health Organization, Geneva, 2004.
4. K. Nordstrom, *Science*, 2002, **296**, 2143.
5. P. L. Smedley and D. G. Kinniburgh, *Appl. Geochem.*, 2002, **17**, 517.
6. F. Pearce, in *New Scientist*, 1995, Sept. 16, p. 14.
7. B. Bearak, in *Globe and Mail*, 1998, Nov. 14.
8. S. Connor and F. Pearce, in *The Independent*, 2001, Jan. 19.
9. K. C. Saha, *Indian J. Dermatol.*, 1984, **29**, 37.
10. K. C. Saha, *Indian J. Dermatol.*, 1995, **40**, 1.
11. B. Sarkar, in *Chem. Eng. News*, 1998, Dec. 14, p. 8.
12. A. Chatterjee, D. Das, B. K. Mandal, T. R. Chowdhury, G. Samanta and D. Chakraborti, *Analyst*, 1995, **120**, 643.
13. D. Das, A. Chatterjee, B. K. Mandal, G. Samanta and D. Chakraborti, *Analyst*, 1995, **120**, 917.
14. R. Banerjee, in *India Today*, 1993, July 15, p. 76.
15. A. H. Smith, E. O. Lingas and M. Rahman, *Bull. World Health Org.*, 2000, **78**, 1093.
16. D. Chakraborti, M. M. Rahman, K. Paul, U. K Chowdhury, M. K. Sengupta, D. Lodh, C. R. Chanda, K. C. Saha and S. C. Mukherjee, *Talanta*, 2002, **58**, 3.
17. D. Rohde, in *New York Times*, 2005, July 17.
18. D. G. Kinniburgh and P. L. Smedley, *Arsenic contamination of groundwater in Bangladesh* WC/00/19, British Geological Survey, Keyworth, UK, 2000.
19. D. Chakraborti, M. K. Sengupta, M. M. Rahman, S. Ahmad, I. K. Chowdhury and M. A. Hossain, *J. Environ. Monitor.*, 2004, **6**, 74N.
20. A. Z. Khan, in *Chem. Eng. News*, 1998, Dec. 14, p. 8.
21. W. L. Lepkowski, in *Chem. Eng. News*, 1999, May 17, p. 45.
22. J. Montaner, in *Globe and Mail*, 2006, Dec. 1, p. A19.

23. B. Archer, in *Globe and Mail*, 2007, July 14, p. F6.
24. P. H. Patel, PhD thesis Harvard University, 2001.
25. P. Bagla and J. Kaiser, *Science*, 1996, **274**, 174.
26. J. Stackhouse, in *Globe and Mail*, 2000, April 22, p. A11.
27. A. M. R. Chowdhury, in *Sci. Am.*, 2004, Aug., p. 87.
28. J. Ahmad, B. N. Goldar, S. Misra and M. Jakariya, *Willingness to pay for arsenic-free, safe drinking water in Bangladesh.*, World Bank Water and Sanitation Program-South Asia, 2003.
29. M. F. Ahmed and C. M. Ahmed, ed., *Arsenic mitigation in Bangladesh*, Local Government Division Ministry of Local Government Rural development and Cooperatives, Government of Bangladesh, 2002.
30. A. v. Geen, K. M. Ahmed and J. Graziano, in *New York Times*, 2005, July 30.
31. W. Lepkowski, in *Chem. Eng. News*, 1998, November 9, p. 12.
32. UNICEF, Media brief. www.es.ucl.ac.uk/research/lay/as/pdf/aslayperson4.pdf. 1999.
33. World Bank, *Q&A facts and figures about the World Bank, Spring 1998*, World Bank, New York, 1998.
34. R. B. Kaufmann, B. H. Sorensen, M. Rahman, K. Streatfield and L. A. Persson, *Assessing the public health crisis caused by arsenic contamination of drinking water in Bangladesh*, World Bank, 2001.
35. S. A. Ahmad, M. H. S. U. Sayed, S. Barua, M. H. Kahn, M. H. Faruquee, A. Jalil, S. A. Hadi and H. K. Talukder, *Environ. Health Perspect.*, 2001, **109**, 629.
36. M. A. K. Barbhuiya, M. H. S. U. Sayed, M. H. Kahn, M. A. Jalil and S. A. Ahmed, *Arsenic contamination and its consequences on human health*, Local Government Division, Ministry of Local Government, Rural Development and Cooperative Government of Bangladesh, Dhaka, 2002.
37. J. M. McArthur, 2006, personal communication.
38. A. A. Meharg, C. Scrimgeour, S. A. Hossain, K. Fuller, K. Cruickshank, P. N. Williams and D. G. Kinniburgh, *Environ. Sci. Technol.*, 2006, **40**, 4928.
39. M. Lee, in *Chem. Brit.*, 1996, March, p. 7.
40. T. Clarke, *Nature*, 2001, **413**, 556.
41. M. Burke, in *Environ. Sci. Technol.*, 2001, Oct. 1, p. 400A.
42. L. J. Kennedy, *Binod Sutradhar and the Natural Environment Research Council*, 2003.
43. (a) H. M. J. Simon, *Binod Sutradhar and the natural environment research council*, Royal Courts of Justice, London, 2004.
    (b) P. J. Atkins, M. M. Hassan and C. E. Dunn, *Trans. Inst. Br. Geogr.*, 2006, **NS31**, 272.
44. Sciencescope, in *Science*, 2004, Feb. 24.
45. M. Day and B. Michalowska, *Success for Bangladeshi villagers in the House of Lords*, Press Release, Leigh, Day & Co, London, 2005.
46. M. O'Sullivan, *Bangladesh claims against the British Geological Society*, News release, British Geological Survey, Keyworth, UK, 2002.

47. W. R. Cullen, 2002–2005, personal experience.
48. BBC, in *BBC News South Asia*, 2006, July 5.
49. M. Hossain, in *SciDev.Net*, 2006, July 6.
50. D. Chakraborti, 2004, personal communication.
51. P. Smedley, *Post conference report: Experts opinions, recommendations and future planning for groundwater problem of West Bengal*. School of Environmental Studies, Jadavpur University, Calcutta, India, 1995.
52. M. Berg, H. C. Tran, T. C. Nguyen, H. V. Pham, R. Schertenleib and W. Giger, *Environ. Sci. Technol.*, 2001, **35**, 2621.
53. D. N. G. Mazumder, A. K. Chakraborty, A. Ghose, J. Dasgupta, D. P. Chakraborti, S. B. Dey and N. Chattopadhyay, *Bull. World Health Org.*, 1988, **66**, 499.
54. J. M. McArthur, 2001, personal communication.
55. U. K. Chowdhury, B. K. Biswas, T. R. Chowdhury, G. Samanta, B. K. Mandal, G. C. Basu, C. R. Chanda, D. Lodh, K. C. Saha, S. K. Mukherjee, S. Roy, S. Kabir, Q. Quamruzzaman and D. Chakraborti, *Environ. Health Perspect*, 2000, **108**, 393.
56. A. H. Smith and M. M. H. Smith, *Toxicology*, 2004, **198**, 39.
57. S. N. M. Abdi, in *Khaleej Times Online*, 2007, Nov. 1.
58. R. Carson, *Silent spring*, Haughton Mifflin, Boston, 1962.
59. K. Christen, *Environ. Sci. Technol.*, 2001, July 1, 286A.
60. M. M. Rahman, D. Mukherjee, M. K. Sengupta, U. K. Chowdhury, D. Lodh, C. R. Chanda, S. Roy, M. Selim, Q. Quamruzzaman, A. H. Milton, S. M. Shahidullah, M. T. Rahman and D. Chakraborti, *Environ. Sci. Technol.*, 2002, **36**, 5385.
61. A. v. Geen, Z. Cheng, A. A. Seddique, M. A. Hoque, A. Gelman, J. H. Graziano, H. Ahsan, F. Parvez and K. M. Ahmed, *Environ Sci. Technol.*, 2005, **39**, 299.
62. B. E. Erickson, *Environ. Sci. Technol.*, 2003, Jan. 1, 35A.
63. A. Mukherjee, M. K. Sengupta, S. Ahamed, M. A. Hossain, B. Das, B. Nayak and D. Chakraborti, *Environ. Sci. Technol.*, 2005, **39**, 5501.
64. P. T. K. Trang, M. Berg, P. H. Viet, N. V. Mui and J. R. v. d. Meer, *Environ. Sci. Technol.*, 2005, **39**, 7625.
65. B. K. Zia, *Speech of the Hon'ble Prime Minister*, International workshop on arsenic mitigation, Dhaka, 2002.
66. NAMIC, *National Arsenic Mitigation Information Centre*, www.bwspp.org/photo_albam.html.
67. B. Bearak, in *New York Times*, 2002, July 14, p. 1.
68. EPA, *Environmental pollution control alternatives: drinking water treatment for small communities* EPA/625/5-90/025, EPA, Cincinnati OH, 1990.
69. ANSTO, *Australian technology could save millions from arsenic poisoning*, Australian Nuclear Science and Technology Organization, 1998.
70. CMRI, in *CMRI Newsletter*, 2000, January, vol. 10, p. 8.
71. A. H. Khan, S. B. Rasul, A. K. M. Munir, M. Habiduddowia, M. Alauddin, S. S. Newaz and A. Hassam, *J. Environ. Health, Part A.*, 2000, **A35**, 1021.

72. S. Murcott, in *Arsenic exposure and health effects IV*, ed. W. R. Chappell, C. O. Abernathy and R. L. Calderon, Elsevier, Amsterdam, 2001.
73. S. Murcott, in *Arsenic exposure and health effects V*, ed. W. R. Chappell, C. O. Abernathy, R. L. Calderon and D. J. Thomas, Elsevier, Amsterdam, 2003.
74. M. Freemantle, in *Chem. Eng. News*, 2005, April 4, p. 12.
75. K. S. Betts, *Environ. Sci. Technol.*, 2001, Oct. 1, 414A.
76. R. K. Boerschke and D. K. Stewart, in *Technologies for arsenic removal from drinking water*, ed. M. F. Ahmed, M. A. Ali and Z. Adeel, BUET, Dhaka, 2001.
77. OCETA, *Annual report 2001–2002*, Ontario Centre for Environmental Technology Advancement, Toronto, 2002.
78. W. R. Cullen and K. J. Reimer, *Technical evaluation of environmental and human health consideration addressed during implementation of the CIDA-funded ETV-AM project*, CIDA, Ottawa, 2004.
79. DFID, *Rapid assessment of household level arsenic removal technologies. Phase II Report*, Bangladesh Arsenic Mitigation Water Supply Project, Dhaka, 2001.
80. BAMWSP, in *News Letter*, 2004, Sept.
81. Staff correspondent, in *The Daily Star*, 2004, Feb. 26.
82. J. Alam, in *Vancouver Sun*, 2006, March 8, p. A10.
83. CIDA, *List of bilateral projects in Bangladesh* BD/32265, Canadian International Development Agency, Ottawa, 2005.
84. Government of Bangladesh, National policy for arsenic mitigation and implementation plan for arsenic mitigation, 2004.
85. K. Hooper, G. Iskander, F. Hussein, J. Hsu, M. Deguzman, Z. Odion, Z. Ilejay, F. Sy, M. Petreas and B. Simmons, *Environ. Sci. Technol*, 1998, **32**, 3825.
86. A. Ghosh, M. Mukiibi and W. P. Ela, *Environ Sci. Technol.*, 2004, **38**, 4677.
87. C. Jing, S. Liu, M. Patel and X. Meng, *Environ. Sci. Technol.*, 2005, **39**, 5481.
88. CIDA, *Environmental technology verification – arsenic mitigation project-Phase II* Project Number BD/32119, Canadian International Development Agency, 2003.
89. J. Wanczycki, 2007, personal communication.
90. G. Howard, *Arsenic, drinking water and health risk substitution in arsenic mitigation: a discussion paper*, World Health Organization, Geneva, 2003.
91. Staff, in *The Independent*, 2005, Jan. 14.
92. WHO, *Water-Related Diseases Diarrhoea*, www.who.int/water_sanitation_health/diseases/diarrhoea/en.
93. M. A. Hossain, M. K. Sengupta, S. Ahamed, M. M. Rahman, D. Mondal, D. Lodh, B. Das, B. Nayak, B. K. Roy, A. Mukherjee and D. Chakraborti, *Environ. Sci. Technol.*, 2005, **39**, 4300.
94. N. S. Haq, in *New Nation*, 2005, Oct. 1.

95. Grainger Foundation, in *Chem. Eng. News*, 2005, Feb. 7, p. 5.
96. National Academy, *Grainger Challenge Prize Gold Award winner*, www.nae.edu/nae/grainger.nsf.
97. B. Haque, *Bangladeshi–American scientist wins National Academy of Engineers' prize for sustainability*, www.bafi.com, bafi@yahoogroups.com, 2007.
98. (a) H. Z. Rahim, in *The American Muslim*, 2007.
(b) S. Alam, in *The Wall Street Journal*, 2008, Jan. 3.
99. BBC World, *Shell launch "The World Challenge 2006"* Indiantelevision.com, 2006, October 14.
100. B. Halford, in *Chem. Eng. News*, 2006, Nov. 13, p. 12.
101. R. Scott, in *Sci. Am.*, 2006, Nov. 9.
102. Economist.com, in *The Economist*, 2006, Dec. 1.
103. W. Ruckelshaus and J. C. Davies, in *Boston Globe*, 2007, July 7.
104. BAMWSP, in *News Letter*, 2004, Sept.
105. J. Whitney, D. Clark, G. Breit, A. Welch, J. Yount, C. Hoard and J. Imes, in *Behaviour of arsenic in aquifers, soil, and plants: Implications for management*, Dhaka, 2005.
106. G. Howard, F. Ahmed, J. Shamsuddin, S. G. Mahmud, D. Deere and A. Davison, in *Behaviour of arsenic in aquifers, soil, and plants: Implications for management*, Dhaka, 2005.
107. W. J. Broad, in *New York Times*, 2005, July 26.
108. P. L. Short, in *Chem. Eng. News*, 2007, April 23, p. 13.
109. (a) M. M. Rahman, U. K. Chowdhury, S. C. Mukherjee, B. K. Mondal, K. Paul, D. Lodh, B. K. Biswas, C. R. Chanda, G. K. Basu, K. C. Saha, S. Roy, R. Das, S. K. Palit, Q. Quamruzzaman and D. Chakraborti, *J. Toxicol., Clin. Toxicol.*, 2001, **39**, 683.
(b) Fact sheet on arsenic No XII, Disaster Forum 2000 June phys4.harvard.edu/~wilson/arsenic/countries/bangladesh/mortoza/FACTSHT5.html.
110. J. Mercola, *Spirulina for arsenic poisoning*, www.mercola.com/2000/jul/2/spirulina_arsenic.htm.
111. P. Mallick, J. Chakrabarti, B. Guha and A. R. Khuda-Bukhsh, *BMC Complem. Alternat. Med.*, 2003, **3**, 7.
112. (a) M. V. Gamble, X. Liu, H. Ahsan, J. R. Pilsner, V. Ilievski, V. Slavkovich, F. Parvez, Y. Chen, D. Levy, P. Factor-Litvak and J. H. Graziano, *Clin. Nutrit.*, 2006, **84**, 1093.
(b) NIH News, 2007, Oct. 10. www.niehs.nih.gov/news/2007/folic.cfm.
113. CanWest News, in *Saskatoon Star Phoenix*, 2006, Nov. 28.
114. W. J. Verret, Y. Chen, A. Ahmed, T. Islam, F. Parvez, M. G. Kibriya, G. Muhammad, J. H. Graziano and H. Ahsan, *J. Occupat. Environ. Med.*, 2005, **47**, 1026.
115. World Bank, *Bangladesh-arsenic public health project*, World Bank, Washington DC, 2003.
116. S. Khan, in *One World South Asia*, 2004, Feb. 24.
117. *5th International conference on arsenic, Developing country's perspective on health, water and environmental issues*, Dhaka, 2004.

118. World Bank, *Bangladesh country brief*, World Bank, 2004.
119. Hospital, *Million dollar arsenic projects in Bangladesh: arsenic situation deteriorated in Eruani village of Laksham P. S. Comilla district from 1997–2005 Are some children lesser to God?*, Dhaka Community Hospital, Dhaka, 2005.
120. BNET Research Centre, Feb. 2006. findarticles.com/p/articles/mi_kmafp/is_200602/ai_n16069875.
121. Correspondent, in *The Independent*, 2005, Jan 16.
122. (a) J. M. McArthur, D. M. Banerjee, K. A. Hudson-Edwards, R. Mishra, R. Purohit, P. Ravenscroft, A. Cronin, R. J. Howarth, A. Chatterjee, T. Talukder, D. Lowry, S. Houghton and D. K. Chadha, *Appl. Geochem.*, 2004, **19**, 1255.
    (b) M. v. Brömssen, P. Bhattacharya, K. M. Ahmed, M. Jakariya, L. Jonsson, L. Lundell and G. Jacks, In *Behaviour of arsenic in aquifers, soil, and plants: Implications for management*, Dhaka, 2005.
123. J. W. Rosenboom, in *Behaviour of arsenic in aquifers, soil, and plants: Implications for management*, Dhaka, 2005.
124. P. Ravenscroft, R. J. Howarth and J. M. McArthur, *Environ. Sci. Technol.*, 2006, **40**, 1716.
125. Opinion, in *The Telegraph*, 2006, Dec. 16.
126. J. H. Graziano and A. v. Geen, *Environ. Health Perspect.*, 2005, **113**, p. A362.
127. M. Hossain, in *SciDevNet*, 2006, Dec. 14.
128. R. Renner, *Environ. Sci. Technol.*, 2004, May 1, p. 155A.
129. WRI, *Global study reveals new warning signals: Degraded agricultural lands threaten world's food production capacity*, News release, World Resources Institute, Washington, 2006.
130. S. Islam, in *All Headline News*, 2006, Oct. 4.
131. A. Rahman, M. Vahter, E.-C. Ekstrom, M. Rahman, A. Haider, M. G. Mustafa, M. A. Wahed, M. Yunus and L.-A. Persson, *Am. J. Epidermiol.*, 2007, **165**, 1389.
132. E. Oziewicz, in *Globe and Mail*, 2007, Jan. 13, p. A12.
133. J. Matschullat, *Sci. Total Environ.*, 2000, **249**, 297.
134. NRC, in *Drinking water and health*, National Academy Press, Washington DC, 1983, vol. 5.
135. EPA, *Special report on ingesting inorganic arsenic. Skin cancer; nutritional essentiality*, US Environmental Protection Agency, Washington DC, 1988.
136. NRC, *Arsenic in drinking water*, National Research Council, Washington, 1999.
137. W. P. Tseng, H. M. Chu, S. W. How, J. M. Fong, C. S. Lin and S. Yeh, *J. Nat. Cancer Inst.*, 1968, **40**, 453.
138. F. J. Lu, *The Lancet*, 1990, **336**, 115.
139. N. Hotta, in *Asia Arsenic Network*, 1995.
140. H. Y. Chiou, S. T. Chiou, Y. H. Hsu, Y. L. Chou, C. H. Tseng, M. L. Wei and C. J. Chen, *Am. J. Epidemiology*, 2001, **153**, 411.

141. K. Christen, *Environ. Sci. Technol.*, 2006, Sept. 1, 5165.
142. News, in *Viet Nam News*, 2006, July 1.
143. J. Bachmann, M. Berg, C. Stengel and M. Sampson, *Environ. Sci. Technol.*, 2007, **41**, 2146.
144. W. Hundley, in *Dallas Morning News*, 2005, March 4.
145. (a) L. Li, *China's rural poor see hope for drinking water*, www.worldwatch.org/node/4827.
    (b) T. Tsuda, T. Ogawa, A. Babazono, H. Hamada, S. Kanazawa, Y. Mino, H. Aoyaam, E. Yamamoto and N. Kurumatani, *Appl. Organomet. Chem.*, 1992, **6**, 333.
146. C. Hopenhayn-Rich, M. L. Biggs and A. H. Smith, *Int. J. Epidemiol.*, 1998, **27**, 561.
147. J. M. Borgono, P. Vincent, H. Venturino and A. Infante, *Environ. Health Perspect.*, 1977, **19**, 103.
148. H. V. Aposhian, A. Arroyo, M. E. Cebrian, L. MD Razo, K. M. Hurlbut, R. C. Dart, D. Gonzalez-Ramirez, H. Kreppel, H. Speisky, A. Smith, M. E. Gonsebatt, P. Ostrosky-Wegman and M. M. Aposhian, *J. Pharm. Exp. Therapeut.*, 1997, **282**, 192.
149. T. W. Gebel, in *Science*, 1999, March 5.
150. A. H. Smith, A. P. Arroyo, D. N. Mazumder, M. J. Kosnett, A. L. Hernandez, M. Beeris, M. M. Smith and L. E. Moore, *Environ. Health Perspect.*, 2000, **108**, 617.
151. A. H. Smith, M. Goycolea, R. Haque and M. L. Biggs, *Am. J. Epidem.*, 1998, **47**, 660.
152. A. H. Smith, G. Marshall, Y. Yuan, C. Ferreccio, J. Liaw, O. v. Ehrenstein, C. Steinmaus, M. N. Bates and S. Selvin, *Environ. Health Perspect.*, 2006, **114**, 1293.
153. (a) C. Ferreccio, C. Gonzalez, V. Milosavjlevic, G. Marshall, A. M. Sancha and A. H. Smith, *Epidemiology.*, 2000, **11**, 673.
    (b) S. Yang, in *UC Berkley News*, June 12. www.berkley.edu/news/media/releases/2007/06/12_arsenic.shtml
154. G. Miller, 2002, personal communication.
155. Reuters Burkina Faso, 2006, Dec. 29.
156. E. Hamamoto, *Jap. Med. J.*, 1955, **1649**, 3.
157. M. Dakeishi, K. Murata and P. Grandjean, *Environ. Health*, 2006, **5**, 31.
158. Y. Shimbun, in *Daily Yomiuri Online*, 2007, July 18.
159. J. White, in *The Budapest Times*, 2006, Jan. 16.
160. A. H. Welsh, S. A. Watkins, D. R. Helsel and M. J. Focazio, *Arsenic in the ground-water resources of the United States*, US Geological Survey, Denver, 2000.
161. D. Foster, *Report shows high arsenic in some southeast NH private wells*, US Geological Survey, 2003.
162. F. H. Wilson and D. B. Hawkins, *Environ. Geology*, 1978, **2**, 195.
163. Drinking water, *Arsenic in drinking water*, Document for public comment, Federal-Provincial-Territorial Committee on Drinking Water, Ottawa, 2004.

164. L. M. Del Razo, G. C. Garcia-Vargas, H. Vargas, A. Albores, M. E. Gonsebatt, R. Montero, P. Ostrosky-Wegman, M. Kelsh and M. E. Cebrian, *Arch. Toxicol.*, 1997, **71**, 211.
165. EPA, in *Federal Register*, 2000, June 22, vol. 65, p. 38888.
166. EPA, in *Federal Register*, 2000, October 20, vol. 65, p. 63027.
167. C. R. Sunstein, *The arithmetic of arsenic*, AEI Brookings Joint Centre for Regulatory Studies, Chicago, 2001.
168. D. Jehl, in *New York Times*, 2001, April 19.
169. (a) C. Hogue, in *Chem. Eng. News.*, 2001, May 21, p. 51.
    (b) K. G. Brown and G. L. Ross, *Reg. Toxicol. Pharmacol.*, 2002, **36**, 162.
170. *Public submits over 95 000 comments in support of 3 ppb arsenic standard*, www.environmentatrisk.org
171. K. Christen, *Environ. Sci. Technol.*, 2000, July 1, p. 291A.
172. R. Woods, *EPA to propose withdrawal of arsenic in drinking water standard; seeks independent reviews*, EPA, Washington, 2001.
173. J. Fialka, in *Wall Street Journal*, 2001, April 24, p. A 20.
174. EPA, in *Federal Register*, 2001, Jan. 2001, vol. 66, p. 6976.
175. NRC, *Arsenic in drinking water. 2001 Update*, National Research Council, Washington DC, 2001.
176. A. H. Smith, P. L. Lopipero, M. N. Bates and C. M. Steinmaus, *Science*, 2002, **296**, 2145.
177. M. J. Mass, A. Tennant, B. C. Roop, W. R. Cullen, M. Styblo, D. J. Thomas and A. D. Kligerman, *Chem. Res. Toxicol.*, 2001, **14**, 355.
178. Research watch, in *Environ. Sci. Technol*, 2001, July 1, p. 281A.
179. P. J. Gurian, M. J. Small, J. R. Lockwood and M. J. Schervish, *Environ. Sci. Technol.*, 2001, **35**, 4414.
180. J. K. Burnett and R. W. Hahn, *A costly benefit*, American Enterprise Institute-Brookings Joint centre for Regulatory Studies, 2001.
181. D. Saunders, in *Globe and Mail*, 2004, Feb. 7, p. F3.
182. B. Thompson, A. Aragon, F. Frost and B. Walters, *Arsenic exposure and health effects, Vol. IV*, ed. W. R. Chappell, C. O. Abernathy, R. L. Calderon and D. J. Thomas, Elsevier, Amsterdam, 2001.
183. Times, in *New York Times*, 2004, April 25.
184. News, in *Helenic News*, 2005 hellenicnews.com/readnews.html?newsid=1300&lang=US
185. M. Scharfenaker, in *J. Am. Water Works Assn.*, 2006, April, p. 16.
186. J. Eilperin, in *Washington Post*, 2006, April 1.
187. B. Donahue, in *Outside Magazine on line*, 2001, Feb.
188. B. Riley, *Las Vegas Review-Journal*, 2005.
189. V. Pearson, in *Lahontan Valley News*, 2006, Nov. 21.
190. M. C. Pierce, 2003, personal communication.
191. D. Myers, in *Reno News and Review*, 2007, Mar. 1, 2007.
192. B. Wasson, in *Lahontan Valley News*, 2006, Jan. 13.
193. M. Walker, M. Benson and W. D. Shaw, *J. Water Health*, 2005, **3**, 305.
194. Associated Press, in *Houston Chronicle*, 2007, April 25.

195. E. Jensen, in *The Arizona Republic*, 2007, July 18.
196. Health Canada, *Arsenic*, www.hc-sc.gc.ca/ewh-semt/pubs/water-eau/doc_sup-appui/arsenic/rationale-justification_e.html
197. www.codebluewater.com
198. Legal Briefs, *OK residents complain of arsenic solicitors*, 2006 Feb. 9.
199. *J. Environ. Sci. Health, Part A*, 2007 **42**, 1093–4529.
200. T. L. Brown, H. E. LeMay, B. E. Burnsten and C. J. Murphy, *Chemistry the central science, 11th edit.*, Pearson Education Inc. USA, 2007.
201. S. Drury, *Water and wellbeing: arsenic in Bangladesh, Science in context, topic 3*. The Open University, Milton Keynes UK, 2006.

Subject Index

Page numbers in *italic* refer to figures and tables. Page numbers in **bold** refer to topics in boxes.

Aborigines 216
  artifacts 86
Abyssinia 241–2
accidental poisoning
  arsenic trade workers 126
  arsine exposure 101–2
  chemical weapons workers 234, 243, 266
  chromated copper arsenate 69
  King George III 90–2
  Luce, Clare Booth 88–9
  Peale family 89–90
  see also drinking water; dust/gas exposure; food adulteration/contamination
accumulators/hyperaccumulators 299–300
acetarsone *23*, 29, 30
acidum arseniosum 17
aconite 167
acute promyelocytic leukemia 6, 20–2, **20–1**, 39
Adamsite 78, *222*, 228–9, *251*
  civilian casualties 234, 259
  cleanup/disposal 265, 271, 274
  post WWI production 240, 244, 245
  tear gas 255, 259
Adulteration of Foods Act 113–14
Afhanistan 275

Africa
  drinking water 376
  Morinaga milk poisoning **377**
Agency for Toxic Substances and Disease Registry (ATSDR) 287
Agent Blue 75, 257–9, 260
Agent Orange 257–60
ague 15
AIDS 30
air, arsenic in 288, 325
Albania 273–4
Albertus Magnus 1
algae
  food products 305–7
  freshwater 299
  marine 292–3
alternative therapies 6
American Civil War 82
American University Experimental Station 271–2
amoebiosis 29, 30
analysis *see* forensic toxicology; speciation; tests
analysis
  arsenic speciation 291–4, **298**
  see also forensic toxicology; tests
anemia 18
aniline dyes 106
animal feed additives 58–60
  see also horses

Subject Index

animal studies 41
animals and plants *see* living organisms
anthrosphere 313–14
anticholinesterase action 136, 249
antidotes
    bezoars and unicorn horn 205
    British Anti-Lewisite 29, 31, 41–2, **246–8**
    chelation 41–2
    clays 205–6
    iron oxide 22
    universal 166–7
antimony compounds 132, 136–7, 138, 139–40, 141, 144
Antofagasta 375
aphrodisiac 5, 6, 184
    and Styrian arsenic eaters 12, 13
Arabians, ancient 3, 19
Arctic, arsenic in **86–7**
Argentina 375
Armenia 309
Armstrong, Herbert 185–7
Arrhenal (medication) 24
arsanilic acid 30, 58
arsenic acid 66, 80, *293*
Arsenic Act (1851) 112, 176–7
arsenic eating *see* eating
Arsenic Gang, The 195
Arsenic and Old Lace 189–90
arsenic pentoxide 78
arsenic sulfide 1–2, 19
arsenic triiodide 10
arsenic trioxide 3
    in agriculture 66, 78
    baby powder 112
    and bubonic plage 63–7
    Chinese medicines 4–5, 20
    Fowlers's solution 10, 15, 16–17
    Indian medicines 6
    and leukemia 6, 20–2
    taxidermy 85, 87
    toxic action 33–4
    white smoke 2, 4, 8
    wood preservative 67
    Yellowstone production/pollution 315–18

Arsenicam Album 14
arsenicosis 34, 35–6, **36–7**
    see also toxicity
arsenilic acid 23
arsenobenzene 36
arsenobetaine 292, *293*, 294–6
    demethylation 334
    metabolites 310
    sources 296–8
arsenocholine 295
arsenolite 3, 4
arsenopyrite 8, 79, 311, 313, 324
arsenosugars 292, *293*, 294–5
    metabolites 310
    sources 296–8
arsenous acid *293*
arsine **101–2**, 109, *293*
    chemical warfare 220, 244, 245
    exposure guidelines 102
    industrial accidents 100–2
    Marsh test 173, 290
    toxicity 35, 80, 99–102, 119
arsphenamine 25, 28, 30, 32
Arthur (war gas) 244, 246
Asarco company 319–22
Asiatic Pills 5
asthma 14, 19
atmospheric arsenic 288
atoxyl 23, *23*, 24, 32
Australia 246, 253
Ayurvedic medicines 6, 205
Aztecs 13

baby powder 112
Back-to-Sleep campaign 135–7
BAL *see* British Anti-Lewisite
balneotherapy **44–5**
Baltic Sea 265
Bangladesh American Foundation Inc 368
Bangladesh Rural Advancement Committee 352–4
Bangladesh/West Bengal water crisis 305
    background and first responses 350–2
    Dipankar Chakraborti's campaign 358–60, 388
    illnesses/symptoms 351, 354–6

lawsuit against BGS **256–7**
mitigation techniques/projects 354, 362–70
  capsules/sachets 364, 368
  deep and dug wells 369–70
  Grainger Prize 352, 367–9
  membrane filtration 370
  nanoparticles 369
  sand filters 353, 368
  three-pitcher filter 353, 364
  verification/evaluation 365–7, 387
political initiatives 352, 362, 365
rainwater harvesting 370
source of the arsenic 355–6
surface water ponds 370
surveys
  ground water 351–2, 356, *361*
  preferred water supply 353–4
tube wells 350–1, 354
  testing/painting 352–3, 360–2
Battle of the Somme 220, 226
Baud's Pills 19
Bay Illness 111
Béchamp, A. 23
beer incident 120–4
beetles 61, 76–7
Begbie, James 17
Besnard, Marie 187–8
Bethel process 68
bezoars 5, 205
Biette's solution 17
bioaccesibility/bioavailability 310–12
  gastrointestinal analogues 312–13
  sequential selective extraction 312
biocide *see* OBPA
biomarkers
  hair 39–40, 148–9
    Napoleon 150–1, 153–7
  liver and kidney 185–6
  nails 40, 154
  saliva 40–1
  urine 38–9, 289, 291–2
biomethylation 116–17, 118, 140
  antimony 138–9
biosphere 289
  see also living organisms

bis(2-chloroethyl)sulfide *see* mustard gas
bitumen 327–8
Black Death *see* bubonic plague
Black Widows 187, 191–4
blackfoot disease 374
Blandy, Mary 172, 188
blistering agent *see* vesicants
Blixen, Karin 46
Blue Cross munitions 223–4, 226, 263
boll weevil 63, 66, 75
Bollstädt, Albert 1
Bolsheviks 239
bone manure **101–2**
book *see* crime fiction
Boothe, Clare 88–9
Borgia family 168–9, 199
Borodin, Alexander 46–7
bottled water 308–9
Bougrat, Pierre 201–2
Bowles Moor 272
Brinvilliers, Marie de 170, 191
British Anti-Lewisite 29, 31, 41–2, **246–8**
British Geological Survey 351, **356–7**, 358, 361–2
British Pharmaceutical Codex 18, 19, 20, 29, 30
brothels 26–7
Brunswick green 88
bubonic plague 7, 9, 13–14
  and arsenic trioxide 64–7
Bunsen, Robert 22
bust pills 18
butarsen *23*, 31

cacodyl 22
cacodyl chloride 230
cacodyl oxide 22, 230
cacodylic acid *see* dimethylarsinic acid
Cactoblastis cactorum 65–6
Cadet's fuming liquor 22, 217
calcium arsenate 62, 63, 106
Cambodia 258
Canada
  chemical warfare 246, 253
  drinking water 379, 387
  forestry 76–7

Subject Index 401

cancer 16, 18, 21, 40
  animal models 41
  and arsenicosis 34, 35–6, **36–7**
  in Bangladesh 351, 354
  lung 36, 330
  nasal 35, 324
  orpiment cure 7
  see also leukemia
candles 114, 199
canec **74–5**
Cangrande della Scala 169
car batteries 101
carbon monoxide 250, 254
carrageen 305, 306
CBR munitions 221
Chagas disease 32
Chakraborti, Dipankar 358–60, 388
Challenger, Frederick 116–18, 118, 138
Challenger pathway 39, **116–17**, 118, 296, 333
Chapman, A. Chaston 289–90
Chapman, George 84
Chatterton, Thomas 176
chelation 41–2, 43
chemical warfare 215–77
  ancient formulations 215–16
  attacks post WWI
    Germany in Poland 250
    Iraq in Iran 274–5
    Italy in Ethiopia 241–2
    Japan in China 242–4, 272–3
    Middle East conflicts 275
    Spain in Morocco 240–1
    tear gas/riot gas 256, 259, 275
    USA in Vietnam 259
  chemical agents
    listed/described *222*, 248–50, *251*, 255–6
    toxicity 232–3, 249
    see also Adamsite; Lewisite; mustard gas; nerve gas; tear gas
  chemical disarmament 261–2
  definition 215
  human trials
    Allies 227, 252–3
    Germany 254

    Japan 243, 254
  stockpile disposal
    Albania 273–4
    China 272–3
    Europe 264–6, 270–1, 272
    Japan 266, 272–3
    post-war dumping/cleanup 263–7
    Russia 265, 267–8
    United States 266–7, 268–70, 271–2
  Vietnam War 257–60
  weapons conventions 217, 261–2
  World War I 217–21
    battlefield tactics 225–6, 228
    casualty figures 233–4
    compensation for victims 237–8
    debates about/public reaction 234–40, 252
    early developments 217–21
    post war cleanup 228, 263–4
    quantities deployed 233
    United States entry 226–7
  World War II, buildup for
    Britain/Europe 244–5
    Canada 246
    Germany 240
    Russia 245
    United States 245–6
Chemical Weapons Convention 261–2, 264, 267, 270, 276
chickens *see* poultry
Children's Safe Water Fund 368
Chile 375–6
China
  drinking water 356, 374–5
  Japanese invasion 242–4
Chinese medicine 3, 4–5, 19, 205
  and acute promyelocytic leukemia 6, 20–2, **20–1**
chlorine gas 218–19, 230, 238
chloroacetophenone *see* CN gas
o-chlorobenzylidinemalanonitrile *see* CS gas
chlorodimethylarsine 22
chlorodiphenylarsine *222*, 223, 240
β-chlorovinyldichloarsine *see* Lewisite

Christy, Agatha 190
chromated copper arsenate 56, 68–72
  alternative treatments 74
  bans and precautions 71–2
  poisoning incidents 69
  treated wood disposal 72–4
chromatography **291**, 292
church picnic 201
Churchill, Winston 244–5, 251
cigarettes 300
circuses **83–4**
Clark, Sally 130–1
clays, protective 205–6
Clemens' solution 17
CN gas 251, 255, 256, 259, 271
coal 287, 324–5
  tar ponds 328–9
coal tar dyes 106
coccidiosis 58
Codex 18, 19, 20, 29, 30
codling moth 61, 62
coffee, poisoned 200, 201
coke, gas 122–3
Cold Lake 327
Colombia 258
Colorado potato beetle 61
Comprehensive Environmental
  Response, Compensation, and
  Liability Act *see* Superfund
condom, poisoned 198
consumption of arsenic, world **56–7**
copper acetoarsenite *see* Paris green
copper arsenite *see* Scheele's green
Cordoba 375
Cordon Rouge 263
cot death *see* sudden infant death
  syndrome
cotton 63, 66, 75
Cotton, Mary Ann 192–3
CR gas 256
crabgrass 76, 321
crib death *see* sudden infant death
  syndrome
crime fiction 8, 34, 84, 176, 190–1
CS gas *251*, 256, 259
curry, poisoned 199–200

cyanodiphenylarsine 225, 234, 240
Cyprus 256

Dalit organization 369
Darwin, Charles 43–4
Davis, Denton 143
defoliants 257–60
Delwar Hosen **363–4**
demons 9
dentistry 19
Department for International
  Development (DFID) 365
DFMO 32, 33
Dhaka Community Hospital 359
dibromomethylarsine 225
dichloromethylarsine *222*, 225, 244
Dick (war gas) *222*
dimercaptopropanesulfonic acid
  (DMPS) 42, 248
2,3-dimercaptopropanol *see* British
  Anti-Lewisite
dimercaptosuccinic acid (DMSA) 42,
  248
dimethylarsine 119, *293*, 310
dimethylarsinic acid 29, 40, 41, 75–6,
  258, *293*
  discovery (cacodylic acid) 24
Dinesen, Isak 46
dioxin 257, 259
Dirofilaria immitis 60
disodium methylarsonate (DSMA)
  75, 76
Donovan's solution 10, 18
Douglas fir 299
Dragon Boat Festival 4
DRES establishment 246
drinking water 14, 36, 40, 41, 76
  Africa 376
  Argentina 375
  Bangladesh/West Bengal *see*
    Bangladesh/West Bengal
  bottled water 308–9
  Canada 379, 387
  and chemical warfare 241, 273
  Chile 42, 375–6
  China 374–5

Hungary 378
India 358, 360
International Standards 349–50
Mexico 379
mitigation technologies 364–70
Nepal 374
Romania 378
United States 36, 302–3, 379
    see also Safe Drinking Water Act
Vietnam 374
Dulce et Decorum Est 237
dumping at sea 263–6
duplicate plate studies 304–5
dust/gas exposure 107–10
    see also Gosio gas
dyes see pigments

eating arsenic
    dose rate 11, 12
    perceived benefits 8, 12–13
    Styrians 7–13, 84
    sudden withdrawal 197–8
    Victorian England 180, 184
Ecklonia radiata 292
ecosystems 76, 287
eczema 6
Edgewood Arsenal 227, 239
EDTA therapy 43
eflorithine 32, 33
Egypt 275
Ehrlich, Paul 24–5, 28, 29
Eisenhower, General D. 251
elemental arsenic 1–2, 57
Eliot, George 43
Elizabeth I of England 198
embalming 3, 81–4
emerald green 88, 90
environmental arsenic
    naturally occurring 287–90
    see also drinking water; living organisms; pollution
erosion 332
Esmarch's solution 17
essential nutrient, arsenic as 300, 303
Ethiopia 241–2
Everett smelter 319–20

Fallon 385–7
Farris, Hazel 84
feed additives see animal feed
female yellow 4
fentanyl 262, 276
ferns 289, 299–300
fertilizer 78–9
fever 15
fiction, crime, see also crime fiction
filters, drinking water 353, 364, 368
fire retardants 132, 136–7
fish 289
    poisoned 5
fishing 266
Flanagan, Catherine 193–4
Flaubert, Gustave 176
Flin Flon smelter 318
flypapers 66–7, 183, 196
food and drink
    adulteration/contamination 38, 104, 113–14
        beer incidents 120–3
        deliberate poisoning 197–8, 199–202
        Morinaga milk powder **377**
        sugar syrups 123
        wine 124
    algal food products 305–7
    edible plants 299
    effect of cooking 304
    essentiality of arsenic 300–1
    market basket surveys 302–4
    recommended arsenic limits 289
    rice 302, 303–4
    seafood 289–90, 302
food tasters 167, 206–7
forensic toxicology **169–70**
    illustrated by chronology of crime 171–87
    neutron activation analysis **148–9**, 154, 156–7
    see also biomarkers
Forest of Dean 115–16
forestry 76–7
Forshufvud, Sten 147–8, 150–2, 154–5, 160
Fort Conger **86–7**
Fort Detrick 257

Foulkes, Charles 218, 231, 235
Fowler, Thomas 15
Fowler's solution 10, 13–19, 20, 35
Franco, General Francisco 240
French poisoners 170–1
Fries, Amos 235
fumo mortale 168
fungus *see* microbial action

Gaddafi, Colonel M. 275
gallium arsenide 57, 79–80
Ganges River 350, 355
garlic smell 16, 110, 111, 115, 119, 173
gas chambers 231, 250, 254
gas mask/respirator 219, 230, 233, 246
gas warfare *see* chemical warfare
gas/dust exposure 107–10
   see also Gosio gas
Gassed (painting) 236
Gay's solution 19
Geneva Convention/Protocol 239, 257, 274
   and tear gas 256, 261
George III of England 90–2
Ghana 376
Giant mine *see* Yellowknife
Gilligan, Amy 189
glass manufacture 80–1, 332
glyphosate 75, 258–9
gold mining 314, 315–17, 376
   micro-organisms 334
   and prospecting 314–15
golf courses 76
gonorrhea 26–8
Gosio, Bartolomeo 110
Gosio gas 110–12, 113
   composition 115, 116
   toxicity questioned 118–20
Grainger Prize 352, 367–9
Grandmothers of Nagyrev 195–6
grapefruit 62
Graves, Robert 27
Great Consols mine 289
Great Slave Lake 316, 317
Greeks 2–3, 167, 215
Green Cross munitions 223, 226

Guizhou Province 326
Gulf War 274–5
gulls 67
gunpowder 215–16
Gutzeit test 173
gypsy moth 61

Haber, Fritz 218, 228, 235
   biography **229–32**
Haber, Ludwig 232, 235, 250
Habla massacre 274
Hague Convention 217, 218, 220
hair 39–40
   King George III 90, 91
   Napoleon 150–1, 153–7
   neutron activation analysis **148–9**
Halabja 241
Haldane, J. B. S. 235
Hall, David W. 270
harassing agent **255–6**
Hardaker, William 114–15
Harle's solution 17
Harmatz, Joseph 250
Harvey, Donald 197
Hassall, Arthur 113
Hata, Sahachiro 25
Hatton, F. A. 83
Hawaiian canec **74–5**
Hayashi, Masumi 200
heartworm 60–1
hemoglobinurea 99, 100
herbal medicine 3, 5
herbicides 56, 57, 61, 67, 78
   crabgrass control 76, 321
   prickly-pear poison 65, *66*
Higgens, Margaret 193–4
hijiki 305–7
Hindle, W.V. 216
hittrich (smoke) 8
Hoch, Johann 195
Hoechst Dye Works 24, 25
Hohenheim, Theophrastus von 6–7
Holmes, Thomas 82
homeopathy 14
homicide *see* poisoners
horses

Subject Index 405

arsenic feeding 9, 10, 47
   Phar Lap **10–11**
Houdini, Harry 207
Hoxha, Enver 273
Huai Nan Tzu 3
human trials
   chemical warfare 227–8, 243, 252–3, 254
   pesticides 255
Humbug Billy 114–15
Hungary 378
Huo Yuan Jia 207
Hussam, Abul 368
Hussein, Saddam 274
Huxley, Thomas 43
hydride generation 290, 291
hydrogen 100, 102
hydrosphere 288
hyperaccumulators 299–300
hyperkeratosis 36, 37

India
   drinking water 358, 360
   medicines 5–6
   use of poisons 167
inorganic arsenic 79–80
   in food *302*
   ingested from environment *303*
   metabolism of 38–9, 310
   minerals 1–6, 78, 313–14, 333–5
   see also mineral medicine
insecticides 56, 57, 61–3, 66, 67, 78
insurance policies 195
International Standards for Drinking Water 349
Iran 274
Iraq 274, 275
iron oxide 22, 114
iron pyrite *see* pyrite
Ironite fertilizer 78–9
Ishii, General Shiro 243, 254
Israeli–Palestinian conflict 256
Italy
   Ethiopian invasion 241–2
   poisoners 168–9
Iwo Jima 251–2

Jack the Ripper 185
Jame's Fever Powder 91
Japan
   attack on China 242–4
   chemical weapon disposal 266
   human testing 243
   and Pacific War 251–2
   poisoned curry 199–200
   Tokyo subway attack 276
JBR munitions 221
Jesuit's bark 14–15
Jewish terrorists **250**
Johnson Atoll 269
Johnson grass 76

kelp 306, 307
Kelvin Commission 111–12, 114
Kempenaar Valley 331
Kenna, W. M. 105–6
khaki disease 120
kharsivan 28
King, Henry 112
King, William 176
King's yellow 87
kombu 305
Kushtay medicines 5–6

La Cantarella 168
La Spara 169
La Toffana 169
La Voisin 170–1
lachrymators *see* tear gas
Lafarge, Marie 174–6
Lake Roosevelt 321–2
lakes and rivers 288
landfill 332
L'Angelier, Emile 178, 181
Laos 258
Lawrence, D. H. 45
lead arsenate 61–3, 330
League of Nations 241, 242
Legge, T. M. 100
legislation 289
   Arsenic Act 112, 176–7
   Safe Drinking Water Act 379–81
   Sale of Food and Drugs Act 123–4

Leonardo da Vinci 168, 216
Leptospirillun ferrooxidans 334
leukemia 6, 18, 19
    treatment for APL 20–2, **20–1**, 39
leukomelanosis 37
Lewis, Winford Lee 224, 227, 235
Lewisite 31, *222*, 224–5, 227–8, 232, 233, 238
    disposal/cleanup 266, 267–8, 270, 271, 273, 274
    post WWI use/production 242, 243, 244–5
Limerick Report 136, 137–8, 139–40
Lincoln, Abraham 83
lips, poisoned 199
liquor arsenii 17
Litvinenko, Alexander 202
liver 185–6
living organisms 289–301
    algae 292, 299
    arsenic accumulators/hyperaccumulators 299–300
    ecosystems 76, 287
    fish 289, 294
    marine invertebrates 289–90, 292, 294
    mechanism of action of arsenicals **30–1**
    terrestrial plants and animals 294–6, 299–300
    see also speciation
Livingstone, David 24
London purple 63, 106
Luce, Clare 88
Lumbicus rubellus 289
Lyford, Benjamin 83

McCurdy, Elmer 83–4
Mace 256, 260
Madame Tussaud's Waxworks 189
magic bullet 24, 29
magical properties 3, 9, 10
Maimonides, Moses 205
malaria 4, 24
male yellow 3
Manchester beer incident 120–4
marine algae 292–3
market basket surveys 302–4

Marsh, James 38, 173–4
Marsh test 12, 38, 100, 173, 290
    Madeleine Smith case 178
    Marie Lafarge case 175
Mason, George 368
mass poisonings 199–202
Materia Medica 3, 5
Maxwell-LeFroy, Harold 238
Maybrick, Florence 179–80, 183–5
Meadows, Roy 130
Medici family 168–70
medicinal uses *see* mineral medicine; organoarsenicals
Mees' Lines 40
melanosis 36, 37
melarsomine *23*, 61
melarsoprol (Mel B) *23*, 30, 31, 32, 33
Mercola, Joseph 143
Merk Index 19
metabolism 292
    arsenobetaine and arsenosugars 310
    inorganic arsenic 38–9, 310
    microbial 333–5
methylarsenicals 22–4, 29, 39, 41, 290
    and Challenger pathway **116–17**, 118–19, 296, 333
    in the environment 290, *293*, 333
methylarsine *293*
Methyldick (war gas) *222*, 225
microbial action 110–13, 116–17, 119, 120
    arsenic oxidation/reduction 333–5
    and cot death theories 131–3
    possible metabolites 138
Midgely, Thomas 191
Minamata incident 117–18
mineral medicine
    acute promyelocytic leukemia 6, 20–2, **20–1**
    historical review 2-20
        Fowler's solution 13–20
        Styrian arsenic eaters 7–13
minerals, arsenic containing 2–6, 313–14
    microbial action on 333–5

Subject Index

mining and smelting 2, 8, 287, 314–15
   claim against the Cobalt Smelting Co. **335–8**
   Everett and Tacoma smelters 319–21
   Flin Flon Manitoba 318
   Giant Yellowknife mine 315–18
   Globe smelter 321
   Great Consols mine 289
   Iron King Mine 78
   mine waste/tailings 78–9, 313, 314, 323
   nickel arsenide 323–4
   Teck Cominco Ltd 321–2
   use of microbes 334
mispickle 8, 313
Mithridates VI 106
monitored natural attenuation 331–4
monomethylarsonic acid 24, 29, 75–7, *293*
monosodium methylarsonate (MSMA) 75–6
Montvoisin, Catherine 170–1
Morocco 240–1
Morris, Malcolm 107, 112
Morris, William 106, 124–6
mosquitoes 60, 61
Mott MacDonald International 351, 361, 388
moulds *see* microbial action
Mozart, Wolfgang Amadeus 208
mummies 2, 81
   and circuses **83–4**
Munster (Germany) 270–1
murder *see* poisoners
museums 85–6, 105, 159
mushrooms 295, 296
music and opera 191
Mussolini, Benito 241
mustard gas 221–3, 232, 233
   cleanup/disposal 263, 265, 266, 268, 271, 273, 274
   post WWI use 241–4
mycophenolic acid 120
myeloma 21, 22

Nagyrev, Grandmothers of 195–6

nails 40, 154
nanoparticles 369
Napoleon I of France 106, 120, 145–61
   autopsy 146–7, 160–1
   hair analysis 150–1, 153–7
   poisoning hypotheses 147–51
      conspiracy theory 151–2, 153
   possible arsenic sources
      arsenical straws 160
      self medication 158
      smoke and preservatives 158–9
      wallpaper 145–6
      wine and water 157–8
   preservation of the corpse 83, 152–3
Natural Environment Research Council **356–7**
natural gas 325
neoarsphenamine 28, 29
neokharsivan 28
neosalvarsan *23*, 26, 28
Nepal 374
Nernst, Walther 218, 230, 231, 232
nerve gas 240, 245, **248–50**, 259
   disposal 269, 270
   in Tokyo 276
   in Yemen 275
neutron activation analysis **148–9**, 154, 156–7, 188, 207–8
nickel arsenide 323–4
Nicotina tabacum 300
Niewland, J. 227A. 227
Niu Huang Jie Du Pian 5, 20
nori 305
Northern Ireland 256
nuclear weapons 276–7
Nuremburg code 254

OBPA biocide 78, 132, 133, 134, 145
oceans and seas 288
   dumping 263–6, 266–7
OCETA 365–7
O'Day, Marie 84
Oetzi body 2
oil 287
oil sands 327–8
Ojo Caliente spa 45

Okunoshima 242, 243
Ontario Centre for Environmental
   Technology Advancement 365–7
opera and musicals 191
Operation Davey Jone's Locker 264
Operation Trail Dust 257
Orfila, Mateu Joseph 175, 176
organoarsenicals
   chemical structures 23, 293
   historical review 22–8
      atoxyl 23–4
      methylarsenicals 22–4, 29
      salvarsan and derivatives 24–8
   method of action **30–1**
   post WWI developments 28–31
      sleeping sickness 31–3
   use as poisons **201–2**
   warfare agents 222, 232–3
   see also speciation
orpiment 2, 4, 87, 313
   in Ayurvic medicines 6
   Paracelsus' cures 7
osmoregulators 296
Owen, Wilfred 237
10,10′-oxybisphenoxarsine see OBPA
   biocide

Pacific War 251–2
paddy fields 305
paints 87–9
Palmer, William 85
Paracelsus 6–7
pararealgar 87
Paris green 61, 88, 103, 111
patents, German 27–8
PAX crabgrass control 321
Peale, Chares Willson 89–90
Peale, Raphael 90
Pearson's solution 17
pedosphere 288
penicillin 29
Penicillium breviacaule 111
Penny, Frederick 178–9
pepper spray 256
peppers, arsenical 326
Periodic Table 1, 79

Perkins, William 217
Peruvian bark 14–15, 16
pesticides 56, 57, 61–3, 65, 66–7, 287
   bans on 69
   demethylation 334–5
   dimethyl arsenic acid 75–7
   Hawaiian canec **74–5**
   monomethylarsonic acid 75–7
   museum objects 85–6
   soil replacement 329–30
   and tobacco 300
   see also chromated copper arsenate
petards 216
PH helmet 219
Phar Lap **10–11**
pharmaceuticals see mineral
   medicine; organoarsenicals
phenarsazine chloride see Adamsite
phosgene 220, 221, 223, 230, 232,
   240, 245
   cleanup/disposal 272
physiological based extraction test
   312–13
phytoremediation 299, 300
picnic, church 201
pigeons 67
pigments and dyes 56, 87–8, 101,
   103–6
   aniline dyes 106
   dust/gas exposure problems 107–10
Pill of Marvelous Application 4
Pill of Six Miraculous Drugs 5
pine beetles 76–7
Pine Bluff 245
pishuang 4
plague see bubonic plague
plants and animal see living organisms
poison book 112, 177
Poison Formulas 169
poisoning and poisoners 166–211
   ancient times 167
   antidotes and protection 205–7
   biggest villains 192
   chronology of crimes/forensic
      science 171–87, 203–4
   delivery systems 197–9

mass/public poisonings 199–201
organoarsenicals **201–2**
political poisoning 202–3
profile of a poisoner 191–2
public perception/media representation 189–91
serial killers 85, 192–7
see also accidental poisoning; chemical warfare
Poisons Affair, The 170–1
pollution
  cleaning up 329–31
    micro-organisms 333–5
    monitored natural attenuation 331–2
  coal 324–5, 328–9
  from mining and smelting 315–24, 334
  from pesticides/herbicides 62, 67, 321, 329–30
    Hawaiian canec **74–5**
    paddy fields 305
    PAX crabgrass control 76, 321
  natural gas 325
  oil sands/bitumen 327–8
  Sydney Tar Ponds 328–9
  see also under chemical warfare
popes 168, 199
Popova, Alexe 194
Porton Down 220, 244, 254, 257
Potter, Dr 16
pottery 88
poultry
  feed 58–9
  litter 59–60
Practice of Pharmacy 17
prickly-pear cactus 65–6
Priyanto, Pollycarpus 202–3
production of arsenic, world **56–7**
psoriasis 6
Pteris vittata 289, 299
PVC biocide *see* OBPA
pyrite 102, 121, 314, 324, 325

rabbits 65
rat poison 64

realgar 2, 20, 87
  in Ayurvic medicines 6
  chemical formula 3, 313
  in Chinese medicine 3–4, 5, 6
  as pesticide 61
  wine 4, 88
red arsenic (sulfide) 3
Redstone arsenal 245–6
registered sales 112, 177
regulation see legislation
Reinsch, Hugo 174
Reinsch test 174, 178, 179
  and impure copper 182
Renczy, Vera 194
resistered sales 177
respirator/gas mask 219, 230, 233, 246
rice 303, 303–4, 304–5
Richardson, Barry 131-4, 136, 141, 144-5
ring, poisoned 199
riot control/vomiting agents *222*, 232, 256, 259
  and Chemical Weapons Convention 262, 276
riots, civil 256, 260
rivers and lakes 288
Rocky Mountain Arsenal 268–9
Roman poisoners 167
Romania 378
Ronaldsay sheep **307–8**
Roosevelt, Franklin D. 251, 257
roxarsone *23*, 58–9
Royal Oak Mines Inc. 317
Russia 245, 267–8

S-adenosylmethionine (SAM) 117, 296, *297*
S-dimethylarsinoglutathione 21
Safe Drinking Water Act 379–81
  setting Maximum Contaminant Levels (MCL) 381–3
    cost/benefit analysis of MCL 383–5
    small systems 387
    Fallon 385–7
    water filters 388–9
Saginaw Plate Glass company 332
Saha, K. C. 350

Sale of Food and Drugs Act 123–4
saliva 40–1
salvarsan *23*, **25–6**, 28
   deliberate poisoning 201–2
   Ehrlich's work on 25
   mechanism of action 30–1
   venereal disease in WWI 26–8
San Pedro 375
Sargent, John Singer 235, 236
sarin 249, 259
   disposal 269, 270
   Tokyo subway 276
Sayers, Dorothy 8, 190
Scheele, Karl 88, 99
Scheele's green 19, 88, 90, 103–4, 105, 114
   microbial action on 111
Schweinfurt green 88, 103, 111
Scopulariopsis brevicaulis 111, 116, 118
   antimony biomethylation 138
   and cot death 132, 134, 135, 139, 145
seafood 289–90, 304
   speciation studies 291–5
seas *see* oceans
seaweed 305–6
   Ronaldsay sheep **307–8**
self-mummification 81
semiconductors 79–80
sequential selective extraction 312
serial killers 85, 192–7
sheep dip 333
sheep, seaweed eating **307–8**
sheep skins 141–3
sheep's bit 299
shellfish *see* seafood
Shipman, Harold 197
sick-building syndrome 120
Sin Lak 5
skin keratosis 34, 35–6
skin treatments 2, 6, 18, 24, 35–6
   cosmetic 180, 181
sleeping sickness 24, 30, 31–3
   arsenical action mechanism **30–1**
smelting *see* mining and smelting
Smethurst, Thomas 181–3
Smith, Alan 351, 376

Smith, Madeleine 178–81, **178–9**
Socrates 167
sodium arsenite 41, 67, 85
sodium borohydride 290
soil
   arsenic content 288
   contamination *see under* pollution; chemical warfare
soman 240, 249
Sommer, Cynthia 189, 203–4
South Africa
   protests 256, 275
   water 275
Spain 240–1
Spara, Hieronyma 169
sparrow 295
spas 43–4, **44–5**
speciation
   analytical techniques 290–1, **298**
   seafood studies 291–4
   terrestrial *vs* marine organisms 294–6
Spilsbury, Bernard Henry 185, 186
Spring Valley 271–2, 276
springs 43–4, **44–5**
Sprott, T. J. 140–1, 143, 144–5
spruce beetles 77
Squyer, Edward 198
S.S.Vaderland 101
Stachibotrys chartarum 120
stibene 132, 134, 137
stockpiles *see under* chemical warfare
Stolzenberg, Hugo 240
Styrian arsenic eaters 7–13, 84
Styrian defense, The 11, **179–80**, 181, 184
Sudan 275
sudden infant death syndrome 130–45
   Back-to-Sleep campaign 135–7
   sheep skins 141–3
   toxic gas hypothesis 131–4, 137
      Limeric Report 136, 137–8
      summing-up of evidence 144–5
      support for 140–1, 143
   TV coverage 136–7, 140
Suffield Defense Research Establishment 246
sugar-cane 74–5

Subject Index

sulfactin 248
sulfanilamide 29
sulfathiazole 29
sulfuric acid 100–1, 102, 121
  Bangladeshi attacks **366**
Superfund 313, 319–20, **320–1**, 322, 329
Sutradhar, Binod **356–7**
Swango, Michael 196–7
Swann Park 330
Sydney Tar Ponds 328–9
symptoms of poisoning *see* toxicity
syphilis 7, 16, 18, 24, 25, 28, 29
  AIDS compared 30
  and Karen Blixen 46
  Tuskegee sharecroppers 29
  in World War I 26–8

tabun 240, 349
Tacoma Smelter 319–20
tailings/mine waste 78–9, 323
Taiwan 373–4
tar ponds 328–9
taxidermy 85–7
Taylor, Alfred Swaine 17, 107, 177, 182
Taylor, Zachary 207–8
Tchaikovsky, Pyotr Ilyic 209–10
tear gas *251*, **255–6**
  civil protests 260
  and Geneva Protocol 261, 276
  WWI 217, 218, 219, *222*, 226, 232
termites 64–5
Terra Sigillata 205
terrorists **250**, 276
tests for arsenic
  Gosio test 111
  Gutzeit 173
  Marsh test 173, 290
  sight and smell 111, 172, 173
  X-ray fluorescence 86
  see also biomarkers; forensic toxicology; speciation
tetramethylarsonium ion *293*
tetramethyldiarsine 22
Thalib, Munir 202–3
Tharix maioni 289
Theriac of Andromachus 166, 170

thiacetarsamide *23*, 61
Thiobacillus ferrooxidans 334
Third Reich 240
three-pitcher filter 353, 364
Tintoretto 87
tobacco 12, 14, 300
Toffana, Giulia 169
Toogood, F. S. 186
toxicity 14, 17, 33–5
  animal studies 41
  arsenicosis and cancer 34, 35–6, **36–7**
  arsine 80, 99–100
  bioavailability/bioaccessibility 310–12
  Gosio gas 118–20
  see also antidotes; biomarkers
Toxicity Characteristic Leaching Procedure 311, 325, 366
toxicology *see* forensic toxicology
Treaty of Strasbourg 217
Treaty of Versailles 231, 239
tree sparrow 295
Treponema pallidum 25, 29
Tridacna maxima 294
Trilon 300 240, 244
trimethylantimony 138, 144
trimethylarsine 116, 119, 139, 144, *293*
trimethylarsine oxide *293*
Trisenox 21
trypanosomes 24, 30, 31–3
tryparsamide *23*, 29, 32
tryparsone *23*, 29
Tschudi, T. von 7, 11
tsetse flies 24, 31
tube wells 350–4, 360–2, 374
tuberculosis 29
Turing, Alan 210
Turner Commission 134–5

ulcers 2, 16, 18
UNESCO 367
UNICEF 350, 354, 376
unicorn horn 205
United Nations Children's Fund 350
urine 38–9
  speciation studies 289, 291–2
uses of arsenic, global **56–7**

vaginitis 29, 30
Vashon Island 299
venereal disease 5
    World War I 26–8
    World War II 29
    see also syphilis
"verdant assassin" 112–13, 114, 120
Versailles Peace Treaty 231, 239
vesicants *222*, 224–5, 232
    ointments against 247
    see also mustard gas
veterinary products 60–1
Vietnam War 257, 261
    Adamsite and tear gases 259
    Agent Blue and Agent Orange 257–9, 259–60
    public reaction 260
Vietnamese drinking water 355, 358
Vincennite (VN) 220
volcanoes 287
vomiting/riot control agents *see* riot control
VX gas 249, 259
    disposal 269, 270

wallpaper 103–4, 105–6, 115, 119
    dust/gas exposure problems 107–10
    and Napoleon's death 145–6
    and William Morris 124–6
warfare agents *see* chemical warfare
Wasserman, August von 25
water *see* bottled water; dinking water
weathering 332
Webster, John 185
weedkiller *see* herbicide
Weider, B. 151–2, 153, 154–5, 160
Weigert, Carl 24
wells
    dug/deep 369–70
    tube 350–4, 360–2, 374

West Bengal *see* Bangladesh
Wheal Jane mine 334
white ants 64–5
white arsenic (oxide) 3
    smoke 2, 4, 8
wig powder 91, 92
Wilson, Maud 180
Wilson, Thomas 15
Wimmer, Karl 254
witches 9
Wolferstan, Thomas 24
Wolmanized wood 68
wood preservatives *see* chromated copper arsenate
Woodpeckers 77
World Bank 353
World Health Organization 349, 360
World Wars
    chemical weapons *see* chemical warfare
    venereal disease 26–8, 29
worms 289

X-ray absorption spectroscopy 296, **298**
X-ray fluorescence 86

yellow arsenic (sulfide) 3
Yellow Cross munitions 221, 224, 226
yellow root tonic 5
Yellowknife
    animals and birds 295–6
    Giant mine 315–17
        cleanup 317–18
Yushchenko, Viktor 202

Zam-Zam water 309
ZIO-101 22
Zyklon B 231, 250, 254
    disposal 265